Passive Solar Buildings

Solar Heat Technologies: Fundamentals and Applications
Charles A. Bankston, editor-in-chief

1. *History and Overview of Solar Heat Technolgies*
Donald A. Beattie, editor

2. *Solar Resources*
Roland L. Hulstrom, editor

3. *Economic Analysis of Solar Thermal Energy Systems*
Ronald E. West and Frank Kreith, editors

4. *Fundamentals of Building Energy Dynamics*
Bruce Humm, editor

5. *Solar Collectors, Energy Storage, and Materials*
Francis de Winter, editor

6. *Active Solar Systems*
George Löf, editor

7. *Passive Solar Buildings*
J. Douglas Balcomb, editor

8. *Passive Cooling*
Jeffrey Cook, editor

9. *Solar Building Architecture*
Bruce Anderson, editor

10. *Implementation of Solar Thermal Technology*
Ronal Larson and Ronald E. West, editors

Passive Solar Buildings

edited by J. Douglas Balcomb

The MIT Press
Cambridge, Massachusetts
London, England

This book was set in Times Roman by Asco Trade Typesetting Ltd., Hong Kong, and was printed and bound in the United States of America.

Library of Congress Cataloging-in-Publication Data

Passive solar buildings / edited by J. Douglas Balcomb.
p. cm.—(Solar heat technologies; 7)
Largely material written from 1983 to 1985.
Includes bibliographical references and index.
ISBN 0-262-02341-5
1. Solar buildings. 2. Solar energy—Passive systems. I. Balcomb, J. Douglas. II. Series.
TJ809.95.S68 1988 vol. 7
[TH7413]
697'.78—dc20
[690'.8370472] 91-46488
CIP

Contents

	Series Foreword by Charles A. Bankston	vii
	Acknowledgments	ix
1	**Introduction** J. Douglas Balcomb	1
2	**Building Solar Gain Modeling** Patrick J. Burns	39
3	**Simulation Analysis** Philip W. B. Niles	111
4	**Simplified Methods** G. F. Jones and William O. Wray	181
5	**Materials and Components** Timothy E. Johnson	199
6	**Analytical Results for Specific Systems** Robert W. Jones	235
7	**Test Modules** Fuller Moore	293
8	**Building Integration** Michael J. Holtz	331
9	**Performance Monitoring and Results** Donald J. Frey	399
10	**Design Tools** John S. Reynolds	485
	Contributors	515
	Index	523

Series Foreword

Charles A. Bankston

This series of ten volumes summarizes research, development, and implementation of solar thermal energy conversion technologies carried out under federal sponsorship during the last eleven years of the National Solar Energy Program. During the period from 1975 to 1986, the U.S. Department of Energy's Office of Solar Heat Technologies spent more than $1.1 billion on research and development, demonstration, and technology support projects, and the National Technical Information Center added more than 30,000 titles on solar heat technologies to its holdings. So much work was done in such a short period of time that little attention could be paid to the orderly review, evaluation, and archival reporting of the significant results.

In response to the concern that the results of the national program might be lost, this documentation project was conceived. It was initiated in 1982 by Frederick H. Morse, Director of the Office of Solar Heat Technologies, Department of Energy, who had served as technical coordinator of the 1972 NSF/NASA study "Solar Energy as a National Resource" that helped start the National Solar Energy Program. The purpose of the project has been to conduct a thorough, objective technical assessment of the findings of the federal program using leading experts from both the public and private sectors, and to document the most significant advances and findings. The resulting volumes are neither handbooks nor textbooks, but benchmark assessments of the state of technology and compendia of important results. There is a historical flavor to many of the chapters, and volume 1 of the series will offer a comprehensive overview of the programs, but the emphasis throughout is on results rather than history.

The goal of the series is to provide both a starting point for the new researcher and a reference tool for the experienced worker. It should also serve the needs of government and private-sector officials who want to see what programs have already been tried and what impact they have had. And it should be a resource for entrepreneurs whose talents lie in translating research results into practical products.

The scope of the series is broad but not universal. It is limited to solar technologies that convert sunlight to heat in order to provide energy for application in the building, industrial, and power sectors. Thus it explicitly excludes photovoltaic and biological energy conversion and such thermally driven processes as wind, hydro, and ocean thermal power. Even with this limitation, though, the series assembles a daunting amount of information. It represents the collective efforts of more than 200 authors and editors. The volumes are logically divided into those dealing with general topics such as the availability, collection, storage, and economic analysis of solar energy and those dealing with applications.

Volume 7 is one of four volumes covering various aspects of solar building design and performance. Volume 4 deals with building energy dynamics; volume 9, with solar building architecture; and volume 8, with passive cooling. This volume covers passive solar heating of residential and commercial buildings. Beginning in 1973, the federal government supported most areas of solar research. The early emphasis was on active solar technology and on the performance of active technologies used in the building sector. Passive solar building design began to receive increasing government support following the first passive solar conference held in Albuquerque in 1976, which revealed widespread national interest. By 1982, the federal R&D budget for passive solar heating and cooling exceeded the budget for active solar.

The information contained in volume 7 is primarily analytical and quantitative. About half of the volume is devoted to quantitative methods for modeling, simulation, and design analysis of passive buildings; the other half summarizes the quantitative results of testing and monitoring of models and buildings. The reader will find that this volume contains analytical methods that range from rigorous application of the laws of heat transfer and thermodynamics to rules of thumb; analytical results for specific materials, components, and systems; test results for systems ranging from small styrofoam models to large commercial buildings, and information on means of integrating passive solar features into building architecture. Volume 9 also covers many of these topics with an emphasis on the building architecture rather than the physical phenomena involved in maintaining the interior environment.

Acknowledgments

My introduction to the "DOEDOC" project, as I came to call it for short, was a meeting in Washington convened by Dr. Fred Morse in 1982. At that time the solar program was in free fall and Fred, with his great perception, realized the importance of documenting the incredible achievements that had been made in less than a decade of solar energy research and development. At that meeting, and at the many that followed, Dr. Morse emphasized the importance of the legacy that these publications would leave. I wish to acknowledge the devotion and tenacity that he, more than anyone else, put into the long drive that has resulted in this MIT series. Many times, when the whole project was in jeopardy due to bureaucratic weak knees or editorial procrastination, he kept it going. Now, well after Dr. Morse's retirement from the Department of Energy, the momentum he instilled in the project has carried yet another volume into print.

Dr. Charles Bankston has served as Editor in Chief of the series. (The official name of the project was the "Solar Thermal Energy Conversion Technology Status and Assessment Program," not DOEDOC, but this was too long for me.) Countless times Charles reminded all the managing editors (and especially me) of the job at hand and kept us going. His enthusiasm, his knowledge of the subject, his skill in working with people, and his perseverance have made the whole series, not just this volume, possible.

In any project of this magnitude there is an unsung hero. For volume 7 this person is Dr. Robert Jones. Robert has been the technical proofreader for the whole volume, a job that he conducted with skill, dedication, and true concern.

This volume, *Passive Solar Buildings*, was originally to have been co-edited by Bruce Wilcox of the Berkeley Solar Group, and myself. Bruce worked hard on the project and contributed substantially in the initial phases, particularly in rounding up the right people for the task, but the pressures of survival in the real world of the mid-1980s forced his resignation before the project could be completed.

Many others have helped to bring volume 7 to print. Those who have helped in the editing of individual chapters include Edward Arens, Charles Barnaby, William Beckman, William Carroll, Joe Carroll, Bing Chen, Craig Christensen, James Clinton, John Duffy, Larry Flowers, and Winslow Fuller.

1 Introduction

J. Douglas Balcomb

1.1 Scope and Content

1.1.1 Scope

This book describes developments in passive solar buildings that took place from the early 1970s through 1989. Much of the work covered was supported with federal funding from the U.S. Department of Energy (DOE) Office of Solar Applications (or its predecessors, the Energy Research and Development Administration or the Atomic Energy Commission); however, some was funded by other federal agencies, state government agencies, or privately. Although the emphasis was to be on work resulting from the DOE's solar program, authors were asked to be inclusive and to reference all relevant research, whether it was federally funded or not and whether it was done within this time frame or not.

Except for this introductory chapter, the material in this book was written from 1983 to 1985 and therefore is current only up to that time. This chapter was written in 1990 and, to a limited extent, references developments after 1985. This should not present a major problem, however, because few developments in passive solar heating took place during 1985 to 1989 as a result of reductions in federal government support that began after 1980. Nonetheless, a considerable number of builders and designers continued to construct passive solar buildings, and technical progress continued, albeit at a low level. The proceedings of the annual Passive Solar Conference sponsored by the American Solar Energy Society document many of the developments that took place during the late 1980s.

1.1.2 How this Book Relates to Others in the Series

Four books in The MIT Press series Solar Heat Technologies: Fundamentals and Applications bear a strong relation to passive solar buildings. Reading all four will give the reader a comprehensive overview of the background and the current state of the art in passive solar buildings. Inevitably, there are overlaps among the books, but we hope that these have been minimized. In any case, the perspectives of the different authors, all of whom participated in the developments of this vital period, should be valuable in themselves.

Volume 9 of this series, *Solar Building Architecture*, edited by Bruce Anderson, is a comprehensive overview of the field and perhaps should be read first. Written largely by architects, this volume is focused on the building itself and not on the evaluation of building performance. The evolution of passive solar buildings from 1975 through 1983 is traced in the context of the modernist movement in architecture. Although the volume emphasizes passive architecture, active solar equipment is also discussed. Also included is a discussion of the influence of the solar tax credits of 1980 to 1985, which accounted for the rise and fall of a major active solar industry. The integration of passive heating, natural cooling, and daylighting is emphasized throughout the volume.

Volume 8, *Passive Cooling*, edited by Jeffrey Cook, is a comprehensive overview of natural cooling, low-energy cooling, or passive cooling—each of these terms is used by different people to describe heat rejection from a building to the environment based on either passive mechanisms or mechanical methods that require minimum operating energy. This subject had originally been slated for inclusion in volume 7, but the large amount of material suggested that it should be a separate volume. Volume 8 does not cover simulation modeling of passive-cooled buildings because many of the same mathematical techniques are used as for passively heated structures. The reader is referred to chapter 2 in volume 7, "Building Simulation Analysis," by Philip Niles, for a discussion of these modeling issues.

Volume 4, *Fundamentals of Building Energy Dynamics*, edited by Bruce Hunn, is a comprehensive overview of mathematical modeling of both buildings and the heating, ventilating, and air-conditioning (HVAC) equipment installed in buildings. The distinction between the discussion of this topic in this volume and in volume 4, which also covers modeling, is a practical one. Volume 4 focuses on large building computer programs, such as BLAST and DOE-2, whereas this volume emphasizes programs that focus on the dynamic response of the building structure. As we will explain later, most of those who evaluated passive solar buildings did not use the large programs, and so a new approach to analysis emerged. Also, the emphasis on mechanical equipment in volume 4 is nearly absent in this volume. Each discussion therefore has its place.

1.1.3 Overview of this Book

This introductory chapter provides an overview of passive solar heating and briefly summarizes other chapters. Chapters 2, 3, and 4 describe work done on the modeling of passive systems. Topics are building solar gain modeling, simulation analysis, and simplified methods. Chapters 2 and 3 are applicable both to passive solar heating and passive cooling. The remaining six chapters describe passive solar heating from an applications perspective. Topics are components, analytical results for specific systems, test modules, subsystem integration into buildings, performance monitoring and results, and design tools.

1.1.4 Overview of this Chapter

The remainder of this chapter consists of three sections. Section 2 presents an overview of passive solar heating. It describes and evaluates the era between 1976 and 1983, when most of the developments took place. Design and development features that have contributed to effective buildings are highlighted. Section 3 deals with modeling of passive systems (chapters 2 through 4). Section 3.1 describes the historical context within which the passive systems analysis has taken place, explaining why development has taken the path it has. Section 3.2 is an overview of the three chapters on analytical modeling. Section 4 deals with passive solar heating (chapters 5 through 10). Section 4.1 is an overview of these six chapters. Section 4.2 includes a brief description of two recent developments not described elsewhere in this book.

1.2 Overview of Passive Solar Heating

1.2.1 The Era of Passive Solar Development

A principal aim of this book is to provide future researchers and practitioners with an entrée into the literature of passive solar heating. A great deal of work has been done that does not have to be redone. Solar energy, like many other fields, is replete with examples of reinvention by succeeding groups of enthusiasts. Lessons learned are forgotten. Mistakes that may have been excusable the first time are made repeatedly; it is a waste of time, energy, and resources—all quantities we are striving to save. Some reinvention is inevitable and even desirable in the evolution of a new discipline, especially one that involves a subject as familiar and personal as the buildings we all live and work in. Each generation must experience and relearn the lessons of the last, either through trial and error or by recourse to the collective wisdom of its predecessors. It is hoped that this book and the extensive list of references cited throughout will save future researchers and practitioners both time and disillusionment.

Historically, interest in solar energy has occurred in waves. The great enthusiasm of the early 1960s died out almost completely by the end of the decade in response to low energy prices. Then, in reaction to the oil embargo of 1972, interest exploded again and lasted until about 1981, when political changes torpedoed the considerable momentum that had built up, the "energy crisis" was proclaimed solved, and nearly all interest in saving energy abated. In the early 1990s we are experiencing another revival of interest in energy conservation and renewable sources of energy as a result of environmental concerns, worries about societal sustainability, and national energy vulnerability. If we can be a bit more practical this time, we can also benefit from the experiences and collective wisdom of those who rode the earlier waves.

Passive solar heating works. Properly designed and constructed, it is cost-effective, practical, comfortable, and aesthetic. These facts, widely disputed fifteen years ago by skeptical engineers, are now a matter of record. But experience shows that if the time between cycles is such that most of those in the new wave did not ride the last, many newcomers do not take the time to look at the record. Those who do will have the advantage in planning for the buildings of the future. Time, energy, and resources will be saved.

A major effort to promote passive solar energy applications in the United States started in about 1976 and focused mainly on passive solar heating of single-family residences; however, work has also been done on multifamily and commercial buildings and on natural cooling and daylighting. It is estimated that there are now more than 200,000 residential and 15,000 commercial passive solar buildings in the United States; passive solar technology is thus an accepted and proven one (Renewable Energy Institute 1986).

Detailed computer simulation models have been used extensively to predict the performance of passive solar buildings. Design tools that architects can use have been developed by correlating the results of numerous computer simulations of different passive solar strategies in different climates. Design guidelines have evolved that balance conservation strategies and passive solar strategies, depending on climate and economics. Design competitions have been used effectively to promote the construction of many buildings throughout the country.

A major government program launched by the Department of Energy to quantify the performance of passive buildings resulted in the detailed monitoring of more than 100 residential structures and 22 commercial ones. In general, performance has been excellent and occupants have been very satisfied.

The most important conclusion to emerge from this decade of work is that we have learned how to design, construct, and operate buildings that use a fraction of the energy of conventional buildings, that provide a more comfortable and livable interior environment, and that cost little or no more than others to construct.

Although there are historical and archaeological examples of the use of passive solar techniques by native Americans, very few buildings were consciously designed to make use of passive solar energy before the 1970s. For example, some direct-gain buildings were constructed in the Midwest in the 1930s, and a group of engineers at the Massachusetts Institute of Technology experimented with passive solar test rooms in 1946. Significantly, although they measured reasonable performance in the cloudy Boston climate, the MIT group abandoned this line of research in favor of work on systems with solar collectors, pumps, and storage tanks, setting the stage for the future. Thus, active solar became the prime strategy promoted for development after the 1972 oil embargo triggered a major solar research and development program.

The first evidence of strong interest in passive solar technology in the United States appeared in 1976 at a passive solar conference held in Albuquerque, New Mexico. National conferences have been held annually since then. Interest continued to be very strong through 1982 but declined somewhat in the last few years, as public concern over all energy issues slackened. Support for research, development, and commercialization of passive solar technology by government agencies lagged behind support for active solar by about four years. This situation was reversed, however, with passive solar receiving the greater emphasis, largely because of the good performance, low cost, ease of maintenance, and acceptance of passive solar approaches by both building designers and the public.

Peak funding for passive solar technology occurred in the 1979 to 1981 period, during which DOE's annual expenditure on passive solar averaged $29 million (Renewable Energy Institute 1986). Subsequently, this funding declined to about $8 million per year for all of solar buildings research. During the peak period, the emphasis was on commercialization activities, such as information programs, design competitions, training programs, and surveys of consumer attitudes. These activities were completely phased out, and the program became almost entirely devoted to research.

A recent and very significant development is the realization that passive strategies can be very effective in institutional and commercial buildings. Usually, the major passive strategy employed is daylighting, since these types of buildings are used mostly during the daytime. However, there is a natural synergy between daylighting and passive solar heating if the windows used for daylighting face south. It is also possible to reduce the cooling requirements of these buildings, since much of the cooling load in them is due to heat generated by artificial lights (Gordon et al. 1986).

1.2.2 Performance Evaluation

1.2.2.1 Passive Solar Performance The net energy benefit of adding passive solar strategies varies with climate, system type, system design, and system size. When the passive system and building conservation parameters are optimized for the climate, the net benefit (defined as a reduction in backup heat compared with a perfectly insulated wall) usually ranges from 50,000 to 100,000 Btu/year per square foot (ft^2) of aperture (150 to 300 kilowatt-hours [kWh]/year per square meter [m^2] of aperture). The added cost of the passive solar features usually ranges from \$5 to \$15/ft^2 of aperture (\$50 to \$150/m^2 of aperture) for residential construction. For the sizes that are usually built, the passive solar added cost ranges from 4% to 8% of the total cost of construction. If the cost of the displaced backup heat is in the range of \$0.02 to \$0.07/kWh, payback time is usually within five to ten years.

1.2.2.2 Performance Analysis Initially, passive solar building performance was estimated most frequently by scientists and engineers using mainframe computer programs to perform hour-by-hour simulation analyses. Normally, these calculations were done for the purpose of research or systems analysis rather than for the design of a particular building.

Two main mathematical approaches have been used to describe the dynamic behavior of buildings. The most straightforward one uses thermal networks, in which the flow of heat from point to point in the building and the storage of heat in massive elements are characterized by a set of ordinary differential equations. The other standard approach uses weighing functions, in which the time response on one side of a wall is described as a convolution of the inputs on both sides. Harmonic analysis is a third technique, but it is used less often. Each method has its strong advocates and its advantages and disadvantages.

1.2.2.3 Passive Solar Design Most passive building design is done without any analysis, simply by emulating other buildings, by intuition, or by using simple rules of thumb. When an analysis is made, the most widely used design tools are simplified monthly methods based on correlations of results of numerous computer simulations. The most common of these is the Solar Load Ratio (SLR) method, in which the correlation parameter is the ratio of solar gain to building load (Balcomb et al. 1982, Balcomb et al. 1984). This analysis can be done by hand or with the aid of one of the numerous microcomputer programs based on the method.

A very effective design tool, targeted specifically to builders, is a set of builder guidelines in which each guideline package is written for a specific locality (Passive Solar Industries Council 1989). This package, which has been distributed throughout the United States, contains three parts: (1) guidelines that give general advice and identify performance potentials; (2) simple fill-in-the-blank worksheets for calculating annual heating and cooling requirements and thermal comfort; and (3) a case-study example. The approach is nonprescriptive; it allows for wide variation in the design of the building—from superinsulated at one extreme to explicitly passive solar at the other. The guidelines are described more fully in section 4.2.2.

Simulation programs that take full advantage of the greater computing power of the current generation of microcomputers are becoming a trend and are normally used in the design of larger structures, such as institutional and commercial buildings, for which energy issues are much more complex than for residential buildings. We can foresee that computer-assisted design (CAD) techniques, now commonly used for drafting, will be expanded to include energy analysis.

1.2.2.4 Test Modules Test modules have been built in three sizes: small test boxes, usually about a 3-ft (1 m) cube; test rooms, usually about 40 ft^2 (4 m^2) in area; and unoccupied larger test buildings. Each size has proved to be very valuable for different types of tests, and each has also been an excellent tool for teaching college students the principles of monitoring, data evaluation, and comparing theory and experiment.

Test modules have played an important role in passive research; quite a large number of them have been built at several institutions to obtain data under carefully controlled conditions. The units serve one or more of the following purposes:

1. Directly comparing competing strategies side by side
2. Obtaining data to validate computer programs
3. Testing components under realistic conditions

Test modules are sometimes free-running, that is, operated without auxiliary heating or cooling. More commonly, they are operated with a thermostatically controlled inside environment. This allows a more direct comparison of the net energy benefit of the units and yields a more realistic operating profile.

1.2.2.5 Monitored Buildings Results of a large-scale DOE program indicate good performance in actual occupied buildings monitored over periods of one or more years using twenty or more sensors connected to hour-by-hour recording equipment. A subsequent computer analysis of the data from 48 homes in various climates showed the following (Balcomb 1983):

• Passive solar homes use 70% less auxiliary heat than conventional homes, with an average solar contribution of 37% of the total heating load or 55% of the net heating load (total minus internal gains from people, light, and appliances). Solar savings are about 27% of the net heating load, on average.

• Among the various systems employed (direct gain, Trombe walls, and sunspaces), there was no noticeable difference in performance. In fact, many of the houses used combinations of two or more solar options to good advantage.

• Building heat loss coefficients per unit floor area of 4 to 6 Btu/°F-day-ft^2 (0.8 to 1.5 W/°C-day-m^2) are routinely achieved, underlining the importance of good conservation practice. Auxiliary heating requirements as low as 1 to 2 Btu/°F-day-ft^2 (0.25 to 0.5 W/°C-day-m^2) are demonstrated in sunny climates, and values of 1.5 times these levels are achieved routinely in all climates.

• Solar fractions of 50% or more are often achieved; however, in a few cases the solar performance is illusory, because losses from the solar elements exceed solar gains.

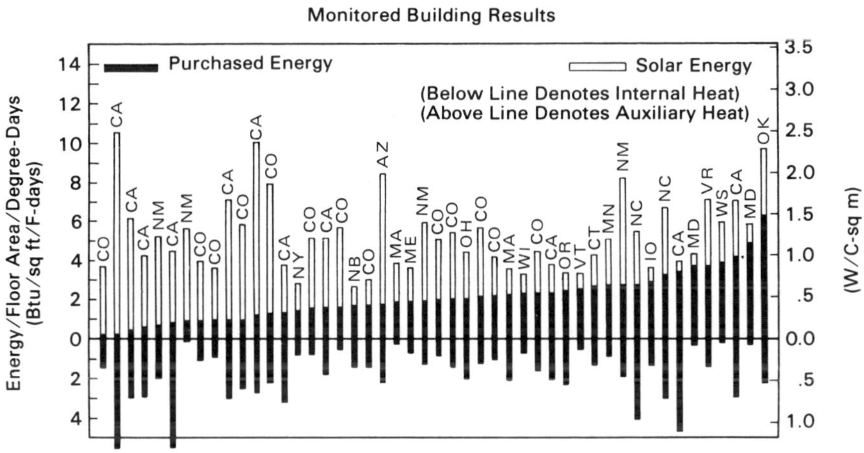

Figure 1.1
Annual energy use of 48 monitored residential buildings (Balcomb 1983). The bars show seasonal energy (usually five or six months) divided by the building floor area and the actual heating degree-days for the season (base 65°F). The black portion of the bar denotes purchased energy; the portion below the axis represents internal gains, the portion above the axis is auxiliary heat. The total length of the bar is the total heat required by the building—essentially the building load coefficient, determined in a coheating test. Thus, by subtraction, the white portion of the bar is the solar gain less any vented heat. The state where the site is located is denoted above the bar. The buildings are rank-ordered according to auxiliary heat. Several of the houses with low internal gains were unoccupied but were thermostatically controlled to normal levels.

The heat *saved* by solar is less than the white part of the bar, because some of the solar energy transmitted is lost through the solar aperture and does not contribute to heating the rest of the house. An analysis was performed to subdivide the white part of the bar; the results vary from house to house, but the average taken over the 48 houses indicates that the two portions are nearly equal: the solar savings are about one-half the white part of the bar, and the remainder is heat transmitted back to the environment through the solar aperture.

- Movable insulation systems intended to reduce heat losses at night have generally not worked out well. In fact, it is questionable whether they saved any energy at all. This is partly due to the fact that there is no tradition of outside roll-down shutters in the United States. Most of the products that were used were of recent design, most were designed to fit inside the window rather than outside, and many were not mechanically reliable. Moreover, occupants who were unaccustomed to the daily operation of such systems tended to resent the extra effort involved.
- Comparisons were made of monitored results with performance estimates based on actual weather, solar, and inside temperature conditions. Deming and Duffy (1986) found that, on the average, both the SLR and the unutilizability-method estimates match the auxiliary heat used quite well (within 4% for SLR and 13% for unutilizability). However, for a given passive solar building, the predictions can vary significantly from measurements (the root-mean-square difference is about 30% for both methods). Much of the discrepancy for individual houses appears to be due to uncertainties in the measured heat-loss coefficient and not necessarily in the passive solar prediction.
- Occupants of the houses are very satisfied with them.

Analysis of the data from twelve monitored institutional and commercial buildings by Gordon et al. (1986) shows the following:

- The overall energy savings compared with base-case buildings of the same size and function in the same location is about 47%. Lighting energy is reduced 65%, heating energy is reduced 44%, and cooling energy is reduced 68%. Only the small category of "other energy" (mostly energy to run equipment and fans) increases.
- Daylighting was a major design strategy in all of the buildings. Post-occupancy evaluation indicates that the most common reaction is a strong appreciation of the quality of the inside environment, especially the light.
- People's actual use of the buildings is higher than estimated. This partly explains why the saving in heating energy is not always quite as large as predicted.
- The construction cost of the buildings was nearly identical to that estimated for the base-case buildings. The added cost of the solar options was partially offset by reductions in the cost of equipment.

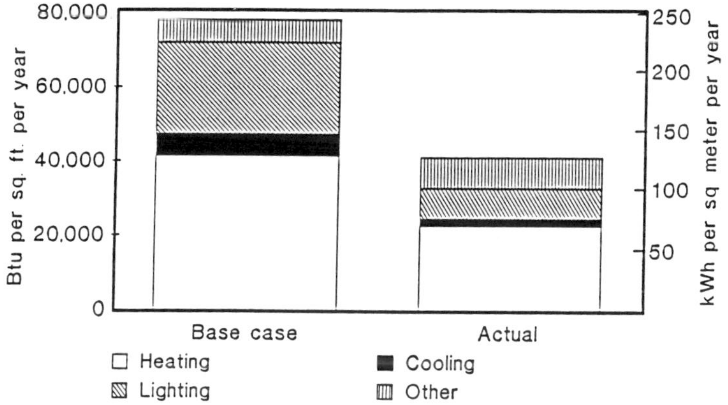

Figure 1.2
Annual energy use averaged over the 12 buildings of the DOE Passive Solar Commercial Buildings program. The bar chart on the left, labeled base case, is the predicted energy use of a building of the same architectural program and same floor area, built with typical contemporary design techniques, that is, without the explicit use of bioclimatic strategies. The bar chart on the right is thc actual monitored energy. The height of the bars is the total energy, measured at the site boundary. This is disaggregated according to four categories, as shown by the hatching. "Other energy" refers to internal heat from fans, equipment, and people. All energies are referred to source energy so that they may be compared fairly.

• The peak demand for electricity was substantially reduced compared with that of the base-case buildings.

• Problems with controls occurred in some of the buildings. Most often this happened when the thermostat had been set back too far at night or the heating system had been started up too late in the morning, and the building was sometimes uncomfortably cool during early winter mornings.

• The cost of *designing* these buildings was greater than the cost of designing conventional buildings because of the extra effort needed to optimize the interrelationships of the passive solar and other systems. This probably will be less noticeable in time, as daylighting and passive solar technology become more routine and as better design tools become available.

• The important lesson stressed by Ternoey et al. (1985) is that the key to achieving good overall energy performance lies in integrating daylighting, passive solar heating, and natural cooling into the overall design.

1.2.3 Passive Practice

In less than one decade, *passive solar* became part of the vocabulary of building designers, buyers, realtors, financiers, and researchers. Coined in the early 1970s, the word *passive* was intended to emphasize an alternative to the then-popular *active* solar technique. Because it employs conventional building materials and because the basic concepts are easily understandable, passive solar technology has found ready acceptance among designers and builders. The term *passive solar* is now used to include the strategies of passive solar heating, natural cooling, and daylighting.

By itself, passive solar technology does not make for a good energy-efficient building, as many of the early examples demonstrate. Many other factors, such as architectural design and siting, must also be sound. However, passive solar technology provides visible testimony that good energy design has been a priority. Some of the factors that have proved to be effective, considering both technical and marketing issues, follow.

1.2.3.1 Residential Buildings

ALL-SOLAR SUBDIVISIONS Purchasing a solar building located among non-solar buildings requires a conviction about the technology that is unlikely in anyone but an innovator. By contrast, clusters of passive solar buildings signal a trend and give evidence of a degree of societal acceptance necessary to most people for undertaking a large financial obligation. Rather than feeling innovative, the buyer becomes part of the trend along with his neighbors.

SITE PLANNING AND DEVELOPMENT With a little forethought, good solar access can be planned into a development with no loss in density. Zoning restrictions can be adopted that guarantee continued solar access to every building.

SUNSPACES Sunspaces have become extremely popular in the United States, both for new construction and for retrofits of existing buildings. They provide a sunny living area and are often used for growing plants. Properly designed, they require no backup heat, furnish as much passive solar heat to the house as the same glazed area of direct gain or Trombe wall, and do not overheat in summer.

BALANCED CONSERVATION AND PASSIVE SOLAR SYSTEMS The economic trade-off between more insulation and more solar gains leads to an easily derived optimum design solution that depends on climate. However, the curve is fairly flat, near-optimum performance can be realized over a reasonably wide range of design choices. But in all cases, good insulation practices and low infiltration are essential. If this is not done, the required solar area will be too large, thermal mass requirements for adequate heat storage will be too great, and control will be difficult.

THERMAL COMFORT Thermal comfort requires stable interior temperatures, which means that a building should have small temperature swings under free-running conditions. This requires adequate thermal mass for heat storage, a proper relationship between the location of solar gains and heat storage, and effective thermal distribution. Without these essentials, a building will not create the kind of internal environment necessary for acceptance of passive solar buildings are adopted on a large scale.

HIGH-QUALITY CONSTRUCTION By adherence to the principles outlined here, the designer will have already integrated good thermal qualities into a building. It is only consistent to follow up with using good materials and high-quality construction practices to ensure that a building will be viable over the long term.

1.2.3.2 Institutional and Commercial Buildings

DAYLIGHTING Good daylighting design, which was well understood and practiced before the advent of the fluorescent lamp, is coming back. The use of natural light in buildings is essential to good architecture and it makes good energy sense. By using windows, light shelves, clerestories, roof monitors, and atria, we can bring daylight into most of the rooms of most buildings. Using natural light not only reduces the need for artificial lighting energy, it reduces cooling loads produced by the lights, since natural light contains more light per unit of heat. Good daylighting design avoids direct-beam sunlight penetration into the building. Instead, sunlight is diffused from strategically placed louvers or baffles that are painted eggshell white to achieve colorless, soft lighting. Glare is avoided by placing all such diffusing surfaces and glazing well above eye level.

By orienting windows used for daylighting to the south, we reduce both heating and cooling loads. However, artificial lighting systems and their controls must be integrated well with daylighting, or energy may not be saved.

ATRIA The atrium has become a very popular design element with architects, because it adds character and vitality to a building. Atria are very complex both thermally and optically and are not well understood. They can have either a positive or a negative energy impact on a building, depending on their design.

BALANCING Thermal balancing of a building must be done on a space-by-space and time-of-day basis, considering orientation, patterns of internal heat generation, and occupancy. Mismatches between the availability of natural energies and the energy needs of various spaces may require transporting heat from space to space. However, this should be minimized, because it invariably increases complexity and cost.

DESIGN PROCESS The design process used was largely iterative, starting from a base-case building and relying on evaluation tools, such as the large computer program DOE-2. Guidance tools would be very valuable in the process, and they will be welcomed by the design community. However, very few of these tools have been developed, because the problem is so multidimensional. As computer-based tools evolve and computers with greater memory and speed become available, we can expect to incorporate more sophisticated analyses earlier in the design process. Some developers of tools predict that CAD-based systems utilizing knowledge-based, expert-system programming techniques will ultimately predominate.

1.2.4 Regionalism

We see a growing trend toward regionalism in architecture. This fits in well with passive solar design, because building types that evolved before this century were often based on accommodation to the climate. This trend is particularly evident in the southeastern United States where several traditional housing types have been identified while making effective use of protection from the sun and natural ventilation. Often, the same designers are concerned with historical preservation, the use of traditional building styles and techniques (updated for modern materials and methods), and passive solar design. We can hope that the current major emphasis on stylistic issues will gradually shift toward a return to a functionalism that pervades the entire building design.

1.2.5 Problems

Passive solar development has not been without its problems, two of which are discussed below.

1.2.5.1 Overheating Overheating by the sun can occur in winter or in summer, and it is usually caused by design errors. Winter overheating can occur if the south glazing is oversized or if there is insufficient thermal storage mass for the amount of direct gain. A good design practice is to limit the collecting area for solar gain to a value that would result in an average inside temperature no greater than 72°F (22°C) on a clear January day. Thermal storage should be sized to limit the inside temperature swing to no more than 11°F (6°C) on a clear winter day. In the case of conventional wood-frame construction with gypsum-board interior sheathing and carpeted floors, this imposes a limit on the south glazing area of about 7% of the floor area. Beyond the 7% glazing area, added mass is recommended. In any case, the area of direct gain should not be greater than about 13% of the floor area of the building, or problems of glare, ultraviolet fading of materials, and loss of privacy will result. If a solar glazing area greater than this amount is desired, then an indirect system, such as a Trombe wall or a sunspace, should be used.

Summer overheating can usually be traced to excessive east-facing, west-facing, or horizontal glass in the design. This is a common problem in sunspaces. The remedy is to emphasize south-facing glass and avoid large glazing areas having any other orientation. If skylights in houses or overhead glazing in sunspaces are used, they should be covered in the summer. Proper natural ventilation is also essential to avoid high summertime inside temperatures.

1.2.5.2 Perception This is listed as a problem because the public, the building industry, and many design professionals do not understand passive solar systems nearly as well as they should. The excellent performance results that have been obtained are not widely recognized, nor is a need for saving energy a priority for everyone. Most people are very conservative about their homes, and they are reluctant to make significant changes in a home's style or appearance.

There are two solutions to the problems of perception. The first is public education. Publicizing effective and attractive passive solar buildings can alleviate many concerns. Exotic concepts, such as underground buildings or extremely modernistic designs, are likely to meet resistance and should be avoided. Certain types of buildings, such as schools and libraries, provide excellent opportunities for passive solar technology and daylighting, and are particularly suitable as demonstrations. Experimentation on new concepts should be confined to the laboratory; only tried and proven systems should be put into buildings that are to be actually used, because errors are always publicized and are very harmful to the image of passive solar technology.

The second solution is to establish a good program of educating architects both within colleges of architecture and in short courses for practicing architects. Education of builders can best be done through builder guidelines packages and seminars.

1.2.6 Design Tools

Because of the importance of good design tools to the implementation of passive solar heating, we digress at this point to describe the evolution of one such design tool and to discuss how some design tools are related to building analysis programs. G. F. Jones and William O. Wray in chapter 4 and John Reynolds in chapter 10 also address design tools.

Many believe that a key factor in the transfer of passive solar technology from the research level to standard practice will be the development of suitable design tools. The term "design tool" means different things to different people. Many architects think of a design tool as an aid in the design process, whereas many engineers (who usually are not designers) think of design tools as computer programs. A computer program can certainly be a design tool, but few are. Design tools sometimes evolve out of practical experience as a codification of conventional wisdom. This may come from an aggregation of experience in design offices or as a result of feedback from the field regarding successful applications of a particular design procedure. Some very effective design tools are simple graphical procedures. However, most of them originate from the repeated application of a complex analysis procedure. The latter type is the subject of the following discussion.

1.2.6.1 Evolution of Design Tools Figure 1.3 shows a logical progression of design tool development, evolving from detailed hourly simulation analysis into a simplified method and then to design guidelines. It describes the development history of many of the simple analysis methods and

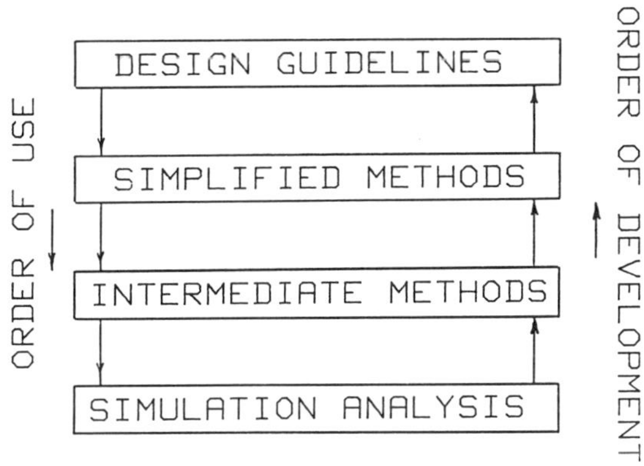

Figure 1.3
The order of development of a design tool (bottom to top) is opposite to the order of use by a practitioner (top to bottom). Development proceeds from complex analysis to simplified techniques. The building designer wants simple techniques first, even if the results are approximate, and may use complex analysis later in the design process, if at all.

design guidelines. The development progression proves to be quite confusing for architects and building designers who are concerned primarily with design guidelines and to a lesser extent with simplified methods, and who are rarely interested in simulations at all. Thus we see a reverse order, shown in figure 1.3 as an arrow from bottom to top, in the use of these techniques as design tools.

For the *developer* of a design tool it makes sense to follow the bottom-to-top order. The simplified methods have most often been developed by correlating the results of multiple simulation analyses. Design guidelines then evolve systematically from the application of analysis results. Designers frequently use simplified methods to perform the analyses required to determine these results rather than use simulation directly, both to save computer time and as a matter of convenience.

Unfortunately, the *users* of the design tool, building designers, often do not understand this progression. It appear backward to them, and indeed it does go backwards, from their perspective. As they attend conferences and read the literature, they are exposed to mathematical treatises that they do not, or even do not want to, understand and that are often presented as if they were intended to be used as design tools. The practitioners want guidelines and do not realize that although guidelines are the simplest to use, they can be the hardest to develop. Likewise, the scientists developing the tools are often not conversant with design practice nor sympathetic and patient enough to see the needs of designers.

1.2.6.2 A Case Study: Conservation and Solar Guidelines The evolution of conservation and solar guidelines during the work on passive systems done at Los Alamos is an example of the bottom-to-top process in figure 1.3; the author has first-hand knowledge of this particular example. It provides an interesting case history and holds some lessons for those who might want to develop other simplified analysis procedures or design guidelines. One lesson is the realization that the evolution from the complex (hourly simulations) to the simple (design guidelines) can be a huge and time-consuming undertaking. In fact, the process described is not yet finished, because many other factors, including thermal comfort, amenity value, and daylighting, should be accounted for (summer cooling loads were included in the Los Alamos work to a limited degree). It is unlikely that this endeavor will ever be finished because of the immense effort required to do a credible job. Nonetheless, many people have used the published results, which justifies the time and effort spent on the process.

The very first analysis work on passive systems done at Los Alamos was a simulation of a water wall, chosen mostly because it was easy to model and had already been built. Figure 1.4 shows the thermal network used. The building simulation consisted of one differential equation to represent the water mass and one algebraic equation to represent the room heat balance. However, the analysis also included the glazing transmittance equations, which were much more complex, but fortunately we had computer subroutines at hand for the detailed analysis of solar glazing. The simulation was driven using hourly solar and temperature data taken from National Weather Service tapes. Figure 1.5 shows some of the simulation results, indicating the heating energy saved because of the water wall. The simulation of a lightweight, passive solar, water-wall building was very straightforward and simple but reasonably realistic. These results were completed in a matter of only two or three weeks in 1975.

It became immediately apparent that the good performance of passive systems claimed by practitioners could be explained based on simulation. (At this point, a similar but more complex thermal-network simulation had already been made for the Skytherm® roof-pond system, but the Los Alamos researchers did not know about it until later.) For several years after this first result was obtained, the Los Alamos simulations evolved

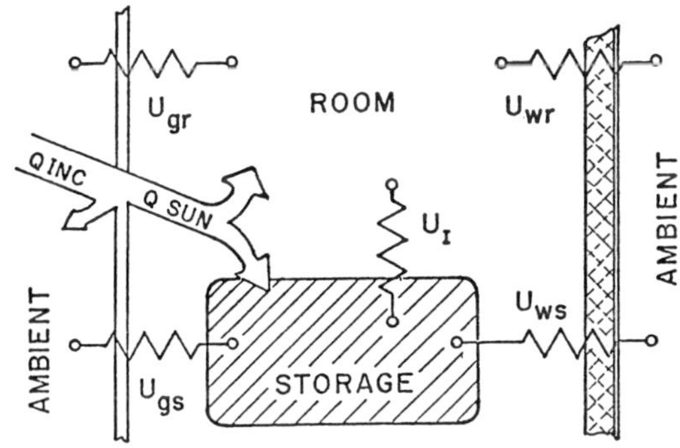

Figure 1.4
Schematic of a thermal network of a passive solar, water-wall building. Incident solar energy is reflected, absorbed in the glass (ignored in this simple model), or transmitted into the space. On absorption at a surface, this energy is converted to heat, some going very quickly to heating the room air, and some going into heating the water. The water tank stores heat and over time loses heat either to the outside through the glass or to the room. The resistor symbol represents linear heat flow (proportional to temperature difference). A backup heating system supplies whatever heat is necessary to maintain the room at the thermostat setting. This schematic is implemented in the computer program as two energy-balance equations—one for the room (an algebraic equation) and one for the water mass (a single-order time differential equation). Inputs to the calculation are incident solar radiation and outside temperature.

into more and more elaborate models but always used the same general thermal-network approach. Trombe walls were modeled extensively, followed by direct-gain systems and then sunspaces. All these models evolved within the general framework of the PASOLE computer program developed by Robert McFarland. Extensive work was devoted to validating the models by comparing the predicted results with data taken under carefully controlled conditions in side-by-side test rooms.

The solar load ratio, or SLR, method was developed to generalize the results to other climates, using a simplified monthly calculation procedure based on correlating results from thousands of month-long, hour-by-hour simulations. This evolution is well documented in the progression of Los Alamos progress reports, technical papers, and the user-oriented series of passive solar design handbooks. This literature is heavy on analysis technique and light on design guidelines.

The next step in the process was to develop economics-based optimizations for balancing the initial investment in conservation features (insulation and air-tightening) with investments in passive solar heating features.

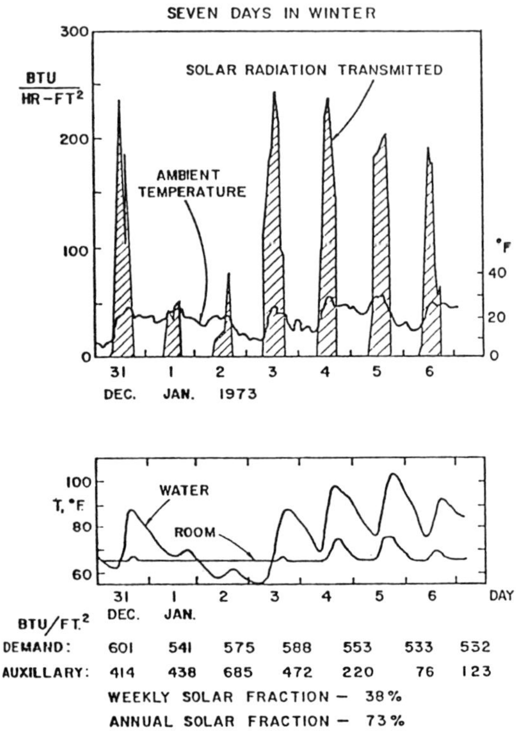

Figure 1.5
Results from the simulation of fig. 1.4 for a week of severe weather. In this case all transmitted solar gains go to the water tank located between the glass and the room. Input data are shown on the top graph. The lower graph shows calculated temperatures. Daily integral heat losses from room to ambient and backup heat are shown. The weekly solar heating fraction is 38%. The annual solar fraction for the same system is 73%.

Introducing life-cycle economics added an unscientific character to the work and certainly added a high degree of uncertainty because of the need to forecast energy prices. However distasteful this was for the scientist, it was a necessary step, because there is no readily *quantifiable* basis other than an economic one for optimizing the design. (Thermal comfort could be used as a basis for optimization in some cases.)

Only after all this work was done was it possible to focus on design guidelines, which took the form of conservation factors and values of load collector ratio for a large number of locations. They enable the designer to determine guideline numbers for insulation R-values and passive solar collection area.

The final result has proved to be quite useful but was eclipsed to some degree during the long time required to reach the end point. Practitioners in the field had evolved design practices by trial and error that worked well in their different climates. They tended not to be overly impressed by the fact that their hard-won results had been vindicated by a theoretical approach. The methodology is nonetheless of value in showing how the optimum mix varies with first cost, fuel cost, and climate. We can generalize as to locations where passive solar practice has not developed, including many locations outside the United States.

1.2.6.3 Evaluation Tools vis-à-vis Guidance Tools Design tools can be divided into two major categories, *evaluation tools* and *guidance tools*. A good example of an evaluation tool is an energy-analysis computer program. A description of the building and the local climate is fed into the program. The computer output is an estimate of monthly and annual energy use and perhaps hourly profiles of temperatures or energy demand. Although quite informative, these results do not actually provide any direct guidance as to how the design should be changed to achieve an improvement. Such guidance may come only from experience, from the intuition of the analyst, or by a "brute-force" rerunning of the program to accumulate results that show the sensitivity of the design to parametric changes. Other examples of analysis tools are the SLR and the unutilizability methods to estimate monthly performance.

Increasingly, such evaluation methods have come to be equated with design tools. The way that they are used in design is illustrated in figure 1.6. The key point is that the time when the tool is used is *after* a design step has been taken. The tool is used to evaluate the consequences of a

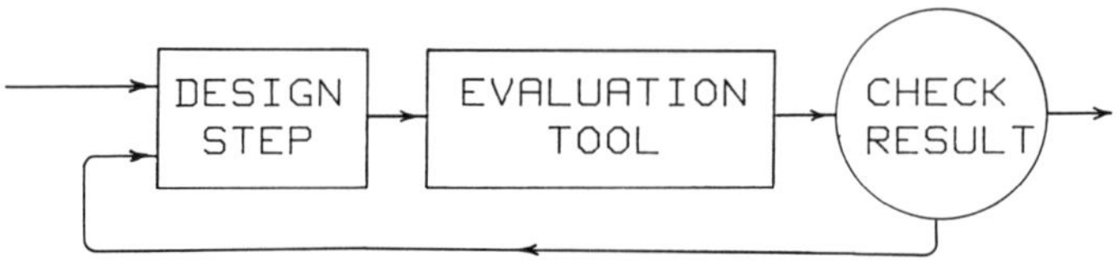

Figure 1.6
Schematic of the use of an evaluation tool in the design process. The tool is used to evaluate the performance of a proposed scheme. The result is then compared to a desired value. If the result is not satisfactory, the design is changed and the process repeated. This iteration is called a *feedback loop*.

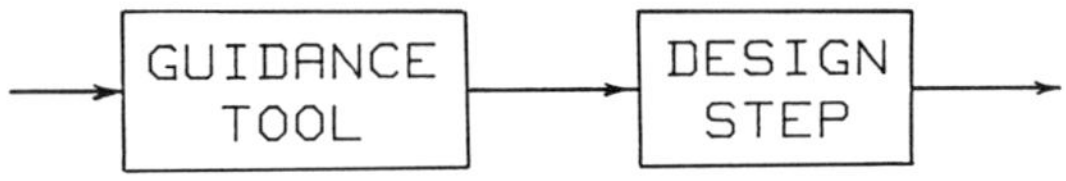

Figure 1.7
Schematic of the use of a guidance tool in the design process. This kind of tool is used to aid in developing a design based on some general criteria known in advance of the design step.

tentative design decision. Corrections can be made if the result is not satisfactory. The process may be iterated until a satisfactory result is obtained. Using an evaluation tool as a design tool may be effective, but it is not very efficient. Billions of numerical calculations may be required within the computer to complete one year of simulation just to produce one useful number. This process may then be repeated several times to study a trend. Of even greater concern is that the evaluation is often not done at all or done only after the last phase of the design process and only to document the consequence of the design. Rarely, and only if the results really look bad, will a redesign be undertaken. It is too costly in terms of design time and tends to upset the traditional relationship between designer and engineer.

The second category of design tool is called a *guidance tool*, as shown schematically in figure 1.7. This tool is used *before* taking a design step. The most widely used guidance tools are "rules of thumb." These rules evolve through experience and are usually quite general. They involve, at most, one input parameter. While they are quite useful, most do not account for climate as a variable and thus do not integrate the essence of bioclimatic design. An example of such a rule might be, "The area of south-facing windows should be 10% of the floor area." This passes the test of simplicity but fails to account for climatic variations or other critical factors, such as building internal heat generation or the need for daylighting.

A good guidance tool should take into account information about both the local climate and the architectural program of the building. Based on these, it should provide recommendations as to the next appropriate design direction. The challenge is to devise guidance tools that are simple enough to be employed early in the design process and yet comprehensive enough to be useful. The fact that so few good guidance tools exist is testimony to the difficulty of this challenge.

In practice, both guidance and evaluation tools can be used effectively in design. Ideally, one would use a guidance tool before each design step and an evaluation tool after the step, to verify that the desired result was indeed obtained. If the guidance tool is effective, then there should be little or no need to iterate as the design evolves. The process will proceed rapidly and smoothly to a building that will be well adapted to its intended use and climate, in short, a comfortable and economical building with low operating costs.

A comprehensive evaluation tool may still be required at the conclusion of the design to ensure that the performance prediction is satisfactory or to satisfy compliance with regulations where required. Note, however, that such a final evaluation gives no clue as to whether the final design is optimum in any way. It provides only a set of performance numbers without any indication that a much better set might be obtained, perhaps even at a lower cost.

One lesson that has been reinforced repeatedly is that design tools must be tailored to a particular class of user *and* to a a particular phase of a design. No single tool can hope to fit every need, any more than one tool can be used to build a house. The three principal user categories—builders, architects and engineers—speak different languages, take different approaches to the problem, and have different expectations. Thus they require different tools.

1.2.6.4 The Use of Design Tools One way of looking at this problem is shown in figure 1.8. Here we emphasize the relationship of different design tools to the various phases of the design process. Guidance tools will be most valuable during the transition from one design phase to another. Of particular importance is the step from the *programming* phase (where relevant data are compiled, such as client needs; economic, regulatory, and

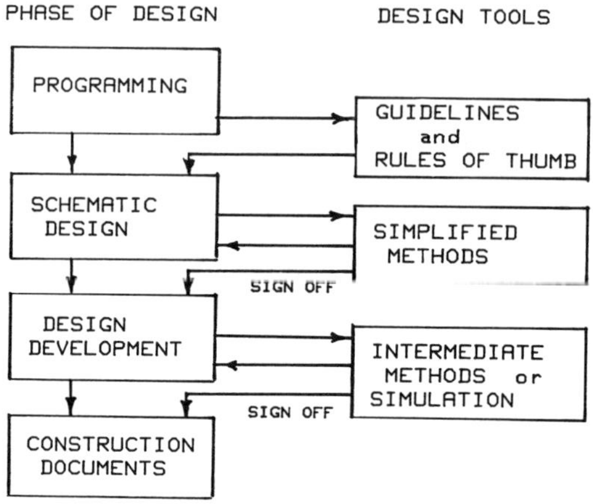

Figure 1.8
The relationship of design tools to the four steps in the normal residential-building design process. For nonresidential buildings, more complex tools would be used earlier in the design process; however, guidelines are still essential at the earliest stages.

site constraints; and climatic records) to the *schematic design* phase (sometimes called *preliminary design*), where most of the key design decisions are made. An example of such a tool for residential design is the conservation and solar guidelines given in the ASHRAE book *Passive Solar Heating Analysis* (Balcomb et al. 1984).

Evaluation tools are needed during each design phase, and they are especially valuable during the schematic design phase to quantify the implications of various trial design suggestions. However, to be useful, such a tool must be easy to employ and must produce results quickly (in perhaps ten minutes). The SLR method (in the annual version) was intended to satisfy this need for design of residential buildings. The monthly form of the SLR method and the unutilizability method take longer to apply but are more flexible. They were developed primarily for use during the *design development* phase, when such details as wall sections are being considered.

Using simulations as evaluation tools for residential design fits into this pattern at the very end of the design process. Simulations are used primarily as means of confirmation.

The situation is much more involved for nonresidential buildings, increasing in complexity with the size of a building. The reader is referred to volume 9 of this MIT series, *Solar Building Architecture*, edited by Bruce Anderson, for a comprehensive discussion of the many issues that should be considered in adapting larger buildings to the environment.

In larger buildings, there is much more leeway for analysis in the design budget and a general tendency to use more complex tools much earlier in the design process than in the design of smaller buildings. In some cases, simulation is used even before the schematic design is completed. Consideration of internal gains plays a major role, and selection of the HVAC system becomes a key issue. The whole process is described in some detail by Ternoey (1985).

1.2.7 Future Research Directions

Currently, government funding for passive solar technology is available primarily for research, with an emphasis on systems analysis and on new products and materials. The objective of the research is both to improve performance and to achieve greater market penetration for passive solar techniques.

Systems analysis is aimed at better understanding of how passive systems work in buildings, at determining the most effective directions for research, and at developing design tools. Daylighting is receiving priority in institutional and commercial buildings. Natural cooling continues to be investigated, although it has thus far proved to be difficult to implement and quantify. New design tools are being developed. Methods of maximizing the information that can be derived from short-term monitoring are being investigated.

The research on new products and materials emphasizes work on improved glazings and other aperture materials with three objectives: reduced heat losses, improved transmittance, and greater controllability. Coatings on glazings that have a low emittance in the infrared are now in common use, and efforts are being made to improve their performance. Coatings with optically switchable properties are being investigated using thermochromic, photochromic, or electrochromic phenomena. Aerogels and evacuated glazings are being investigated to try to reduce heat transmission coefficients to as low as 0.1 Btu/°F-h-ft^2 (0.6 W/°C-h- m^2) while maintaining at least 50% optical transmittance. Work is also being done on improving the thermal storage capacity of building materials by impregnating them with a solid-solid phase-change material.

1.2.8 Energy and Cost-Savings Potential

Residential buildings in the U.S. consume approximately 15 quads of energy annually (4.4×10^{12} kWh), and commercial and institutional buildings consume 10 quads (2.9×10^{12} kWh). The sum of these represents 36% of U.S. energy consumption at an annual cost of $160 billion. Of these 25 quads, electricity represents 62%, while natural gas and oil represent 36% and 10%, respectively. The major uses are space heating (9.6 quads), lighting (3.7 quads), cooling/air conditioning (3.7 quads), hot water (3 quads), and appliances/other (5.4 quads).

The potential for displacing these energy requirements by using conservation and passive solar strategies depends on several factors: the potential savings for each building, the fraction of new buildings that use these strategies coupled with the rate of new building construction, and the rate of retrofit of existing buildings. Using reasonable estimates for the maximum credible market capture for these strategies, one can predict an annual savings of perhaps 10 quads per year (2.9×10^{12} kWh) in 2010, corresponding to an annual savings of at least $64 billion, in current dollars. It is important to note that this savings would accrue each year at no additional expenditure of capital. This enormous potential should easily justify a much greater emphasis on these strategies; however, this potential will probably not be realized if current attitudes continue.

In addition to building owners, there would be many beneficiaries if we achieve the full potential of conservation and passive solar strategies. Our national security would be enhanced with less dependence on energy imports. The reduced need for new electric-generating capacity would also free vast capital resources for investments in modernizing U.S. industry. Environmental quality would be enhanced as a result of reductions in extracting and burning conventional fuels that pollute both the air and water. The private building industry and its suppliers would benefit from more construction, especially to retrofit existing buildings. Building designers would benefit from a greater need for their services. Unemployment would be reduced in response to increases in the demand for local construction skills.

1.2.9 Conclusions

Passive solar development has been at least partially successful in the United States. But current public complacency about energy issues in general has greatly slowed the technology's penetration of the market. This complacency is clearly associated with decreases in oil prices, even though the cost of energy for heating, cooling, and lighting buildings is no lower than it was in 1977 to 1979 when interest in the technology was increasing.

We have learned how to build much more efficient buildings and how to use passive systems to effectively harness environmental energies. The challenge now is to regain lost momentum. The best way to do this is to emphasize the fact that passive solar techniques lead to a higher-quality interior environment. The energy issue has already received ample attention. We should stress the thermal comfort possible with passive heating and cooling and the delightful character of natural light. Since a major purpose of a built environment is to provide shelter and comfort, we should let it be known that passive strategies can do this very well indeed—both reliably and economically.

1.3 Passive Systems Analysis

1.3.1 Historical Context

The analysis of passive systems followed in a natural way from the great upsurge in interest in passive systems that accompanied and followed the first Passive Solar Heating and Cooling Conference, which took place in Albuquerque, New Mexico, in May 1976. Some work had preceded that conference, most notably the evaluations that were done for the Skytherm® roof-pond house at Atascadero, California. The scientists and engineers who were involved in the analysis of passive systems had generally little background in buildings science per se, but they were physicists or engineers well versed in the simulation of other physical systems ranging from electrical circuits to nuclear reactors. This led to a rich mix of modeling approaches as well as to some confusion in terminology. Eventually, as the field matured and a nomenclature was developed for passive systems, the approaches tended to come together and the terminology tended to normalize.

Before 1976, a comprehensive approach to the thermal modeling of buildings had already been developed. This was documented in the literature of organizations such as the American Society of Heating, Refrigeration, and Air-Conditioning Engineers (ASHRAE), the Army Construction Engineering Research Laboratory (CERL), and the Building Technology

Center at the National Bureau of Standards (NBS, now the National Institute for Standards and Technology). The techniques had been committed to large computer programs, such as BLAST and NBSLD, and generally used the thermal-response-factor, time-domain approach. This work is described in volume 4 of this MIT series, edited by Bruce Hunn.

Most of the scientists involved in passive systems modeling did not use these existing computer programs but, instead, developed their own methods and wrote new programs. One reason was that the large, existing building-analysis programs had not been designed for buildings that were strongly solar driven, and, in one way or another, they were not programmed to simulate systems or operating modes of greatest interest. For example, the programs did not (at that time) allow the building inside temperature to "float" between upper and lower thermostat settings, as necessary to simulate the normal behavior of direct-gain passive heating systems (or most any other building, for that matter). Nor could the programs handle indirect approaches, such as floating-temperature sunspaces with convective coupling to the building and thermal storage walls with thermocirculation.

A second reason for this departure was that many of the passive-system modelers had already been involved in the simulation of active solar systems, for which new models were required anyway. In many of these programs the building was often considered as a simple resistive load on the thermal storage (that is, without dynamics). Thus, it was natural for the scientists to extend these techniques to passive systems rather than to rely on the existing large, inflexible programs.

Many other computer programs for building thermal analysis existed before 1976, but they were generally designed for sizing mechanical HVAC equipment. The programs were not designed to simulate performance but only to estimate peak building loads under design-day conditions. Most of the programs assumed the building to be in steady state and ignored dynamic effects. Furthermore, many of these programs were proprietary to commercial HVAC companies, were poorly documented, and were regarded with some suspicion by a community seeking to minimize its reliance on mechanical systems. Thus, the existence of programs such as NBSLD and the proprietary equipment-manufacturers' programs had little impact on the development of passive system analysis tools.

The net result was that passive systems modelers found themselves in a new field in which methods developed for other purposes were applied to passive solar buildings. It was a rich and exciting evolutionary period. At the beginning, the focus was on passive solar heating, and only later (around 1980) did cooling considerations receive much attention. Most of the early attention to cooling was not directed to methods of enhancing the natural rejection of heat by the building but rather to an evaluation of the summer-cooling consequences of winter passive-heating design strategies. To a large extent, this mind-set persists; accurate, validated methods do not exist for evaluating buildings with windows open for natural ventilation or with design features that would enhance radiative cooling. This is partly because such an analysis is technically a much more difficult job than a heating analysis and partly because such work was postponed until funding for development had greatly diminished.

1.3.2 Overview of Chapters 2–4

Three chapters of volume 7 deal with passive systems analysis. They address in some detail methods of analysis of passive buildings and outline some of the principal conclusions that have been drawn from the systematic use of these techniques to explore the nature of passive building performance.

In chapter 2, "Building Solar Gain Modeling," Patrick Burns deals with the thorny subject of converting information typically supplied about solar gains—namely, global horizontal solar radiation—into the appropriate heat inputs within a simulation model. He presumes that the reader knows how the atmosphere diffuses direct-beam solar radiation entering it into a mixed field of beam and diffuse radiation impinging on the building site. This subject is covered in detail in volume 2 of this series, *Solar Resources*, edited by Roland L. Hulstrom. But this analysis, complex as it is, is only the beginning of the story. Various site effects further modify the solar radiation, such as ground reflectance onto building walls and shading by neighboring buildings or trees. Building elements such as overhangs, projecting walls, or louvers can dramatically change the solar gains impinging on the windows. Then, window transmission must be taken into account, depending on the number of panes of glass (or other glazing material), refractive index, thickness, absorption, etc. Once inside the building, the solar radiation can be reflected or absorbed, or it can even exit the building through the same or another window. Altogether, it is a complex and somewhat uncertain process, but a critical one if the succeeding thermal analyses are to be at all accurate.

In chapter 3, "Simulation Analysis," Phillip W. B. Niles picks up where Burns' chapter leaves off. He explains both the principles and details of dynamic heat transfer and storage in buildings. Various levels of detail are appropriate for various applications, from a simple "degree-day" model that treats the building as a resistive load, to more subtle and complex interactions, such as radiation to a sky that is generally colder than the air temperature. The messy subject of infiltration is discussed, as is the more tractable but mathematically complex subject of direct radiative heat exchange between interior surfaces. Types of models discussed include the popular thermal-network and weighting-functions approaches, and also the powerful, though sometimes confusing, harmonic analysis technique. Various models of thermal comfort are described. Throughout, there is an attempt to present methods that will be directly useful to persons developing thermal models of buildings.

In chapter 4, "Simplified Methods," G. F. Jones and William O. Wray describe the process used to develop simplified methods for estimating passive building performance based on monthly or annual analysis rather than hourly simulation. The purpose of these methods is to generalize the hourly simulation result of most interest, annual energy savings, in terms of the principal variables, namely, the building load coefficient, the solar collection aperture size, and the climatic variables of temperature and solar radiation. Simplified methods developed by different research teams are all based on correlating results obtained by simulations. The precision that is sacrificed is acceptably small and easily justified by the enormous gains in calculation speed and convenience. Other simplified methods, such as the diurnal heat capacity method for estimating temperature swings, are also described as well as a method for accounting for mass effects on passive system performance based on effective heat capacity.

1.4 Passive Solar Heating

1.4.1 Overview of Chapters 5–10

The subjects of these six chapters are somewhat diverse, but they all relate to the implementation of passive solar heating in buildings.

In chapter 5, "Materials and Components," Timothy E. Johnson reviews several research efforts that have come out the laboratory and are now making their way into the marketplace. He correctly places the greatest emphasis on window products that reduce heat loss or increase transmittance to solar heat or daylight, since these have had the most profound effect on passive solar applications. As an architect, Johnson gives equal weight to issues of architectural practice and energy performance; here again, the window is the key element. Low-emittance coatings and gas fills, both firmly established in the marketplace, are discussed as well as the emerging concepts of aerogel windows, evacuated windows, and switchable glazings. Other components reviewed include movable insulation products for windows and both conventional and phase-change materials for heat storage.

In chapter 6, "Analytic Results for Specific Systems," Robert W. Jones reviews and summarizes the considerable volume of material that has been published based on calculations made with hour-by-hour computer simulation programs, such as those described by Niles in chapter 3. The emphasis here is on showing the predicted results of passive systems rather than on calculation technique. Much of this material has now been interpreted and made its way into guidelines for designers. Time spent studying this material and its many references will save future researchers and designers the time it takes to replicate most of the many thousands of annual computer simulations already made. Many sensitivities will change somewhat with the specific building, but the trends will generally be the same. The various useful performance indices that have been devised to present the results are reviewed. Parameter studies showing how performance is affected by design decisions are evaluated. This chapter gives a good general overview of the performance that we can expect from passive systems.

In chapter 7, "Test Modules," Fuller Moore presents a comprehensive review of the many test modules that have been constructed and monitored to evaluate the performance of passive solar concepts. These experimental structures vary in size from test boxes that are roughly a meter on each side, to full-height, single-zone test rooms, to nearly full-size houses, some with multiple rooms. Test modules allow data to be taken economically under carefully controlled conditions and provide a useful intermediate step between analysis and the construction and monitoring of actual passive solar buildings. The results are used to demonstrate proof of concept, to compare by direct observation how different concepts work, and to provide data useful for validating computer simulation programs.

Test modules can also be effective teaching tools for students. Monitored results from several test modules are summarized.

In chapter 8, "Building Integration," Michael J. Holtz reviews the evolution of demonstration projects and design competitions concerned with the integration of passive solar concepts into actual occupied buildings. Residential and commercial buildings are discussed separately. Design elements and how they are integrated are elaborated. A historical review describes the extensive effort undertaken, both with government support and wholly in the private sector, to move passive solar systems into the marketplace. Many major programs are discussed, such as the Passive Solar Commercial Building Program and the Passive Solar Manufactured Building Program. Lessons learned are summarized. Design competitions that were intended to foster dissemination are described, and several noteworthy buildings are reviewed. Recommendations are made for future directions. A major contribution of Holtz's work is in documenting the history of the vital and exciting evolution of passive solar technology from an obscure and misunderstood curiosity into a mainstream, proven approach to saving energy in the buildings sector.

In chapter 9, "Performance Monitoring and Results," Donald J. Frey presents a comprehensive review of monitoring and results, primarily in the United States, from 1969 to 1987, for full-size residential and commercial passive solar buildings. A variety of data-collection methods are described, and five different approaches to data analysis are discussed. Monitoring activities that have been reported in the literature range from individual buildings, such as the Wallasey School in England, to aggregations of several buildings, such as the National Solar Data Network (20 buildings), the SERI Class B (60 houses) and Class C (335 houses) monitoring programs, the Tennessee Valley Authority solar program (35 houses), and the DOE Passive Solar Commercial Buildings Program (19 buildings). Generally, the results show very good performance; savings of 50% to 95%, compared with the energy use and energy cost of conventional construction, are typical. It is usually less clear how to accurately apportion the sources of savings; for example, whether the good performance can be attributed to load reduction (conservation) or to passive solar systems. Nonetheless, the results, taken individually or in the aggregate, show remarkable performance. It is clear that conservation, passive solar heating, natural cooling, and daylighting all play important and complementary roles. A major contribution of this chapter is in clarifying the various methods of evaluating monitored data, a difficult and sometimes ambiguous task that is usually underestimated.

In chapter 10, "Design Tools," John S. Reynolds presents a short but precise review of design tools from the perspective of an architect. The categorization of design tools is discussed, and a number of specific tools are described. Reynolds points out that in recent years we have seen a tendency toward sophisticated, numeric-based design tools that are strong on evaluation but weak on guidance. It is a good note to end the book on: major work is needed to develop tools that will provide assistance in conceiving a good design in the first place rather than just quantify the shortcomings of a proposed design.

1.4.2 Recent Developments

The authors of this book have presented a comprehensive review of both the analysis and realization of passive solar buildings up through about 1987. As we pointed out earlier, the phasedown of research on all solar energy topics has led indirectly to a dramatic reduction in new published material. Many groups were presenting material in reports, journals, and conferences in 1981, but then the number of institutions and individuals working in the field and the volume of research and presentations declined, roughly in proportion to the funding available for research. Thus, there are not very many recent developments. Two of these, though, are discussed briefly in the paragraphs that follow: short-term energy monitoring and guidelines for home builders.

1.4.2.1 Short-Term Energy Monitoring Short-term energy monitoring (STEM) is briefly described by Frey in the chapter on performance monitoring (chapter 9). The basic approach has been developed by Subbarao (Subbarao 1988; Subbarao et al. 1988). The most recent developments include an increase from 4 to 21 in the number of buildings that have been monitored using the method, the validation of the method, and the use of this method to evaluate building cooling strategies.

There is potential for confusion in the multiplicity of acronyms used during the evolution of this monitoring method. The original work was called BEVA, which stands for Building Energy Vector Analysis. This refers to Subbarao's use of harmonic analysis to evaluate building performance (see chapter 3). The term BEVA, however, has been phased out. Subbarao used the BEVA mathematical formalism in developing a *method* for analyzing buildings being monitored, called PSTAR, which stands for Primary and Secondary Terms Analysis and Renormalization. This acronym accurately describes the basic procedure. The *project* at the Solar Energy Research Institute that puts this method to use in evaluating monitored buildings data is called STEM, which stands for Short-Term Energy Monitoring. Thus, there is a distinction between the method (PSTAR) and the project (STEM).

In the PSTAR method, a building hourly simulation model is calibrated based on data taken during a test that is usually conducted over three days. First, a simulation model is developed based on the building plans or on site observations made by the person doing the test, or both. This is called the audit model; it serves as a first approximation to the real building. Then, using another model derived by harmonic analysis and the audit model, all significant heat-flow terms are disaggregated according to a specific and unique categorization. The three primary heat-flow terms are renormalized (by multiplying by three constants) in order to force agreement between the observed data and the model during selected time periods when the heat-flow term is dominant or at least large. These heat flows are the steady-state heat loss, the solar gains, and the heat discharges from storage as a result of changes in inside temperature. Minor heat flows that are not renormalized include discharges resulting from changes in outside temperature, depression of the sky temperature below outside air temperature, changes in infiltration, and heat transfers to adjacent spaces such as a garage or basement.

During a STEM test, the heat required to maintain a desired inside temperature level is supplied from electric-resistance convection heaters controlled by the data-acquisition computer. This heat, plus any other casual electric heat introduced into the building, is accurately measured using a Hall-effect power meter attached to the main electric junction box.

Once the audit model has been renormalized, it can be used to make long-term predictions of building performance based on hourly data from historical records, such as a TMY (typical meteorological year) data set. Peak building demand or other quantities of interest can also be predicted.

The renormalized building model can also be used to infer energy flows due to other causes. The building, in effect, is used as a dynamic calorimeter. For example, the overall efficiency of the normal backup heating or cooling system in the building can be measured by comparing the heat predicted by the model over a period of a few hours to the total measured fuel input to the system. This is normally done routinely on the last night of the STEM test. The same approach has been used to measure heat removal by natural ventilation and the shading coefficients of miniblinds.

PSTAR has been validated in a number of ways (Burch 1990). One is to demonstrate that the three renormalization constants are repeatable from test to test. Another is to verify that long-term energy use is accurately predicted. And another is to accurately predict other known heat flows, such as those due to a fan-coil cooling unit (when the water flow rate and temperature difference are known) and forced ventilation of the building (when air flow and temperature difference are known). Other consistency tests have lent researchers confidence in the method.

1.4.2.2 Guidelines for Home Builders A difficult problem facing the author of design guidelines in the United States is how to deal effectively with the incredible variety of weather situations. Most authors either present a general method that can be used with monthly weather data or attempt to regionalize the weather. The first approach invariably leads to either an unacceptable loss of accuracy and specificity, or it is far too unwieldy to present and too complicated to apply, thereby losing its desired audience. The second approach does not work well because there are just too many climatic variations to be considered. For example, there are seven important climate zones in Arizona and ten in California.

The 58-page package, *Passive Solar Design Strategies: Guidelines for Home Builders*, takes a different approach (Passive Solar Industries Council 1989). Each version of the booklet is based on weather data for one specific place, and thus it addresses only that locality. This complicates the production and distribution of the booklets but makes the user's life much easier.

The guidelines booklets are produced by a special computer software package that combines text and numbers to create site-specific packages, including a ready-to-print copy of the guidelines books and viewgraphs for workshop presentations. Thus it is possible to produce individual guideline booklets as they are requested. The packages can be generated for 205 U. S. locations, or they can be modified and customized to apply to adjacent sites, as long as long-term monthly weather data are available. As a result, guidelines for more than two thousand sites can potentially be created. These booklets are available from the Passive Solar Industries Council, which also presents seminars based on the material.

The basic approach in the guidelines is to reference everything to a base-case house design. This provides a logical starting point for the designer. The base case is a standard 1,500-sq-ft house built in conformance with typical insulation levels used in the location. The guidelines present the annual heating and cooling loads of the base-case house in the location and clearly indicate the effectiveness of various conservation, passive solar, and natural cooling strategies. Several example house designs are described that will save 20%, 40%, and 60% compared with the base-case energy requirement.

A second key part of the guidelines is a set of four one-page, fill-in-the-blanks worksheets. These enable designers to quickly evaluate the energy and comfort characteristics of a proposed design in their location. Annual heating and cooling energy, and both conservation and comfort indices are calculated with the aid of a set of location-specific tables. The values obtained can be compared with the base-case house presented in the guidelines; the design can then be adjusted as necessary to meet design goals. These worksheets take the guidelines beyond the realm of the typical write-up that gives only general advice and make them a really useful tool by providing a design procedure capable of quantitative evaluation. This is essential if the designer is serious about energy efficiency.

An example house illustrates how to use the worksheets. This employs improved insulation, a sunspace, shading, some added mass, and a ceiling fan to reduce heating and cooling loads. The annual energy reduction is typically more than 50%, depending on location. Filled-out worksheets included in the guidelines are intended to be used as a case study during the workshop presentation.

A parallel effort has resulted in a computerized version of the guideline worksheets. Known as *BuilderGuide*, this program allows designers to complete the worksheet calculations in a fraction of the time needed for the handwritten version. The Passive Solar Industries Council disseminates *BuilderGuide* as an integral part of its overall guidelines program.

References

Balcomb, J. D. 1983. Evaluating the performance of passive solar heated buildings. *Proc. 6th Technical Conference of the ASME Solar Division*, Orlando, FL, April 19–21, 1983. New York: American Society of Mechanical Engineers.

Balcomb, J. D., R. W. Jones, R. D. McFarland, and W. O. Wray. 1982. Performance analysis of passively heated buildings: Expanding the SLR method. *Passive Solar Journal* 1 (2): 67–90.

Balcomb, J. D., R. W. Jones, R. D. McFarland, and W. O. Wray. 1984. *Passive Solar Heating Analysis.* Atlanta, GA: American Society of Heating, Refrigerating, and Air-Conditioning Engineers.

Balcomb, J. D. 1984. Passive solar research and practice. *Energy and Buildings,* vol. 7. Elsevier-Sequoia, pp. 281–295.

Burch, J. D., K. Subbarao, C. E. Hancock, and J. D. Balcomb. 1990. *Repeatability and Predictive Accuracy of the PSTAR Short-Term Building Monitoring Method.* SERI/TR-254-3606. Golden, CO: Solar Energy Research Institute.

Deming, G., and J. J. Duffy. 1986. Un-utilizability design tool predictions compared to auxiliary measurements of passive solar residences. *Proc. 11th National Passive Solar Conference,* June 7–11, 1986. Boulder, CO: American Solar Energy Society, pp. 88–93.

Gordon, H. T., P. R. Rittleman, J. Estoque, G. K. Hart, and M. Kantrowitz. 1986. Passive solar energy for non-residential buildings. *Advances in Solar Energy,* vol. 3. New York: Plenum Press, pp. 171–206.

Passive Solar Industries Council. 1989. *Passive Solar Design Strategies: Guidelines for Home Builders.* Booklets for specific localities available from PSIC, 1090 Vermont Ave. NW, Suite 1200, Washington, DC 20005.

Renewable Energy Institute. 1986. *Annual Energy Technology Review, Progress Through 1984.* REI, 1001 Connecticut Ave., Suite 719, Washington, DC 20036.

Subbarao, K. 1988. *PSTAR—Primary and Secondary Terms Analysis and Renormalization.* SERI/TR-254-3175. Golden, CO: Solar Energy Research Institute.

Subbarao, K., J. D. Burch, C. E. Hancock, A. Lekov, and J. D. Balcomb. 1988. *Short-Term Energy Monitoring (STEM): Application of the PSTAR Method to a Residence in Fredricksburg, Virginia.* SERI/TR-254-3356. Golden, CO: Solar Energy Research Institute.

Ternoey, S., L. Bickel, C. Robbins, R. Busch, and K. McCord. 1985. *The Design of Energy Responsive Commercial Buildings.* New York: Wiley-Interscience.

2 Building Solar Gain Modeling

Patrick J. Burns

2.1 Introduction

2.1.1 Overall Perspective

This chapter deals with the calculation of solar heat gain of buildings. Such energy gain can occur through transparent surfaces, such as windows or skylights, or through opaque surfaces, such as walls or roofs. The perspective of this chapter begins with the solar radiation near the building. This radiation is regarded as the sum of the direct normal radiation, the sky diffuse radiation, and the ground-reflected radiation (see also volume 2). These radiation components are followed through their interaction with surfaces to obtain the instantaneous rate of heat gain to the building (or to specific building surfaces if this level of detail is desired).

The objective of this chapter is to describe how these heat transfer rates may be determined and the results incorporated into a thermal model of a building. From such a thermal model, the thermal response of the building may be predicted from the environmental driving functions: the rate of solar heat gain, the ambient temperature, the wind speed, and other variables. The thermal characteristics of buildings (solar absorptances of surfaces, thermal conductivities of materials, etc.) are usually independent of the solar gains; thus calculations of solar heat gains are most efficiently done as input calculations independent of the specific thermal model. Exceptions include shading control algorithms and shading materials that depend on the solar radiative transfer—the methods presented in this chapter are sufficiently general to cover either case. Thermal models are not discussed in this chapter (see Niles, chapter 3).

In general, three applications exist for solar gain calculations: 1. simplified design calculations that require very approximate data, 2. less detailed thermal network computer models in which the total solar gain is required, and 3. more sophisticated thermal network computer models that require the solar heat gain to be apportioned to individual surfaces. This chapter briefly reports the available methods for these applications. Thus, the emphasis is on methods of obtaining the solar data required for input into a computer-based design method, for example from a solar preprocessor.

2.1.2 Conventions and Definitions

In this chapter, the terms *glass, glazing, solar aperture*, and *windows* encompass all types of transparent and translucent materials used in the building industry. Areas of glass refer to the net area. That is, if the entire projected areas of the mullions and sill are subtracted from the projected area roughed out for the window (the gross area), the net window area is obtained. The net area may usually be assumed as constant with little absolute error.

The formulas are developed with surface i as the surface of interest and surface j as an ancillary surface. Any number of surfaces i can exist, but the surface specifier i is omitted from the formulas unless it is absolutely essential. The calculations assume actual solar radiation data are available. The asymmetry (around solar noon) inherent in such data can significantly affect the time response in buildings of large thermal capacitance. Therefore, these data are preferred over artificial data that are symmetric about solar noon. In this context, the data are integrated (averaged) over the hour and reported at the end of the hour. The instantaneous calculations of rates are then valid, on the average, over the previous hour. Whenever relevant, it is assumed that the actual composition of buildings is approximated by series and parallel resistance networks containing lumped values of heat capacitance and thermal resistance.

Primed time quantities indicate values in hours; all other time quantities are in seconds. Angular values are in degrees, not radians. A "barred" quantity indicates an average value. A primed radiative material property indicates a directional dependence; a superscript s indicates a specular value (no superscript indicates a diffuse value).

2.1.3 Organization of the Chapter

The chapter is arranged logically as solar energy travels from its origin in the sun to its absorption by building surfaces (first internal, then external). Section 2.2 briefly discusses solar radiation at the earth's surface, including the directional and spectral distributions and the intensity. Section 2.3 describes the geometry of the building surface–sun system, and in section 2.4 a method is presented for calculating the incident energy in the absence of external shading. Section 2.5 covers external factors such as the terrain, trees, and other surfaces that affect the incident energy. Simplified methods are briefly discussed in section 2.6. Section 2.7 presents the types and properties of glazing materials. Section 2.8 discusses internal factors, including furniture, plants, and shading devices. Section 2.9 addresses solar effects on opaque exterior surfaces. In section 2.10, strategies are presented for effecting efficient computation. Final observations and comments in section 2.11 complete the chapter.

2.2 Solar Radiation at the Earth's Surface

The important features of solar radiation at the earth's surface are its directional distribution, spectral distribution, and intensity. The directional distribution is of paramount importance because it allows the calculation of the incident energy on building surfaces. The spectral distribution is important in cases where the radiative properties (transmittance, absorptance, and reflectance) depend on wavelength. See also volume II.

A great deal of information is available in the literature (ASHRAE 1981, Threlkeld 1970, Henderson 1970, Frazier 1980, Thekaekara 1973, Moon 1940, Threlkeld and Jordan 1958, Parmelee and Aubele 1952, Bliss 1961, Watt 1978, Goody 1964, Coulson 1975) on the nature of solar radiation at the earth's surface. The physics of the interaction of the solar radiation with atmospheric constituents is extremely complicated—even if the components of the atmosphere were well defined. The actual case is even more complex because atmospheric components are known only approximately at best. A great deal of fundamental research is still being conducted on this topic.

The simplified engineering approach that circumvents these difficulties is measuring the solar radiation incident at the earth's surface. No atmospheric modeling is necessary if these data are measured at the location of interest. However, models for the directional and spectral distributions must still be applied to the data, if these details are not measured, to perform calculations of the energy incident on building surfaces.

Figure 2.1 shows the various idealized components inherent in the energy incident on the earth's surface. Outside the atmosphere, the solar radiation is undisturbed and arrives in a direct line from the 1/2-degree cone, which is the angle occupied by the disk of the sun. On entering the atmosphere, this radiation is scattered by air molecules, water droplets, ice crystals, dust, and chemical pollutants. A significant amount of the scattered radiation is only slightly diverted from its original path and arrives in a 5-degree cone. This slightly scattered radiation is termed the

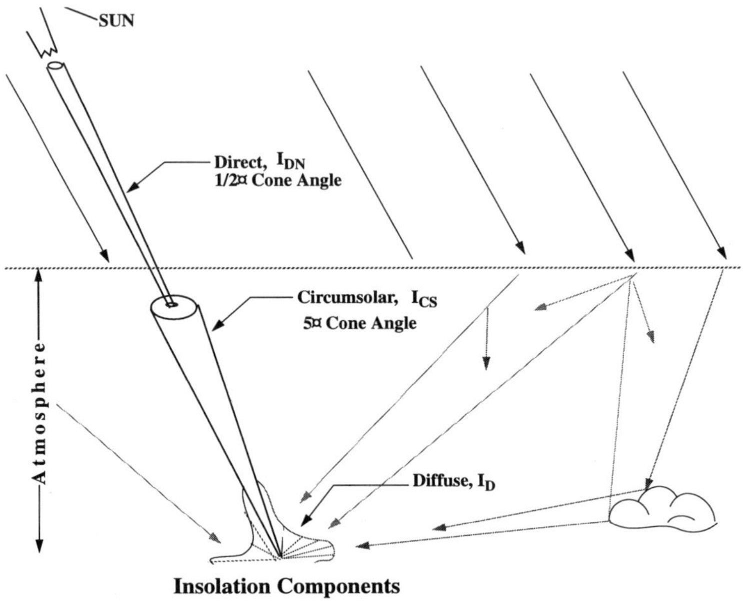

Figure 2.1
Directional distribution of solar radiation at the earth's surface. Source: Watt 1978.

circumsolar radiation because it arrives centered around the direct solar radiation. The sum of these two components is the quantity measured by pyrheliometers and is termed the direct normal, or beam radiation. It is accurately modeled to arrive in a path coincident with the ray from the center of the sun to the center of the earth.

The remaining radiation, called the diffuse component, is scattered in all directions with a distribution that depends on the specific state of the atmosphere. However, the distribution of the diffuse radiation typically is heavily weighted around the direction of the direct normal radiation because of the preferential forward scattering inherent in the upper atmosphere. Also, there is usually a hump in the distribution at angles near the horizontal, known as "horizon brightening," caused by reflection from clouds and the lower-level atmospheric constituents of both the incident and the ground-reflected radiation.

The exact directional dependence of the solar radiation and the distribution of energy between the direct normal and diffuse components depends on wavelength. The total incident energy is obtained by including all incident wavelengths, and the total energy measured on the ground implicitly contains all wavelengths.

Finally, existing solar radiation data are valid for the atmospheric conditions of the past. As atmospheric pollution and associated photochemical changes continue to occur, the incident solar radiation will change. This change will be very location dependent.

2.2.1 Spectral Distribution

The spectral distribution of solar radiation refers to its relative amount at each wavelength or frequency (Leckner 1978, Hatfield, Giorgis, and Flocchini 1981, Brine and Iqbal 1983, Bird, Hulstrom, and Lewis 1983, Guzzi et al. 1983). Since different wavelengths of solar radiation interact differently with the atmospheric constituents, the spectral distribution is different at the earth's surface than outside the earth's atmosphere. In fact, the distribution changes continuously as the atmosphere is traversed. The amount of atmosphere the radiation encounters is quantified as the air mass m, a dimensionless measure proportional to the number of air molecules along the path length through the atmosphere, defined as 0 outside the atmosphere and 1 when the sun is directly overhead at sea level. The air mass may be smaller than 1 at elevations above sea level or greater than 1 for other angles.

Figure 2.2 shows the spectral dependence of solar radiation for air masses of 0, 1, and 5. Outside the atmosphere, the radiation is characteristic of that emanating from a blackbody source at 6000 K. Preferential absorption by atmospheric constituents at selected wavelengths causes the spectral distribution to shift as it travels to the surface. It is apparent from the figure that the energy shifts to higher (longer) wavelengths as the air mass increases. The total amount of energy, characterized by the area under the curves, decreases as the air mass increases, to about one half the extraterrestrial value at $m = 1$ and to about one quarter at $m = 5$.

Although the reduction in energy is directly accounted for when using measured data, the shift in wavelength is normally neglected. Indeed, most material radiative properties reported in the literature are those corresponding to the solar spectrum at $m = 1$ and are only approximately valid for higher air masses. As the air mass increases, the radiative properties should be weighted more toward higher wavelengths, resulting in larger values of the absorptance for dielectrics and smaller values for metals.

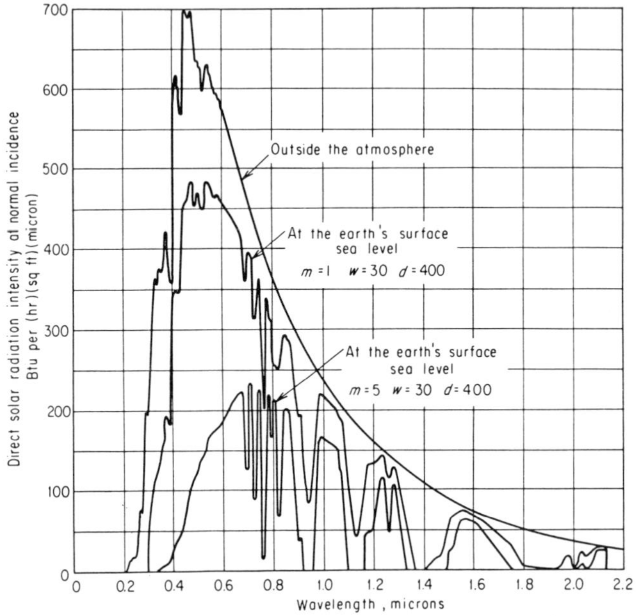

Figure 2.2
Spectral variation of solar radiation at the earth's surface ($m = 1$ and $m = 5$) and extraterrestrial solar radiation.

These differences can result in errors of up to 10% for building materials with highly wavelength-dependent properties, such as some metals and light-colored paints.

2.2.2 Directional Distribution

The solar radiation at a building surface is normally approximated as the sum of three contributions: 1. direct normal radiation, 2. sky diffuse radiation, and 3. radiation reflected from the ground. In this approximation, the surface is considered isolated; that is, no account is taken of radiation reflected from other surfaces. The standard measurement—the total radiation incident upon a horizontal surface—contains only the direct normal and sky diffuse components. It is much easier to separate the total horizontal radiation into components because it does not contain the ground-reflected radiation (which is determined separately). The next step is to correlate the direct normal radiation (a well-defined, measurable quantity), using data sets that contain separate measurements of direct normal radiation, with the total horizontal radiation and weather variables. Typically, the direct normal solar radiation is correlated with the clearness index, K_T, defined by

$$K_T = I_h/H_e, \tag{2.1}$$

where H_e is the extraterrestrial solar radiation incident on a horizontal surface,

$$H_e = S_c[1 + 0.034 \sin(2\pi\overline{N}/365)] \sin \beta, \tag{2.2}$$

where $\overline{N}$ is the midday of the month, S_c is the solar constant, and I_h is the solar radiation incident on a horizontal surface.

An alternative approach is to form a correlation of the diffuse component with K_T. This is generally not as effective because the diffuse component is not a well-defined, easily measurable quantity. Indeed, an absolute measurement would require integration over the sky hemisphere (excluding the circumsolar direction), and this is not done in practice. (Existing correlations of the sky diffuse sometimes result in direct normal values that are far greater than the solar constant and should be used with caution.)

Even when the direct normal component is the object of a correlation, it is necessary to assume a directional distribution function for the sky diffuse in order to calculate the radiative flux on surfaces of orientations other than horizontal. The assumption often made is that the sky diffuse component is isotropic. The nature and effects of this assumption have been investigated by Sowell 1978, Erbs, Stauter, and Duffie 1980, Davies and McKay 1982, Ma and Iqbal 1983, LeBaron and Dirmhim 1983, Morris and Lawrence 1971, and Dave 1977.

An effort to improve the modeling is evident in the recent work of several investigators (Puri et al. 1980, Steinmuller 1980, Healey, Near, and Boston 1980, Iqbal 1981, Thomalla et al. 1983). In this work, a fraction of the diffuse is included with the direct normal radiation through a weighting factor proportional to the area of the sky projected in the beam direction. This method is known as the moment method, because moments (or powers) of the weighting factor are included in the directional modeling. Conceptually, this model represents the first two terms of a truncated, double Fourier series expansion in the altitude angle and the circumferential angle, which would, if untruncated, represent any sky diffuse distribution. It seems probable that it will be worthwhile to include only a very small number of these terms before the lack of data and insufficient measurement resolution introduce error as large as does the truncation.

From an engineering viewpoint, it is desirable to assess the difference between the two models. Annual differences range to as high as 13% in absorbed energy under a typical overhang and are highest in clear, sunny areas, and hourly differences are as high as 100% in the summertime when the absorbing surface is entirely shaded from beam radiation (Reuth 1984).

2.2.3 Polarization

The radiation striking a surface from a particular direction is most fundamentally considered in terms of not only a single wavelength but also a single component of polarization. The radiation striking the outside of the atmosphere has approximately random (equal components of) polarization. Observations of the polarization at the earth's surface indicate approximately random polarization except at angles nearly 90 degrees away from the direction of the sun. However, there is very little energy in solar radiation incident from these directions, and polarization effects may be neglected with good accuracy. This is fortunate because the radiative properties of common building materials are measured under random polarization.

2.3 Local Solar Radiation Geometry

The local solar radiation geometry (and, to a lesser extent, material properties) determines at each instant the amount of the direct normal component of the incident energy that will be transmitted through or absorbed by a building surface. The temporal integration of this energy provides the direct normal contribution to the building solar gain.

2.3.1 Coordinate System

It is convenient to define a coordinate system with the origin located at the center of the base of the collecting surface (that is, the building surface of interest). Discussion of such a coordinate system is widely available in the literature, e.g., in Threlkeld 1970, ASHRAE 1981, Kreith and Kreider 1978,

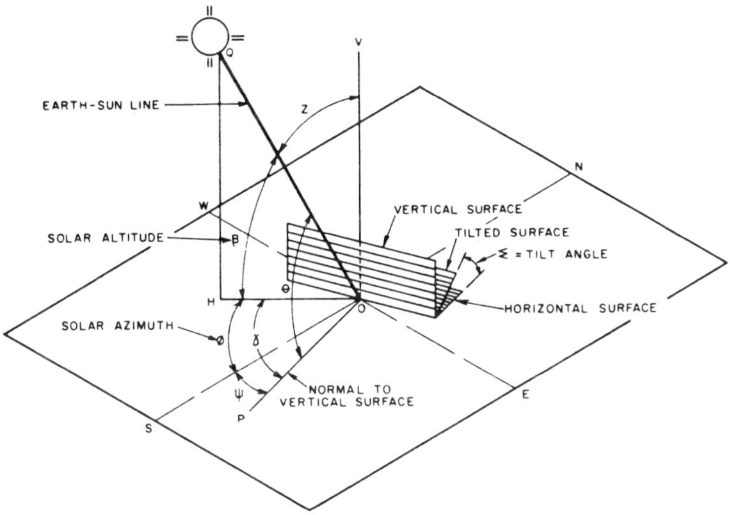

Figure 2.3
Solar angles for vertical and horizontal surfaces.

and Duffie and Beckman 1980. The notation used by ASHRAE (1981) is used here because the *ASHRAE Handbook of Fundamentals* is probably the most widely used reference on this material.

The building surface, the sun, and the associated angular definitions are shown in figure 2.3. The surface orientation is defined by the surface azimuth Ψ that lies in the horizontal plane and advances counterclockwise (looking from above) from due south, and the surface slope Σ that lies in the plane normal to the collecting surface. The position of the collecting surface is defined relative to the earth's centroid by the latitude LAT, the longitude LON, and the local time t. The earth's position is defined relative to the sun by the distance from the sun, d, and the day of the year, N, which determines the declination δ. The declination is the angle between the plane of the ecliptic (that is, the plane of orbit of the earth) and the plane of the earth's equator. Approximately (Cooper 1969),

$$\delta = 23.45 \sin[360(284 + N)/365]. \tag{2.3}$$

The angle of incidence θ is the angle between a normal to the surface and the vector to the sun.

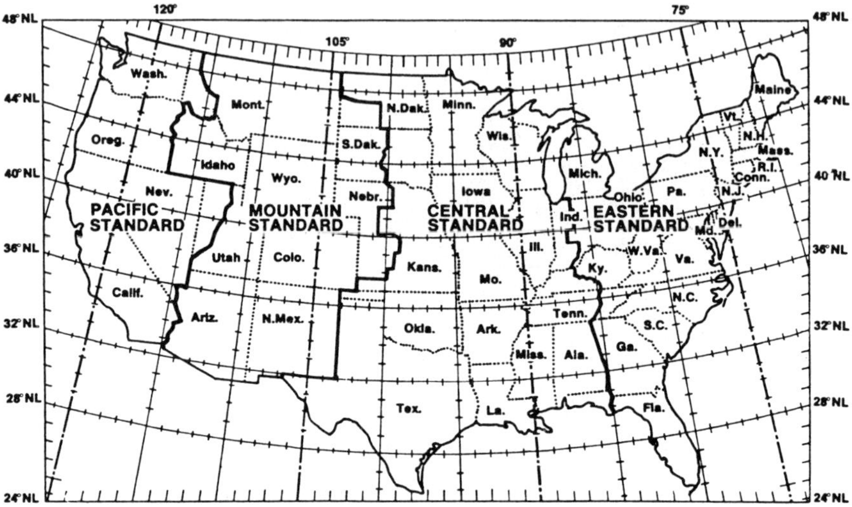

Figure 2.4
Standard longitudes and time zones for the United States. Source: Mazria 1979.

2.3.2 Local Solar Time

It is necessary to use a time scale that consistently reflects the position of the sun relative to the surface of interest (Pitman and Vant-Hull 1978). That time scale is local solar time in which the sun is directly overhead at solar noon. The corrections to ordinary local time are generally not significant in predicting the amount of incident energy, but they are important in determining the local sunrise and sunset times. Some solar processors are sensitive to these times of day.

The first correction to local time arises because of time zones. The time zones are nominally centered at the standard longitudes LON_s as shown in figure 2.4. A time zone associated with these standard longitudes nominally encompasses $\pm 1/2$ hour of revolution of the earth corresponding to ± 7.5 degrees of longitude. The correction for any longitude LON is as follows:

$$\Delta t_L = (LON_s - LON)3600/15 = 240(LON_s - LON). \tag{2.4}$$

The other correction arises because the earth's orbit is elliptic rather than circular, causing the speed with which the earth travels around the sun to change accordingly. This correction to the time is depicted in figure 2.5.

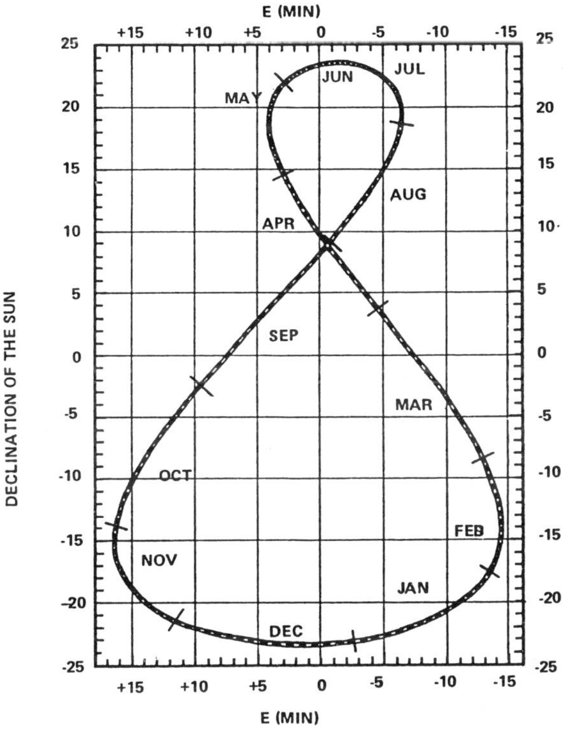

Figure 2.5
The analemma. Source: Watt 1980.

The curve is represented by the following approximation, which is known as the equation of time:

$$\Delta t_k = -(445.0 \sin D - 15.44 \cos D + 554.0 \sin 2D + 217.0 \cos 2D), \quad (2.5)$$

where

$$D = 2\pi N/365. \quad (2.6)$$

The total correction is obtained by summing the corrections given by equations (2.4) and (2.5). The result is the local solar time

$$t_s = t + 240(LON_s - LON) + (445 \sin D - 15.4 \cos D + 554 \sin 2D + 217 \cos 2D), \quad (2.7)$$

where t is the local standard time (in seconds).

2.4 Local Solar Radiation Estimation

In this section are presented the three components—beam, sky diffuse and ground-reflected—of solar radiation transmitted through transparent (or translucent) building surfaces. For the incident components, the transmittances may be set to 1. For the components absorbed on opaque exterior surfaces, the transmittances may be replaced with absorptances. For components absorbed by interior surfaces, the transmittances may be replaced with transmittance-absorptance products. Solar properties of various materials are discussed in sections 2.5, 2.7, 2.8, and 2.9.

2.4.1 Beam Radiation

The angle of incidence for beam radiation θ determines the key radiative properties (transmittance and reflectance) for beam radiation of the collecting surface. This angle is depicted in figure 2.3 as the cone angle between the incoming ray and the surface normal. Several ancillary angles are also included on the figure. These include the solar azimuth ϕ, which is measured clockwise in the horizontal plane from due south to the line defined by the vertical projection of the incoming ray onto the horizontal plane (east is negative, west is positive). The solar altitude β is the angle between this line and the incoming ray. The zenith angle Z is the complement of the solar altitude. The relative azimuth γ is the difference between the surface and the solar azimuths, $\phi - \Psi$.

It is most convenient to perform calculations in terms of the solar altitude β and the relative azimuth γ. These angles are computed in terms of the latitude L, the declination δ, and the hour angle H as follows:

$$\sin\beta = \cos L \cos\delta \cos H + \sin L \sin\delta, \tag{2.8}$$

where the hour angle H is defined as

$$H = 15(t_s' - 12), \tag{2.9}$$

and

$$\cos\phi = (\sin\beta \sin L - \sin\delta)/(\cos\beta \cos L). \tag{2.10}$$

The cosine of the incident angle is given by:

$$\cos\theta = \cos\beta \cos\gamma \sin\Sigma + \sin\beta \cos\Sigma. \tag{2.11}$$

For a vertical surface ($\Sigma = 90$ degrees), equation (2.11) reduces to:

$$\cos\theta = \cos\beta\cos\gamma. \tag{2.12}$$

The reader is urged to consult Threlkeld (1970), ASHRAE (1981), Kreith and Kreider (1978), and Duffie and Beckman (1980) for alternate forms of the equation for $\cos\theta$. The simplifications for other surfaces of specific orientations possess the advantage of ease of application. It is not necessary to extract the actual angle θ from equation (2.11) or (2.12) except in instances where the radiative material properties are to be calculated from θ and not from $\cos\theta$.

The beam radiation entering the building $\dot{Q}_b$, through a transparent surface may now be expressed as

$$\dot{Q}_b = f_b I_{DN} \cos\theta\, A_g \tau_b, \tag{2.13}$$

where f_b is the unobstructed fraction, or transmittance, of beam radiation passing through external shading obstructions (more detail will be given in section 2.5), I_{DN} is the direct normal flux, A_g is the net glazing area, and τ_b is the transmittance of the glazing for beam radiation. Equation (2.13) is applicable in the absence of exterior reflections. Where a well-defined transmittance-absorptance product exists, this may be substituted for τ_b to yield the net absorbed energy. For an opaque surface, the absorptance may be used in lieu of the transmittance to yield the absorbed energy.

2.4.2 Sky Diffuse Radiation

The next step in the process is to estimate the amount of sky diffuse radiation incident on a collecting surface. In the winter, the sky diffuse radiation incident on a vertical south surface is usually much smaller than the beam radiation. However, for other surface orientations, or in the summer, the incident sky diffuse and beam energies may be of approximately equal magnitude. The calculation depends on the specific model for the sky diffuse radiation.

Considering the purely isotropic sky diffuse model, the transmitted diffuse solar radiation is calculated as

$$\dot{Q} = f_s^0 I_s^0 F_{is} A_g \tau_d, \tag{2.14}$$

where f_s^0 is the unobstructed fraction of diffuse sky radiation passing through shading obstructions, I_s^0 is the diffuse sky solar flux calculated with an isotropic sky diffuse model, F_{is} is the diffuse view factor to the sky, and τ_d is the transmittance of the transparent surface for diffuse radiation. F_{is} is the view factor from the surface to the sky, defined as (Siegel and Howell 1982)

$$F_{is} = (1/A_g) \int_{A_g} \int_{\omega} \cos \chi \, d\omega \, dA_g, \tag{2.15}$$

where ω is the solid angle from a point on the surface to the sky. F_{is} is most conveniently conceptualized as the percentage of sky diffuse energy that is incident upon the surface of interest in the absence of shading. For no overhang, the view factor is

$$F'_{is} = \frac{1 + \cos \Sigma}{2}. \tag{2.16}$$

When an overhang exists, this value is reduced due to the occlusion of a portion of the sky by the overhang. The view factor from the aperture to the overhang, F_{io}, is readily available from many texts in formula (Siegel and Howell 1982, Howell 1982) or graphical (Reuth 1984) form. The actual sky view factor can then be computed as

$$F_{is} = F'_{is} - F_{io}. \tag{2.17}$$

When the moment method of handling the sky diffuse radiation is employed, the total diffuse energy transmitted through a transparent surface is calculated as

$$\dot{Q} = (f_s^0 I_s^0 F_{is} + f_s^2 I_s^2 F_{is}^2) A_g \tau_d, \tag{2.18}$$

which reduces to equation (2.14) if the diffuse radiation is considered to be totally isotropic $I_s^2 = 0$).

The second moment of the diffuse view factor F_{is}^2 (note that F_{is}^2 is not F_{is} squared) is defined as (Puri et al. 1980 and Davies and McKay 1982)

$$F_{is}^2 = (1/A_g) \int_{Ag} \int_{\langle \omega \rangle} \cos \chi \cos^2 \chi' \, d\omega \, dA_g, \tag{2.19}$$

where χ' is the angle between the direction of viewing and the line of incidence. The brackets enclosing ω indicate that the integral is to be performed over only values of χ' that are less than 90 degrees. However, this method presents some difficulty of application. The integral must be performed numerically over the glazing area and the solid angle, an excessive computational requirement. Fortunately, the double integral may be approximated by the following single integral, evaluated at the centroid of the glass:

$$F_{is}^2 \approx \int_{\langle\omega\rangle} \cos\chi \cos^2\chi' \, d\omega. \tag{2.20}$$

Such an approximation represents an error on a seasonal basis of about 3% of the total absorbed energy (Reuth 1984). The details of the procedure are given in Davies and McKay (1982) and in Puri et al. (1980). It is noted that this component of the sky diffuse energy is generally small so that reasonable approximations may be made.

2.4.3 Ground-Reflected Radiation

The final component of solar radiation incident on a collecting surface is the ground-reflected component. In the summer, for building surfaces that are shaded from the beam radiation and part of the sky diffuse radiation, this component may be dominant. If the total horizontal solar radiation is assumed to be diffusely reflected from the ground, the ground-reflected radiation transmitted through a transparent building surface is

$$\dot{Q} = \rho_r f_r I_h F_{ir} A_g \tau_d, \tag{2.21}$$

where ρ_r is the diffuse ground reflectance, f_r is the unshaded fraction of ground-diffuse radiation passing through shading obstructions, I_h is the total horizontal flux, and F_{ir} is the diffuse view factor between the surface and the ground.

One may account for diffuse reflection of ground-reflected radiation from the underside of an overhang (see figure 2.6) by formulating an effective surface-to-ground view factor F_{ir}:

$$F_{ir} = F'_{ir} + \rho_0 F_{io}, \tag{2.22}$$

where F'_{ir} is the actual glazing-to-ground view factor, defined by

$$F'_{ir} = \frac{1 - \cos\Sigma}{2}. \tag{2.23}$$

ρ_0 is the diffuse reflectance of the underside of the overhang and F_{io} is the view factor from the glazing to the overhang.

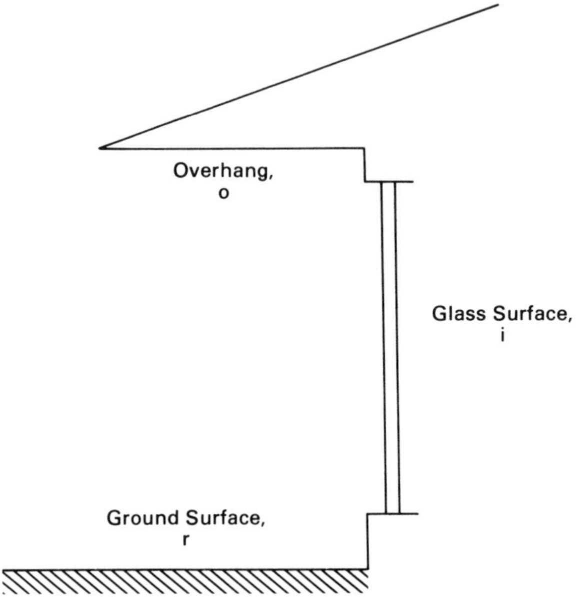

Figure 2.6
Geometry for reflection from overhang.

Section 2.5 addresses the obstruction of both the sky diffuse and the ground diffuse radiation by external obstructions such as trees, giving methods to estimate the unobstructed fractions f_s^0, f_s^2, and f_r.

2.5 External Shading and Reflection

This section establishes the relationship between the reference solar radiation that would strike a simple external surface in the absence of shading and the radiation actually reaching the surface. External obstructions are categorized as site obstructions, such as plants, terrain, and other buildings, and architectural elements such as self-shading and reflection by parts of the building structure, and architectural shading devices, such as overhangs and reflectors. The combined effects are presented in calculational form in section 2.5.3. Simpler, graphical tools are described in section 2.6.

2.5.1 Site Shading—Trees

The judicious placement of certain deciduous trees has long been lauded as the clever symbiosis of man and nature (Olgyay and Olgyay 1957,

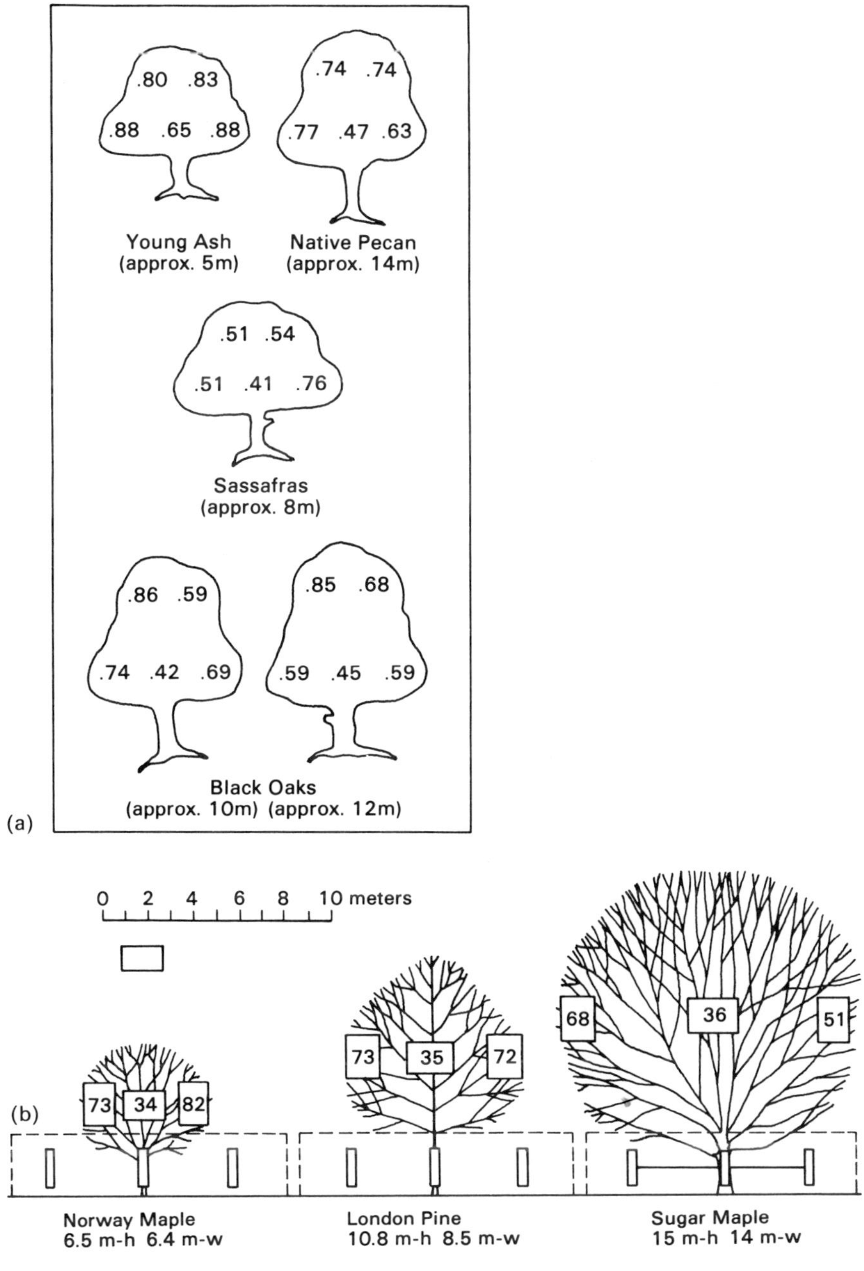

Figure 2.7
Transmittance of various trees. (a) Ash, pecan, sassafras, and oak. Source: Holzberlein 1979. (b) Maples and London pine. Source: Montgomery, Keown, and Heisler 1982.

Kohler and Lewis 1981) on the supposition that the leaves block the unwanted summer sun and fall away during the heating season, thus allowing the sun clear access to the solar aperture. Subsequently, these assumptions have been the object of quantitative experimentation (Holzberlein 1979, Erley and Jaffe 1979, Montgomery, Keown, and Heisler 1982). These more recent findings have indicated that bare, leafless trees transmit only about 1/2 to 2/3 of the solar radiation incident upon them. These results are shown in figures 2.7(a) and (b) and figure 2.8, which present transmittances of various sets of trees. A more rigorous approach is cited in Hottel and Sarofim (1967), but to apply this approach, more information is required than is typically available. In a climate with an appreciable heating load in the winter, the enhanced shading in the summer should not be obtained at the cost of losing the full solar availability during the heating season.

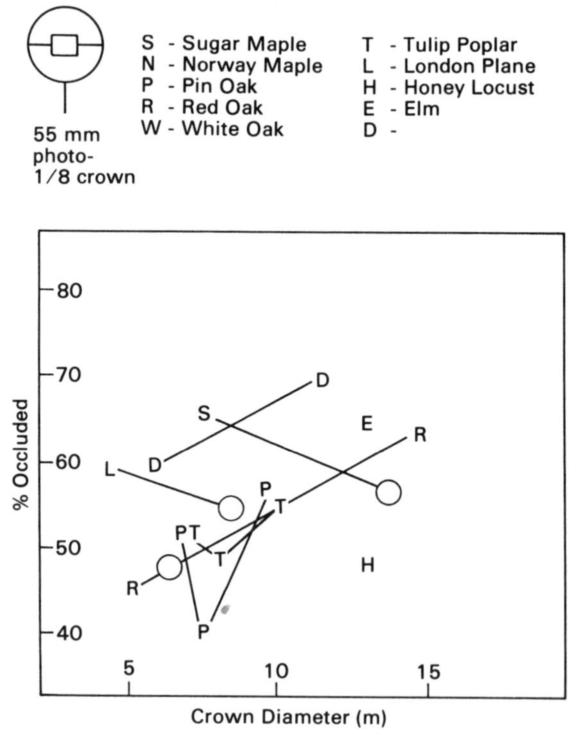

Figure 2.8
Occlusion by various trees. Source: Montgomery, Keown, and Heisler 1982.

Therefore, trees on the south side should not shade any part of the passive solar heating aperture in the winter. However, placing deciduous trees on the east and west sides of houses is generally beneficial. Trees placed in these locations effectively block the summer sun and do not appreciably affect the winter solar gain.

In the event that trees do cause shading, their transmittances should be multiplicatively included in the calculation of incident energy [equations (2.13), (2.14), and (2.18)]. It is recommended that the fraction occluded in figure 2.8 be used as follows:

$$\tau_{trees} = 1 - (\text{fraction occluded}). \tag{2.24}$$

The tree transmittance must be applied to all forms of the incident energy including beam, sky diffuse, and ground reflected—all with appropriate area-averaged weighting factors. A specific procedure is given in the following section.

2.5.2 Calculation of Tree Transmittance

The literature is scant on the subject of estimating the effective transmittance of site features such as trees. The following procedure, developed by the author and presented here for the first time, assumes a line of trees as shown in figure 2.9(a) where surface *1* is the glazing, surface *2* is the overhang, surface *3* is the ground before the trees, surface *4* is the ground to the rear of the trees, and surface *5* is the trees.

The procedure is valid only when the line of trees lies between the sun and the surface of incidence. For other situations the reflectance of the trees would be required, and some of the formulas would have to be modified. Many cases exist, but data on the reflectance of trees are not available. Due to these difficulties and the fact that we do not have established models for situations of this kind, the following development may be considered illustrative rather than comprehensive.

The first component to be considered is the beam radiation. Given the close-up view shown in figure 2.9(b), the fraction partially shaded by the trees (if no end effects exist) is L_{trees}/L_1, and the fraction completely exposed to the sun is $1 - L_{trees}/L_1$. The length L_{trees} is easily obtained by methods to be covered in section 2.5.3.2. Thus, the fraction of beam radiation entering the enclosure can be represented as

$$f_b = [(L_{trees}/L_1)\hat{\tau}_{trees} + (1 - L_{trees}/L_1)]. \tag{2.25}$$

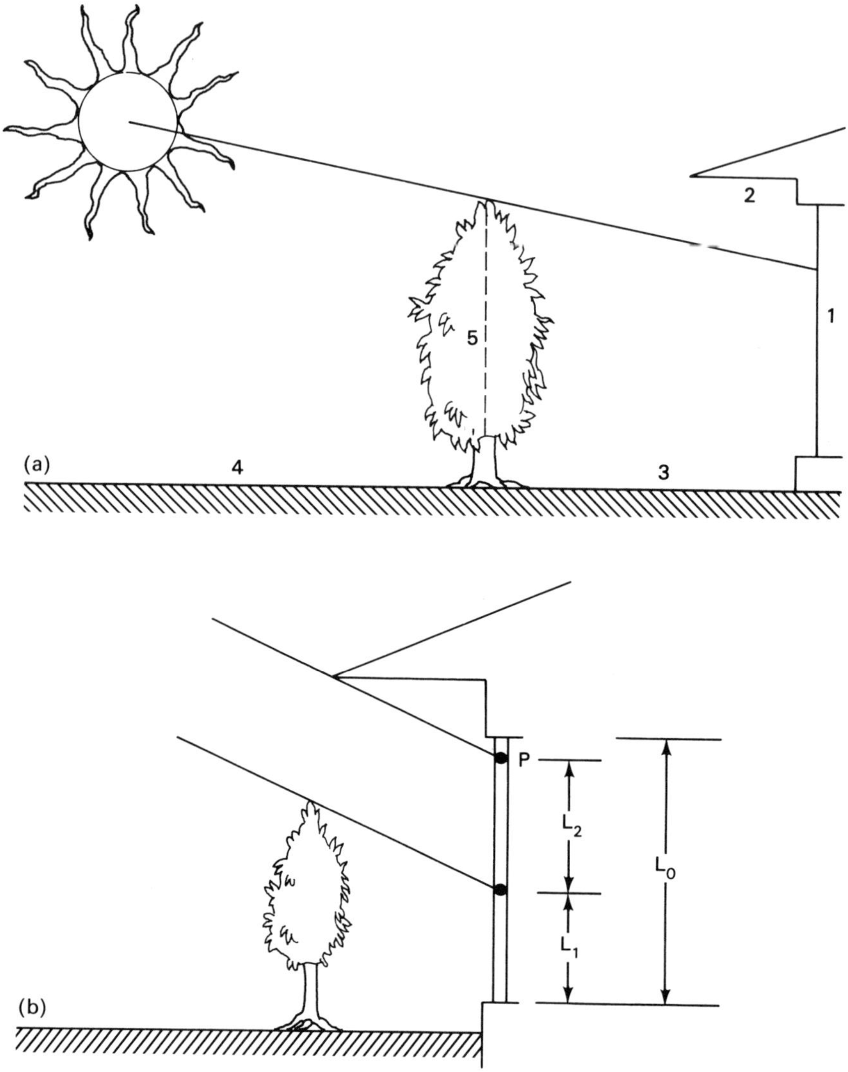

Figure 2.9
(a) Elevation view of tree/overhang geometry; (b) close-up of overhang geometry.

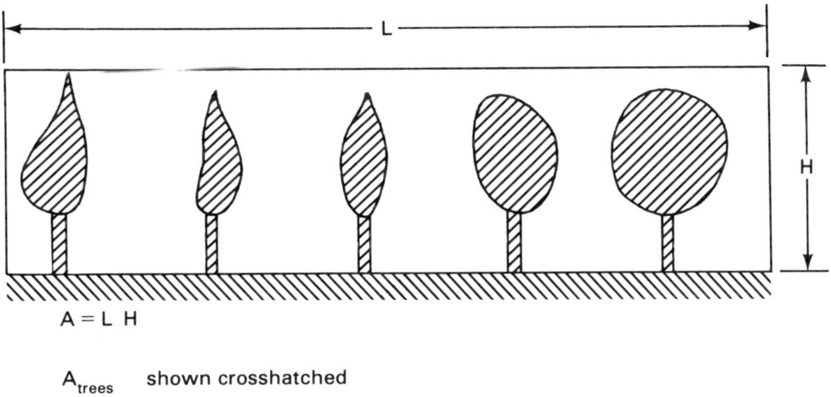

Figure 2.10
Tree geometry.

In all of these equations, the effective transmittance of a row of trees, τ_{trees}, has been utilized. This includes the projected areas of the trees themselves and the projected clear spaces between the trees. As illustrated in figure 2.10, this quantity may be computed as follows:

$$\hat{\tau}_{trees} = [\tau_{trees} A_{trees} + (A - A_{trees})]/A. \tag{2.26}$$

The sky radiation is handled next. For an unshaded vertical surface, $F'_{1s} = 1/2$. However, the direct sky radiation will be diminished by the overhang and the trees, and there will be some sky radiation reflected from the overhang onto the aperture. Thus, the fraction of diffuse radiation entering a vertical aperture with an overhang will be as follows for isotropic sky radiation:

$$f_s = \tfrac{1}{2}\{[\hat{\tau}_{trees} F_{15} + (1 - F_{12} - F_{13} - F_{14} - F_{15})] + \rho_2 [\hat{\tau}_{trees} F_{25} + (1 - F_{21} - F_{23} - F_{24} - F_{25})] F_{12}\}. \tag{2.27}$$

where F_{15} and F_{25} are evaluated over the portion of region 5 in figure 2.9 that occludes the sky.

When sky diffuse radiation moments are being considered, the transmittance of the trees must be included in the argument of equation (2.19) as the integral is being numerically evaluated.

Finally, the ground-reflected radiation must be accounted for. This situation however, is somewhat complicated due to the varying occlusion of the different components of radiation as they pass through the trees to strike the ground and the window. First, total horizontal radiation strikes the ground on the sunward side of the trees and, upon reflection, is occluded as it passes through the trees:

$$\dot{Q}_{r4} = \rho_r I_h F_{14} \hat{\tau}_{trees} \tau_d A_g. \tag{2.28}$$

Next considered is the radiation striking the ground in area *3*. This radiation consists of beam radiation passing through the trees and sky diffuse radiation that is directly incident and occluded by the trees. For isotropic sky radiation, this will be as follows:

$$\dot{Q}_{r3} = \rho_r \{(I_h - I_s^0)\hat{\tau}_{trees} + I_s^0[\hat{\tau}_{trees} F_{35} + (1 - F_{35} - F_{31} - F_{32})]\} F_{13} A_g \tau_d. \tag{2.29}$$

It is probably not worthwhile to account in these calculations for non-isotropic sky radiation. It is therefore recommended that the sky diffuse radiation be considered as isotropic.

The total fraction of radiation reflected from the ground, f_r, may be obtained by summing equations (2.28) and (2.29) with the result

$$f_r = [\hat{\tau}_{trees} I_h (F_{14} - F_{35} F_{13}) + I_s^0 (1 - F_{31} - F_{32} - F_{35})]/(F_{ir} I_h). \tag{2.30}$$

The reflection from the overhang can be added to this total by employing the techniques resulting in equation (2.27). The radiation reflected from the trees may also be included in this fashion, but its magnitude is usually small enough to be neglected.

Except for the beam-radiation portion of these energies, all other parameters in these equations may be precalculated during the initialization phase of the program. Unfortunately, it is necessary to track the beam radiation in section *3* of figure 2.9(a) if the lower ray shown in figure 2.9(b) strikes this area. It is best to perform this calculation in conjunction with the shading calculations for the overhang and the mullions.

Once the topology has been idealized, the remaining task is to characterize the foreground surface. This is well represented by the directional hemispherical solar reflectance ρ' as shown in figure 2.11. Except for bright green grass and snow (not shown), most surfaces may be assigned a constant (independent of the direction of the incident rays) diffuse reflectance with very little error. Additional radiative properties may be found in Threlkeld (1970); ASHRAE (1963); Gubareff, Janseen, and Torberg

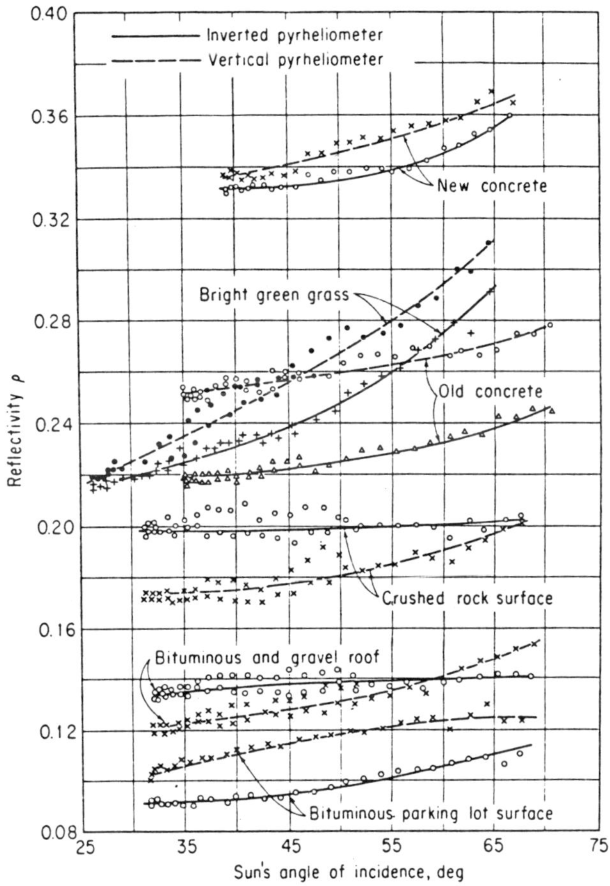

Figure 2.11
Directional solar reflectance of various materials. Source: Threlkeld 1970.

(1960); Touloukian and DeWitt (1972); Mabinton and Goswami (1980); and Kwentis (1967).

2.5.3 Architectural Shades and Reflectors

The significant impact of solar gain through windows is graphically presented in Ewing and Yellott (1976). Efforts to reduce this load during the cooling season often center on the design of effective shading devices. Furthermore, inadvertent reduction or enhancement of solar radiation incident on a surface due to shading or reflection by adjacent surfaces may be significant. Since the geometrical position of the sun is well defined, the problem is simply one of geometrically relating the position of the sun, the collecting surface, and the shading or reflecting surface. Many methodologies have been presented (Sun 1968, Walton 1979, Utzinger and Klein 1979, Jones 1980, Jones and Yanda 1981, Budin and Budin 1982, Sassi 1983, Johnston 1983). The most efficient rely heavily on the use of direction cosines and the projection of end points (Walton 1979, Budin and Budin 1982). Nearly all modeling is based on the specification of surfaces as polygons. The more sophisticated routines include shading by means of fixed shading devices, attached building surfaces other than the collecting surface, and completely separate objects such as hills or detached buildings (Walton 1979). The inclusion of such effects can substantially increase the calculation time, with up to half of the calculation time of a simulation being required for shading calculations. However, even such sophisticated algorithms do not always adequately describe the solar energy input, for example, because they exclude the transmittances of various trees.

2.5.3.1 Reflection from External Surfaces External surfaces reflect radiation to other surfaces. The methods described so far may be used to calculate the incident energy, $\dot{Q}_j$, on a detached surface j. If this radiation is diffusely reflected, it may then be incorporated in the energy transmitted through surface i as follows:

$$\dot{Q}_{i,\mathrm{ext}} = \rho_j \dot{Q}_j F_{ij} A_g \tau_d. \tag{2.31}$$

2.5.3.2 Shading From a Simple Overhang The shading of beam radiation by a simple rectangular overhang is shown in figure 2.12. Extensions to other geometries are relatively straightforward: lengths along the building are represented by L, heights by H, and depths by D. The geometry is symmetric with the overhang protruding from the wall a depth of D_1, in a

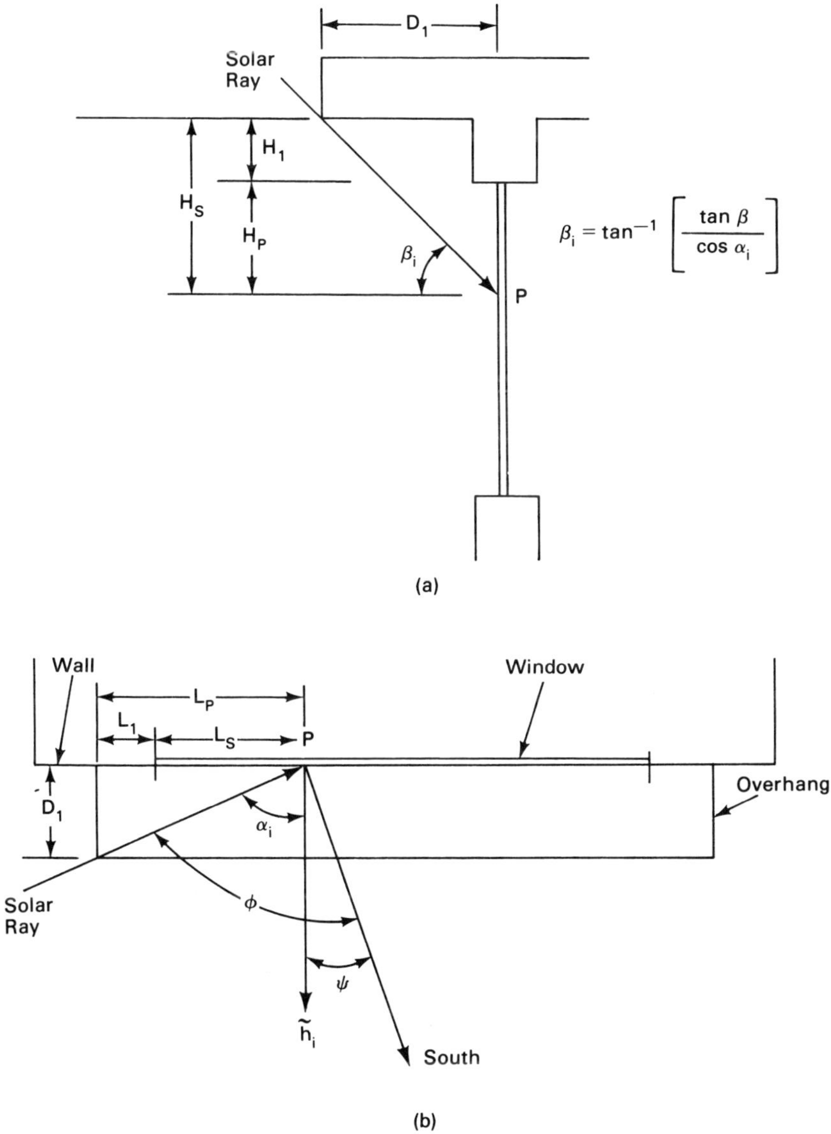

Figure 2.12
Overhang geometry: (a) side elevation of shading geometry.

plane located a height H_1 above the vertically oriented window of dimensions $L \times H$. The overhang extends along the wall a length of L_1 (which may be negative) away from both edges of the window.

The shadow cast by the sun is shown in figure 2.12. For this type of overhang the shading is obtained by determining the location of the point P, which represents an intersection of the limiting ray (grazing the corner of the overhang) with the window surface. The direction of the limiting ray relative to the window is defined by the solar altitude β and the surface azimuth γ_i (see sections 2.3.1 and 2.4.1).

First, the vertical shading line below the overhang must be determined. The relevant angle is the surface altitude β_i (see figure 2.12[a]). The tangent of this angle may be computed as

$$\tan \beta_i = \frac{\tan \beta}{\cos \gamma_i}. \tag{2.32}$$

One may then compute the distance of the point P below the top of the window as

$$H_p = H_s - H_1 = D_1 \frac{\tan \beta}{\cos \gamma_i} - H_1. \tag{2.33}$$

Figure 2.12(b) shows a plan view of the shaded geometry. The distance of the point P from the edge of the window is

$$L_p = L_s - L_1 = D_1 \tan \gamma_i - L_1. \tag{2.34}$$

Once these coordinates are computed, the logical set of situations to be considered in computing the fraction unshaded is shown in figure 2.13. The figure is drawn for the case where the overhang is wider than the window. These numerous decisions and the subsequent calculation of the shaded area can significantly increase the computer execution time.

This discussion applies only to beam radiation. The interaction of the overhang with the sky diffuse and ground-reflected radiation is handled via the view-factor methods described in sections 2.4.2 and 2.4.3.

Finally, a completely integrated approach would require further mapping of the beam radiation onto the interior surfaces as input into the radiosity network to be developed in section 2.8.3. The extension is straightforward and proceeds similarly to the present approach but will not be developed here (see Siegel and Howell 1982).

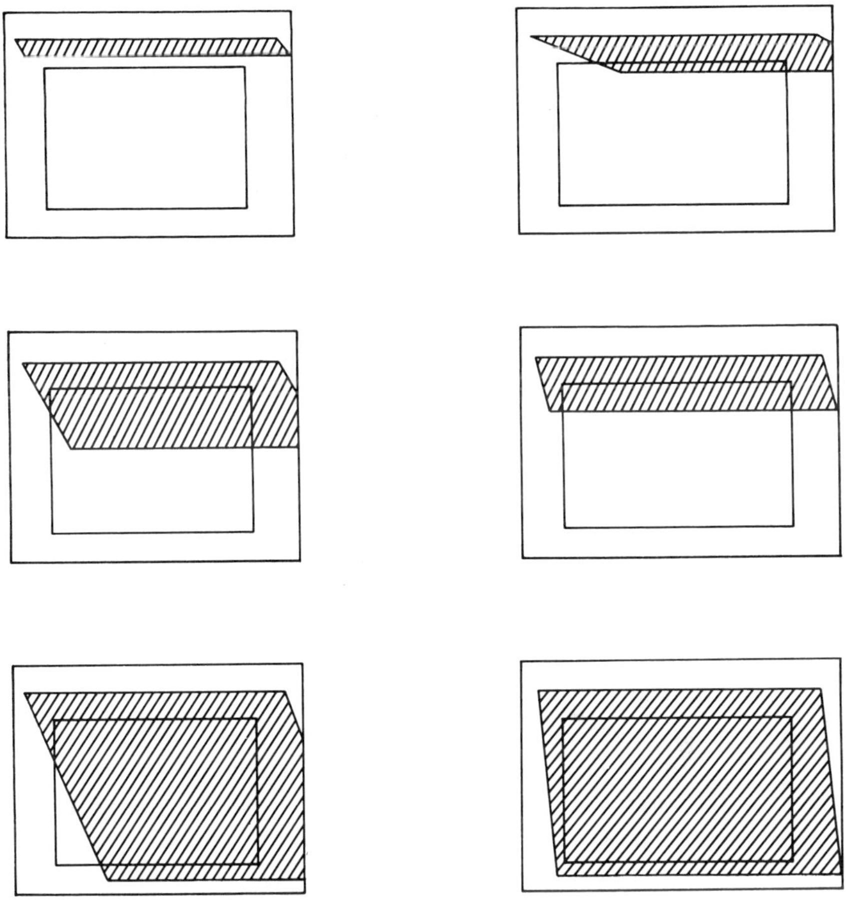

Figure 2.13
Overhang shading cases.

2.5.4 Coordinate Mapping

It is possible to perform the calculations in section 2.5.3 using coordinate mapping (Walton 1979, Eringen 1980). These calculations are most efficient when formulated in terms of direction cosines and may be performed directly for each hour of the simulation, or the shading fractions versus incident angle may be precalculated and stored for later use. The choice between these two methods of calculation depends on the balance between run-time cost and memory cost, and calculations of this type are performed only in very detailed algorithms. A detailed discussion of these techniques is beyond the scope of this chapter, and the interested reader is referred to the abundance of literature on this subject, of which Walton (1979) and Eringen (1980) are representative.

2.6 Simplified Methods—Sun Charts

This section concludes with a discussion of some simplified methods that are available to analyze the solar radiation entering buildings. These methods involve the use of simplifying assumptions as to the character of the beam radiation, charts that geometrically describe the position of the sun in the sky, charts that combine geometry with intensity to allow quantification of the load, and three-dimensional scaled physical models that allow the direct determination of the shading patterns. Since daylight is part of the solar spectrum and the analysis thereof falls primarily into these categories, procedures pertaining to daylighting will also be briefly addressed. Indeed, the geometrical behavior is identical, and the spectral dependence of the material properties is the only distinguishing factor.

An elegant alternative to the hour-by-hour calculations is the method developed by Sharpe (1980). This method entails the analytical integration of the beam radiation entering an aperture with simple shading. The day is divided into periods of no shading, partial shading, and full shading for the aperture of interest. The implicit assumption behind such a method is that the direct normal radiation is constant with time. This implies no correction for air mass or varying cloud cover. However, the method is very useful as input to simplified design tools. Other investigators have developed tables that easily allow the designer to calculate the shadow position and dimensions (Ewing and Yellot 1976, Yago 1982, Lau 1982).

At the earliest stage of design, the most pertinent information is the sun's position in the sky. This information allows the designer to assess, in a rudimentary fashion, the effects of building orientation, overhang geometry, and encroachment of the skyline on the solar window. It is a simple process to construct "sun charts" that represent these effects. An excellent, though dated, discussion of these topics is presented in Olgyay and Olgyay (1957). The simplest of these sun charts is the shading protractor (Olgyay and Olgyay 1957, Libby Owens n.d., Dean 1979, Scofield and Moore 1981, Englund 1982). The development given here follows Mazria (1979), which is perhaps the clearest reference on the subject.

The position (relative to the observer) of the sun in the sky is determined by the solar altitude angle β and the solar azimuth angle ϕ. These angles, which can be calculated using equations (2.8) and (2.10), define the position of the sun in the "skydome," which is an imaginary hemispherical dome covering the extent of the sky as shown in figure 2.14. The grid constructed on the skydome represents constant values of β and ϕ. The cylindrical sun chart, also shown in figure 2.14, is the projection of the grid on the skydome onto such a cylindrical surface as shown. The path of the sun as it travels across the sky (dome) is then represented on this chart. Separate charts must be constructed for each value of the latitude. Longitude corrections are not required as each chart is referred to solar noon. True south, not magnetic south, is used.

Figure 2.15 is a chart for 40 degrees north latitude. The solar altitude is represented on the vertical axis, and the solar azimuth is represented on the horizontal axis. The heavy dark lines represent the trajectory of the sun for the particular days shown. Note that except for June 21 and December 21, each of these lines represents two days, as the sun's path is identical for both the fall and spring seasons on these days. The dashed lines represent the hour of the day from sunrise (on the left) at a solar altitude angle of 0 degrees to sunset (on the right) at a solar altitude angle of 0 degrees, with noon shown as a vertical line in the middle of the chart. The sun's path proceeds from left to right on the chart.

As an example, consider April 21 at 10:00 A.M. The solar altitude is read as 51 degrees, and the solar azimuth is read as 52 degrees east of south. This is identical to the position of the sun at 10:00 A.M. on August 21.

The great utility of these charts lies in the capability to predict when shading of the solar aperture occurs. Thus, if a very high row of hills

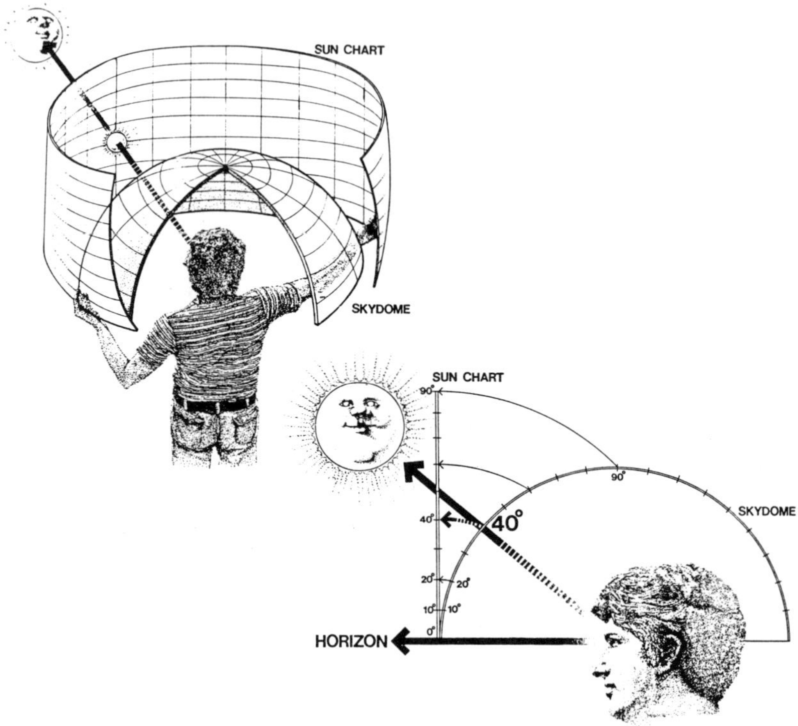

Figure 2.14
Skydome and associated cylindrical sun chart. Source: Mazria 1979.

obstructs the view from our position to a height of 51 degrees and at angles east of 52 degrees east of south, we may determine that our aperture is shaded from direct beam radiation until 10:00 A.M. on the days of April 21 and August 21. In fact, shading exists until 10:00 A.M. for all days when the sun is lower in the sky than 51 degrees, which is the entire period from August 21 to April 21.

The extension of this procedure to locating the physical position of obstructions on the sun chart is known as construction of the "shading mask" for the south-facing surface of interest. Other buildings and the terrain will obstruct the sun from below, overhangs will obstruct the sun from above, and side fins will obstruct the sun from the side. Trees and other significant objects may also be included in the shading mask. The ultimate concern is to have good solar access during winter months and to

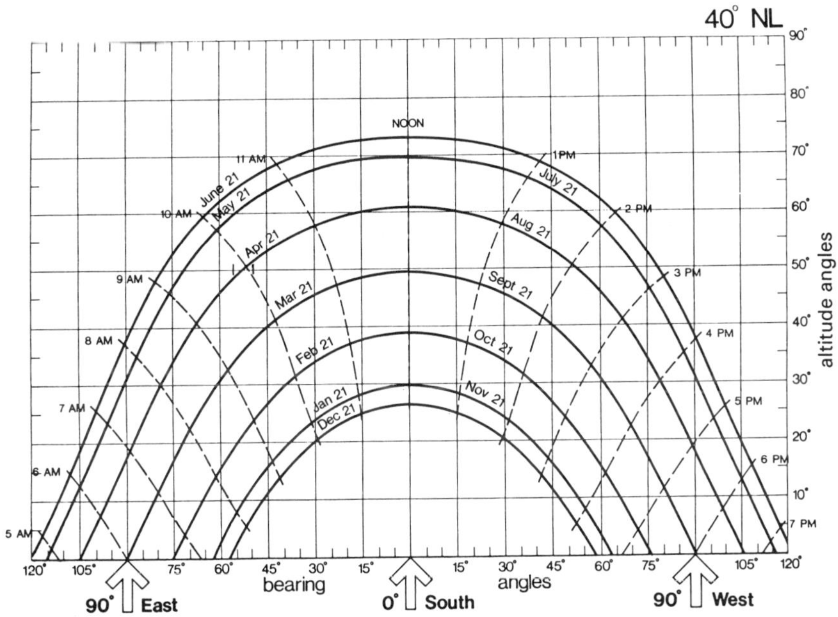

Figure 2.15
Sun chart for 40 degrees north latitude. Source: Mazria 1979.

obstruct solar access during the summer months. Figure 2.16 shows a plot of a site skyline.

The next step is to plot the shading caused by overhangs and side fins. The procedure is somewhat involved, and the reader is referred to Mazria (1979) for details. A utilitarian device known as the shading calculator is provided by Mazria for constructing the shading plot. Figure 2.17 is a completed plot for a south-facing surface.

The final step in obtaining quantitative solar loads is to assess the incident solar flux at the unobstructed times of the day. Mazria again has developed a graphical tool for doing so, known as the solar radiation calculator. Overlayed on figure 2.17, the device contains contours of constant incident intensity. This results in figure 2.18. For the day of interest, values of incident intensity may then be read for each hour. As an example for the south-facing surface under consideration, at 2 P.M. on September 21, the incident intensity is 205 Btu/h ft^2 (646 W/m^2). This value may then be multiplied by the glazing transmittance and unshaded area to obtain the solar energy entering the building.

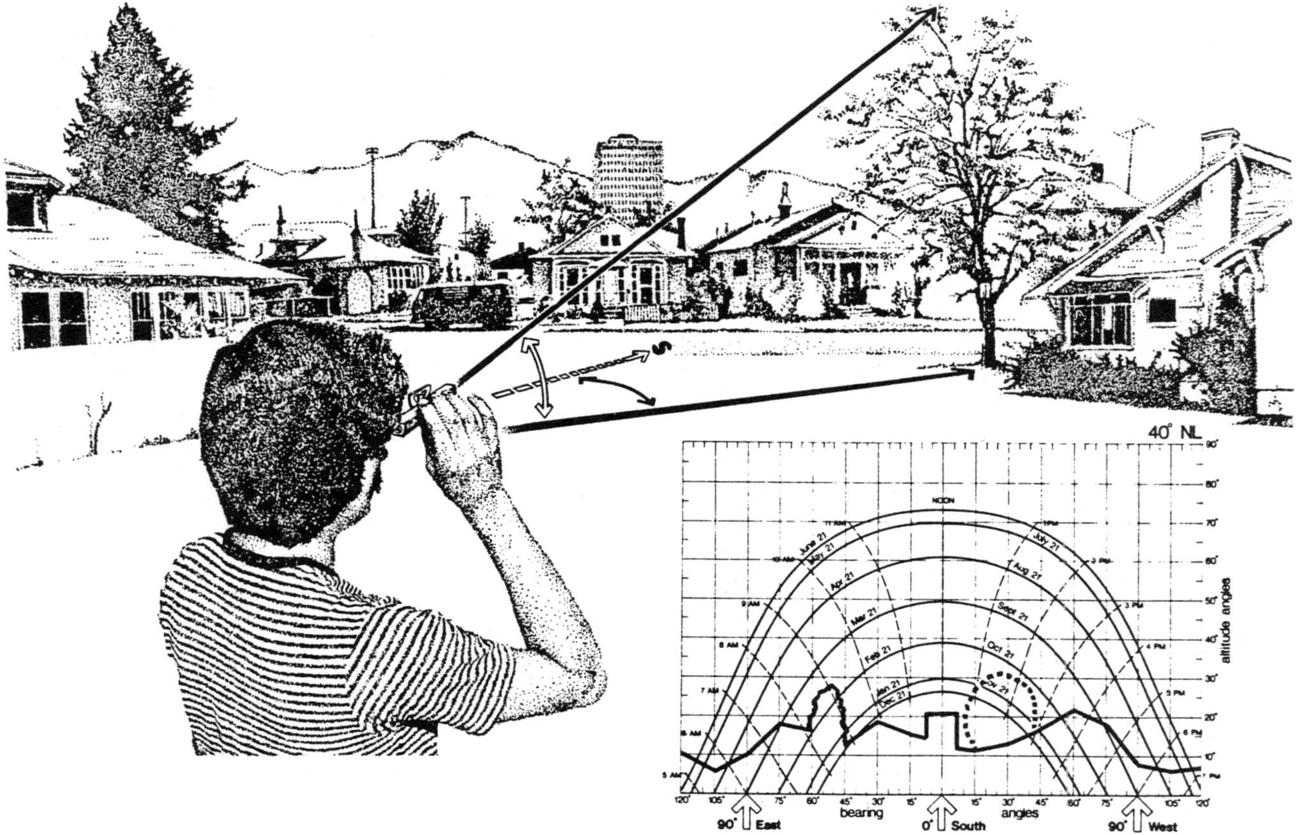

Figure 2.16
Construction of a site skyline for a south-facing surface. Source: Mazria 1979.

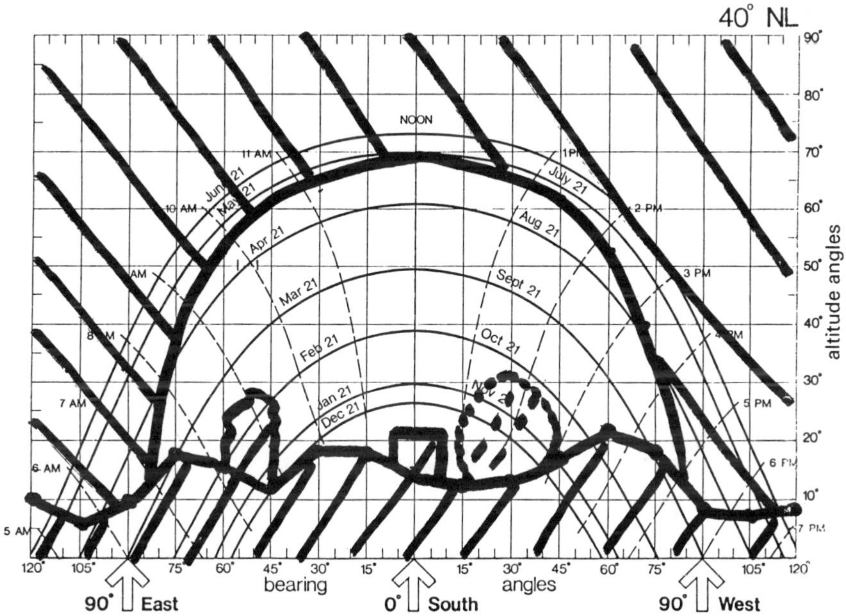

Figure 2.17
Graphical solar radiation chart overlayed on solar window.

Thus, a simple graphical technique exists for calculating the incident solar energy. The technique accounts for surface location (latitude), surface orientation (slope, surface azimuth), and obstructions (terrain, foliage, buildings, and architectural shading devices). However, the technique is based solely on beam radiation with some resulting inaccuracy in winter months and serious error from April through September when sky diffuse and ground-reflected radiation become more important. The solar radiation calculator is based on ASHRAE clear-day modeling of solar radiation. The method does not account for local cloudiness, which must be incorporated via tabulated clearness factors (Jones et al. 1982, Balcomb et al. 1984). Note, however, that clearness factors are monthly average values applied to instantaneous data. The method is simple, appropriate for approximate design.

Other methods have used the geometrical and energy relationships for solar radiation to balance the average building load during both the heating and cooling seasons to enable the designer to determine the opti-

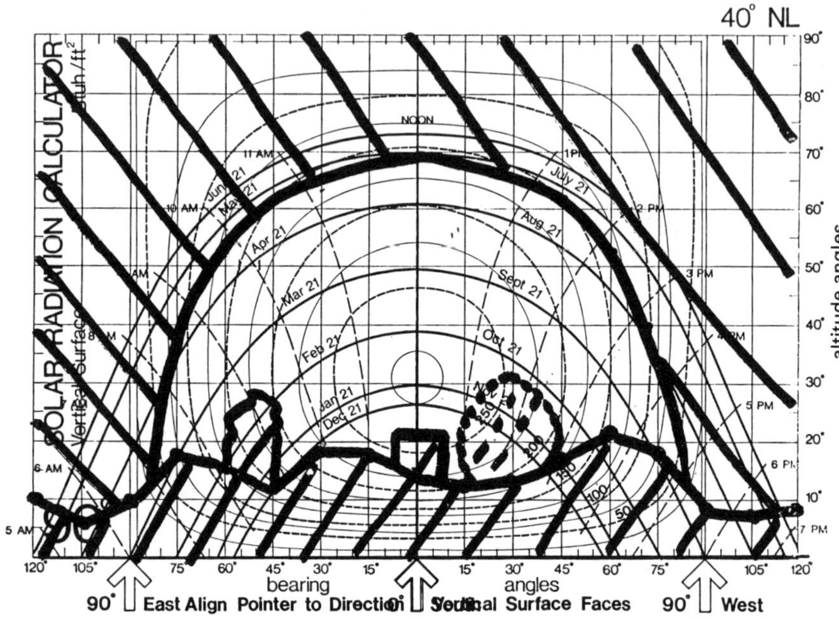

Figure 2.18
Completed sun chart showing "solar window." The upper limit is produced by an overhang.

mal overhang dimensions versus location (Jones 1981, Lau 1982). The most useful methodology, which entails monthly hand calculations, is presented in volume 3 of the *Passive Solar Design Handbook* (Jones et al. 1982) and in *Passive Solar Heating Analysis* (Balcomb et al. 1984). Simple curve fits are provided for the transformation of the total horizontal solar radiation to total solar radiation incident on surfaces of various orientations, the total incident to the total transmitted radiation, and the transmitted solar radiation to the solar radiation absorbed in the passive system. Correction factors are given for several overhang geometries (Balcomb et al. 1984).

It is possible to enhance the amount of incident radiant energy by using reflecting side fins or horizontal plates. The curve fits presented in Balcomb and McFarland (1978) and Balcomb et al. (1984) are especially useful in estimating these enhancements for monthly calculations. Note, however, that the reflection of diffuse radiation from surrounding surfaces (except for the ground) is typically neglected in most calculation procedures and this might, in some extreme instances, result in large relative errors in the calculation of the cooling load for buildings with south overhangs. Chiam (1983) presents a simple and direct method for performing these calculations under the assumption of isotropic diffuse incidence from the reflecting surface.

2.7 Glazings

Glazings are the transparent covers (windows) that admit solar energy into the passive system. This section presents material that focuses on the detailed calculations involving glazings. Glass is singled out for discussion, but the same principles apply to nonglass glazings.

2.7.1 Types of Glass

Types of glass are characterized by function (ASHRAE 1981, Threlkeld 1970, Olgyay and Olgyay 1957, Selkowitz 1978, Selkowitz and Berman 1978, Yellott 1979). The most desirable type of glass to maximize solar energy gain is low-iron glass. Typical soda-lime glass is less effective in transmitting solar radiation. Low-iron glass contains less than 0.02% of iron oxide (Fe_2O_3). This glass is easily distinguishable when viewed from the edge: it appears clear instead of green as does ordinary glass. The other extreme is represented by heat-absorbing glass, which contains an excess of iron oxide and typically absorbs about half the incident solar energy. A view through this glass will appear quite dark. Finally, a variety of new coatings can be applied to glass surfaces, rendering them reflective to either solar or infrared radiation as desired.

2.7.2 Material Properties

The two important radiative properties of glass are transmittance, τ, and absorptance, α. Transmittance and absorptance are broadly defined as the fractions of the incident energy that pass through and get trapped within the glass, respectively. Each quantity depends on both the direction of the incoming ray and its wavelength. The properties (ASHRAE 1981, Threlkeld 1970, Olgyay and Olgyay 1957, Yellott 1979, Parmelee and Aubele 1950, Pettit 1978, Hsieh and Coldwey 1975, Hsieh 1976) are functions of the type of glass, the coatings applied, and the thickness, and they are so represented (or tabulated).

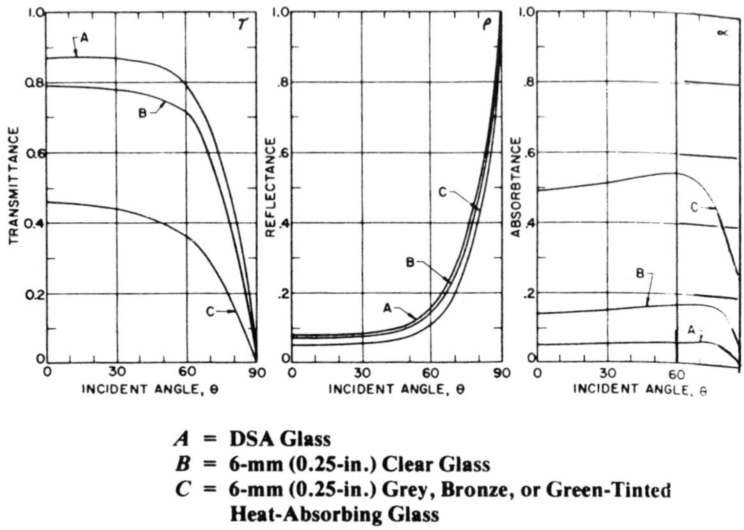

Figure 2.19
Directional properties of glass. Source: ASHRAE 1981.

Property variations with direction are shown in figure 2.19 for three types of glass where $\theta = 0$ *degrees* represents normal incidence and $\theta = 90$ *degrees* represents grazing incidence. These properties have been averaged over all solar wavelengths. As the incident angle increases from the normal, the transmittance decreases, the reflectance increases, and the absorptance increases slightly and then diminishes as the incident ray becomes completely reflected. For isotropic diffuse radiation, the properties are close to those of directional radiation incident at $\theta = 68$ *degrees.*

The variation of the transmittance with the wavelength of the radiation is shown in figure 2.20 for the three types of glass specified. In the visible range (0.4–0.7 μm), the transmittance is very high and remains high to about 3 μm, which is the wavelength range wherein most of the solar energy exists. However, virtually no ultraviolet energy (wavelengths less than 0.3 μm) is transmitted. The data shown are for normal incidence.

A single, 1/8-in.-thick pane of double-strength, grade A glass has been taken as a standard by ASHRAE. This is unfortunate because this glass is actually too thin for many applications; a 3/16-in. thickness is preferred for the larger sizes typical of solar apertures. The increased thickness reduces the transmittance of a single pane by about 5% and of two panes by about

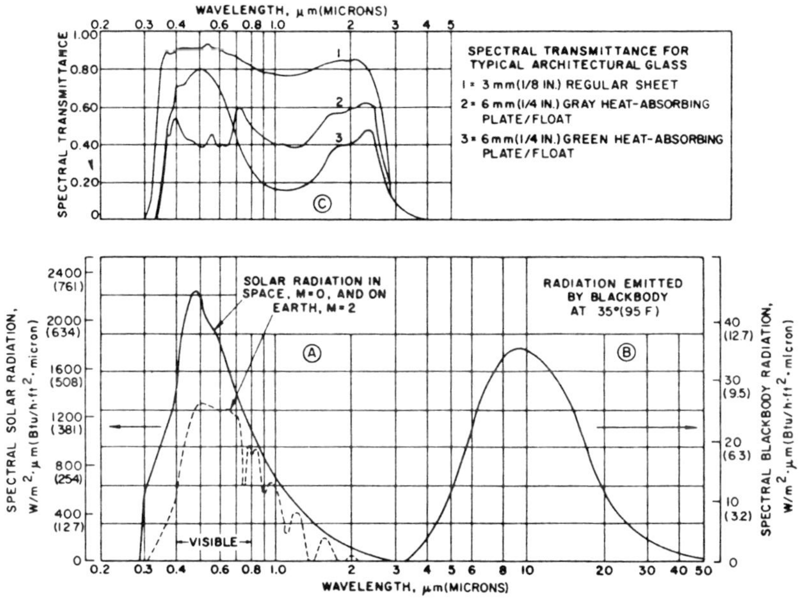

Figure 2.20
Spectral properties of glass. Source: ASHRAE 1981.

9%. In many commercial applications, 1/4-in. thicknesses (and greater) are used for the larger areas to obtain the requisite strength to resist wind loads.

It has been common practice to measure the solar-optical properties of glass (ASHRAE 1981, Yellott 1979, Parmelee and Aubele 1950, Pettit 1978, Hsieh 1976, Christian and Shatynski 1982); however, there has been recent success with calculating these quantities when the electromagnetic properties, especially the index of refraction, are known (Hsieh and Coldwey 1975, Hsieh 1976). Electromagnetic theory may be applied for a light wave incident from air ($n = 1$) at an incident angle, θ, onto a single surface (Siegel and Howell 1982, Born and Wolf 1965) to yield the following reflectivity formulas for dielectric surfaces:

$$\rho_l'(\theta) = \left[\frac{n^2 \cos\theta - \sqrt{n^2 - \sin^2\theta}}{n^2 \cos\theta + \sqrt{n^2 - \sin^2\theta}}\right]^2 \tag{2.35}$$

$$\rho_d'(\theta) = \left[\frac{\sqrt{n^2 - \sin^2\theta} - \cos\theta}{\sqrt{n^2 - \sin^2\theta} + \cos\theta}\right]^2, \tag{2.36}$$

where the subscripts l and d refer to, respectively, the parallel and perpendicular components of polarization. The primes signify a directional dependence.

2.7.2.1 Single Glazing For a single pane of glass there are two air-glass interfaces. An infinite number of internal reflections occur, each being diminished by the internal absorption in the glass and an amount being refracted through the interface. The situation is depicted in figure 2.21(a). The transmissivity of one beam through the interior of the glass is

$$\tau' = e^{-KL}, \tag{2.37}$$

where K is the extinction coefficient and L is the path length of the beam in the glass. K is either a measured quantity or may be calculated from the following formula:

$$K = 4\pi k/\lambda_o, \tag{2.38}$$

where k is the imaginary part of the complex index of refraction, i.e.,

$$\bar{n} = n + ik, \tag{2.39}$$

and λ_o is the wavelength of radiation.

These material properties are available in the literature for aperture materials, or they may be calculated from the electromagnetic material properties using the formulas in Hsieh (1976). The result of summing the infinite series depicted in figure 2.21(a) is:

$$\bar{\rho}'_1 = \rho'\left[1 + \frac{\tau'^2(1-\rho')^2}{1-\rho'^2\tau'^2}\right], \tag{2.40}$$

$$\bar{\tau}'_1 = \tau'\left[\frac{(1-\rho')^2}{1-\rho'^2\tau'^2}\right], \tag{2.41}$$

and

$$\bar{\alpha}'_1 = 1 - \bar{\rho}'_1 - \bar{\tau}'_1, \tag{2.42}$$

where the subscript *1* refers to a single glazing.

Equations (2.40) through (2.42) are valid for only one component of polarization. Each total optical property is a convex combination (weighted by the component of polarization) of the optical properties of each component of polarization. As suggested in section 2.2.3, it is a good ap-

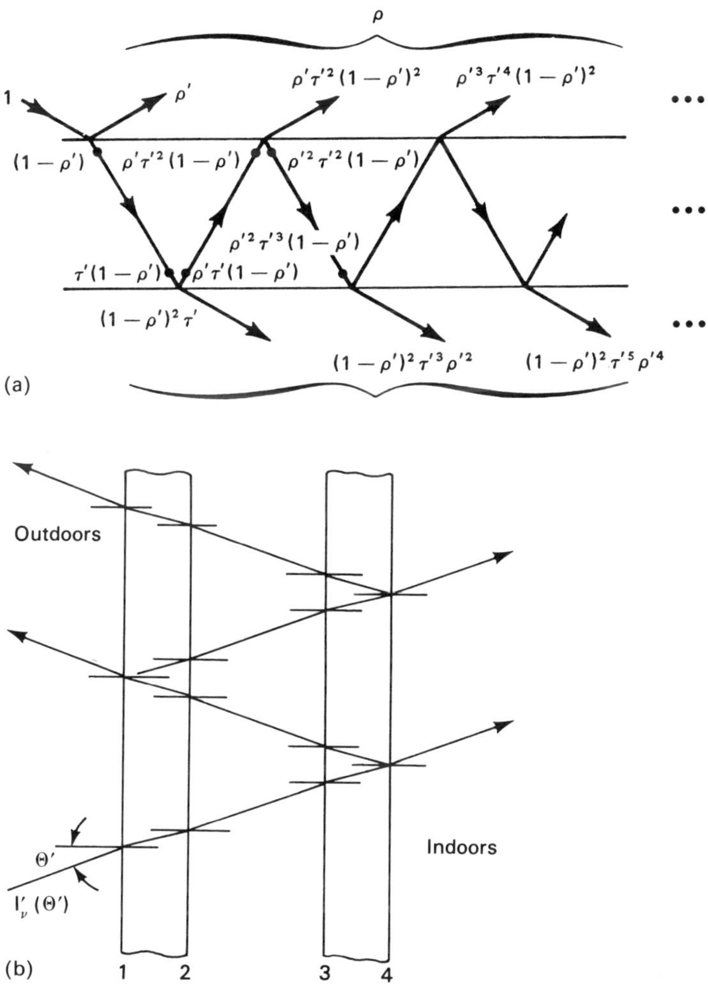

Figure 2.21
Multiple reflections for air-glass interfaces: (a) multiple reflections for a single pane, (b) multiple reflections for two panes.

proximation to assume the polarized incoming rays have equal energies in each component of polarization. Then, the components may simply be averaged:

$$\rho'_1 = \tfrac{1}{2}[\bar{\rho}'_{1l} + \bar{\rho}'_{1d}], \tag{2.43}$$

$$\tau'_1 = \tfrac{1}{2}[\bar{\tau}'_{1l} + \tau'_{1d}], \tag{2.44}$$

$$\alpha'_1 = \tfrac{1}{2}[\bar{\alpha}_{1l} + \bar{\alpha}_{1d}]. \tag{2.45}$$

In equations (2.43)–(2.45), the first subscript *1* refers to single glazing and the second subscript *l* or *d* refers to the component of polarization. Thus, optical properties may be obtained from electromagnetic properties. However, it is a relatively simple and inexpensive matter to measure the optical properties directly. For engineering purposes, it is sufficient to have the properties available; it is unimportant whether they are calculated or measured as long as the end results are correct. However, it is important to ascertain the exact quantities reported by the manufacturer since the optical data may be given in a wide variety of ways.

2.7.2.2 Multiple Glazings The properties of a single pane of glass can be extended to multiple panes by using Stokes' procedure (Stokes 1860). Consider two uncoated panes of glass as shown in figure 2.21(b) with the transfer of radiation from left to right and the surfaces numbered as shown. If τ'_1 and ρ'_1 are the properties for one pane of glass (where the primes indicate a directional dependence), then the following relations may be obtained by summing the series shown in figure 2.21(b) for two identical panes:

$$\tau'_2 = \frac{\tau'^2_1}{1 - \rho'^2_1}, \tag{2.46}$$

$$\rho'_2 = \rho'_1 + \frac{\rho'_1 \tau'^2_1}{1 - \rho'^2_1}, \tag{2.47}$$

and the following for three panes:

$$\tau'_3 = \frac{\tau'_2 \tau'_1}{1 - \rho'_2 \rho'_1}, \tag{2.48}$$

$$\rho'_3 = \rho'_1 + \frac{\rho'_2 \tau'^2_1}{1 - \rho'_2 \rho'_1}. \tag{2.49}$$

It is important to use both components of polarization in the right-hand sides of equations (2.46) through (2.49) and then to take the average values of the left-hand sides of the results as was done in equations (2.43) through (2.45).

An alternative approach is to take the directional properties for multiple panes from curves of tables supplied by the manufacturer. In any event, for computational purposes, the results are ultimately available in the form of a table of properties versus incident angle. Note that, because the three properties sum to *1*, only two must be stored and the third can be obtained subtractively. The table is generally used every time step to obtain the appropriate properties for the prevailing angle of incidence. It is more efficient computationally to form a curve fit to the data during the initialization phase of the computation. This type of curve fit can be accomplished with good accuracy for the transmittance and the absorptance. The normal procedure is to use the cosine of the incident angle ($\cos \theta$) as the curve fit independent variable as described in ASHRAE (1981). However, when the inversion is performed so that the absolute angle θ is determined each time step, this quantity may be used as the curve fit independent variable. The usual procedure is to compute "best-fit" coefficients based on some global criterion as is done in multiple regression analysis. The results will always weave in and out of the data, as can be seen in figure 2.22. For this reason, it is not uncommon for some correction to be necessary near the grazing angle. Since so little energy is incident at these angles, there is little overall error in this approach.

A property of more general use for average hand calculations is the shading coefficient as defined by ASHRAE (ASHRAE 1981, Yellott 1979). The shading coefficient is defined as the ratio of the solar gain through the particular glazing type of interest to the solar gain through ASHRAE's reference glazing, which is called the solar heat gain factor (SHGF). The SHFG is tabulated by latitude in ASHRAE (1981) for a horizontal surface and for unshaded vertical surfaces at the sixteen compass points for each hour of the twenty-first day of each month. The reference glazing has properties characterized by curve A in figure 2.19.

2.7.3 Energy Absorbed in Glazing Material

Radiant energy is transmitted directly through glazed surfaces into the passive system. Additionally, part of the energy absorbed in the glazing flows indirectly into the passive system by long-wave radiation and con-

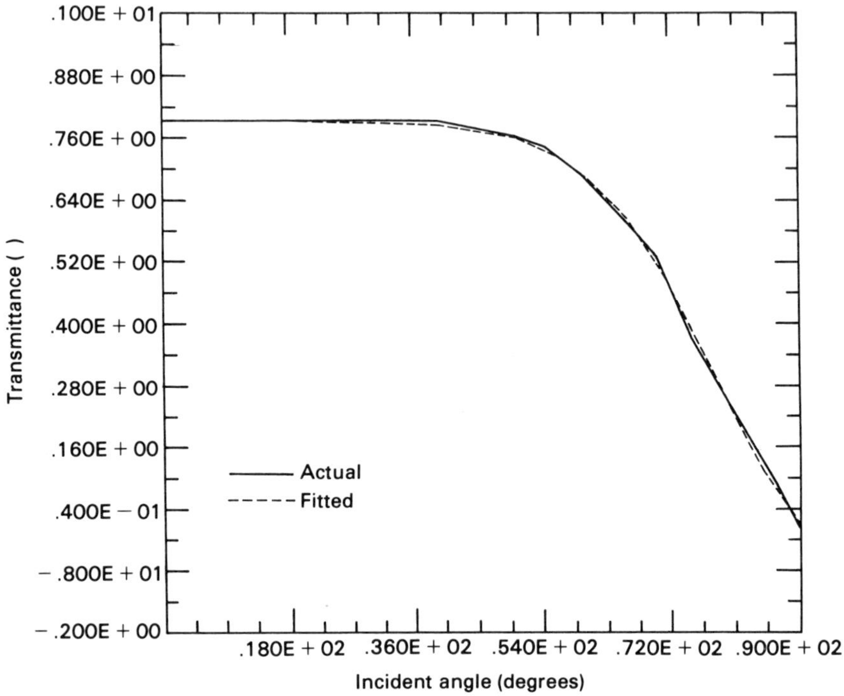

Figure 2.22
Actual and best-fit solar transmittance.

vection. Although this energy constitutes a small fraction of the direct transmission, it can cause substantial changes in the temperature of the glazing, which increases comfort by raising the mean radiant temperature (MRT). This situation is most pronounced when there are multiple glazings because the innermost glazing is relatively insulated from the environment.

The methods of section 2.4 may be applied to the calculation of the energy absorbed in the glazing, resulting in

$$\dot{Q} = [f_b \alpha_{DN} \cos\theta_i I_{DN} + \alpha_d(f_s^0 I_s^0 F_{is} + f_s^2 I_s^2 F_{is}^2 + \rho_r f_r I_h F_{ir})] A_g. \tag{2.50}$$

Other methods consistent with the desired level of accuracy may be used to calculate this quantity.

The thermal balance on a layer of glazing is depicted in figure 2.23. For purposes of illustration, the room is taken as very large and isothermal at

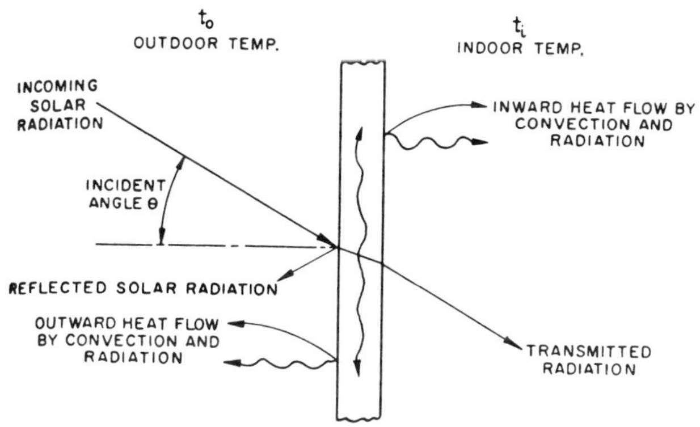

Figure 2.23
Heat balance on a single pane of glass.

Table 2.1
Glazing temperatures for typical conditions

	Temperature (°C)	
Glazing type	With absorption	Without absorption
Single glazing	5.0	4.0
Inner pane of double glazing	18.8	15.0
Inner pane of triple glazing	20.5	16.4

T_R. The temperature of the glazing (neglecting capacitance effects) is as follows:

$$T_g = \frac{\dot{Q}_a/A_g + T_\infty(h_c + h_r)_\infty + T_R(h_c + h_r)_R}{(h_c + h_r)_\infty + (h_c + h_r)_R}, \tag{2.51}$$

where T_∞ and T_r are the outdoor and room temperatures, respectively, and $(h_c + h_r)_\infty$ and $(h_c + h_r)_R$ are the sums of the radiative and convective heat transfer coefficients to the outdoors and to the room, respectively.

Table 2.1 gives values of T_g for both single, double, and triple pane systems, where, $\dot{Q}_a/A_g = 50\ W/m^2(17\ Btu/h\ ft^2)$, $T_\infty = 0°C\ (32°F)$, and $T_R = 20°C\ (68°F)$. Typical values have been used for the convective and radiative heat transfer coefficients, and overall conductances have been used where appropriate. The numbers indicate the importance of including solar absorption in the inner glazings of multiple glazing sets when comfort is being considered.

2.8 Internal Factors

Once radiant solar energy enters the passive system, a fraction is absorbed and a fraction is reflected back out through the glazings. Calculations of these fractions entail the analysis of specific surfaces for which separate energy balances are performed at each time step. Situations encountered in practice include radiative interaction with plants, furniture, window-shading materials, and interior surfaces (walls, ceiling, and floor). The first two categories result in absorption throughout the volume of the space, or *distributed* absorption, while the third category results in radiative transfer that is *localized* behind the glass. These topics are presented in sections 2.8.1 and 2.8.2, respectively. Section 2.8.3 presents a detailed method for calculating the absorption on specific interior surfaces and concludes with a discussion of the effect on comfort of the solar radiation within the space.

2.8.1 Distributed Absorption

Distributed absorption occurs throughout the space, such as at the surfaces of plants and furniture. The absorbers are low-mass objects in the building as opposed to the primary absorption surface, such as a Trombe wall or mass floor. Since these former elements possess small thermal capacitance, the solar energy absorbed is normally considered to be directly transferred to the air as the time constants of these objects are assumed to be much less than the simulation time step increment. Note, however, that a significant portion of the energy absorbed in low-mass objects is transferred to surrounding surfaces by infrared radiation and not to the room air, which occurs by convection. Also, some objects in the space, such as planters, bookcases, and heavy furniture, may well have time constants of one or more hours, so they cannot be characterized accurately as an extension of room air heat capacitance.

The difficulty in modeling distributed absorption lies in the complexity of accurately characterizing the percentage of radiation absorbed by these objects. A detailed analysis (Kwentis 1967) yields the following formula for the transmittance of volumetrically distributed particles:

$$\tau^* = e^{-K_V L/4}, \tag{2.52}$$

where L is the path length and K_V is a volumetric absorption coefficient defined for this situation as

$$K_V = \alpha_s A/V, \tag{2.53}$$

where A is the total projected surface area, V is the volume of the space, and α_s is the solar absorptance of the surfaces. Equations (2.52) and (2.53) are strictly valid only for convex, black surfaces ($\alpha_s = 1$); however, they may be used with good accuracy when $\alpha_s \gtrsim 0.8$. The volumetric absorptance is, then, approximately

$$\alpha^* = 1 - \tau^*. \tag{2.54}$$

These equations represent an accurate approximation for furniture and fairly transparent plants but are invalid when there is significant self-shading. In this case, an analysis along the lines of that presented in section 2.5.2 should be applied. A node should also be included in the simulation to which this solar gain is to be added.

In practice, a typical method of dealing with distributed absorption is to specify the percentage of the radiation entering the space that is absorbed by lightweight objects (SERI 1980, Balcomb et al. 1980, Jones et al. 1982, Balcomb et al. 1984), and this fraction remains constant for all time. Indeed, this assumes that the projected area stays constant (which is only true for a spherical object) and that there is no reflection of radiation. Furthermore, there is no mechanism in such a model to account for the directional redistribution of the reflected radiation. This type of model will thus represent an acceptable approximation where the distributed absorption area is small or where it is randomly oriented. Fortunately, this is representative of most situations encountered in practice.

2.8.2 Local Absorption

Local surfaces are planar surfaces that exist immediately behind the glazing (ASHRAE 1981, Moore and Pennington 1967, Keyes 1967, Pennington 1969, Morrison, Wheeler, and Farber 1976, Siminovitch, Bergeson, and McCulley 1982, Mabinton and Goswami 1980, Yellott 1972), including draperies, venetian blinds, and roller shades. An excellent review of analysis procedures for these devices is presented in ASHRAE (1981) together with the appropriate shading coefficients. Unfortunately, the shading coefficient alone gives no indication of the spatial distribution of heat that remains in the indoor space. In passive solar applications, it is useful to separate this heat gain into a portion that is transmitted through the shading assemblage and a remainder that is absorbed by it. The transmitted portion may be computed from the relation given in appendix A of Moore and Pennington (1967) as:

$$\frac{\text{direct transmitted}}{\text{total transmitted}} = \frac{\tau_g}{1 - \rho_g \rho_s}[1 + 0.215\,\alpha_g \rho_s + 0.75\,\alpha_s] + 0.215\,\alpha_s, \quad (2.55)$$

where the subscript g refers to the glazing and the subscript s refers to the interior shade.

These relations are all approximate, and the described situation does not arise in most passive solar heating applications. However, as the emphasis shifts toward reducing the solar heat input in the summer, the effects are more important, and more detailed analysis procedures, such as those presented in the next section, should be applied.

2.8.3 Radiation Networks

A more detailed analysis may be easily applied by using the net radiation method (Oppenheim 1956), which is valid for diffusely emitted and reflected radiation. Although these procedures may be used anywhere in the analysis, their application is most convenient after the radiation has been traced to the interior surfaces; i.e., ray-tracing techniques are applied until the radiation penetrates the glazing, then radiation networks are applied. Once the incident radiation is calculated for the surfaces under analysis, the net radiation method may be applied to calculate the energy absorbed at these surfaces. The surface heat fluxes are then included in the thermal simulation for the time step.

2.8.3.1 Definition of Radiosity Radiation networks find their application in the analysis of the overall incident energy flux (irradiation) upon a surface i and within wavelength band k, G_i^k, and the overall outgoing energy flux (radiosity) emanating from surface i, J_i^k. The irradiation includes only the emission and reflection from other surfaces, with the incident solar I_i^k handled separately. It is common to define two wavelength bands: a solar band ($k = 1$) from 0 to 3.5 μm and a longwave band ($k = 2$) from 3.5 μm to infinity. The tabulated properties for the materials being considered are then used (solar absorptance α_i^1, solar transmittance τ_i^1, and longwave emittance ξ_i^2). With these bands defined for building materials,

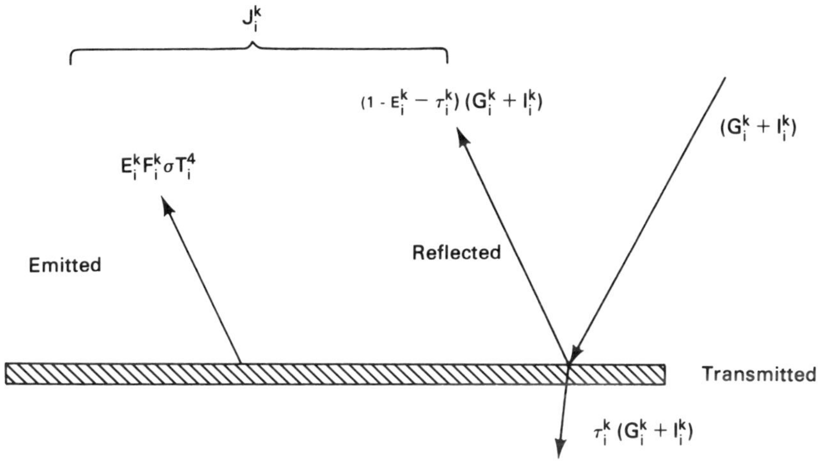

Figure 2.24
Radiation balance on surface i in wavelength band k.

the longwave transmittances are all zero, and the fraction of the blackbody energy existing in the solar band for all surfaces i, F_i^2 may be taken to be zero.

A radiative energy balance for surface i in wavelength band k is graphically depicted in figure 2.24. The directly incident flux I_i^k exists only in the solar band and is computed from methods presented earlier. If there are M surfaces, there are $2M$ unknowns for each band: G_i^k and J_i^k. The following relations may be used to eliminate the G_i^k:

$$J_i^k = \varepsilon_i^k F_i^k \sigma T_i^4 + (1 - \varepsilon_i^k - \tau_i^k)(G_i^k + I_i^k), \tag{2.56}$$

$$G_i^k = \sum_{j=1}^{M} J_j^k F_{ij}. \tag{2.57}$$

Substituting equation (2.57) into equation (2.56) yields

$$J_i^k = \varepsilon_i^k F_i^k \sigma T_i^4 + (1 - \varepsilon_i^k - \tau_i^k)\left(\sum_{j=1}^{M} J_j^k F_{ij} + I_i^k\right), \tag{2.58}$$

where

$$F_i^k = \begin{cases} 0 & (k = 1) \\ 1 & (k = 2). \end{cases} \tag{2.59}$$

Equation (2.58) represents a system of M linear algebraic equations that may be solved by any of the standard subroutines or by iteration. The net energy rate absorbed by surface i then follows as

$$Q_i = A_i \sum_{k=1}^{2} [(G_i^k + I_i^k)(1 - \tau_i^k) - J_i^k]. \tag{2.60}$$

Note that this formulation accounts for reflection of the solar radiation back out the glazing and for the variation of radiative properties with wavelength. Both of these factors are important where light-colored materials are encountered (this is typical in direct-gain structures and especially sunspaces).

Since the material properties and the view factors are constant for the computer run, the two coefficient matrices of equation (2.58) may be precomputed, triangularized, and reduced during the input phase. The solution (back-substitution) then requires only M^2 operations at each time step, which considerably reduces computer time for the run. An approximate alternative approach is presented in Carroll (1980) and Walton (1980). Their technique requires the order of $3M$ operations per time step and is much more efficient for large systems. However, it does not allow reflection of radiation back out the glazings, nor does it allow the material properties to vary with wavelength.

Finally, the net radiation method presented here handles local absorption only by averaging material properties over the local absorption areas and neglects distributed absorption. Distributed absorption can be handled by the "participating medium" method presented in Hottel and Sarofim (1967). It requires one additional node for each distributed absorption surface, which must then be convectively coupled to the air thermal node.

2.8.3.2 Calculation of Shape Factors The above formulations assumed the availability of the radiation shape factors, F_{ij}. Such shape factors can be calculated from the formulas available in Siegel and Howell (1982), Howell (1982), and Reuth (1984). Using shape-factor algebra and reciprocity can reduce the calculation time of this matrix to less than half the time required for the full matrix. However, the calculation of shape factors can still consume significant computer time if the surfaces are finely subdivided.

Finer and finer subdivisions of physical surfaces can increase the solution time drastically. In view of this, it is worthwhile to examine the issue

Table 2.2
Accuracy in the radiation solution versus the number of subdivisions of each physical surface

Grid on each physical surface	Number of radiation surfaces per physical surface	Error (percentage)
1	1	22
2 × 2	4	9
3 × 3	9	3

of spatial discretization error. This issue is significant only for direct-gain and sunspace systems. W. O. Wray at Los Alamos used the SUNSPOT program (Wray, Schnurr, and Moore 1980) to perform detailed radiation analyses including many subdivisions. He concluded that one physical surface may be adequately modeled with one radiation surface. Unpublished work by Burns for a rectangular parallelpiped has enabled the construction of table 2.2, which presents the error in the radiative calculations versus the number of subdivisions per physical surface. This work indicates that the floor should be modeled as three equally divided strips running east and west, but that the remaining physical surfaces may each be modeled as one radiation surface. Arumi-Noe subdivides all physical surfaces into nine (3 × 3) equal radiation surfaces. It must be questioned whether the finer subdivision for all surfaces is merited given the uncertainty in the radiative properties and the large increase in execution time.

2.8.3.3 Specular Reflections The above formulations for diffusely emitting and reflecting surfaces apply to most building materials. However, some enamel paints and some tiles add a significant specular component to the reflectance, which is apparent upon viewing an image of a bright object (e.g., a light) in the surface. The sharper the image, the more specular the surface. Since there are virtually no data of this detail for building materials, one must guess the specular reflectance ρ_s and the diffuse reflectance ρ after visual inspection.

Two cases of special interest to passive solar designers arise in practice. The first is external radiation enhancement by a specular reflector near the south-facing glazing. Common cases are a horizontal reflector below the collecting surface and vertical "wing wall" reflectors beside the collecting surface. These situations have been analyzed (Grimmer et al. 1978), and

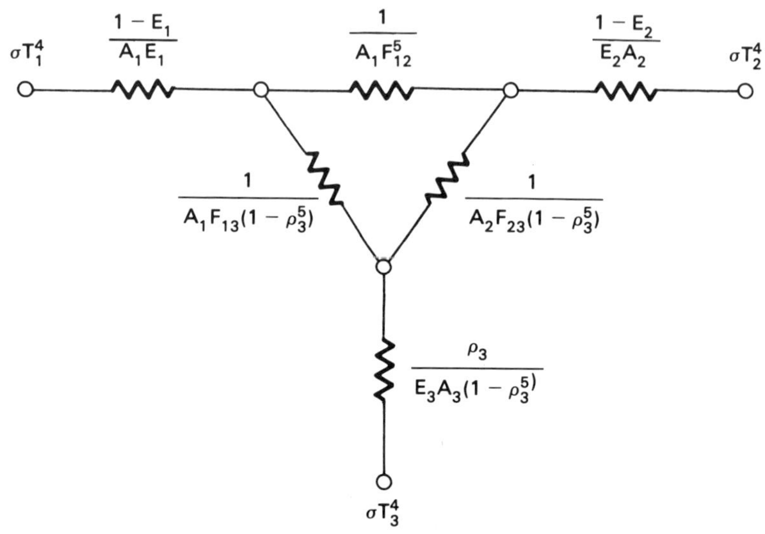

Figure 2.25
Specular radiosity network for three surfaces: 1 and 2 diffuse, and surface 3 specular.

multiplicative factors for monthly calcualtions are presented in Balcomb and McFarland (1978) and Balcomb et al. (1984).

The other instance where it is important to model the radiative material behavior in more detail is where the floor has a significant specular component of reflectance and is relatively light in color. Orders-of-magnitude calculations by the author indicate that errors of up to 20% in the absorbed solar radiation can occur if such surfaces are modeled as diffuse reflectors.

2.8.3.3.1 IMAGING One technique of modeling specular reflectors is the method of imaging where beam tracing methods are used to generate specular shape factors, F^s_{ij}. The method strictly applies only where the specular reflectance is independent of the incident direction (Siegel and Howell 1982, Bobco 1964, Sparrow and Lin 1965, Sarofim and Hottel 1966). When the reflectance has both a specular and a diffuse component, some algebraic manipulation allows construction of radiation networks with modified resistances, such as that shown in figure 2.25. In this instance, the solution techniques mentioned before may be applied.

2.8.3.3.2 MONTE CARLO TECHNIQUES When even greater detail is desired, Monte Carlo techniques may be applied. These techniques can easily account for directional emittances and reflectances, and specular and diffuse components of the reflectance (McNall and Biddison 1970). However, since many rays must be traced, these solution techniques tend to be extremely lengthy. The properties of building materials are frequently not known in sufficient detail to warrant such an approach.

2.8.4 Effects on Comfort

Passive buildings have been purported to be superior to conventional construction because they provide radiant comfort not available in ordinary buildings. An insulated wall that would ordinarily have a surface temperature close to the room air temperature could be replaced by a warm thermal-storage surface in a passive building. The resulting source of radiant heat enables one to lower the thermostat, thereby decreasing the energy usage for space heating with no loss of comfort. The standard method for quantifying this effect of radiant heat is through the definition of the mean radiant temperature, MRT. Other effects, as described below, also influence comfort.

2.8.4.1 Definition of MRT MRT is the temperature of a blackbody sphere that has no convective exchange with the air. (MRT can be defined for any location in any shape space.) For a sphere placed in the geometric center of a rectangular parallelpiped, MRT can be approximated by the geometric average of the area-weighted wall temperatures, and some have included the emittance of the surfaces in the definition (ASHRAE 1981, Fanger, Angelius, and Kjerulf-Jensen 1970, Wray 1979, Gubareff, Janseen, and Torberg 1960):

$$\mathrm{MRT} = \frac{\sum_{i=1}^{n} \varepsilon_i^2 T_i A_i}{\sum_{i=1}^{n} \varepsilon_i^2 A_i}, \tag{2.61}$$

where the superscript *2* refers to the longwave band. Since the longwave radiative emittances of all building materials are all nearly equal to $\varepsilon_i^2 = 0.9$, the emittances may be omitted from the calculation with little error.

The temperature representative of the thermal comfort is the effective temperature, *ET*. It combines the effect of MRT, dry-bulb temperature, wet-bulb temperature, humidity, and air movement to yield the sensation of warm or cold (ASHRAE 1981). The effective temperature may be approximated (Carroll 1980) as

$$ET = 0.42\ \text{MRT} + 0.51\ T_R + 0.04\ T_\infty + 1.1°\text{C}, \tag{2.62}$$

where T_R is the room air temperature and T_∞ is the outdoor ambient temperature. From this formula, it is apparent that one experiences about equal comfort from the room air temperature and from the MRT. The ambient-temperature term incorporates the effects of air infiltration.

2.8.4.2 Radiation Energy Density Although the MRT is a useful concept, it does not account for the comfort effect of solar radiation in the room. Using radiation networks as described in section 2.8.3, one may calculate the MRT in an enclosure with longwave and solar radiative flux. Comparing the longwave radiation from one wall at 32°C (90°F) and five walls at 20°C (68°F) to the shortwave radiation entering through a window, the following aperture areas provide an energy flux from solar radiation equivalent to the energy flux from 1 m^2 (11 ft^2) of wall area:

1. For diffuse solar radiation of *intensity* = *100 W/m² (34 Btu/h ft²)*; A_g = 0.121 m^2 (1.3 ft^2)

2. For beam solar radiation of *intensity* = *800 W/m² (272 Btu/h ft²)*; A_g = 0.0015 m^2 (0.016 ft^2)

It is apparent that radiant effects on comfort may be due primarily to the solar radiation entering the space rather than the elevated temperature of the surrounding surfaces.

To quantify all radiant effects on comfort, it is useful to consider the radiation energy density (Howell 1982). This quantity represents the radiation passing through a point in space and is properly formulated to include both solar and longwave bands, multiple reflections from surfaces, and the directly incident solar radiation. Once a radiation network is solved, the calculation of the radiation energy density is straightforward. This concept is worthy of further detailed investigation.

2.9 Solar Absorption on Opaque External Surfaces

Solar energy can enter the building directly through transparent surfaces. The amount absorbed provides a direct source of heat to the living space. However, the solar radiation absorbed on opaque exterior surfaces provides an indirect source of heat because some of this energy flows into the living space. In the winter, this amount of energy is usually small compared to the large amount of solar radiation entering the living space through windows. In the summer, when the overhang provides shading, the solar radiation absorbed on exterior surfaces can cause an indirect heating load that is greater than the direct heating load of the diffuse solar radiation entering through the windows. Exterior surface temperatures can approach 90°C (194°F) (Kahwaji 1987) during periods of high sunshine and low wind. Indeed, neglecting this amount of energy as a heat source has caused large cooling-load calculation errors (Wray 1980).

2.9.1 Instantaneous Energy Balance on a Node

On an exterior wall, all types of heat transfer occur, including absorption of solar radiation, longwave radiative transfer with the surroundings, convective transfer with the surrounding air, and conductive transfer into the wall. A thermal network may be drawn for the wall system as shown in figure 2.26(a). An energy balance may be determined on the outside wall node (ASHRAE 1981, Threlkeld 1970).

The thermal network shown does not include the significant thermal capacitance of the wall and, thus, it is only valid in the steady state. In this case, all time-varying potentials (temperatures) must be averaged over a suitable time interval, such as the period of the weather data. The steady-state energy balance, under this constraint, is

$$\overline{\alpha_o I} - h_s(\overline{T}_o - \overline{T}_s) + h_\infty(\overline{T}_\infty - \overline{T}_o) + \frac{k}{L}(\overline{T}_i - \overline{T}_o) = 0, \tag{2.63}$$

where the bars indicate an average over 24 hours. The heat transfer into the wall may be calculated as

$$\frac{\dot{Q}}{A} = \frac{k}{L}(T_o - T_i). \tag{2.64}$$

Solving for T_o from equation (2.63) and substituting into equation (2.64) yields, after some rearrangement,

$$\frac{\dot{Q}}{A} = \frac{\overline{\alpha_o I}}{h} + \frac{h_s \overline{T}_s + h_\infty \overline{T}_\infty}{h} - \frac{\overline{T}_i}{\frac{L}{k} + \frac{1}{h}}, \tag{2.65}$$

where

$$h = h_s + h_\infty. \tag{2.66}$$

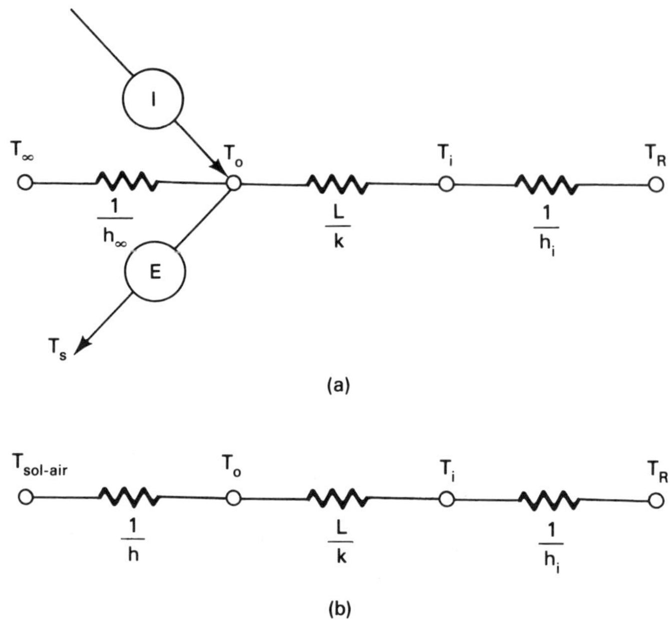

Figure 2.26
(a) Thermal network for a building element; (b) equivalent thermal network for a building element.

Defining

$$T_{\text{sol-air}} = \frac{\overline{\alpha_o I}}{h} + \frac{h_s \overline{T}_s + h_\infty \overline{T}_\infty}{h} \tag{2.67}$$

as the driving potential for heat transfer allows the construction of the equivalent thermal network shown in figure 2.26(b). The sol-air temperature explicitly includes the effects of solar radiative absorption and long-wave radiative emission.

It can be shown that the sol-air temperature approach yields exact answers for linear problems where both the interior room temperature and the heat transfer coefficients are constant. However, recent experimental work at NBS has shown that the approach is inaccurate when the interior temperature is maintained within a deadband. In this instance, the sol-air temperature approach overpredicts the solar load on the building (Burch et al. 1984).

Note that h_∞ includes the effects of both convective and radiative transfer with the surroundings:

$$h_\infty = h_c + F_{o\infty}\bar{\varepsilon}_o\sigma(T_\infty^2 + T_o^2)(T_\infty + T_o). \tag{2.68}$$

Furthermore, longwave radiative transfer with the sky is expressed in terms of h_s:

$$h_s = F_{os}\bar{\varepsilon}_o\sigma(T_s^2 + T_o^2)(T_s + T_o). \tag{2.69}$$

2.9.1.1 Solar Absorptance and Longwave Emittance To accurately calculate the heat transfer coefficients, it is necessary to know the solar absorptance α_o and the longwave emittance ε_o of the surface. Wray (1979) and Gubareff, Janseen, and Torberg (1960) include the properties for many building materials, and Mabinton and Goswami (1980) include the properties for many engineering materials. ASHRAE (1981) and Threlkeld (1970) provide a subset of these data.

In practice, it is sometimes useful to account for the surface roughness since the material properties apply only to smooth surfaces. Unfortunately, most of the analyses done to account for surface roughness (Porteus 1963, Beckman and Spizzichino 1963, Torrance and Sparrow 1966 and 1967, Kanayama 1972, Birkebak and Abdulkadin 1976, Rowley, Algren, and Blackshaw 1930) apply to very idealized surfaces and are applicable only at a single wavelength and only for a single incoming direction. A simple correction for random surface roughness that can be applied to both the solar absorptivity and the longwave emissivity is

$$\bar{\varepsilon} = \frac{4}{\dfrac{3}{\varepsilon} + 1}. \tag{2.70}$$

Equation (2.70) is valid only when the mean surface roughness is much greater than the wavelength of the radiation. This is true for all building materials. Under this restriction, equation (2.70) may also be applied to the solar absorptance.

2.9.1.2 Convective Heat Transfer Coefficient The total heat transfer coefficient, h_∞, used in equations (2.63) through (2.68) includes both convection and radiation. Since the radiative portion is given directly by equation (2.68), the remainder of this section is devoted to a discussion of the convective portion, h_c. This quantity is also germane when considering the heat loss from glazing surfaces, since the resistance of the exterior film is significant in terms of the total resistance through the glass. (See also Niles, chapter 3 of this volume.)

At typical air velocities, the surface conductance depends only slightly on the surface roughness (ASHRAE 1981, Threlkeld 1970, Parmelee and Huebscher 1947, Sparrow and Tien 1977, Lloyd and Moran 1974). Unfortunately, the dependence increases with increasing wind velocity. Indeed, there are times when the wind is nearly stagnant and mixed convection occurs from exterior surfaces. In the rare simulations where the attempt is made to model this accurately, the observations and correlations presented in Metais and Eckert (1964), Brown and Gauvin (1965), Lloyd and Sparrow (1970), Mori (1961), and Carslaw and Jaeger (1973) may prove helpful.

The flow zones around a building are identified with reference to the incident direction of the wind. Positive pressure is experienced by the windward surfaces, and negative pressure is experienced by the leeward surfaces. The extent of these regions depends on the geometry of the building and the magnitude and direction of the incident wind. The direction of flow over the surfaces may be primarily parallel (for surfaces parallel to the incident wind direction), primarily normal (on the windward side), or mixed (on the leeward side). All these situations cause the magnitude of the heat transfer coefficient to change radically (ASHRAE 1981, Parmelee and Huebscher 1947, Sparrow and Tien 1977, Lloyd and Moran 1974). Unfortunately, the situation is generally too difficult to model in detail. The only definite statement that can be made is that the ASHRAE conductance value recommended for design is far too high compared to the actual average value. This is because a winter design wind velocity of 15 mph (7 m/s) is recommended (ASHRAE 1981), which results in values for the convective coefficient that are typically two to three times the actual average values. However, this is consistent with ASHRAE's conservative approach that tends to recommend design values on the high side in predicting the load.

The following simplified equation is recommended as an approximate approach to calculating the external film coefficient (Sparrow and Tien 1977):

$$h = h_o + k_w V_w, \tag{2.71}$$

where $h_o = 5.1$ W/Km^2 (0.9 Btu/h °F ft^2) and includes both free convection and longwave thermal radiation, and an appropriate value for k_w is 8.17 Ws/Km^3 (0.00012 Btu/°F ft^3). Note that the unit of V_w in equation (2.71) is m/s (ft/h).

2.9.2 Effective U-Value Approach

An alternative approach is to include the solar and longwave energy sources by modifying the overall wall heat transfer coefficient (ASHRAE 1981, Threlkeld 1970, Mackey and Wright 1944, Stewart 1948, Bullock 1961). The effective U-value is defined as the net energy transfer through the wall over a specified time period divided by the inside-air to outside-air temperature averaged over the same period. From a fundamental standpoint, this approach is not as desirable as the conduction transfer function or sol-air temperature approaches because the energy sources modify the potential, not the conductance. The problem with the approach is that the results apply only to a particular location, wall construction, wall color and orientation, and time period. However, the approach is useful when approximate answers are being sought or where incident solar radiation data are not available for the surface of interest.

2.10 Efficient Computation

At the cost of a small degree of inaccuracy, some of the above quantities in a detailed solar processor (as presented in sections 2.3 and 2.4) may be calculated only occasionally. In fact, almost a 50% reduction in computation time results from some simple approximations that cause only a small change (<5%) in the amount of absorbed energy. This is normally considered to be entirely acceptable in view of the inaccuracy inherent in the approximations involved in the modeling of the diffuse and ground-reflected components of the solar radiation.

2.10.1 Sunrise, Sunset Calculation

The most obvious method of reducing calculation time is to ensure that no calculations of solar radiation are performed at night. Thus, if the horizontal solar radiation is less than or equal to zero, the entirety of solar calculations should be bypassed. Furthermore, all measured radiation may be considered as diffuse before sunrise and after sunset. These times may be determined by setting the solar altitude β to zero in equation (2.8), with the result

$$H_s = \cos^{-1}[-\tan(L)\tan\delta]. \tag{2.72}$$

Converting to local time in hours yields

$$t_s = (H_s/15\ \mathrm{deg}/h) + 12h - (\Delta t_k + \Delta t_L)/3600. \tag{2.73}$$

It is also advantageous to calculate the "sunrise" and "sunset" hour angles on the particular surfaces of interest, that is, the times between which the surface can be exposed to direct normal radiation. The maximum exposure times are then taken between actual sunrise and sunset on horizontal surfaces, and surface "sunrise" and "sunset" on other surfaces. No beam radiation and shading calculations need to be done outside this exposure time. An alternative approach is to exclude beam radiation calculations when $\cos\theta$ is negative. From an efficiency standpoint, the calculation times are nearly identical.

Furthermore, it is frequently advantageous to take all the incident radiation as diffuse for some time after sunrise and some time before sunset. This avoids potential mismodeling of the radiation, which occurs only when the diffuse has been the object of correlation, that is, when I_s^0 is obtained from I_h and K_T. The beam radiation is then calculated as

$$I_{DN} = (I_h - I_s^0)/\cos\beta. \tag{2.74}$$

When $\cos\beta \ll 1$, equation (2.74) may yield $I_{DN} > H_e$.

The criterion to circumvent such circumstances recommended by Chapman (1980) is to set $I_{DN} = 0$ (that is, $I_s^0 = I_h$) for $\beta < 8$ degrees. Since the radiation is mostly diffuse and weak at these times (because m is large), this approximation represents very little error in the total incident energy during a day. This situation does not arise if I_{DN} is correlated versus K_T; therefore this approach is recommended.

2.10.2 Temporal Quantities

The calculation time may be substantially reduced by selecting time periods in which certain quantities are held constant. There are four categories of such time periods:

1. The entire run (usually one year)

The constant quantities are those associated with the local position, the surface angles, and the glazing properties as follows:

a. the time correction for the longitude [equation (2.4)]
b. the sine and cosine of the latitude [equations (2.8) and (2.10)]
c. the sines and cosines of surface slopes and azimuths [equation (2.11)]
d. all surface view factors [equations (2.14–2.18), (2.21–2.23)]
e. the curve fits for the glazing transmittance and absorptance
f. the diffuse transmittance and absorptance

2. Each month

The constant quantities include the extraterrestrial solar radiation, the declination, the orbital correction for the time, and all quantities involving only these variables and those that are held constant for the run as follows:

a. the midday of the month $\overline{N}$
b. the extraterrestrial solar radiation at the midday of the month [see equation (2.2)]
c. the declination at the middle of the month and the sine and cosine thereof [equations (2.8) and (2.10)]
d. the orbital correction for the time at the middle of the month [equation (2.4)]
e. all terms in the determination of cos θ [equation (2.11) or (2.12)] involving only the quantities

$$T_1 = \cos L \cos \delta \tag{2.75}$$

$$T_2 = \sin L \sin \delta \tag{2.76}$$

f. ground reflectance (see figure 2.11)

The time period for which these quantities are held constant (that is, one month) is arbitrary. One month was chosen because this time period occurs naturally in the problem.

3. Each day

The constant quantities include all those associated with sunrise and sunset as given by equations (2.72) and (2.73).

4. Each hour

The finest subdivision of time for solar calculations should be one hour because this is the period over which the data are averaged. Moreover, the level of accuracy of the calculations does not warrant a further subdivision. Should values be needed at a smaller time interval, they should be obtained by interpolation and not by calculation.

The values calculated within the hour are as follows:

a. the solar time [equation (2.7)]
b. the hour angle and its sine and cosine [equation (2.9)]
c. the sines and cosines of the solar azimuth and altitude [equations (2.8) and (2.10)]
d. the direct normal components and the diffuse components of the incident radiation
e. the incident angles or their cosines for the direct normal radiation [equations (2.11) or (2.12)]
f. the glazing transmittances and reflectances for the angles of the direct normal radiation
g. the fractions of the surfaces exposed to the beam radiation f_b
h. the components of the incident solar radiation for the various surfaces given by equations (2.13), (2.18), and (2.21)
i. the radiation absorbed in the glazing [equation (2.50)]

In principle, all surfaces that exchange energy radiatively may be coupled through a diffuse radiation network that requires a solution for each time step, although some of these effects may be accounted for in the definitions of the view factors as, for example, in equation (2.19). The degree of detail used in performing radiative network calculations varies widely in simulations. The inherent inaccuracies in solar processors must be balanced against the expenditure of excessive computer time. The level of detail at which this occurs depends on computation costs and the accuracy required for a given problem.

2.11 Summary

The chapter began with a discussion of the nature of solar radiation at the earth's surface. Both spectral and directional variations were covered. It was elucidated that due to the spectral shift with air mass, solar properties are not absolute but may change by up to about 10% under extreme conditions. However, directional variations are the cause of much greater concern, as errors of up to 100% in the sky diffuse can occur when overhangs are present. It was noted that there have been no investigations of solar radiation under overhangs, so the matter remains inconclusive for the present.

The solar radiation was traced past external objects, through glazings, and to absorption on interior surfaces. An issue of primary concern was the occlusion caused by various bare trees. A second-order direct method was proposed to correct for the effect of trees on the beam, sky diffuse, and ground reflected radiation. Overhang shading concluded the discussion of external effects.

Types of glass, the material properties thereof, and the effect upon energy balances were addressed next. The effect of the utilization of both beam and diffuse transmittances was addressed. The discussion proceeded with an analysis of internal distributed absorption on surfaces such as plants and furniture. Local absorption on building surfaces was covered next. Radiation networks, imaging, and Monte Carlo techniques were presented with an indication of the substantial calculation times involved therein. The section on internal absorption concluded with a discussion of shortwave and longwave radiative effects upon comfort. It was pointed out that neglecting shortwave radiation can lead to gross errors in the assessment of comfort.

The chapter then proceeded with a discussion of the effect of solar absorption upon external opaque surfaces. Important aspects to be considered include: the radiative material properties and the value of the external film coefficient. The sol-air temperature approach was recommended as a method for analysis, whereas the effective U-value approach was not. Finally, a section on efficient computation was provided.

In summary, I feel that while solar radiation calculations can be formulated with great complexity and result in a very precise method of calculation, accuracy must always be questioned. Factors not accurately accounted for typically include 1. effects of directional and spectral properties of materials; 2. transmittance and reflectance of external objects; 3. effects of environmental coatings such as dust and dirt; 4. transmittance, absorptance, and reflectance of distributed internal surfaces; and 5. perhaps the most significant of all, directional distribution of the sky diffuse radiation. No comprehensive investigation of the errors caused by these interdependent effects has been conducted to date. In view of this situation, the overall goals of performing detailed solar radiation calculations should perhaps be viewed from a relative rather than an absolute perspective.

Acknowledgment

I wish to express gratitude to Dr. J. Douglas Balcomb for his sagacious review of the manuscript. It is the best review I have ever received.

References

Alford, J. S., J. E. Ryan, and F. O. Urban. 1939. Effect of heat storage and variation in outdoor temperature and solar intensity on heat transfer through walls. *ASHVE Transactions* 45:393.

American Society of Heating, Refrigerating, and Air-Conditioning Engineers. 1963. *ASHRAE Transactions* 69:31.

American Society of Heating, Refrigerating, and Air-Conditioning Engineers. 1981. *ASHRAE Handbook: 1981, Fundamentals.* Chap. 27. Atlanta, GA: ASHRAE, pp. 27.1–27.48.

Balcomb, J. D., and R. D. McFarland. 1978. A simple empirical method for estimating the performance of a passive solar heated building of the thermal storage wall type. *Proc. 2d National Passive Solar Conference*, Philadelphia, PA, March 16–18, 1978. Newark, DE: American Section of the International Solar Energy Society, p. 377.

Balcomb, J. D., D. Barley, R. McFarland, J. Perry, W. Wray, and S. Noll. 1980. *Passive Solar Design Handbook. Vol. 2: Passive Solar Design Analysis.* DOE/CS-0127/2. Washington, DC: U.S. Department of Energy.

Balcomb, J. D., R. W. Jones, R. D. McFarland, and W. O. Wray. 1984. *Passive Solar Heating Analysis: A Design Manual.* Atlanta, GA: ASHRAE.

Beckman, P., and A. Spizzichino. 1963. *The Scattering of Electromagnetic Waves from Rough Surfaces.* New York: Macmillan.

Bird, R. E., R. L. Hulstrom, and L. J. Lewis. 1983. Terrestrial solar spectra data sets. *Solar Energy* 30:563.

Birkebak, R. C., and A. Abdulkadin. 1976. Random rough surface model for spectral, directional emittance of rough metal surfaces. *Int. J. Heat Mass Transfer* 19:1039.

Bliss, R. W. 1961. Atmospheric radiation near the surface of the ground. *Solar Energy* 5:103.

Bobco, R. P. 1964. Radiation heat transfer in semigray enclosures with specularly and diffusely reflecting surfaces. *J. Heat Transfer* 86:123.

Born, M., and E. Wolf. 1965. *Principles of Optics.* 3d. ed. New York: Pergamon Press.

Brine, D. T., and M. Iqbal. 1983. Diffuse and global solar spectral irradiance under cloudless skies. *Solar Energy* 30:447.

Brown, C. K., and W. H. Gauvin. 1965. Combined free and forced convection. *Can. J. Chem. Engr.* 43:306.

Bryan, H., et al. 1981. The use of physical scale models for daylighting analysis. *Proc. 6th National Passive Solar Conference*, Portland, OR, September 8–12, 1981. Newark, DE: American Section of the International Solar Energy Society, p. 865.

Budin, R., and L. Budin. 1982. A mathematical model for shading calculations. *Solar Energy* 29:339.

Bullock, C. E. 1961. *Periodic Heat Transfer in Walls and Roofs.* Master's thesis, Department of Mechanical Engineering, University of Minnesota.

Burch, D. M., D. F. Kritz, and R. S. Spain. 1984. The effect of wall mass on winter heating loads and indoor comfort—an experimental study. *ASHRAE Transactions* 90(I):94.

Cannon, T. W., and L. D. Dwyer. 1981. An all-sky, video-based luminance mapper for daylighting research. *Proc. 6th National Passive Solar Conference*, Portland, OR, September 8–12, 1981. Newark, DE: American Section of the International Solar Energy Society, p. 855.

Carroll, J. A. 1980. An MRT method of computing radiant energy exchange in rooms. *Proc. Systems Simulation and Ecomonic Analysis*, San Diego, CA, January 23–25, 1980. Golden, CO: Solar Energy Research Institute, p. 343.

Carslaw, H. S., and J. C. Jaeger. 1973. *Conduction of Heat Solids.* 2d ed. U.K.: Oxford University Press, p. 105.

Chapman, J. 1980. User's guide for FREHEAT. Master's thesis, Dept. of Mechanical Engineering, Colorado State University, Fort Collins.

Chiam, H. F. 1983. Transmittance of reflected diffuse radiation. *Solar Energy* 30:75.

Christian, K. D. J., and S. R. Shatynski. 1982. Passive solar windows produced by physical vapor deposition. *Proc. 7th National Passive Solar Conference*, Knoxville, TN, August 30–September 1, 1982. Newark, DE: American Section of the International Solar Energy Society, p. 855.

Coulson, K. L. 1975. *Solar and Terrestrial Radiation: Methods and Measurements.* New York: Academic Press, p. 322.

Dave, J. V. 1977. Validity of the isotropic-distribution approximation in solar energy estimations. *Solar Energy* 19:331.

Davies, H. 1954. The reflection of electromagnetic waves from a rough surface. *Proc. Inst. Elec. Engr. London* 101:209.

Davies, J. A., and D. C. McKay. 1982. Estimating solar irradiance and components. *Solar Energy* 29:55.

Dean, E. T. 1979. Graphic methods for determining the solar access design envelope with irregular topography. *Proc. 4th National Passive Solar Conference*, Kansas City, MO, October 3–5, 1979, Newark, DE: American Section of the International Solar Energy Society, p. 287.

Duffie, J., and W. A. Beckman. 1980. *Solar Engineering of Thermal Processes.* New York: Wiley.

Englund, J. S. 1982. Solar access to residential buildings. *Solar Engineering*, 1982, ASME, p. 357.

Erbs, D. G., R. C. Stauter, and J. A. Duffie. 1980. The basis and effects of inaccuracies in diffuse radiation correlations. *Proc. 1980 Annual Meeting of the American Section of the International Solar Energy Society*, Phoenix, AZ, June 2–6, 1980. Newark, DE: American Section of the International Solar Energy Society, p. 1429.

Eringen, A. C. 1980. *Mechanics of Continua.* 2d ed. Huntington, NY: Krieger, p. 27.

Erley, D., and M. Jaffe. 1979. *Site Planning for Solar Access.* HUD-PDR-481. Washington, DC: U.S. Department of Housing and Urban Development.

Ewing, W. B., and J. I. Yellot. 1976. Energy conservation through the use of exterior shading of fenestration. *ASHRAE Transactions* 82(I):703.

Fanger, P. O., O. Angelius, and P. Kjerulf-Jensen. 1970. Radiation data for the human body. *ASHRAE Transactions* 76:323.

Frazier, K. 1980. *Our Turbulent Sun.* Englewood Cliffs, NJ: Prentice Hall, pp. 1–31.

Goody, R. M. 1964. *Atmospheric Radiation: Theoretical Basis.* New York: Oxford University Press-Clarendon Press.

Grimmer, D. P., et al. 1978. *Augmented Solar Energy Collection Using Various Planar Reflective Surfaces: Theoretical Calculations and Experimental Results.* LA-7041. Los Alamos, NM: Los Alamos National Laboratory.

Gubareff, G. G., J. E. Janseen, and R. H. Torberg. 1960. *Thermal Radiation Properties Survey—A Review of Literature.* 2d ed. Minneapolis, MN: Honeywell Research.

Guzzi, R., G. Lo Vecchio, R. Rizzi, and G. Scalabrin. 1983. Experimental validation of a spectral direct solar radiation model. *Solar Energy* 31:359.

Hatfield, J. L., R. B. Giorgis, Jr., and R. G. Flocchini. 1981. A simple solar radiation model for computing direct and diffuse spectral fluxes. *Solar Energy* 27:323.

Healey, J., R. Near, and J. Boston. 1980. Solar irradiance data for the passive designer. *Proc. 5th National Passive Solar Conference*, Amherst, MA, October 19–26, 1980. Newark, DE: American Section of the International Solar Energy Society, p. 201.

Henderson, S. T. 1970. *Daylight and Its Spectrum.* New York: American Elsevier, pp. 1–115.

Holzberlein, T. M. 1979. Don't let the trees make a monkey of you. *Proc. 4th National Passive Solar Conference*, Kansas City, MO, October 3–5, 1979. Newark, DE: American Section of the International Solar Energy Society, p. 416.

Hottel, H. C., and A. F. Sarofim. 1967. *Radiative Transfer.* New York: McGraw-Hill, pp. 201, 202.

Howell, J. R. 1982. *A Catalog of Radiation Configuration Factors.* New York: McGraw-Hill.

Hsieh, C. K. 1976. Calculation of thermal radiative properties of glass. *ASHRAE Transactions* 82:734.

Hsieh, C. K., and R. W. Coldwey. 1975. Thermal radiative properties of glass. *ASHRAE Transactions* 81:260.

Iqbal, M. 1981. The influence of collector azimuth on solar heating of residential buildings and the effect of anisotropic sky-diffuse radiation. *Solar Energy* 26:249.

Johnston, S. A. 1983. A vector computer graphic procedure for calculating the irradiated area of collector surfaces in buildings. *Proc. 8th National Passive Solar Conference*, Santa Fe, NM, September 7–9. 1983. Boulder, CO: American Solar Energy Society, p. 71.

Jones, R. E., Jr. 1980. Effects of overhang shading of windows having arbitrary azimuth. *Solar Energy* 24:305.

Jones, R. E., Jr., and R. F. Yanda. 1981. Finite width overhang shading of south windows. *Proc. 1981 Annual Meeting, American Section of the International Solar Energy Society*, Philadelphia, PA, May 27–30, 1981. Newark, DE: American Section of the International Solar Energy Society, p. 867.

Jones, R. W. 1981. Summer heat gain control in passive solar heated buildings: fixed horizontal overhangs. *Passive Cooling: Proc. International Passive and Hybrid Cooling Conference*, Miami Beach, FL, November 11–13, 1981. Newark, DE: American Section of the International Solar Energy Society, pp. 402–406.

Jones, R. W., J. D. Balcomb, R. D. McFarland, and W. O. Wray. 1982. *Passive Solar Design Handbook, vol. 3: Passive Solar Design Analysis.* DOE/CS-0127/3. Washington, DC: U.S. Department of Energy.

Kahwaji, G. Y. 1987. Heat transfer in building enclosures. Ph.D. diss., Dept. of Mechanical Engineering, Colorado State University, Fort Collins.

Kanayama, K. 1972. Apparent directional emittance of V-groove and circular-groove rough surfaces. *Heat Transfer Jpn. Research* 1:11.

Keyes, M. W. 1967. Analysis and rating of drapery materials used for indoor shading. *ASHRAE Transactions* 73:VIII.4.1.

Klein, S. A., W. W. Beckman, and J. A. Duffie. 1976. A design procedure for solar water heating systems. *Solar Energy* 18:113.

Kohler, J., and D. Lewis. 1981. Let the sun shine in. *Solar Age* 6:45.

Kreith, F., and J. F. Kreider. 1978. *Principles of Solar Engineering.* New York: McGraw-Hill, pp. 37–84.

Kwentis, G. K. 1967. Sc. D. diss., MIT, Cambridge, MA.

Lau, A. 1982. Design and effectiveness of fixed overhangs; development of a novel design tool. *Proc. 7th National Passive Solar Conference*, Knoxville, TN, August 30–September 1, 1982. Newark, DE: American Section of the International Solar Energy Society, p. 393.

LeBaron, B., and I. Dirmhim. 1983. Strengths and limitations of the Liu and Jordan model to determine diffuse from global irradiance. *Solar Energy* 31:167.

Leckner, B. 1978. The spectral distribution of solar radiation at the earth's surface—elements of a model. *Solar Energy* 20:143.

Libby Owens. N.d. *Sun Angle Calculator.* Toledo, OH: Libby Owens Ford Glass Company.

Lloyd, J. R., and W. R. Moran. 1974. Natural convection adjacent to horizontal surfaces of various planforms. *J. Heat Transfer* 96:443.

Lloyd, J. R., and E. M. Sparrow. 1970. Combined forced and free convection on vertical surfaces. *Int. J. Heat Mass Transfer* 13:434.

Ma, C. C. Y., and M. Iqbal. 1983. Statistical comparison of models for estimating solar radiation on inclined surfaces. *Solar Energy* 31:313.

Mabinton, R. C., and D. Y. Goswami. 1980. Compilation of thermal radiative properties of building and furnishing materials and other interiors for use in energy conserving designs. *Proc. Annual Meeting of the American Section of the International Solar Energy Society*, Phoenix, AZ, June 2–6, 1980. Newark, DE: American Section of the International Solar Energy Society, p. 1131.

Mackey, C. O., and L. T. Wright, Jr. 1944. Periodic heat flow—homogeneous walls or roof. *ASHVE Transactions* 50:293.

Mazria, E. 1979. *The Passive Solar Energy Book.* Emmaus, PA: Rodale Press.

Metais, B., and E. R. G. Eckert. 1964. Forced, mixed and free convection regimes. *J. Heat Transfer* 86:295.

McNall, P. E., Jr., and R. E. Biddison. 1970. Thermal and comfort sensations of sedentary persons exposed to asymmetric radiant fields. *ASHRAE Transactions* 76:123.

Montgomery, D. A., S. L. Keown, and G. M. Heisler. 1982. Solar blocking by common trees. *Proc. 7th National Passive Solar Conference*, Knoxville, TN, August 30–September 1, 1982. Newark, DE: American Section of the International Solar Energy Society, p. 473.

Moon, P. 1940. Proposed standard solar radiation curves for engineering use. *J. Franklin Institute* 230:583.

Moore, G. L., and C. W. Pennington. 1967. Measured and application of solar properties of drapery shading materials used for indoor shading. *ASHRAE Transactions* 73:VIII.4.1.

Mori, Y. 1961. Buoyancy effects in forced laminar convection flow over a horizontal flat plate. *J. Heat Transfer* 83:479.

Morris, C. W., and J. H. Lawrence. 1971. The anisotropy of clear sky diffuse solar radiation. *ASHRAE Transaction* 77:136.

Morrison, C. A., J. M. Wheeler, Jr., and E. A. Farber. 1976. An experimental determination of shading coefficients for selected insulating reflective glasses and draperies. *ASHRAE Transactions* 82:38.

Muniz, F. A. 1982. In search of the tropicool: a shading device calculator. *Proc. 7th National Passive Solar Conference*, Knoxville, TN, August 30–September 1, 1982. Newark, DE: American Section of the International Solar Energy Society, pp. 917–922.

Olgyay and Olgyay. 1957. *Solar Control and Shading Devices*, Chap. 8. Princeton, NJ: Princeton University Press.

Oppenheim, A. K. 1956. Radiation analysis by the network method. *ASME Transactions* 65:725.

Oretskin, B. L. 1982. Studying the efficacy of light wells by means of models under an artificial sky. *Proc. 7th National Passive Solar Conference*, Knoxville, TN, August 30–September 1, 1982. Newark, DE: American Section of the International Solar Energy Society, p. 459.

Parmelee, G. V., and W. W. Aubele. 1948. Solar and total heat gain through double flat glass. *ASHVE Transactions* 54:407.

Parmelee, G. V., and W. W. Aubele. 1952. Radiant energy emission of atmosphere and ground. *ASHVE Transactions* 58:85.

Parmelee, G. V., and R. G. Huebscher. 1947. Forced convection heat transfer from flat surfaces. *ASHVE Transactions* 53:245.

Pennington, C. W. 1969. Solar heat gains through 3/8 in. and 1/2 in. grey plate glass with indoor shading. *ASHRAE Transactions* 75:44.

Pettit, R. B. 1978. Solar averaged transmittance properties of various glazings. *Proc. Annual Meeting of the American Section of the International Solar Energy Society*, Denver, CO, August 28–31, 1978. Newark, DE: American Section of the International Solar Energy Society, p. 294.

Pitman, C. L., and L. L. Vant-Hull. 1978. Errors in locating the sun and their effect on solar intensity predictions. *Proc. 1978 Annual Meeting of the American Section of the International Solar Energy Society*, Denver, CO, August 28–31, 1978. Newark, DE: American Section of the International Solar Energy Society, p. 701.

Porteus, J. O. 1963. Relation between the height distribution of a rough surface and the reflectance at normal incidence. *J. Opt. Soc. Am.* 53:1394.

Puri, V. M., R. Jiminez, M. Menzer, and F. A. Costello. 1980. Total and non-isotropic diffuse insolation on tilted surfaces. *Solar Energy* 25:85.

Reuth, J. G. 1984. The solar load ratio-solar fraction design method for cylinder water walls. Master's thesis, Dept. of Mechanical Engineering, Colorado State University, Fort Collins.

Rowley, F. B., A. B. Algren, and J. L. Blackshaw. 1930. Surface conductances as affected by air velocity, temperature and character of surface. *ASHVE Transactions* 36:444.

Sarofim, A. F., and H. C. Hottel. 1966. Radiative exchange among non-Lambert surfaces. *J. Heat Transfer* 88:37.

Sassi, G. 1983. Some notes on shadow and blockage effects. *Solar Energy* 31:331.

Scofield, S. H., and F. Moore. 1981. Climatological sundial. *Proc. 6th National Passive Solar Conference*, Portland, OR, September 8–12, 1981. Newark, DE: American Section of the International Solar Energy Society, p. 382.

Selkowitz, S. 1978. Transparent heat mirrors for passive solar heating applications. *Proc. 2d National Passive Solar Conference*, Philadelphia, PA, March 16–18, 1978. Newark, DE: American Section of the International Solar Energy Society, p. 329.

Selkowitz, S. 1981. A hemispherical sky simulator for daylighting model studies. *Proc. 6th National Passive Solar Conference*, Portland, OR, September 8–12, 1981. Newark, DE: American Section of the International Solar Energy Society, p. 850.

Selkowitz, S., and S. Berman. 1978. Energy efficient windows program activities. *Proc. 2d National Passive Solar Conference*, Philadelphia, PA, March 16–18, 1978. Newark, DE: American Section of the International Solar Energy Society, p. 335.

Selkowitz, S., M. Rubin, and R. Creswick. 1980. Average transmittance factors for multiple glazed window systems. *Proc. 4th National Passive Solar Conference*, Kansas City, MO, October 3–5, 1979. Newark, DE: American Section of the International Solar Energy Society, p. 383.

Solar Energy Research Institute. 1980. *SERIRES User's Manual*. Golden, CO: SERI.

Sharp, K. 1980. Analytical integration of the insolation on a shaded surface at any tilt and azimuth. *Proc. 5th National Passive Solar Conference*, Amherst, MA, October 19–26, 1980. Newark, DE: American Section of the International Solar Energy Society, p. 166.

Siegel, R., and J. R. Howell. 1982. *Thermal Radiation Heat Transfer*. 2d ed. New York: McGraw-Hill.

Siminovitch, M. J., D. E. Bergeson, and M. T. McCulley. 1982. Thermal protection of the solar aperture in the cooling season—a quantitative performance based study. *Proc. 7th National Passive Solar Conference*, Knoxville, TN, August 30–September 1, 1982. Newark, DE: American Section of the International Solar Energy Society, p. 883.

Sowell, E. F. 1978. The use and limitations of ASHRAE solar algorithms in solar energy utilization studies. *ASHRAE Transactions* 84:77.

Sparrow, E. M., and S. H. Lin. 1965. Radiation heat transfer at a surface having both specular and diffuse reflectance components. *Int. J. Heat Mass Transfer* 8:769.

Sparrow, E. M., and K. K. Tien. 1977. Forced convection heat transfer at an inclined and yawed square plate—application to solar collectors. *J. Heat Transfer* 99:507.

Steinmuller, B. 1980. The two-solarimeter method for insolation on inclined surfaces. *Solar Energy* 25:449.

Stewart, J. P. 1948. Solar heat gain through walls and roofs for cooling load calculations. *ASHVE Transactions* 54:361.

Stokes, G. G. 1860. On the intensity of light reflected from a transmitter through a pile of plates. *Proc. Royal Soc. London* 11:546–556.

Sun, T. Y. 1968. Shadow area equations for window overhangs and side fins and their application in computer calculation. *ASHRAE Transactions* 74:I.1.1.

Thekaekara, M. P. 1973. Solar energy outside the earth's atmosphere. *Solar Energy* 14:109.

Threlkeld, J. L. 1970. *Thermal Environmental Engineering*. 2d. ed. Chap. 13. Englewood Cliffs, NJ: Prentice Hall, pp. 279–311, 320.

Threlkeld, J. L., and R. C. Jordan. 1958. Direct solar radiation available on clear days. *ASHVE Transactions* 64:45.

Thomalla, E., et al. 1983. Circumsolar radiation calculated for various atmospheric conditions. *Solar Energy* 30:563.

Torrance, K. E., and E. M. Sparrow. 1966. Off-specular peaks in the directional distribution of reflected thermal radiation. *J. Heat Transfer* 88:223.

Torrance, K. E., and E. M. Sparrow. 1967. Theory for off-specular reflection from roughened surfaces. *J. Opt. Soc. Am.* 57:1105.

Touloukian, Y. S., and D. C. DeWitt. 1972. *Thermal Radiative Properties*. Vols. 7, 8, and 9 (1970–1972). TPRC Data Series. New York: IFI Plenum.

Utzinger, D. M., and S. A. Klein. 1979. A method of estimating monthly average solar radiation on shaded surfaces. *Proc. 3d National Passive Solar Conference*, San Jose, CA,

January 11–13, 1979. Newark, DE: American Section of the International Solar Energy Society, p. 295.

Walton, G. N. 1979. The application of homogeneous coordinates to shading calculations. *ASHRAE Transactions* 85(I): 174.

Walton, G. N. 1980. A new algorithm for radiant interchange in room loads calculations. *ASHRAE Transactions* 86: 190–208.

Watt, A. D. 1978. On the nature and distribution of solar radiation. *Proc. Annual Meeting of the American Section of the International Solar Energy Society*, Denver, CO, August 28–31, 1978. Newark, DE: American Section of the International Solar Energy Society, p. 656.

Watt, A. D. 1980. The sun and its radiation. *Introduction to Meteorological Measurements and Data Handling for Solar Applications.* DOE/ER-0084. Washington, DC: U.S. Department of Energy.

Windheim, L. S., K. V. Dury, and L. A. Daly. 1981. The substitution of daylighting for electric lighting in a large office building. *Proc. 6th National Passive Solar Conference*, Portland, OR, September 8–12, 1981. Newark, DE: American Section of the International Solar Energy Society, p. 875.

Wray, W. O. 1979. A simple procedure for assessing thermal comfort in passive solar heated buildings. *Sun II: Proc. American Section of the International Solar Energy Society Annual Meeting*, Atlanta, GA, May 1979. Newark, DE: American Section of the International Solar Energy Society, p. 327.

Wray, W. O. 1980. A quantitative comparison of passive solar simulation codes. *Proc. 5th National Passive Solar Conference*, Amherst, MA, October 19–26, 1980. Newark, DE: American Section of the International Solar Energy Society, pp. 121–125.

Wray, W. O., N. M. Schnurr, and J. E. Moore. 1980. Sensitivity of direct gain space heating performance to fundamental parameter variations. *Proc. 5th National Passive Solar Conference*, Amherst, MA, October 19–26, 1980. Newark, DE: American Section of the International Solar Energy Society, pp. 92–95.

Yago, J. R. 1982. Shadow factors. *Proc. 7th National Passive Solar Conference*, Knoxville, TN, August 30–September 1, 1982. Newark, DE: American Section of the International Solar Energy Society, p. 71.

Yellot, J. I. 1972. Effect of louvered sun screens upon fenestration heat loss. *ASHRAE Transactions* 78: 199.

Yellot, J. I. 1979. Glass—an essential component in passive heating systems. *Proc. 3d National Passive Solar Conference*, San Jose, CA, January 11–13, 1979. Newark, DE: American Section of the International Solar Energy Society, p. 95.

Nomenclature

A	area (m^2)
A_g	glazing area (m^2)
D	depth (m), or day parameter given by Equation 2.6 (rad)
d	distance from the sun (m)
ET	effective temperature, defined by Equation 2.62 (°C)
F	radiation view factor

F_i^k fraction of black body radiation in wavelength band k, for surface i
f fraction of radiation
H height (m), or hour angle defined by Equation 2.9 (degrees)
h heat transfer coefficient (W/m^2 °C)
H_e extraterrestrial solar radiative flux (W/m)
H_s sunrise hour angle (degrees)
I incident solar radiative flux (W/m^2)
I_{DN} direct normal solar radiation flux (W/m)
I_h solar radiative flux incident on horizontal surface (W/m^2)
I_i^k radiation in wavelength band k, incident on surface i (W/m^2)
I_s^0 zeroth moment of the sky diffuse radiative flux (W/m^2)
I_s^2 second moment of the sky diffuse radiative flux (W/m^2)
K extinction coefficient (m^{-1})
K_T clearness index defined by Equation 2.1
k thermal conductivity (W/m °C), or imaginary component of complex index of refraction
k_w coefficent in Equation 2.71 (Ws/°C m^3)
L latitude (degrees) or length (m)
LON longitude (degrees)
LON_s standard longitude (degrees)
M number of surfaces
m air mass
MRT mean radiant temperature, defined by Equation 2.61 (°C)
N day of the year
n index of refraction
$\dot{Q}$ heat transfer rate (W)
S_c solar constant, 1353 Wm^{-2}
T temperature (°C)
$\overline{T}$ average daily temperature (°C)
T_1, T_2 terms constant for the month given by Equations 2.75 and 2.76
t time (s)
t' time in hours (h)
t_s local solar time (s)
V volume (m^3)
V_w wind velocity (m/s)
z zenith angle (degrees)

Greek Symbols

α	solar absorptance
α_t	thermal diffusivity (m^2/s)
β	solar altitude defined by Equation 2.8 (degrees)
γ	solar-surface azimuth defined by Equation 2.10 (degrees)
δ	declination (degrees)
ε	longwave emittance
λ_0	wavelength of radiation (m)
θ	angle of incidence of beam radiation (degrees)
ρ	solar reflectance
Σ	surface slope (degrees)
σ	Stefan-Boltzmann constant ($Wm^{-2} K^{-4}$)
τ	solar transmittance
ϕ	solar azimuth (degrees)
Ψ	surface azimuth (degrees)
ω	solid angle (steradians)
χ	angle between direction of viewing and surface normal (degrees)
χ^1	angle between direction of viewing and incident beam direction (degrees)

Superscripts

k	denotes wavelength band
S	denotes a specular component
$'$	denotes a directional quantity
*	denotes an "effective" quantity
—	denotes an average quantity, or a complex quantity

Subscripts

b	denotes beam
C	denotes convective
d	denotes diffuse, or perpendicular component of polarization
ext	denotes external
g	denotes glazing
h	denotes horizontal

i	denotes surface i
j	denotes surface j
k	denotes Kepplarian (from the analemma)
L	denotes longitude
l	denotes parallel component of polarization
o	denotes outside, or overhang
P	denotes the point defining the extent of shading
R	denotes inside room air
r	denotes ground reflected, radiative
S	denotes solar, or diffuse sky, or (in Equation 2.54) interior
∞	denotes ambient

3 Simulation Analysis

Philip W. B. Niles

3.1 Scope of Chapter

This chapter reviews recent developments in simulation analysis as used to predict the thermal performance of passively heated and cooled buildings. The chapter will consider how thermal systems are represented by thermal networks and will review efficient methods used for solving the network equations. The finite-difference method of solution to the resulting differential equations will be emphasized, but analytical solutions by frequency-domain approaches will also be reviewed. The heat transfer algorithms used throughout passive systems simulations will be reviewed. Neither solar input algorithms, nor ventilation algorithms will be covered in this chapter—see Burns, chapter 2 of this volume and Chandra, volume 8 of this series. Other methods of transient heat transfer analysis are discussed in the chapter by Busch, volume 4.

3.2 Thermal Network Modeling

3.2.1 Nature of Simulation Analysis

A significant part of the rapid development of the field of passive solar buildings has been due to the feasibility and optimization results afforded by simulation-based research studies. The use of simulation allowed researchers to build and operate systems in the computer, where weather and building parameters can be varied much more easily and with more generality than is economical to do experimentally. To perform a thermal simulation of a building system, the building is first translated into formal mathematical terms, usually a system of equations, with each equation describing a discrete thermal process in the system. The equations are then solved in concert to give the building's response over some time period, typically a year. Because of the huge number of calculations involved, early building simulation programs were only economical to use for building research (Buckberg 1971, Willcox 1954). Computer code designers have thus been challenged to develop strategies to make the simulations run more efficiently and, at the same time, to make them friendlier to use, more flexible, and more accurate. As a result of these efforts and concurrent improvements in computer technology, simulations have become economical to use in routine design work.

A number of simulation methodologies have been based on analytical frequency-domain methods. Though limited, they are a valuable adjunct to time-domain methods used to provide network reduction techniques, code validation solutions, and simple analytical solutions.

3.2.2 Thermal Networks

In the thermal-network approach, the thermal components of a building are represented by lumped parameters, which are analogous to lumped electrical circuit elements and frequently use the same symbols. Before the advent of digital computers, thermal networks representing buildings were built of electrical components, and thermal outputs were determined by making electrical measurements. The basic principles of representing thermal systems by networks are discussed in elementary heat transfer texts (e.g., see Holman 1981). Vemuri (1981) gives a more in-depth treatment of the principles underlying network analysis of field problems. Peikari (1974) discusses the fundamentals of electrical networks, and Muncey (1979) and Kimura (1977) treat building thermal networks.

In practice, each element of a building thermal system can be represented by a lumped thermal conductance or a lumped thermal capacitance or a combination of these, interconnected to represent the energy pathways, which are usually idealized to be one-dimensional. Points connecting any two elements are called nodes. Time-varying inputs, such as outdoor temperature (analogous to electrical voltage) and insolation (analogous to current), can be applied to any node. Some circuit elements, usually variable resistances, can represent passive or active thermal controls by taking on parameter values that are a function of time or temperature.

Figure 3.1 depicts a thermal network using standard electrical symbols. It shows a simple circuit representing the direct-gain building indicated. Conductance U_0 represents the conductance between the outside and inside air, accounting for all of the building envelope except for the portion U_4 adjacent to the heat storage. Node T_1 represents the room air, which has a heat capacitance of C_1. U_1 is the conductance between the room air T_1 and the storage wall surface at T_2. The storage is lumped into one "T-circuit," represented by two equal conductances U_2 and U_3, and centered heat capacitance C_2. The system has two solar inputs, Q_1 and Q_2.

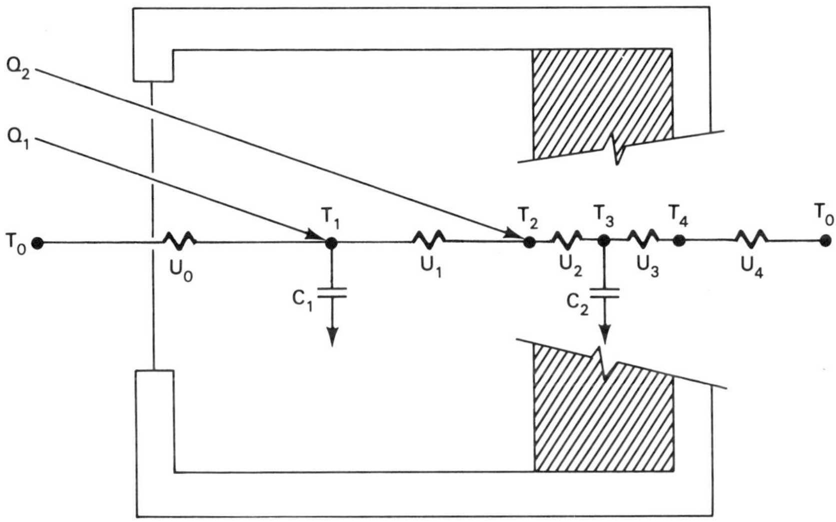

Figure 3.1
Circuit of direct-gain building.

Conductances U_0, U_1, and U_4 are assumed to include long-wave radiation transfers as well as conductive and convective transfers.

Although the elements of figure 3.1 are easily described with electrical symbols, many thermal network elements, such as forced convection, have no direct electrical counterpart. Carroll and Clinton (1980) suggest a set of symbols tailored for thermal networks. Their notation is particularly useful in helping to make network analysis assumptions explicit. The normal capacitance symbol, for example, which is an artifact of the two-terminal nature of electrical capacitors, is simply replaced by a circle at the storage node.

3.2.3 System Equations

Sebald (1981) develops the general form of the thermal system equations by generalizing the equations derived for an example similar to figure 3.1. Following Sebald, an energy balance is expressed at each node for which the temperature is to be determined. Since the temperature at node 4 is not of interest, node 4 is eliminated (dissolved) by combining U_3 and U_4 into a single conductance $U_5 = U_3U_4/(U_3 + U_4)$. The energy balance gives

$$C_1(dT_1/dt) = (T_0 - T_1)U_0 + (T_2 - T_1)U_1 + Q_1, \tag{3.1}$$

$$C_2(dT_3/dt) = (T_0 - T_3)U_5 + (T_2 - T_3)U_2, \tag{3.2}$$

$$0 = (T_1 - T_2)U_1 + (T_3 - T_2)U_2 + Q_2. \tag{3.3}$$

As can be seen here, and as is the case in general, the system of coupled equations for the nodal temperatures consists of first-order differential equations for the mass nodes and of algebraic equation(s) for the massless node(s). These equations can be expressed in matrix form as

$$\begin{bmatrix} dT_1/dt \\ dT_3/dt \end{bmatrix} = \begin{bmatrix} -(U_0 + U_1)/C_1 & 0 \\ 0 & -(U_5 + U_2)/C_2 \end{bmatrix} \begin{bmatrix} T_1 \\ T_3 \end{bmatrix} + \begin{bmatrix} U_1/C_1 \\ U_2/C_2 \end{bmatrix} T_2 + \begin{bmatrix} 1/C_1 & U_0/C_1 & 0 \\ 0 & U_5/C_2 & 0 \end{bmatrix} \begin{bmatrix} Q_1 \\ T_0 \\ Q_2 \end{bmatrix}, \tag{3.4}$$

$$0 = [U_1 \; U_2] \begin{bmatrix} T_1 \\ T_3 \end{bmatrix} - (U_1 + U_2)T_2 + [0 \;\; 0 \;\; 1] \begin{bmatrix} Q_1 \\ T_0 \\ Q_2 \end{bmatrix}. \tag{3.5}$$

Generalizing to any number of nodes, equations (3.4) and (3.5) can be written in more compact matrix notation as

$$dT_m/dt = A_1 T_m + A_2 T_n + B_1 E, \tag{3.6}$$

$$0 = A_3 T_m + A_4 T_n + B_2 E. \tag{3.7}$$

Equation (3.6) represents a set of m differential equations for the m mass nodes, and equation (3.7) a set of n algebraic equations for the n massless nodes. A_1, A_2, A_3, A_4, B_1, and B_2 are matrices containing heat conductances and capacitances, and T_m and T_n are vectors of the temperatures of the massive and nonmassive nodes. E is a vector of the system inputs. This set of equations constitutes an initial-value problem. Initial conditions, consisting of the starting storage mass temperatures, must be specified.

The numerical methods discussed in the following section can be used to solve this set of $m + n$ equations, but the solution is more efficient if the number of equations is reduced. As Sebald (1981) shows, these equations can be reduced to a single matrix differential equation by solving equation (3.7) for T_n:

$$T_n = -A_4^{-1} A_3 T_m - A_4^{-1} B_2 E. \tag{3.8}$$

Using this to eliminate T_n in equation (3.6) gives:

$$dT_m/dt = AT_m + BE, \tag{3.9}$$

where

$$A = A_1 - A_2 A_4^{-1} A_3 \tag{3.10}$$

and

$$B = B_1 - A_2 A_4^{-1} B_2, \tag{3.11}$$

with initial conditions at zero time of $T_m = T_{m0}$.

Thus, equations (3.6) and (3.7) have been reduced to equation (3.9), consisting of m first-order differential equations. If nonmassive node temperatures need to be determined—for example, a mass surface temperature may be required to determine the mean radiant temperature—they can be found from equation (3.8) after equation (3.9) is solved.

In the case of the figure 3.1 example, it should be noted that equation (3.9) would result directly from the energy balances on the mass nodes if the massless node 2 had initially been dissolved as was done with node 4. In this case, Q_2 would have had to be split appropriately between nodes 1 and 3. For more complex networks, methods of reducing the number of equations using equivalent circuits determined by use of frequency-domain analysis are discussed in section 3.3.1.

Thermal systems for buildings can usually be represented with a set of matrix equations of the form of equations (3.8) and (3.9). The form of equation (3.9) is also applicable in modeling single-phase or two-phase rockbeds (see section 3.3.4).

Frequently the circuit elements and resulting matrix coefficients are not constants. Movable insulation used over the windows at night, for example, would make conductance U_0 time-varying—either on a schedule, or dependent on current temperatures. If the placement of the movable insulation depended on the values of inside and/or outside temperature, U_0 would be temperature-dependent, whereby the heat transfer would be non-linear, that is, not proportional to the first power of the temperature difference. Other causes of temperature-dependent and/or time-varying coefficients are fan or natural convection-driven ventilation and backup heating and cooling equipment.

Network equations represented in the form of equation (3.9) happen to take the convenient form of *state-variable equations*, for which there is an extensive circuit and systems analysis literature. Peikari (1974) and Reid (1983) discuss the fundamentals of solving the common forms of state-variable equations, including the cases in which the matrix coefficients are time-varying and/or temperature-dependent.

3.2.4 Solution of Network Equations

The heart of a simulation program concerns the method of solution of the initial-value problem posed by the system equations—such as equations (3.6), (3.7), or (3.9). Although analytical solutions are possible, they are most suitable when the coefficients of the matrices are constant (Peikari 1974). Because thermal networks result in time-varying or temperature-dependent coefficients, numerical methods of solution are used in most of the thermal network programs that have been developed. The method of finite differences is the most commonly used numerical approach. It consists of replacing the derivatives in equation (3.9) by finite differences, rendering the equations into algebraic form that can be solved iteratively or simultaneously.

There is a multitude of finite-difference algorithms that can be used to solve networks, and, as pointed out by Nogatov (1978), it is impossible to choose any one best method for general use. Each has its advantages and disadvantages relative to a particular application. The choice of a particular procedure depends on such factors as the type of building to be modeled, the desired accuracy, the capabilities of the available computer, and the experience of the individual. Since the computer time required for an analysis is inversely proportional to the time step, it is expedient to utilize numerical schemes that allow large time steps to be used without sacrificing accuracy.

The finite-difference methods usually fall into two categories: *explicit* and *implicit*, depending on whether one can predict the temperature at a node independent of the predictions made at other nodes, or whether all node predictions must be made simultaneously.

3.2.4.1 Explicit Methods Perhaps the simplest way to understand and perform explicit method is forward-differencing, also known as Euler integration. This is used in the SEA-PAS program (Clinton 1979) and in SERI-RES (Palmiter and Wheeler 1983). In this case, as elaborated on by Sebald (1981), the time derivatives in equation (3.9) are approximated by the ratio of a finite temperature difference to a finite time interval:

$$dT_m/dt \approx (T_{m,k+1} - T_{m,k})/\Delta t, \tag{3.12}$$

where $T_{m,k+1}$ is the mass-node temperature vector at time $t_{k+1} = t_k + \Delta t$, and $T_{m,k}$ is the vector at time t_k. If all other terms in equation (3.9) are evaluated at time t_k and the terms are regrouped, equation (3.9) becomes

$$T_{m,k+1} = [(I + A\Delta t)T_m + BE\Delta t]_k. \tag{3.13}$$

Since all of the factors on the right side of equation (3.13) are evaluated at time t_k, one can solve for all of the mass temperatures at time t_{k+1} by solving one nodal equation at a time, thus marching forward in time indefinitely as long as the inputs are defined and as long as the numerical procedure is well behaved.

Two main types of error occur in using finite-difference algorithms: truncation errors and instabilities. Instability errors cause the calculated solution (node temperatures) to oscillate with increasing error in each step. Sebald shows that for the forward-difference algorithm to be stable, the time step, Δt, must be smaller than the smallest time-constant of any node in the system, where the time-constant is defined as the heat capacity of the node divided by the total conductance between the node and all the mass nodes to which it is connected. As can be seen by tracing the development of equation (3.9), the diagonal elements of matrix A in equations (3.9) and (3.13) are the reciprocals of the time-constants of each of the mass nodes.

While setting an upper limit on the time step, the stability limit does not guarantee that the solution will be of desired accuracy; this depends on the truncation error. The error in the predicted temperatures or heat flows caused by replacing a derivative by a finite difference can be considered to be the error due to dropping all but the first term of a Taylor-series representation of the derivative and is called the truncation error. Reducing the truncation error may require yet a smaller time step and will also depend on the spacial discretization implicit in the mass lumping process—see section 3.3.1.1. In SERI-RES (Palmiter and Wheeling 1983), the time step is specified by the user in the input. After scanning the mass lumping specified by the user, the program outputs the value of the minimum time step required for stability. The user can then use this minimum, specify a smaller time step, or change the mass lumping specifications.

There are numerous explicit differencing methods besides forward differencing. The modified Euler method (Euler Corrector-Predictor) is slightly more complicated than the Euler method. It entails first using the Euler method to estimate the temperature of each node at time t_{k+1}—call it $T'_{m,k+1}$. Letting $f(T_m, t)$ stand for the right hand side of equation (3.13), the predicted temperature is found from

$$T_{m,k+1} = T_{m,k} + [f(T'_{m,k+1}, t_{k+1}) + f(T_{m,k}, t_k)]\Delta t/2. \tag{3.14}$$

This equation is also a special case of what is known as the second-order Runga-Kutta method. The Runga-Kutta methods are characterized by probing out in front of the present solution to determine the approximate slope of the temperature function. Different-degree Runga-Kutta methods probe various fractions of a time step in front and use various weights on the final average. Chapman et al. (1980) uses the fourth-order Runga Kutta algorithm in the building simulation program FREHEAT to update mass temperatures.

3.2.4.2 Implicit Methods The most common implicit methods are the central and backward difference methods. The central difference method is more accurate than the forward difference for a particular time discretization, and although its solutions can oscillate, they cannot be unstable. Central differencing is used to update mass temperatures in PASOLE (McFarland 1978) and CALPAS (Niles 1980, 1981).

With all the methods, equation (3.12) is used to approximate the time derivative in equation (3.9). However, whereas in the forward difference method the right-hand side of equation (3.9) is evaluated at the beginning of the time interval expressed in equation (3.12), in the central difference method the right-hand side of equation (3.9) is evaluated at the middle of the time interval—at $(t_{k+1} + t_k)/2$. Equation (3.9) thus becomes

$$(T_{m,k+1} - T_{m,k})/\Delta t = [(AT_m + BE)_{k+1} + (AT_m + BE)_k]/2. \tag{3.15}$$

Rearranging, this can be written as

$$RT_{m,k+1} = PT_{m,k} + \Delta t(Q_1 + Q_2)/2, \tag{3.16}$$

where $R = [I - \Delta tA/2]_{k+1}$, $P = [I + \Delta tA/2]_k$, $Q_1 = [BE]_{k+1}$, $Q_2 = [BE]_k$. The subscripts k and $k + 1$ mean that the expressions are evaluated at t_k and t_{k+1}, respectively, and I is the identity matrix.

Because of the presence of the matrix R, each of the m equations in equation (3.16) contains more than one unknown temperature, so the system of equations given by equation (3.16) must be solved simultaneously. For this reason the method is called an implicit method. When applied to the nodes in a lumped representation of a homogeneous storage mass, the central difference solution shown here is known as the Crank-Nicolson method.

The backward difference method is also a commonly used implicit method. The programs UWENSOL (Emery 1978) and TEANET (Kohler 1980) use backward difference algorithms. In the backward difference case, the right-hand side of equation (3.9) is evaluated only at the future time. Replacing the time derivative by equation (3.12), equation (3.9) becomes

$$(T_{m,k+1} - T_{m,k})/\Delta t = (AT_m + BE)_{k+1}. \tag{3.17}$$

Combining like terms yields

$$(I - \Delta t A) T_{m,k+1} = T_{m,k} + \Delta t (BE)_{k+1}. \tag{3.18}$$

Thus, like equation (3.16), this system of equations must be solved simultaneously. Since the right-hand side of equation (3.17) only involves conditions at time t_{k+1}, the backward difference method is sometimes called the full-implicit method.

There is a considerable variety of methods used to solve the simultaneous system of equations represented by either equation (3.16) or (3.18). In general, nonlinearities are handled by iteration, so that standard linear simultaneous equation subroutines are generally utilized. For example, at each time step, TEANET uses the Gauss elimination routine supplied with the TI-59 programmable calculator. CALPAS also uses a Gauss elimination algorithm. UWENSOL (Emery 1978) is made to handle large matrices resulting from multi-zone buildings and uses a linear equation routine made for large sparse matrices. Although FREHEAT (Chapman 1980) uses an explicit difference method to update mass temperatures, they solve the massless node temperatures simultaneously using a library subroutine from IMSL (1984) that does an L-U decomposition by the Crout algorithm (Gerald and Wheatley 1984).

Symbolically, it is convenient to write the solutions to equations (3.16) and (3.18) by using inverse matrices. Multiplying equation (3.16) through by the inverse of matrix R yields

$$T_{m,k+1} = R^{-1}[PT_{m,k} + \Delta t(Q_1 + Q_2)/2]. \tag{3.19}$$

Similarly, Equation (3.18) can be written as

$$T_{m,k+1} = (I - \Delta t A)^{-1}[T_{m,k} + \Delta t(BE)_{k+1}]. \tag{3.20}$$

In the case of equation (3.19), matrix R need only be inverted once during the simulation if the system coefficients in A are constants, but at every time step if the parameters vary with time. Frequently a system has only a few configurations—with or without night insulation, for example. In this case, the inverted R matrices can be stored and reused repeatedly for that configuration. As pointed out by Sebald (1980), computer time is saved in this process if instead of inverting the R matrix, R is decomposed into an upper triangular matrix, and back substitution is used everytime the solution to equation (3.16) is performed. Sometimes, the system has varying coefficients for a limited time—during natural ventilation, for example. In this case, the system of equations must be solved for each time step during the ventilation period.

White et al. (1980) compared the accuracy and computation time of the following four finite difference solution techniques: the backward difference implicit method, the forward (Euler) scheme, the modified Euler scheme, and the fourth-order Runga-Kutta solution method. The authors used these methods to analyse a double-glazed unvented mass wall system. They split the wall into between four and sixteen nodes, with the massive edge node configuration of figure 3.2(b). They varied the time step between three and twenty time steps per hour. While not definitive, their comparison indicates that the explicit methods in general, and the forward difference method in particular, was perhaps the best choice under the circumstances.

It is also possible to use a mixture of difference algorithms in solving one network. CALPAS, for example, uses a mixture of backward, forward, and central differences in order to speed up the network solution. Given the temperature of the air of a zone, the masses coupled to the air are updated using a forward difference, eliminating the need to solve simultaneous equations. Internally, the mass nodes of a given homogeneous mass element are updated with a central difference. The air, in turn, which is assumed to have the heat capacity of all the short-time constant materials in the room, is updated by a backward difference. The motivation for this hybrid method is to speed up solution of building networks that contain masses with a wide disparity in time-constants.

3.3 Heat Storage Modeling

3.3.1 Solid Phase Storage

3.3.1.1 Spacial Discretization of One-Dimensional Distributed Mass In order that a building thermal network be reducible to a system of linear differential or difference equations, the distributed mass elements must be rendered as lumped mass elements. Typically, a mass element is represented as one-dimensional and is sliced into a number of layers of thickness Δx, each represented by a T-circuit consisting of a thermal capacitance in the center and a thermal resistance on each side, as indicated in figures 3.1 and 3.2. The slices must be thin enough not to cause undue inaccuracies in the solution. McAdams (1954) suggests that in general, lumping will yield reasonable estimates when the Biot number is less than 1/10. The Biot number is defined as $h\ \Delta x/k$, where k is the thermal conductivity, and h is the conductance coefficient from the surface. A value of $Biot = 1/10$ indicates that the surface resistance is at least ten times the internal resistance. The FREHEAT program (Chapman 1980) uses this criterion.

A more intrinsic characteristic of the mass than the Biot number is also useful in determining the appropriate lump size. The process of representing a distributed mass as a series of lumps is essentially the process of representing the second-order spatial derivative in the heat conduction (diffusion) equation[1] by a finite-difference expression. This causes a truncation error in the solution to the heat conduction equation. Using Fourier series, Vemuri (1981) shows that if the heat flow is periodic, the truncation error due to using a central difference expression for the second derivative of temperature with respect to distance is proportional to the square of the ratio of lump size, Δx to the wavelength, λ, of the temperature distribution in the material.

Thus, to reduce trunction error, the lump thickness should be much smaller than λ. This assures that the temperature changes will be small over the width of the lump so that temperature can justifiably be assumed to be constant in the lump as is assumed in the finite difference solution. This criterion is similar to that expressed by Chirlian (1973) to determine the appropriate lump sizes in electrical circuits.

For periodic heat flow with a period, P, in a semi-infinite material of thermal diffusivity, v, Carslaw and Jaeger (1959) give the wavelength as $\lambda = (2\pi d)$, where d is the characteristic depth equal to $(2v/\omega)^{1/2}$, where ω is the radian frequency. Goldstein (1978) considers three cycles/day as the highest frequency essential in representing the environmental driving functions. This is confirmed for winter weather by Anderson and Subbarao (1981). By the above equation, for this frequency the wavelength in heavyweight concrete is about 60 cm (24 in.). Ten percent of this would give a lump size of about 6 cm (2.4 in), which is consistent with the common practice of lumping concrete into 5- to 10-cm (2- to 4-in.) thick slices as done by Sebald et al. (1979), Monsen et al. (1979), Palmiter and Wheeling (1983), and McFarland (1978).

For a forward finite difference representation of the heat conductance equation, Ozisik (1968) shows that the truncation error is given by

$$0.5\ \Delta t \partial^2 T/\partial t^2 - (v\Delta x^2/12)\partial^4 T/\partial x^4 + \text{terms of order } (\Delta t)^2 \text{ and } (\Delta x)^2, \tag{3.21}$$

where v is the thermal diffusivity and T is the exact solution. The first term of the truncation error is seen to be proportional to the time step (i.e., of order Δt) and the second truncation error term is proportional to the square of the spatial-step (order $[\Delta x]^2$). Equation (3.21) shows that the error also depends on the higher-order rates of change of the temperature in time and space. Thus, the truncation error depends partly on the input-boundary conditions, indicating why the error associated with a particular finite-difference algorithm and discretization cannot be determined explicity from the system equations. Clearly, small enough temporal and spatial steps could reduce the truncation error to any desired level. Although very small time steps could introduce roundoff errors, generally, the smaller the time step, the more accurate the solution. Indeed, a multitude of finite-difference flaws can be overcome with a small enough time step, though at the expense of computation time.

A shrewd method of reducing the truncation error in the forward difference solution is given by Ozisik (1968). By differentiating the heat conduction equation, it can be seen that $\partial^2 T/\partial t^2 = v^2 \partial^4 T/\partial x^4$, showing that the derivatives in equation (3.21) have the same sign. In fact, both terms in equation (3.21) can be made to cancel if such values of Δt and Δx are chosen that the Fourier number, $Fo = v\Delta t/\Delta x^2$, is chosen to equal 1/6. This strategy was employed by Sebald (1981) in the SEA-PAS program by using a 20-minute time step and 3-in (7.6-cm) thick subdivisions of concrete elements, and the results were found to yield acceptable truncation errors. Note that this time step is much smaller than that required for stability, which is easily shown to require $Fo < 1/2$ for an internal node of a homogenous mass (Croft and Lilley 1977).

It should be noted that it is common to subdivide as indicated either in figure 3.2(a), where the edge node is massless, or in figure 3.2(b), where the edge node has half the mass of interior nodes. It is easy to see that since the edge mass lump in figure 3.2(b) has half the mass of the edge lump of figure 3.2(a), it would have only half of the conductance to adjoining nodes in order to have as large a time constant as the interior nodes. This is not possible unless the surface conductance h were zero. Thus, the more stable solution technique is that of figure 3.2(a). For an interior node spacing of Δx, Croft and Lilley (1977) show that the limiting Fourier numbers in cases (a) and (b) are $(2 + Bi)/(2 + 3\ Bi)$ and $1/[2(1 + Bi)]$, respectively.

The lump size need not be uniform throughout a uniform storage mass. Balcomb (1983a) shows that the admittance of storage mass (adiabatic on the unexposed surface) is maximum when the thickness of the mass is about $1.2\ d$, where d is the characteristic depth. Thus, at a given frequency, heat does not penetrate significantly into a material beyond the depth d, sometimes referred to as the penetration depth. This explains why a number of researchers use thin lumps near the surface of the mass so as to represent the mass correctly for the short wavelengths that only penetrate the surface of the mass, and progressively larger lumps farther from the surface in correspondence with the longer wavelength heat that penetrates there.

3.3.1.2 Network Reduction The nonuniform lump size networks discussed above are an example of various schemes that are used to reduce the number of lumped elements below that required with uniform lumping. The motivation to reduce the elements is, of course, to reduce the number of calculations required to analyze the system.

Network reduction can only succeed if the response of the network needs to be accurate for just a limited range of frequencies of exitation. For example, if only the diurnal frequency response is of concern, a slab of mass insulated on the back can be represented by one lumped resistor connected to one lumped capacitor. The size of the resistor and capacitor can be determined by forcing the admittance of the lumped circuit to match the admittance of the real mass. For masses somewhat thinner than the characteristic length, even the resistance can be neglected. This latter approximation was used by Niles (1979) to determine closed-form analytical

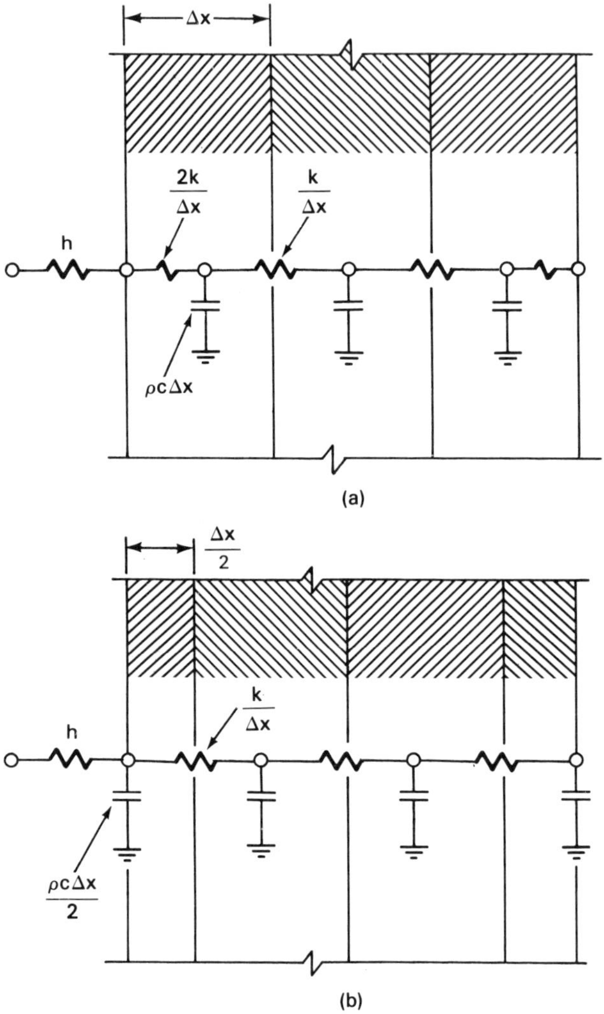

Figure 3.2.
Subdivision of mass with (a) massless edge node, and (b) with massive edge nodes.

solutions of the temperature swing in direct-gain houses. Balcomb (1983a) uses similar approximations, including reducing layered walls to one lump to obtain temperature swings with his diurnal heat capacity procedure. Besides simplifying each mass, he aggregates different masses into one capacitance by vectorially adding their admittances.

The admittance of a mass wall is essentially a frequency-domain response function (see section 3.5) relating the heat transferred into the wall to the surface temperature swing of the wall. Alternate response functions can also be used as the basis of equivalence between lumped and continuous walls. Goldstein (1978) considers the response function that relates the mass surface temperature to the room temperature (or room sol-air temperature to include internally absorbed sun), and he determines this function for both a one-lump RC circuit and the continuous mass. Instead of forcing them to be equal at a particular frequency, he tries to match the poles and zeros of the functions. The simple one-lump circuit assumed has only one pole and one zero (which occur at the time-constants of the circuit when it is coupled to the room air), whereas the real material has an infinite number of poles and zeros to be matched. This is similar to the fact that the admittance was matched at only one frequency, yet the real mass has the correct admittance at all frequencies. His procedure matches the lumped circuit poles and zeros with the lowest frequency pole and zero of the continuous media. This procedure is limited to material thinner than the characteristic length (at the diurnal frequency) because beyond this point, the second pole of the continuous media occurs at a frequency of interest in the environmental forcing functions. Thus, it should be considered, yet cannot be without a more complex lumped circuit. For materials thicker than the characteristic thickness, Goldstein uses the admittance-matching scheme formerly discussed.

It is likely that when the dominant forcing frequency is known, such as the diurnal frequency in passive buildings, the admittance method will be superior since it yields an exact match between response functions at this important frequency. Goldstein's response functions usually underestimate the response function at the diurnal frequency. The pole-and-zero procedure would seem to work best when the dominant frequency is not known. It would assure that until the driving frequencies are high enough to be near the second pole of the continuous medium (at which time the second pole will affect the response), there will be some faithfullness to the lumped representation.

Goldstein also obtains lumped parameters for replacing Trombe walls, but these are probably the least successful, and not surprisingly so, since the Trombe wall response functions involve both sides of a thick mass. As pointed out by Goldstein, a T-circuit can, at most, cause a time lag of six-hours, whereas a continuous Trombe wall can have phase shifts up to twelve hours.

Although a one-lump admittance-matching method works well for design day types of analysis, hour-by-hour simulations over extended periods will not be accurate for thick mass unless the admittance of the storage is accurate for cycles whose periods are on the order of a few days. That is, if the real mass can be effective in storing heat over periods of a few days, the reduced mass should have the same ability. Balcomb (1983a, 1983b) used a two-lump network reduction scheme to represent masses, including layered walls. Basically, he forces the admittance of the lumped circuit to match the admittance of the real wall at two different frequencies—he picked the diurnal frequency and 1/3 the diurnal frequency (corresponding to a three-day period). A match at two frequencies requires twice as many arbitrary lumped parameters; thus the two T-circuits. To determine the values of the two-frequency lumped parameters, Balcomb solves a set of simultaneous nonlinear algebraic equations; he includes a program listing to accomplish this.

Figure 3.3 shows an example from Balcomb (1983a) of a detailed thermal network for a room with a 12-in (30-cm) concrete back-insulated wall. He made the first lump thinner, and the last thicker, somewhat arbitrarily. Figure 3.3(b) shows the reduced network obtained by the two-frequency matching procedure. The conductance and heat capacity values shown have the units of Btu/h °F, and Btu/°F, respectively. The relative solar gains to the air and surface nodes are circled. Using actual weather during a severe winter transient to drive the two circuits, the room temperature responses were determined using a finite-difference procedure. Figure 3.4 shows that the room temperatures predicted by both networks differ by less than the width of the line in the plot.

3.3.1.3 Heat Transfer through the Ground As illustrated in the above sections, heat conduction through most building components in the building envelope is treated as one-dimensional in the majority of the current simulation programs. Most programs ignore altogether the complex multidimensional heat flow process for building corners. The determina-

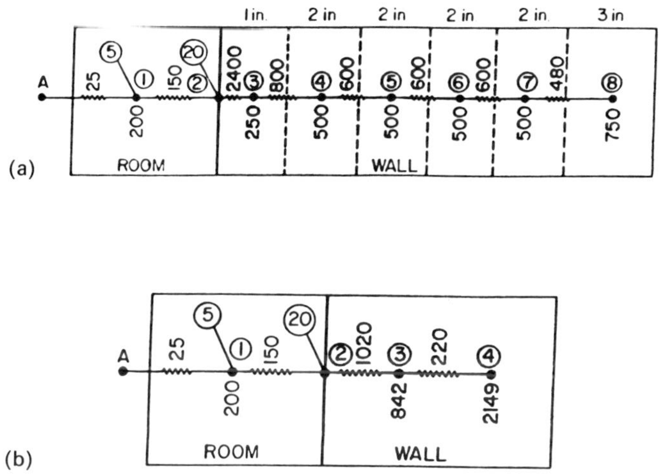

Figure 3.3
(a) Detailed, and (b) reduced thermal network of a 12-in. (30-cm) concrete wall with a surface area of 100 ft^2 (30 m^2). The network is put in the context of a 100-ft^2 (30.5-m^2) direct-gain room with 25 ft^2 (7.6 m^2) of south glazing. Source: Balcomb 1983a.

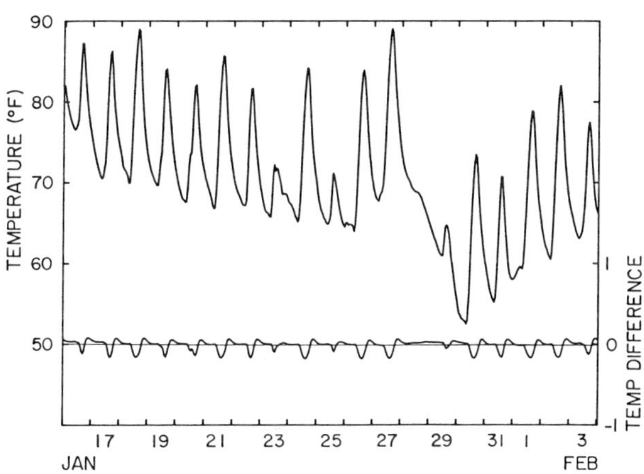

Figure 3.4
Comparison of room temperature simulations made with the networks of figure 3.3. The lower curve shows the temperature difference between the two simulations. Source: Balcomb 1983a.

tion of the heat transfer through multidimensional basement, bermed walls, and slab floors is the bane of simulation programs.

Although multidimensional analysis for any type of building geometry can be handled by finite-difference or finite-element techniques, generalized routines are difficult to implement, and computer running times and memory limitations are frequently prohibitive (Kusuda 1980).

It is recognized that extreme algorithm accuracy is pointless unless the thermophysical properties of the soil are known better than is currently the practice. Soil density, specific heat, and conductivity all have a significant influence on the heat transfer rates. The effects on heat transfer of moisture transport in the ground are just beginning to be investigated for applications to building simulation (Meixel and Bligh 1983).

Obtaining the rate of heat transfer through the ground with comparable accuracy to that through plane walls requires at least a two-dimensional, and, preferrably, a three-dimensional, network. Numerous authors have discussed two- and three-dimensional finite-difference and finite-element solution methodologies for various slab and basement configurations (Andrews 1979, Ambrose 1981, Metz 1983, Meixel and Bligh 1983, Yard 1984, Shipp and Broderick 1984, Perry et al. 1985). Such routines are time consuming because the heat conduction domain influenced by the ground heat flow is extremely large, requiring a large number of grid points, and because of the large time-constants of the ground, the simulation may need a long "warm-up" period before the effect of the assumed initial mass temperatures die out—this may be on the order of a few months.

Although justifed for research programs, these solutions usually require too much computer time to be useful as regular routines in design simulation programs intended for building design or compliance analysis. For design purposes, it is unlikely that local variations in slab or basement wall temperature are needed, as long as the model gives an estimate of the correct total heat transfer rate from the surface. As pointed out by Carroll and Clinton (1980), a useful algorithm for design programs should represent the ground heat flow in three important respects: its responses to inside and outside transients, and its steady-state heat losses. Its other properties may differ somewhat from those of real floors.

Clearly, one of the main needs facing the designers of simulation programs is to develop accurate but fast procedures to be incorporated conveniently into finite-difference-based simulation programs intended for design purposes. Various schemes have been utilized to satisfy this need; none are entirely satisfactory so far.

Perhaps the most common scheme used in finite-difference-based programs is to use an equivalent one-dimensional RC circuit approximation to the actual three-dimensional heat transfer network. One of the first formal methods of determining a one-dimensional equivalent circuit was proposed by Muncey and Spencer (1977). They obtain a Fourier series solution to the heat conduction equation applied to the problem of a zero thickness rectangular slab resting on a semi-infinite ground of known conductivity. This solution determines the rate of heat transfer for steady periodic exitation at any frequency, including zero (yielding the steady-state conductance). Their solution is confirmed by the Fourier transform solution of an uninsulated rectangular slab given by Delsante et al. (1983). They show how their solution can be used to determine four complex response functions analagous to the elements from the matrix method (see section 3.5) characterized by the admittances shown in figure 3.14. They then propose two different distributed parameter RC circuits that are intended to approximate these transfer functions, presumably with the expectation that these circuits could be implemented in a finite-difference scheme for rapid solution. Such a circuit is useful because it essentially expands their analytical solution, limited to steady-periodic weather conditions, to one applicable to arbitrary weather.

The simpler of their proposed circuits is based on the proposition that the ground can be approximated by a one-dimensional slab with the same area as the actual slab but thick enough to give the correct overall conductance of the ground. This conductance can be obtained from their Fourier series solution. They show that at the diurnal frequency, the driving point admittance of this equivalent circuit is in fair agreement with the same admittance of the corresponding transfer function. This is important so that variations in room temperature induce the correct amount of diurnal storage. This match is not surprising since the penetration depth is small at the diurnal frequency, so the slab acts as if it were one-dimensional. However, the transfer admittance, expressing the heat flow to the indoors due to the annual cycle in outdoor temperature, does not compare well with that obtained from the analytical solution. Their more complex one-dimensional equivalent circuit agrees somewhat better with the analytically derived transfer functions.

Other reasonable but essentially ad hoc one-dimensional RC ground loss approximations are given by Carroll and Clinton (1980), Chapman et al. (1980), and Balcomb (1983a). Carroll and Clinton use a lumped RC circuit with small lumps near the surface to handle the "fast transient" occuring there. The deepest node has a time constant of two months. The deeper conductances are adjusted upwards to approximate the three-dimensional effects that increase heat transfer and storage with depth.

A number of methods have been developed to obtain monthly average heat transfer values for application to simplified load analysis procedures. Yard (1984) and Akridge (1983) give methods that produce monthly average heat transfer results. MacDonald et al. (1985) review these and other methods developed for use with simplified analysis techniques, and give a brief discussion of each method and its limitations. They show that there is a wide disparity in answers when the various methods are applied to an example basement. One of the most sophisticated and general of these methods is given by Mitalis (1982, 1983). Mitalis's method also has the potential of being adaptable to use in finite-difference programs.

Mitalis used two- and three-dimensional finite-difference programs to determine the heat transfer rates through basement walls and floors (Mitalis 1982, 1983), and through slab floors (Mitalis 1987) subject to steady periodic conditions. In both situations, geometries with various insulation configurations were analysed. These solutions were obtained for a constant indoor temperature, and inside surface diurnal heat storage effects were not considered. Largely based on these results, Mitalis obtained simple empirical formulations that gave the steady periodic heat transfer rate at the inside surface in the form of a steady-state and sinusoidal component, where the sinusoidal component depends on an amplitude attenuation factor and phase-lag terms. Some of these equations were modified by Mitalis (1982) so that the results would agree more closely with experimental data. Yard et al. (1984) show how dimensionless relationships offer an alternate method to that of Mitalis to correlate results obtained from large multidimensional programs. Although the specific results of Yard et al. are limited to a fixed basement configuration, their methodology may be useful for numerous configurations with some promise of being faster than Mitalis's.

Mitalis's empirical equations have the form of a transfer admittance expressing the effect of the outdoor temperature cycle on the heat transfer at the inside surface. Since his solution assumes a constant inside temperature, the driving point admittance (self-admittance) is not addressed. Because of its generality and form, Mitalis's solution may be adaptable to the needs of finite-difference programs. This could be done in a manner analagous to what was done by Muncey and Spencer. That is, it is likely that a one-dimensional equivalent circuit can be matched to the annual frequency transfer admittance given by Mitalis so that the circuit will correctly portray the influence of transient weather conditions on inside heat transfer. The inside part of this circuit could then be modified to correctly account for the driving point admittance, or else this admittance could be handled by a separate parallel circuit not connected to the oudoors, essentially as the circuit shown in figure 3.14.

Kusuda (1984) compares the Mitalis solution with the exact solution given by Delsante et al.(1983) for uninsulated slabs, showing that Mitalis's solution may not be too accurate in some cases. For instance, the timing of the winter peak heat loads differed by months between the two methods. Since this timing is likely to be very sensitive to the slab edge geometry, it is possible that the difference in time lag could be explained by diffences in assumed geometry in this regard. Kusuda also suggests that Mitalis's solution doesn't scale properly with slab area. MacDonald et al. (1985) shows that Mitalis's solution scales poorly with basement depth.

There are several areas where improved ground heat transfer algorithms are needed. Expanded soil conductivity options and the addition of moisture transport effects are needed. The heat transfer is strongly affected by the short portions of the path between the inside and outside conditions, but there is a deficiency of results sensitive to this effect—for example, how the heat transfer depends on the thickness of walls resting on slab-on-grade floors. More options on possible insulation thicknesses and floor plans with complex shapes are generally needed. The heat transfer through berms and earth-covered walls also needs to be addressed. With some improvements in the state of the art, it should be possible to incorporate a two- or three-dimensional finite-difference or finite-element routine in a simulation program preprocessor, from which one-dimensional circuits can be derived for rapid hourly processing. This is currently being done in DOE 2.1B to generate ground heat transfer response factors (Sullivan et al. 1985). Possibly, the rapid multidimensional solution procedure suggested by Shen and Ramsey (1983) has use in such a preprocessor.

3.3.2 Liquid Storage

Liquid storage has largely been limited to water wall and roof-pond type applications. In the water wall case, drums, vertical cylinders, or rectangular parallelepiped tanks are situated in front of the solar aperture. These configurations have been used in both opaque and partially transparent form.

3.3.2.1 Water Walls Rectangular opaque steel water tanks are the most easily simulated configuration. Because natural convection mixes the water, it is usually assumed to be isothermal, so, in effect, it can be treated as one lumped capacitance. There is some question about just how isothermal the water is in tanks. Vertical stratification has been reported—10°F (6°C), for example, in the Star Tannery house (Sandia 1979)—but probably has little influence on performance.

Modeling semitransparent water walls is more challenging. Hull et al. (1980) give validated modeling information on the transparent version of the rectangular tank where the water is contained between two glass walls with a partially transparent glass barrier vertically dividing the water. They simulate the water on each side of the divider by a two-lump network because they found the horizontal resistance to be higher than expected, corresponding to a Nusselt number of 2.5. Fuchs and McClelland (1979) discuss determination of the solar absorption as a function of the position between the glass walls so that the solar input can be assigned to the correct nodes.

Van der Mersch et al. (1980) have extensively studied the more complicated situation of partially transparent vertical water tubes that are separated so as to allow direct gain penetration to the room behind them and to allow room air circulation around the tubes. They assume isothermal water—their experimental data indicate that stratification is less than 2°C (4°F)—and a one-dimensional thermal network. The principal difficulty in modeling the tubes is to determine the destination of the insolation admitted through the aperture. They found that the room solar gain was noticeably affected by intercylinder shading and cylinder surface reflections (they assumed specularly reflecting Tedlar covered cylinders). They discuss the solar algorithms they developed to determine the distribution of beam and diffuse insolation. Another major part of their effort was to compare simulation results with test-cell data in order to determine the best values to use for surface conductance. Interestingly, the cylinder surface to room conductance determined by Burns et al. (1979) was 8 W/m2^{2} °C (1.4 Btu/h ft^{2} °F), close to ASHRAE's vertical wall combined coefficient values of 8.5 W/m^{2} °C (1.5 Btu/h ft^{2} °F)—see section 3.4.1.1.

3.3.2.2 Roof Ponds The water ponds of roof-pond heating and cooling systems, like water walls, are virtually always modeled with a one-dimensional network (Niles 1975, Miller and Mancini 1977, Jones 1982). As with water walls, the main problem is knowing appropriate thermal parameter values to use in the network. A few of the problem areas are discussed below. Evaporation and sky radiation algorithms are also related to roof-pond modeling and are discussed in section 3.4.3 and 3.4.2, respectively. An extensive review of the roof-pond literature and algorithms is given in Marlatt (1984).

Determining the appropriate solar absorption coefficient of the water ponds is not as easy as it at first appears. The low winter sun angles on horizontal ponds makes reflections losses large and sensitive to the angle. Absorptivity is thus difficult to predict, especially when the pond surface may consist of multiple layers of plastic that can be folded, wrinkled, yellowed, dirty, or can harbor air bubbles. Los Alamos experiments (Jones 1982), for example, circumstantially indicated anomalously low water bag absorptivities possibly due to these factors.

Two simple empirical equations are reported in Marlatt (1984), giving solar absorptivity of clear water layers as a function of water depth. A roof-pond simulation developed by Tavana (1980) models absorption in the water by considering separate extinction coefficients in five different wavelength bands. There appears to be a lack of experimental data to validate these models, particularly as regards showing the effect of water turbidity.

Many roof-pond simulations assume that all of the sun is absorbed at the bottom liner of the water container. Due to overturning, the water is assumed to be well mixed and at a uniform temperature (e.g., see Miller 1977). These assumptions are probably reasonable for clear water and highly transparent top surfaces. However, with less ideal conditions, there is evidence that some ponds have temperature stratification (Niles 1975). While not necessarily affecting the overall absorptivity, stratification can seriously affect the heat transfer to the building and the environment. When the water stratifies (and it seems clear that it will, at least when the insulation panels are closed and the water is hotter than the room), the heat transfer through the insulation panels and to the room will be inaccurate if its calculation is based on the average water temperature.

During the heating mode of operation, the air in the room probably stratifies near the ceiling. Little good experimental data is available on the magnitude of convection under these circumstances. In any case, since convection will be small or negligible, radiation heat transfer becomes the dominant mode of delivering heat to the house during heating. For this reason, it is important that the radiative heat tranfer algorithms accurately account for real room conditions, as do the simulations of Miller (1977) and Jones (1982), which use surface-to-surface radiation networks.

As discussed in section 3.4.1.1, Faultersack and Loxsom (1982) give data on ceiling heat transfer coefficients for the cooling mode when the water is colder than the room. No stratification is expected in this mode. Their results are based on experimental data from their corrugated-ceiling test building, and may not be applicable to flat ceilings.

As discussed in sections 3.4.1.2 and 3.4.3, the forced convection coefficient at the outside surface of exposed ponds is not well known either, yet has a fair influence on performance, especially if the ponds are unglazed. A principal problem with determining an appropriate correlation relating heat transfer to wind velocity is knowing the relationship between weather-bureau wind velocities and the appropriate wind velocity to use on the roof. The situation is complicated by microclimate influences, such as wind shadows and turbulence levels, and by roof conditions, such as parapets and reflectors.

Although the roof-pond simulation programs have largely been successful for sensitivity studies, it is doubtful that they are very accurate in general application. In most cases, the unknown parameters required in the models have not come from isolated controlled experiments but have been acquired by the intuition-based tuning of the model until it agreed with a particular test house's performance. Much more fundamental data is needed for accurate predictive modeling of roof ponds.

3.3.3 Phase-Change Storage

Phase-change storage materials (PCMs) are becoming more prevalent, and simulation programs have been largely successful in predicting their performance (Grimmer et al. 1977). Liquid-to-solid phase-change materials have been the predominant choice to date, and they offer some challenging simulation problems (Solomon 1979).

In the program PASOLE (McFarland 1978), phase-change materials are simulated by representing the heat of fusion as an increase in heat capacity over a given temperature range of about 20–40°F (10–20°C) wide. Thus, heat capacity becomes temperature dependent.

SERI-RES (Palmiter and Wheeling 1983) incorporates a phase-change algorithm that simulated the phase change more rigorously. The program allows the user to specify a one-dimensional multilayer lumped network just as with sensible heat storage but with any number of phase-change layers. These layers have the same nodal equations as a pure capacitance layer, but differ in their ability to store heat without change in temperature. That is, the phase-change material layer behaves as a pure capacitance layer until its temperature reaches the user-specified melting point. Then the temperature of the layer is held constant, while latent energy is stored up to the total latent storage capability of the layer. When the total latent storage capability is reached, the layer again behaves as a pure capacitance layer. Ideally, such a simulation would allow for differences in density, specific heat, and conductivity of the two phases. For transparent PCMs, internal solar heating should be simulated.

The main drawback of such models is the lack of sensitivity to three-dimensional effects that influence some phase-change configurations. For example, when the phase-change material is used as a directly sunlit thermal wall and is separated from the glazing by a vertical air space, vertical natural convection currents in the air space allow a vertical stratification that subjects the uppermost phase-change material to higher temperatures than the material below. This stratification can cause the PCM to melt at the top and result in overheating. Natural convection currents in the liquid phase of the material itself can also occur, including separation of phases.

Bourdeau (1980) describes modeling phase-change material in a manner that apparently allows for some three-dimensional effects to be considered. In order to describe more accuately the effects of natural convection in the air next to the PCM containment, which result in faster melting of the upper containers, the wall is considered as a pile of one-dimensional elements exposed to different energy inputs. Each element consists of a one-dimensional network consisting of several nodes connected by thermal capacitors, like the one in SERI-RES. Two variables are associated with each node: its temperature and its proportion of melted material. A heat balance is kept, and the amount melted is updated at each time step. Details are not given on how the convection is modeled.

It is apparent that more phase-change simulation algorithm development would be useful for liquid/solid PCM simulation. For PCMs with a solid-to-solid phase change (and consequently no separation of phases), the modeling should offer problems similar to those encountered when modeling sensible storage.

3.3.4 Rockbed Storage

Rockbed storage is commonly used in passive/active hybrid buildings. Usually placed under concrete floors, rockbeds are charged either from a sunspace or collectors, or by outside air when used for cooling. They exchange heat with the living space either passively through the floor or actively by forced convection. Given a well defined rockbed configuration, the air temperature, direction, and flow rate entering the bed, the principle thermal problem is to determine the outlet temperature of the bed at any time. Because the solution to this problem can involve a significant fraction of the total building simulation run time, efficient algorithms have been fairly vigorously pursued.

Rockbeds are just a special case of packed beds, which have been widely used in the chemical industry. An early mathematical solution to the pebble-bed problem was given by Shumann (1929). He solved essentially the following coupled set of partial differential equations, one expressing an energy balance on the rock and the other an energy balance on the air. A one-dimensional rockbed is assumed, with axial dimension x:

$$\partial T_a/\partial x = (NTU/L)(T_r - T_a), \tag{3.22}$$

$$\partial T_r/\partial t = (NTU/\tau)(T_a - T_r), \tag{3.23}$$

where t is the time; T_a is the air temperature; T_r is the rock temperature; L is the rockbed length; A is the rockbed cross-sectional area; NTU is the number of transfer units, L/λ; λ is the characteristic length for the rockbed, $(\dot{m}c_p)_a/(Ah_v)$; h_v is the volumetric heat transfer coefficient; τ is the characteristic time for the rockbed, $(mc)_r/(\dot{m}c_p)_a$; $(mc)_r$ is the total rock heat capacity; and $(\dot{m}c_p)_a$ is the product of the mass flow rate and specific-heat of the air.

For long-term performance analysis under varying weather conditions, numerical solutions of equations (3.22) and (3.23) can be easily obtained. Duffie and Beckman (1974) give equations (3.22) and (3.23) with the term $U(T_o - T_r)/(mc)_r$ added to equation (3.23) to account for conduction through the sides of the bed to the environment at T_o. They give a forward finite-difference scheme for solving these equations. The stability condition of the finite-difference form of equation (3.22) requires that the bed be broken into slices of thickness Δx, less than λ, typically comparable to a few rock diameters. Similarly, equation (3.23) shows that the time step, Δt, must be less than τ/NTU, that is, smaller than the time it takes the air to travel a distance λ. This is on the order of a few minutes. As a result of these limitations, the Duffie and Beckman model yields a rather time-consuming computer code.

Mumma and Marvin (1976) solved the same differential equations with an algorithm that affords much larger discretizations. Essentially, they solve equation (3.22) analytically over each bed segment by assuming a constant rock temperature in the segment. Putting the resulting exponential expression for air temperature change into equation (3.23) uncouples the equations and allows a forward difference solution of equation (3.23). They show that they can obtain the same accuracy solution as Duffie and Beckman while using much larger space and time discretizations. Their technique of uncoupling the equations by assuming how the rock temperature varies in a segment is discussed in general by von Fuchs (1981), who presents an improved version of this method and gives an excellent review of recent rockbed models.

If NTU becomes infinite, the rock and air temperature can be assumed to be identical, and equations (3.22) and (3.23) reduce to a single equation, the so-called single-phase formulation of the rockbed model—see Hughes et al. (1976). The single-phase equation takes the form of the advection equation (Vemuri 1981) that characterizes a wave traveling without shape change at a wave velocity of L/τ and indicates that no wave front dispersion occurs. In reality, the wave front should smear due to finite rockbed axial conductivity and finite temperature difference between the rock and air. Hughes et al. solved the equation using finite differences. They use a forward difference to represent the time derivative. To represent the spacial derivative, they use an "upwind" difference: the temperature difference between the node being updated and the node upstream, divided by the node spacing. This is commonly used in numerical fluid-flow analysis to make the finite difference form of the advection equation stable but artificially produces front smearing. Although an artifact of the numerical procedure, and influenced by the choice of discretization sizes, this smearing is an advantage in the present application. They found that for the simulation of typical active system rockbeds, choosing the spacial discretization so that the bed was broken into five segments resulted in a reasonable compromise between accuracy and calculation effort. Their solution can be obtained much faster than the two-phase model solution.

Both the University of Wisconsin's TRNSYS (1981) and SERI-RES (Palmiter and Wheeling 1983) implement the single-phase solution in their rockbed subroutines. In both cases, five bed segments are nominally used. Kohler et al. (1979) implement the single-phase equation using a backward time difference with the upwind spacial difference.

Carroll and Clinton (1980) found a way to choose the spacial and temporal discretizations so that the smearing that results from the discretization accurately accounts for the real temperature wave dispersion. They argue that the smearing of a pulse is initially asymmetrical but quickly approaches a Gaussian distribution. Thus, the smearing is adequately characterized by its dispersion rate: the rate of growth of variance in axial position of a pulse. They propose a simple equation relating the appropriate time and spacial discretizations. To account for temperature gradients within the rock itself, a corrected characteristic length proposed by Jeffreson (1972) is used. This method seems to give quite accurate results while retaining the speed of the single-phase methodology.

In most of the models discussed above, a term is also incorporated to account for finite thermal conduction in the bed, particularly when the air is not flowing. Dietz (1979) reviews the literature, and considering only conduction effects, he develops an equation to predict the effective conductivity of rockbeds. Carroll and Clinton (1980) propose an extention of Dietz's result to include the radiation heat transfer effects. It is shown that the radiant heat transfer is comparable in magnitude to the conduction heat transfer determined by Dietz. Internal natural convection can increase the effective conductivity in otherwise static rockbeds if there is a sufficient vertical temperature gradient in the rock. This may have particular relevance to passive discharge underfloor rockbin performance. Katto and Masuoka (1967) determined the critical Rayleigh number for the onset of convection. The magnitude of the expected increase in heat transfer as a result of free convection has not yet been reported in the literature.

A number of analytical simulation models have been developed for rockbed analysis. Riaz (1978), Sowell and Curry (1980), and White and Korpela (1979) give solution methods based on convolution methodologies and may be useful, particularly when very precise results are desired. Recent comparisons between rockbed simulations and test data tend to confirm that the simulations are fairly accurate. The papers of Persons et al. (1980), von Fuchs (1981), and Coutier and Farber (1982) are particularly interesting.

3.4 Heat Transfer Algorithms

In the modeling discussion in this chapter, it has been assumed that one has access to accurate values for the various heat transfer coefficients involved. For example, the coefficients U_0, U_1, and U_4 in figure 3.1 all incorporate knowledge about the magnitude of convection and radiation coefficients used to represent the heat transfer between the interior and exterior surfaces of the building and their environment.

In this section we will explore the resources available to determine the values of the convection and radiation coefficients. Related radiation network simplifications will also be discussed. In addition, evaporation coefficients are discussed for application to the modeling of buildings with roof ponds.

3.4.1 Convection

This section covers recent research related to determining convection heat transfer rates to building surfaces both on the inside and outside of buildings. Although air movement between rooms is discussed here, infiltration and natural ventilation through the exterior building envelope is covered in another chapter.

3.4.1.1 Convection between the Air and Surfaces, Inside of Buildings Natural convection and thermal radiation are responsible, in roughly equal measure, for the transfer of heat between surfaces inside of buildings. Thus, the determination of appropriate and simple ways to model convection is important.

Essentially all current state-of-the-art building analysis programs assume that convection occurs between room surfaces that are at uniform temperature over their surface to room air at a single temperature. With few exceptions (Andersson 1980), the convection coefficients between the surfaces and the air are assumed to be constant, independent of temperature difference,[2] and, at most, dependent on surface orientation and direction of heat flow. Probably the most widely used values for convection coefficients are the constant values of the combined convection/radiation

Table 3.1
Selected ASHRAE convection heat transfer coefficients (h)

For vertical surfaces	3.08 W/m^2 °C (0.54 Btu/h ft^2 °F)
For horizontal surfaces	
With heat flow upward	4.04 W/m^2 °C (0.71 Btu/h ft^2 °F)
With heat flow downward	0.95 W/m^2 °C (0.17 Btu/h ft^2 °F)

coefficients recommended by ASHRAE (1981, table 23.12-1). Table 3.1 gives the convection portion of the ASHRAE combined coefficients, determined by subtracting the thermal radiation portion of ASHRAE's values, which for surfaces with an emissivity of 0.9 is 0.92 W/m^2 °C (0.16 Btu/h ft^2 °F). The development of the combined coefficient concept is discussed in section 3.4.2.1.

It has long been recognized that the use of constant convection coefficients oversimplifies reality, but it has been less clear just how much error their use has been causing in the results, or how to make major improvements on these assumptions. In order to introduce temperature dependence, ASHRAE's correlations for free convection from flat plates are sometimes used (e.g., Walton 1983), although these equations are meant for plates with free edges, not connected to perpendicular walls, and so would not be expected to apply well. In effect, the use of temperature-independent convection coefficients assumes that air currents or temperature stratification, if they exist, are of secondary significance or can be accounted for by appropriate choice of room temperature and/or convection coefficient, and that heat transfer based on average surface temperature is equivalent to that based on the actual surface temperature distribution.

A number of studies have looked at the sensitivity of performance to the magnitude of convection coefficient. For example, Mitalis (1965) found the cooling load in a typical office (with solar gains) to be relatively insensitive to h if the cooling thermostat was governed by the air temperature. However, he found that the room MRT, and thus comfort conditions, were very sensitive to the assumed convection coefficients. Andersson et al. (1980) concluded that to correctly predict air temperature as a function of time would require coefficients that were sensitive to changing surface temperature conditions.

It is clear that compared to the conductance of a normally insulated building envelope element, the convection coefficient usually has little influence on the overall thermal resistance of the wall. However, at windows and other highly conductive envelope elements, the convection coefficient can significantly influence the magnitude of the conductive gains and losses. It also has a strong effect on the transfer of heat to and from mass storage elements.

Recent research results imply that in these important locations, where the rates of heat transfer tend to be high, both the constant and temperature dependent ASHRAE coefficients are generally too large, even when the there are significant natural convection air currents. For example, Gadgil et al. (1982) used a finite-difference solution to the Navier-Stokes equations, in what we might call a numerical experiment, to determine the convection coefficients on the window and solar heated slab floor in a strongly solar heated direct-gain room. With the room air at 25°C (77°F), he determined that the convection coefficient on the 35°C (95°F) floor to be 1.7 W/m^2 °C (0.30 Btu/h ft^2 °F), compared to ASHRAE's 4.04 W/m^2 °C (0.71 Btu/h ft^2 °F) for this situation. On the south window, which had a surface temperature of about 15°C (59°F), the coefficient was found to be 2.7 W/m^2 °C (0.48 Btu/h ft^2 °F), compared to ASHRAE's 3.03 W/m^2 °C (0.53 Btu/h ft^2 °F). A similar analysis of the room with a warm north wall opposite the window showed a comparable disparity between the numerical solution and ASHRAE values. Carroll (1980) also argues for using smaller coefficients than given by ASHRAE, this despite the fact that Carroll adds an effective 0.45 W/m^2 °C (0.08 Btu/h ft^2 °F) to the convection coefficient due to the finite emissivity of room air (see section 3.4.2.1).

Although a numerical convection model, like the one used by Gadgil, could conceivably be incorporated into a building analysis program, the computing time and storage required will be prohibitive for design type simulation programs for some time. Thus, the principal research effort by Gadgil et al. has been to use the numerical model, along with experimental validation, to develop simple correlations that could be used directly in building analysis programs. As an example, using the same numerical convection program, Altmayer et al. (1982) (see also Kammerud et al. 1982, Bauman et al. 1980, 1983) have developed correlations that predict the heat transfer rates in a two-dimensional room similar to that for which the above data were given. They considered the heat transfer in the room shown in figure 3.5.

There is a hot surface, h, of area A_h and temperature T_h, a cold surface, c, of area A_c and temperature T_c, with the rest of the subsurfaces isothermal at a intermediate temperature T_n, and total surface area A_n. The total wall

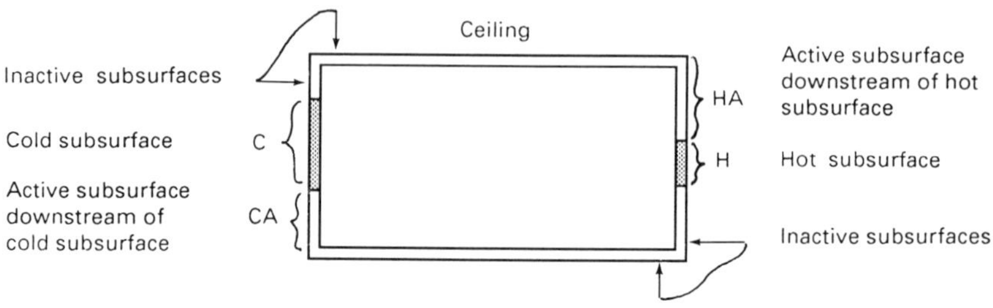

Figure 3.5
Elevation view of room with natural convection.

area is A. In general, the local convection heat transfer rate at any point on a surface will be dependent on the temperature of the surface and the adjacent air. The temperature of the adjacent air is in turn dependent on the temperature of all the surfaces the air has exchanged heat with, particularly those immediately upstream. Because of this, Altmayer et al. found it useful to assume a constant convection coefficient and develop a correlation for an effective air temperature adjacent to the surface of interest, rather than assume a constant room air temperature and a varying convection coefficient.

They found that approximately 90% of the heat transfer occurred at surfaces h and c, and at surfaces ha and ca immediately downstream of h and c. They thus directed their attention to getting correlations for determining the net transfer to and from these four "active" subsurfaces. They found that the rate of heat transfer from surface x (either c, h, ca, or ha) could be obtained from

$$Q_x = K(T_x' - T_x), \tag{3.24}$$

where K is 2.42 W/m^2 °C (0.426 Btu/h ft$^{2\circ}$ F), T_x is the surface temperature of surface x, and T_x' is the effective air temperature adjacent to surface x given by

$$T_x' = K_1 T_h A_h/A + K_2 T_c A_c/A + K_3 T_n A_n/A + K_4 T_h + K_5 T_c. \tag{3.25}$$

Table 3.2 gives the dimensionless correlation constants depending on which of the four active subsurfaces is being considered. Sensitivity studies done by Altmayer et al. showed that this correlation held for a wide range of hot and cold subsurface areas, and inactive surface temperatures.

Table 3.2
Correlation constants for equation (25)

Subsurface	K_1	K_2	K_3	K_4	K_5
c	1.49	1.38	0.89	0	0
h	1.49	1.38	0.89	0	0
ca	0.76	0.7	0.46	0	0.49
ha	0.76	0.7	0.46	0.49	0

For rooms with approximately two-dimensional heat flow, this correlation could be quite useful since it gives the heat transfer rate as a function of the surface temperatures, which are usually known during a simulation analysis. However, the above correlation is the only one available and is, as far as is known, limited to the two-dimensional geometry and temperature distribution, for which it was developed, and to laminar flow. It does not account for furnishings, and the effects of real-wall roughnesses and protuberances are largely unknown. Although the authors point out experimental evidence to support their use of a laminar flow model, they recognize that the onset of turbulence is being approached in their studies.

Also, their correlation does not account for the effect of room-to-room convection, which may significantly alter the airflow and air temperature patterns in a room. There is evidence (Ruberg 1979, Nansteel 1982) that when three-dimensional effects are included, strong turbulence can be introduced. Although the numerical experiment results generally show that the convection currents are constrained to a narrow boundary layers near the surfaces, Nansteel's flow visualizations clearly illustrate the complexity of the convection currents when partitions exist in the enclosure.

There has been a wide disparity between the convection coefficient values used by various researchers for the convection between the room air and the floor, particularly when the floor is colder than the average room air temperature. Carroll and Clinton (1980) argue for a value of zero unless the slab is considerably warmer than the air. This would probably be the case if the walls were adiabatic and all of the slab was colder than the air and ceiling. However, if convection loops are set up in the building, as by differentially heated walls, then it seems possible that significant convection could occur between the room air and slab. In the heated slab floor example from Gadgil et al. (1982), the unheated part of the slab is 40°C colder than the average room air temperature, yet is being cooled convectively—by cold air draining off the cold south wall glass. He found the convection coefficient to be almost as big as the ASHRAE value of 0.95 in table 3.1. Balcomb (1983a) measured surprisingly warm slab floors in low northern non-sunlit locations of the Accardo house, caused by strong convective air flow loops originating in the sunspace.

The heat transfer coefficient between the room air and ceiling is particularly important to the performance of roof-pond buildings. Faultersack and Loxsom (1982) measured the heat transfer coefficient on the inside surfaces of an unpartitioned building with a roof pond and a slab floor. The ceiling was made of ribbed metal with an actual surface area about twice the projected plan area. They used a heat flow-meter and radiometer measurements to determine the average convection rates from the inside surfaces to the air. During a time when the ceiling was on the average cooler than the room air by about 2°F (1°C), cooler than the slab by about 3°F (2°C), and cooler than the walls by about 5°F (3°C), they measured ceiling, walls, and slab floor convection coefficients of 3.3, 2.6, and 4.0 $W/m^2\,°C$ (0.58, 0.46, and 0.71 $Btu/h\,ft^2\,°F$), respectively, based on ceiling projected area. These values are all somewhat lower than the respective constant ASHRAE coefficients of 4.0, 3.1, and 4.0 $W/m^2\,°C$ (0.71, 0.55, and 0.0.71 $Btu/h\,ft^2\,°F$) for these surfaces, despite the ribbing. They did not report on the nature of the inside convection currents, but the surface temperature distribution would indictate that the currents were favorable for good convection on all the surfaces. Other estimates of ceiling heat transfer coefficients for roof-pond systems have been made by Loxsom et al. (1981), Tavana et al. (1980), Miller and Mancini (1977), Niles et al. (1976), and Balcomb et al. (1979).

Clearly, convection is the least understood mode of heat transfer in buildings. The numerical experiment approach promises to be a valuable tool to determine convection correlations, but it is only in its infancy, and much more work needs to be done. In addition, more full-scale experiments need to be performed to validate this work.

3.4.1.2 Convection on the Outside of Buildings The art of determining the appropriate convection coefficients to use on the outside of buildings is not well established, but the problem is less severe than that of the convection coefficients on the inside walls, because the outside surface conductances are usually wind induced and tend to be larger than the inside coefficients, thus having less influence than the inside coefficients on the overall thermal conductance of the building. However, the outside

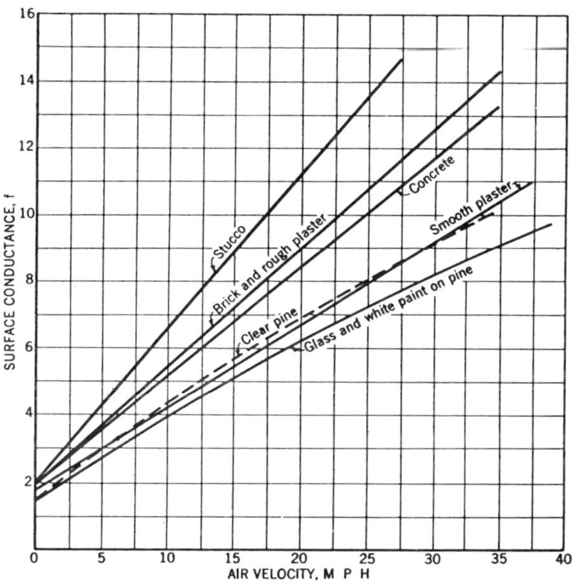

Figure 3.6
Surface conductances for different 12-in.² (30-cm²) surfaces as affected by air movements. Source: ASHRAE 1981.

coefficients do significantly influence the magnitude of the nonsolar heat transfer through windows and the solar gains through walls. Note that the wall solar gains are essentially inversely proportional to the outside surface conductance and are much more sensitive to the assumed value of surface conductance than is the nonsolar window heat transfer.

ASHRAE's (1981) suggested surface conductances are shown in figure 3.6; they include a radiation coefficient portion amounting to 4 W/m^2 °C (0.71 Btu/h ft^2 °F). These values are recommended by ASHRAE (1975) for computer calculations. The ASHRAE summer and winter design conductances of 23 and 34 W/m^2 °C (4.1 and 6.0 Btu/h ft^2 °F) for wind velocities of 7.5 and 15 mph (12 and 24 km/h) are fairly consistent with these values.

The accuracy of these numbers has been discussed frequently (Cooper and Tree 1973, Cole and Sturrock 1977, Carroll and Clinton 1980), with concern that they were measured under conditions that weren't representative of those encountered on the exterior of buildings. Among other things, building surface lengths are bigger than those that were used, and yet average turbulent heat transfer is known to decrease with increasing surface length. Cooper and Tree (1973) show that accepted flat-plate boundary layer theory predicts lower coefficients for flow on windward building surfaces. Leeward surfaces, in the separated flow region, were not considered.

There is a lack of experimental data on full-size house heat transfer coefficients. Since the flow configuration around a house-mounted solar collector would possibly be dominated by the house it is on, Duffie and Beckman (1980) suggest using the average house heat transfer value for the collector. For houses, he advises using the results of Mitchell (1976), intended to determine the heat transfer between animal forms and the ambient air. Largely using wind-tunnel data, Mitchell shows that animal shapes have approximately the same average heat transfer coefficient as a sphere if the diameter of the sphere is taken as the cube root of the volume of the animal. Mitchell also took data on spheres near the ground in a rural location to determine the effect of atmospheric turbulence on the results and found an average heat transfer enhancement of 23% due to atmospheric turbulence. Based on this enhancement and Mitchell's wind-tunnel data correlation, Duffie and Beckman suggest the equation

$$Nu = 0.42(Re)^{0.6}, \tag{3.26}$$

where Nu and Re stand for the dimensionless Nusselt and Reynolds numbers, respectively; see Holman (1981) for their definitions. The characteristic length appearing in both the Nusselt and Reynolds numbers should be taken as the cube root of the building volume as discussed above; the velocity appearing in the Reynolds number should be evaluated at the mid-height of the building.

The average winter U.S. wind velocity (Climatic Atlas 1977) is near the world average wind speed of 11 mph (18 km/h)—Duffie and Beckman (1980). Assuming this is measured at the standard 30-ft (10-m) height, then neglecting other microclimate corrections, the logarithmic law for vertical velocity distribution (Aynsley 1977), shows that the wind would be about 6 mph (10 km/h) at a 6-ft (2-m) height on a house. A 1200 ft^2 (112 m^2) house 10 ft (3 m) high would have a 23-ft (7-m) characteristic dimension. For these conditions, equation (3.26) yields $h = 7\ W/m^2\,°C$ ($1.2\ Btu/h\ ft^2\,°F$), considerably below the radiation-coefficient corrected value of 14 $W/m^2\,°C$ (2.5 Btu/h ft^2 °F) given for the smooth plaster surface in figure 3.6.

The Reynolds number in this case is about 1.2 million, an order of magnitude above that for which Mitchell's data were obtained. Although the flow characteristics around a sphere don't suddenly change in character above a Reynolds number of 100,000, using Mitchell's equation to extrapolate above this value is likely to underestimate the heat transfer coefficient because the exponent on the Reynolds number typically increases with the Reynolds number for flow around bodies.

Kind (1983) used a 1 : 32 scale model in a wind tunnel to measure the convection coefficient on solar collectors that were centered on one side of a pitched roof and were approximately 1/6 the total roof area. Although not explicity whole house data, his results may shed some light on the average house heat transfer coefficient. He determined the heat transfer with various house orientations and under simulated atmospheric boundary layer conditions with regard to velocity profile and turbulence. As with Mitchell's data, the Reynolds numbers were an order of magnitude below those likely for real houses. Also like Mitchell, he found the house coefficients to closely follow those for a sphere and suggested using sphere correlations to extrapolate his data to higher Reynolds numbers.

Kind found the heat transfer from the collectors to be relatively insensitive to wind direction, varying $\pm 20\%$ from the average value. The highest value occurred when the wind was parallel to the collector surface, while the lowest value occured when the collector was on the lee side of the building with the wind parallel to the building diagonal. Heat transfer coefficients were almost the same for windward and leeward wind perpendicular to the collectors. Kind found the heat transfer to be about 15% higher for the more turbulent urban location tests, as compared to the rural tests.

Kind's results yield a heat transfer coefficient with the same Reynolds number dependence as Mitchell but with the leading coefficient about 35% smaller. This is possible to explain by the fact that the heat transfer measured does not include the surface of the roof near the edges, only on the collectors centered on the roof. However, the results of Sparrow et al. (1982) imply that this would underestimate the coefficients by only a few percent.

Sparrow et al. (1981, 1982) also determined correlations for the heat transfer from surfaces on pitched roofs, both facing the wind and facing away from the wind. Below a Reynolds number of 60,000, the windward side had the biggest coefficient. At the highest Reynolds number tested, 100,000, the leeward side was about 15% higher. They suggest that this would increase to 40–50% in the 200,000 range. They correlated their results with Reynolds number to the 1/2 power, typical of laminar flow. However, the lee side heat transfer at the higher Reynolds numbers began to have a higher power dependence, more consistent with recirculating and turbulent flow. In general, Sparrow's results predict smaller heat transfer coefficients than either Kind's or Mitchell's. With the small 1/2-power dependence, the difference would be accentuated if one attempted to extrapolate to higher Reynolds numbers. Part of the reason for these differences could be that Sparrow made no attempt to account for atmospheric turbulence effects, and the wind tunnel used by Sparrow had a very low turbulence level.

Walton (1983) uses Sparrow's correlation for getting the convection coefficient of smooth glass surfaces and modifies it to account for other surfaces by using the relative roughness effects given in figure 3.6, increasing the Sparrow result by a factor of 2.1, 1.7, 1.5, and 1.1 for stucco, brick, concrete, and smooth plaster, respectively.

For translating weather-tape wind data to the local environment, Aynsley (1977) gives the following logarithmic law:

$$VH = V_{ref}(\ln(H/H_{ruf})/\ln(H_{ref}/H_{ruf})), \tag{3.27}$$

where VH is the mean wind velocity at height H; V_{ref} is the known mean wind velocity at the reference height, H_{ref}, usually 30 ft (10 m); H is the height where the wind is to be determined; and H_{ruf} is the roughness height, or approximately 5 to 10% of the average height of the ground roughness elements that effect the vertical velocity profile. Table 3.3 gives roughness heights for various types of terrain. When the wind speed is very low, free-convection effects may dominate. For free convecting rectangular blocks in the house range of Rayleigh numbers, McAdams (1954) gives the correlation

$$Nu = 0.13(RA)^{1/3}, \tag{3.28}$$

Table 3.3
Roughness heights for various terrains

Ground description	Roughness height, ft (m)
Flat, no trees	0.003 (0.001)
Flat or rolling with some trees or low bushes	0.1 (0.03)
Wooded or suburban, low-rise towns	1.0 (0.3)
Urban, high-rise buildings	10.0 (3.0)

with characteristic length defined as $L = 1/(1/L_{horiz} + 1/L_{vert})$. For air and a typical size house, this reduces to

$$h = K(\Delta T)^{1/3}, \tag{3.29}$$

where h is the convection heat transfer coefficient, K is approximately 1.6 $W/m^2\,°C^{4/3}$ (0.23 $Btu/h\,ft^2\,°F^{4/3}$), and ΔT is the house-to-ambient temperature difference. The temperature difference is small except on dark surfaces in direct sun where the difference can approach 50°C (90°F). Light colored walls and glass surfaces will rarely exceed 15°C (27°F) above ambient, for which h is 4 $W/m^2\,°C$ (0.71 $Btu/h\,ft^2\,°F$). When both free and forced convection effects exist, McAdams suggests calculating both and using the larger of the two values.

In conclusion, recent research suggests that the ASHRAE outside surface convection conductances are too large. Largely because of the small Reynolds numbers considered, most of the correlations developed for flat-plate collector heat transfer are suspect. For lack of better data, equation (3.26) is recommended for general use on exterior surfaces of houses, perhaps with figure 3.6 used to include roughness effects. There is clearly a need for building surface heat transfer measurements—if not on full-size buildings in a natural environment, then at least on models at the Reynolds numbers experienced on real buildings. An appraisal of the importance of the outside surface coefficient on solar gains through opaque surface of the building may indicate that solar gain effects should be included in such an experimental measurement program.

3.4.1.3 Convection between Rooms—through Doors and Vents If there are two openings in a building envelope, vertically displaced by h feet, the flow rate through these openings due to stack effect is given in dimensional form by ASHRAE (1977) as

$$V = 9.4Ak(h\Delta T)^{1/2}, \tag{3.30}$$

where V is the volumetric air flow rate (ft^3/min), A is the area of the smallest opening (ft^2), h is the height difference between the centers of the openings (ft), ΔT is the temperature difference between the indoor and outdoor air (°F), and k is a factor to account for the increase in flow caused by the excess in size of one opening over the other. If A_R is defined as the smaller opening area divided by the larger opening area, then ASHRAE's graph for k can be replaced by

$$k = (2/(1 + A_R^2))^{1/2}. \tag{3.31}$$

Equation (3.30) can also be used to give the volume flow rate of air between two rooms connected by two openings vertically displaced by h feet. This equation essentially gives the flow through a series of two orifices, each with a discharge coefficient of about 0.65.

Aynsley et al. (1977) derive this equation and give suggested discharge coefficients for various room geometries. For Reynolds numbers above 10,000, which is usually the case for stack effect driven flow through openings, the minimum possible discharge coefficient for sharp-edged orificies that are small compared to the wall they penetrate is about 0.6 (ASHRAE 1981). Walton (1983) uses 0.6 for ventilation openings. Smaller discharge coefficients are possible at low Reynolds numbers or if there are additional flow restrictions between the openings, as would be the case if the openings are screened or barred. Coefficients larger than 0.6 are possible when the opening is large compared to either the downstream or upstream flow area, being unity when either the upstream or downstream room is the same size as the opening. The ASHRAE value of 0.65 is seen to represent a fairly unrestricted flow situation.

Equation (3.30) is also used to determine flow rates through Trombe wall vents, with the discharge coefficient adjusted to account for additional losses up the face of the wall (McFarland 1978).

The rate of heat transferred, Q, through a vent or doorway due to stack effect can be found by multiplying equation (3.30) by the change in enthalpy of the air per unit volume, yielding

$$Q = KAk\Delta T^{3/2}h^{1/2}, \tag{3.32}$$

where Q is in Btu/h (W), K is 10.3 Btu/h ft$^{5/2}$ °F$^{3/2}$ (142 W/m$^{5/2}$ °C$^{3/2}$), and the other variables are as previously defined. If a doorway of width w by height d is approximated as two vertically adjacent openings with $h = d/2$, then one gets

$$Q = \frac{K}{2\sqrt{2}} w(d\Delta T)^{3/2} \tag{3.33a}$$

or the dimensional equation

$$Q = 3.6w(d\Delta T)^{3/2}, \tag{3.33b}$$

where Q is in Btu/h, w and d are in ft, and ΔT is in °F. That equation (3.33b) is essentially correct has been verified experimentally by Weber and Kearney (1980), who used similitude modeling and full-scale testing to arrive at the equation

$$Q = 4.6w(d\Delta T)^{3/2}. \tag{3.34}$$

Bauman et al. (1983) and Nansteel (1982) report on research aimed at developing detailed correlations between the interzonal heat transfer through openings in partition walls and the temperatures of the surfaces in the zones. Using numerical models to develop a data base, their results so far are less general than those of Weber and Kearney but in the long run may provide more exacting algorithms for room-to-room heat transfer. They have so far studied rooms with differentially heated end walls similar to the building of figure 3.5, but the rooms were three-dimensional and with a partition wall and doorway dividing the room into two parts. For a fixed temperature difference between the end walls, they found that although the heat transfer through the door was sensitive to the height of the door, it was relatively independent of the width of the doorway. In fact, the door opening could be extended all the way to the sidewalls without much effect on total heat transfer.

Although these results seem to contradict those of Weber and Kearney, they are, in fact, compatible: apparently, when the door width is increased, the air flow velocity through the doorway increases along with a decrease in the room-to-room air temperature difference, while the end wall temperature difference remains constant. In fact, the experiments done by Weber and Kearney also utilized differentally heated endwalls and apparently showed the same effect when the heat transfer was correlated with end wall temperature rather than average room air temperature (Nansteel 1982).

The heat transfer correlations developed by Bauman et al. would be useful if extended to a variety of differentially heated wall geometries, but at present, equation (3.34) is more general and is likely to be sufficiently accurate for design programs.

In order to determine the room-to-room air temperature difference necessary to supply heat to an otherwise unheated room via a doorway, Balcomb (1983a) equated the heat transfer through the doorway according to equation (3.34) to the heat lost to the outside from the colder zone. His result in figure 3.7 shows the average temperature difference between the driving room and the unheated room as a function of the average tempera-

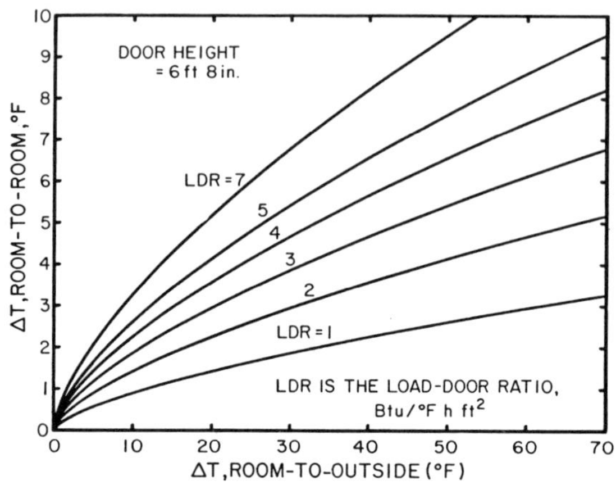

Figure 3.7
Steady-state results for air flow through a doorway to a remote room. Source: Balcomb 1983a.

ture difference between the unheated room and the outdoors. LDR is the ratio of the heat loss coefficient of the unheated room to the door area in Btu/ft^2 h °F.

Balcomb shows that the air flow and heat transfer rates predicted using equation (3.34) agree with measured values in two residences. His measurements clearly illustrate the sizable amounts of heat that can be transferred from sunlit zones to unsunlit zones via natural convection.

3.4.2 Radiation

3.4.2.1 Longwave Radiation Inside of Buildings Longwave radiation is an important mode of heat transfer between surfaces on the inside of buildings and between the outside of buildings and the environment. Inside, the walls transfer heat directly to each other by radiation, and some heat is radiantly exchanged with the air. The net radiant heat transfer is comparable in magnitude to that convected between the walls. Generally speaking, it is possible to calculate the inside radiant energy exchanges quite accurately, certainly more accurately than the convective exchanges discussed in section 3.4.1. However, the exact radiant exchange between surfaces inside a room and between the surfaces and the air is extremely complex, and virtually all of the simulation programs simplify the problem by assuming uniformly radiated grey diffuse surfaces with constant surface

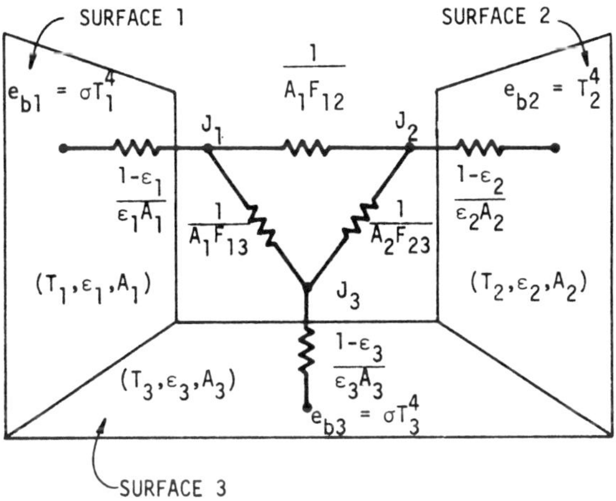

Figure 3.8
Radiation network for three surfaces. Source: Chapman et al. 1980.

temperatures. As is discussed below, a few large simulation programs have view factor calculation algorithms. Other programs use the very simple method of adding a constant radiation coefficent to the convection coefficients of each wall. There are also some intermediate approaches that combine good accuracy with rapid analysis.

Most heat transfer texts (e.g., Holman 1981) present the radiation-network methodology of Oppenheim (1956) for determining the radiant transfer between the walls of an enclosure. As illustrated in figure 3.8, with radiant flux as the potential, each surface has a "surface resistance," $(1 - \varepsilon_i)/(\varepsilon_i A_i)$, that depends on its area and emissivity. The J-nodes, representing the radiosity of each surface, are coupled to each other by a view-factor dependent resistance, $1/A_iF_{ij}$.

The radiosity nodes, J_i, can be dissolved by Y-Δ transformations to yield the equivalent circuit shown in figure 3.9, with resistances $1/A_iF'_{ij}$. With a generalized Y-Δ transformation, this can be done regardless of the number of legs meeting at the radiosity nodes.

The term F'_{ij} is the radiant interchange factor, which is the ratio of the flux emitted by i and ultimately absorbed by j (both directly and after reflections off other surfaces) to the flux that would be emitted by i if it were a blackbody. The total net heat transfer from any surface i is then given by

$$Q_i = \sigma \sum_j A_i F'_{ij}(T_i^4 - T_j^4), \tag{3.35}$$

where σ is the Stefan-Boltzmann constant. The radiative heat transfer given by equation (3.35) would normally be incorporated into an energy balance equation at the surface node for this surface—for example, into equation (3.3) for surface node 2 of figure 3.1. However, to avoid making that equation nonlinear, it is common to first linearize equation (3.35) by approximating $(T_i^4 - T_i^4)$ as $4\bar{T}^3(T_i - T_j)$, yielding

$$Q_i = 4\sigma \sum_j A_i F'_{ij} \bar{T}^3(T_i - T_j), \tag{3.36}$$

where $\bar{T}$ is fixed at a typical average room temperature or is reevaluated for each pair of surfaces. Since $\bar{T}^3$ changes by 1% per degree C, Q_i can easily be off a few percent when $\bar{T}$ is fixed at one value throughout a simulation. Carroll (1981) shows that for typical conditions, these linearization errors are not likely to cause more than a 5% error in any surface heat flux.

Equation (3.36) shows that the effective linearized radiation coefficient between walls can be written as

$$h_{ij} = 4\sigma F'_{ij} \bar{T}^3, \tag{3.37}$$

so that the net radiant heat loss from wall i is

$$Q_i = \sum_j A_i h_{ij}(T_i - T_j). \tag{3.38}$$

Thus, with the resistances changed to $1/(4\sigma F'_{ij}\bar{T}^3)$, figure 3.9 would represent a circuit with temperature as the potential instead of its fourth power. This circuit is shown in figure 3.16(a) in section 3.5, which includes the room convection circuit. Although the F'_{ij} values could be found from the figure 3.8 data, in the program NBSLD, Kusuda (1976) approximates F'_{ij} equation (3.37) as $F'_{ij} = \varepsilon_i F_{ij}$, so that

$$h_{ij} = 4\sigma\varepsilon_i F_{ij} \bar{T}^3. \tag{3.39}$$

This common approximation fails to properly account for reflections between the walls. Carroll (1981) points out that it also does not satisfy reciprocity, unless $\varepsilon_i = \varepsilon_j$, so it can produce heat balance errors (first law violations) in the room. He shows that such heat balance errors can amount to approximately 10% of the net flux for fairly realistic room

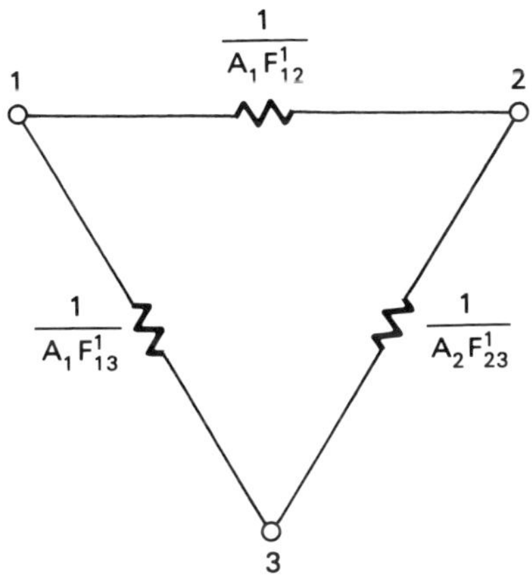

Figure 3.9
Alternate radiation network for figure 3.8, defining F'_{ij} factors.

shapes and are a much larger source of heating/cooling load errors than flux inaccuracies.

Because of geometrical complexities, F_{ij} values are difficult to determine. They also require a tedius three-dimensional description of the building in the simulation input. The DEROB program (Arumi-Noe and Northrup 1979) solves for the view factors by a double numerical integration over the solid angles subtended from each surface. Alternately, a scheme is presented in ASHRAE (1975) that used standard formulas to determine the view factors in rectangular rooms and in gabled attic spaces. The program UWENSOL (Emery et al. 1978) uses a line integral approach (see Sparrow and Cess 1970), which reduces the solution to that of a single integral. Recent discussions of the view factor problem are given by Emery et al. (1981) and Lipps (1983).

To simplify the view factor determination, the program BLAST (Hittle 1977) approximates F_{ij} on the basis of surface area, independent of relative position of the surfaces:

$$F_{ij} = A_j \Big/ \sum_j A_j, \tag{3.40}$$

where the sum is over all j seen by i. Though this relationship is fairly accurate for the facets of rectangular rooms that are not too far from cubical, Carroll (1981) points out that it also lacks reciprocity.

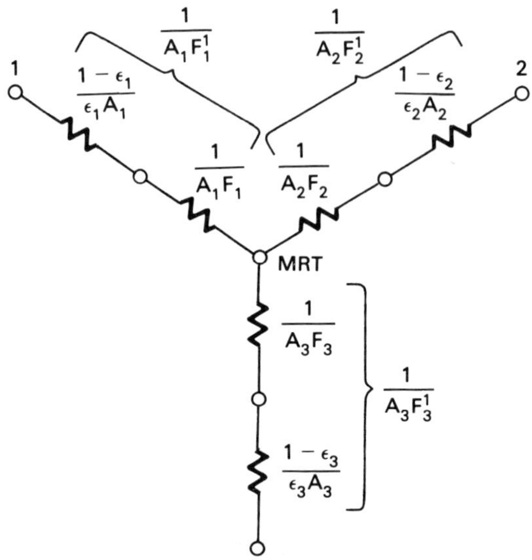

Figure 3.10
MRT network for three surfaces.

An interesting way to model the interior radiation interchange is used by Walton (1983) in the National Bureau of Standards program TARP. This procedure was developed by Carroll (1980, 1981), who calls it the MRT network method. Like equation (3.40), this method avoids a complex view factor calculation. Carroll assumes that each wall exchanges heat radiantly with the mean radiant temperature of the room, as shown in figure 3.10.

Besides the surface resistances of figure 3.8, figure 3.10 has the resistances $1/(A_iF_i)$. It can be argued that the net heat transfer from say, surface 1, should be driven by the difference between its temperature, T_1, and some mean temperature of all the surfaces "seen" by surface 1. This would require that the heat transfer from each surface be calculated using a different mean radiant temperature, each determined as a weighted mean of the temperature of all the other surfaces. In figure 3.10, however, Carroll uses a common MRT to determine the heat transfer from each surface. To eliminate the error due to the self-weighting of each surface in the MRT, the values of F are adjusted. Carroll uses a heuristic argument to develop the following equation for F_i:

$$F_i = 1/\left(1 - A_iF_i \Big/ \sum_i (A_iF_i)\right) \tag{3.41}$$

Given the area, A_i, of each surface and starting with $F_i = 1$, this equation can be solved iteratively for each F_i, yielding the resistances A_iF_i in figure 3.10. This figure essentially defines the MRT, which is itself partly influenced by the size of the F factors. This has the effect of slightly exaggerating the heat transfer from larger surfaces, more than warranted by their larger area.

A major advantage of this approach is that it inherently results in no heat balance errors. Also, by having performed what is essentially a generalized Δ-to-Y transformation, the number of circuits to be analyzed has been reduced. Although not as accurate as a full Oppenheim network solution, Carroll makes a good case that the MRT network method is generally subject to smaller errors than the other common methods used for computing radiant interchange in buildings, even when coplanar surfaces get no special treatment. In addition, this method appears to be simple to use because it does not require as detailed a description of room geometry.

Used in a simulation, the temperature at the MRT node would be determined at each time step similarly to the way the air temperature would be found in the room. As pointed out by Carroll (1981), knowing the MRT in addition to the air temperature also allows a more accurate assessments to be made of comfort conditions.

Most of the building simulation programs treat the indoor air as a transparent medium. Carroll (1980) shows that the air emissivity is typically 0.05 to 0.15 in residences. Thus up to 15% of the radiation leaving a wall may be absorbed by the air. In addition, the air radiates to the walls. In Carroll's MRT network scheme, the air effects can be accounted for by connecting the MRT and air nodes with a conductance that depends on the air emissivity and wall area. This connection is indicated in figure 3.11, which includes the convection resistance circuits.

Note that the MRT and air nodes in figure 3.11 could be shorted if the air temperature and MRT were always equal. Shorting would only be correct if the emissivity of the air were unity. In this case, the heat radiated to the air from the wall would be transported around the room convectively, as would happen if the room were full of a radiantly opaque fluid, such as water or Freon®. In this case, the convection conductance h_i could

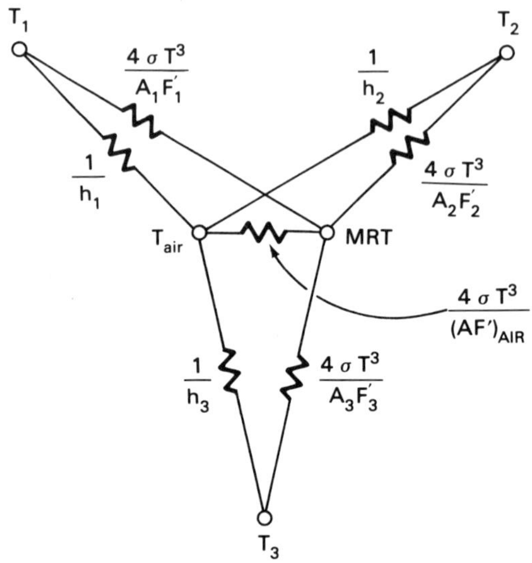

Figure 3.11
MRT network and convection networks connected due to radiant interchange between the air and the walls.

be added to the radiation conductance A_iF_i'. Although not completely accurate physically, this assumption simplifies the calculations. For load calculations, ASHRAE (1972) recommends using a similar but even simpler combined coefficient in which A_iF_i' is replaced by a fixed radiation coefficient equal to $4\sigma\varepsilon_i(\overline{T}^3)$. The corresponding ASHRAE convection coefficients are discussed in section 3.4.1.1.

Since the temperature at the combined (shorted) air and MRT node is a mixture of both air and MRT temperatures, it is sometimes given the name "globe temperature," although it is not the same as the globe temperatures defined by ASHRAE (1972). Hutchinson (1964) discusses some of the major causes of error in ASHRAE's combined coefficient model. For example, if all the walls were equally insulated external walls, their inside surface temperatures would be the same and no net radiant heat transfer would occur. Thus, if the combined ASHRAE surface conductances were assumed, the heat transfer to the air from the external walls would be exaggerated. Carroll and Clinton (1980) point out that the combined coefficient method typically results in a house loss coefficient about 5% higher than does Carroll's MRT method. They also discuss inaccuracies that the combined coefficient method causes when comparing different types of heating systems in evaluating comfort and the effectiveness of ventilation cooling.

An additional method to determine a combined coefficient is discussed in section 3.5, wherein the conductances determined by equation (3.37) are added directly to the convection coefficients. This is shown in figure 3.16 (section 3.5), resulting in each surface's total conductance being given by $[h_i + \sum h_{ij}]$. Subbarao and Anderson (1982a) give a frequency-domain method of determining the error caused by this procedure.

3.4.2.2 Longwave Radiation on the Outside of Buildings Radiant exchanges inside of buildings can be determined as accuately as necessary. However, on the outside of the buildings, the situation is not quite as satisfactory due to the less adequate knowledge of the temperature of the surroundings.

To determine the radiation loss from the outside surfaces, the program FREHEAT (Chapman et al. 1980) employs an equation based on a three-surface network similar to figure 3.8, where one surface represents the wall, one the sky, and one the ground. Assuming the ground and sky to act like blackbodies, the net radiant heat loss from a wall or window, Q_w can be found from an energy balance on this circuit as

$$Q_w = A_w h_r (T_w - T_o), \tag{3.42}$$

where

$$h_r = \varepsilon_w \sigma (T_w^4 - F_s T_s^4 - F_G T_G^4). \tag{3.43}$$

T_w, T_s, T_G, and T_o are the wall, sky, ground, and ambient air absolute temperatures. F_s and F_G are the view factors from the wall to the sky, $(1 + cos(\phi))/2$, and to the ground, $(1 - cos(\phi))/2$, ϕ is the tilt angle of the surface. σ is the Stefan-Boltzmann constant, and ε_w is the emissivity of the wall.

The problem remains of determining the ground and sky temperatures. The ground, or more generally the opaque terrestrial surroundings, are typically hotter than the ambient air during the day and colder at night. The simplest assumption is that both the sky and ground temperatures are always equal to the ambient air temperature. This is least accurate at night, when both the ground and effective sky temperatures are usually below the air temperature. Besides depending on the conditions it is exposed to, the ground temperature is very material dependent.

Good information is available for determining the effective clear sky temperature, at least as seen by a horizontal surface. This is particularly useful for determining the radiant losses from roofs and roof ponds, in which case $F_G = 0$, and $F_s = 1$ in equation (3.43). The Lawrence Berkeley Laboratory's latest sky radiation studies (Berdahl and Martin 1982) include data taken over eighteen months at six cities in the southern part of the U.S. They measured the wavelength and angular dependence of the sky radiance to determine the total hemispheric radiance of the sky on clear days, that is, the total longwave (5 to 50 μm) radiant flux received from the sky hemisphere by a horizontal surface. They give the following one-parameter quadratic equation for the total clear sky emissivity, ε_s:

$$\varepsilon_s = 0.711 + 0.56(T_{dp}/100) + 0.73(T_{dp}/100)^2, \tag{3.44}$$

where T_{dp} is the dew point temperature at ground level, in degrees Celsius. This correlation is the result of an integration over the hemisphere of the emissivity data taken at various zenith angles (Berdahl and Fromberg 1982), with the integrand weighted by the cosine of the zenith angle so that the effective emissivity is determined for a horizontal surface. The corresponding clear sky radiance is given by the equation

$$R = \varepsilon_s \sigma T_o^4. \tag{3.45}$$

An effective sky temperature can be determined from equations (3.44) and (3.45) as

$$T_s = (\varepsilon_s)^{1/4} T_o, \tag{3.46}$$

so that equivalent to equation (3.45),

$$R = \sigma(T_s)^4. \tag{3.47}$$

The net radiant heat loss from a horizontal surface, r, is then given by

$$N = \varepsilon_r \sigma(T_r^4 - T_s^4). \tag{3.48}$$

The above formulations can be applied during cloudy weather by replacing ε_s by the following empirical equation for cloudy sky emissivity given by Martin and Berdahl (1982):

$$\varepsilon_{cs} = \varepsilon_s(1 - Ln) + Ln, \tag{3.49}$$

where ε_s is the clear day emissivity given by equation (3.44), n is the estimated fraction of sky cover by clouds, and L is a parameter that depends on the cloud height and type. They give values for L of: 0.16 for cirrus clouds (height 12.2 km), 0.66 for altocumulus (3.7 km), and 0.88 for stratocumulus (1.2 km). NOAA weather tapes supply n but not cloud type, so regional/seasonal cloud types would have to be known to use this correlation.

Clear and cloudy sky emissivity correlations have also been obtained by Clark and Allen (1978) for the weather of San Antonio, Texas. Blanpied et al. (1982a) give a comparison between clear and cloudy San Antonio data, and a number of correlations. Equation (3.49) fits their data well. Berdahl and Fromberg (1982) compare a number of correlations for clear sky emissivity.

The spectral radiometer measurements made by Martin and Berdahl (1982) show that the emissivity is the smallest at the zenith, and as one approaches the horizon, all the measured emissivities increase rapidly. This is indicated by the LOWTRAN computer model results shown in figure 3.12 and results from the long air path at low altitudes where the air is near the surface temperature.

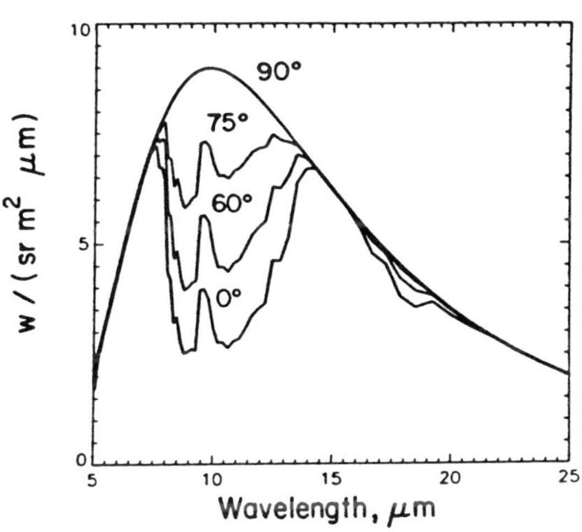

Figure 3.12
Spectral radiance of clear skies. Estimated spectral radiance of the cloudless sky for zenith angles of 0, 60, 75, and 90° surface, dewpoint temperatures of 21 and 16°C (70 and 61°F), respectively, and for mid-latitude, summer atmosphere. Source: Berdahl and Fromberg 1982.

As a consequence of the higher radiance near the horizon, the part of the sky seen by tilted surfaces is not as cold as given by equation (3.46). Based on the spectral radiance curves given in figure 3.12, Blanpied et al. (1982b) developed the following empirical equation giving the ratio of the sky radiance on a surface tilted by θ radians to the sky radiance on a horizontal surface. For clear skies,

$$B_c = 1 + 0.027250\theta - 0.25421\theta^2 + 0.03372\theta^3. \tag{3.50}$$

For skies with cloud cover, n, ranging from 0 to 1, they propose the following equation to determine the B factor:

$$B = (B_c - B_0)(1 - n) + B_0, \tag{3.51}$$

where B_0 is given by

$$B_0 = (1 + \cos\theta)/2. \tag{3.52}$$

Their validation measurements taken thus far over winter months show a fairly good correspondence with the equations.

Other methods for determining tilted surface radiance are given by Walton (1983), Sweat and Carroll (1983), and Kamada and Flocchini (1984).

3.4.3 Evaporation

Evaporation will be discussed here with respect to its application to roof pond cooling. Open water roof ponds are usually only exposed nocturnally. They are cooled primarily by radiation to the sky and evaporation to the ambient air. Frequently, net cooling takes place despite ambient air temperature being higher than the water temperature. When this occurs with a sealed pond (i.e., dry surface) in a still-air environment, the air above the water can stratify and eliminate convection. With a wet pond however, moisture-induced buoyancy (moist air is lighter than dry air) can overcome the temperature-induced stability, so the surface air becomes destabilized and significant evaporation can occur. In order to have an evaporation correlation sensitive to these phenomena, the Trinity University study on roof-pond cooling potential used the evaporation and convection equations recommended by Niles et al. (1978). These evaporation equations were based on the analogy between heat and mass transfer given by Eckert and Drake (1959). Marlatt et al. (1984) review a number of roof-pond evaporation algorithms. The roof-pond simulation program developed by Tavana et al. (1980) used equations (3.53) and (3.54) below, developed by Ryan and Harleman (1973), and Jirka et al. (1978). These equations were developed to determine evaporation and convection from heated ponds, with particular application to cooling basin design. For $T_{VS} > T_{VA}$ (defined below), the rate of heat transfer due to evaporation is given by

$$Q_E = 3128[V + 3.13(T_{VS} - T_{VA})^{1/3}](p_w - p_A). \tag{3.53}$$

Related to this by the Bowen ratio, the convective heat transfer is given by

$$Q_C = 6.04 \times 10^{-4}(T_W - T_A)Q_E/(p_W - p_A), \tag{3.54}$$

where Q_C and Q_E have the units of kJ/h m^2, and

$$T_{VS} = T_S/(1 - 0.378p_S/p_B), \tag{3.55}$$

$$T_{VA} = T_A/(1 - 0.378p_A/p_B), \tag{3.56}$$

where p_W is the partial pressure of saturated water vapor at the water temperature (in atmospheres); p_A is the partial pressure of water vapor in the air at T_A, 2 m above the pond (in atmospheres); p_B is the barometric pressure (in atmospheres); T_W is the temperature of the water (K); T_A is the temperature of the air 2 m above the water (K); and V is the wind velocity at 2 m above the pond (km/h).

The temperatures T_{VA} and T_{VS} are the virtual temperatures that would give dry air the same buoyancy as the air under the actual conditions. As seen in equation (3.53), these equations have a free convection part, where the flow is driven by the temperature differential, and a forced convection part, where the wind velocity determines the flow. The free convection portion is based on the same analogy as Niles's equation and predicts approximately the same answers. The free convection portion also compared favorably to heated-pond laboratory experiments done at MIT, using a 1-m^2 pond and a 7-by-12-m pond. The forced convection portions of equations (3.53) and (3.54) are based on a number of heated-pond and lake measurements, extensively documented in the report of Ryan and Harleman (1973).

Notice that equations (3.53) and (3.54) have no pond length dimension as did the Trinity equations. For surfaces with a 20-ft characteristic length, under forced-convection-dominated conditions, equations (3.53) and (3.54) predict evaporation and convection rates well below those of the Trinity equations. For lake-scale ponds, the two sets of equations are more in agreement. The Trinity equations used a rough flat plate forced-convection correlation. Interestingly, if they had instead based their forced convection coefficient on Mitchell's equation, recommended in Section 3.4.1.2, their results would have been close to the MIT results for house-scale ponds.

Although Ryan and Harleman lacked good validation data for small heated ponds under forced convection, their equations did very well in predicting heated swimming pool losses in measurements made by Klotz (1977). As a consequence, they were used in the Lawrence Berkeley Laboratory heated swimming pool program, POOLS, by Wei et al. (1979). Wei et al. also found the predictions to compare favorably with actual swimming pool data. They used factors of 5–10 to reduce weather bureau wind speeds to what they expected at the pool measurement sites. The velocity correction procedure given by equation (3.27) gives results comparable to their wind correction factors.

A review of heated pool correlations by Shah (1981) also corroborates the accuracy of the MIT equations. Shah predicts results reasonably close to the MIT equation for typical pond conditions. Shah compares his equation with a number of other correlations, including the well-known Carrier equation, which he shows is not reliable except at low velocities and with high temperature differences. Generally, it was found that the MIT equations predicted evaporation rates well below those predicted by the Carrier equation.

In general, as with convection (section 3.4.1), the state of evaporation modeling could be advanced if there was more good data for algorithm validation. This probably means controlled roof-pond experiments under a variety of conditions, with both wet and dry ponds.

3.5 Frequency Domain Methods

The harmonic method, sometimes called the frequency-response method or the admittance method, is generally limited to a building that can be represented by linear equations (with temperature- and time-independent parameters) and that is subjected to steady periodic environmental conditions. As discussed below, these restrictions can be overcome to some extent. However, the method has usually been applied to the simulation of buildings for periodic design-day weather conditions, excluding the effects of weather fluctuations. Because of the second restriction, the method is unable to handle the nonlinear nature of thermostat-controlled backup heating and cooling, and so has generally been limited to determining the "floating" temperature in buildings.

While the solutions available by this method are limited, they have the advantage of being rapid and can be made exact, which is useful for validation of other more approximate solution methodologies. Frequency-domain methods can also give considerable insight into the fundamental thermal characteristics of a building and help determine the parameter groups that characterize it. Frequency-domain analysis is almost indespensible when dealing with network reduction techniques, as discussed in section 3.3.1.2.

In its simplest form, the harmonic method used in building analysis is the same as the AC circuit analysis used in linear electrical circuit theory (for example, see Hayt and Kemmerly 1962). With only sinusoidal exitations, the differential equations describing the systems can be reduced to complex algebraic equations. The temperature (voltage) and heat flow (current) are replaced by their phasor form (giving magnitude and phase angle) in the frequency domain, and the circuit elements by their equivalent complex admittances. Niles (1979) essentially took this approach to develop a graphical procedure for analyzing the performance of direct-gain buildings, like that shown schematically in figure 3.13.

In Niles's approach, the building inputs of insolation and outside temperature are both assumed to be composed of their 24-hour average (zero-frequency) components and of a sinusoidal component at a frequency of $\omega = (2\pi/24) rad/h$. A steady-state energy balance on the air node of figure 3.13 yields the equation

$$\overline{T} = \overline{T}_o + \overline{S}/U, \qquad (3.57)$$

where $\overline{T}$ and $\overline{T}_o$ are the average inside and outside air temperatures, $\overline{S}$ is the total daily insolation admitted into the room, and U is the overall building conductance. An energy balance of the diurnally periodic energy components gives

$$\tilde{T} = A\tilde{S} + B\tilde{T}_o, \qquad (3.58)$$

where $\tilde{T}$ and $\tilde{T}_o$, and $\tilde{S}$ are the sinusoidal temperature and sun components, usually written in phasor notation. A and B are given by $A = 1/(U +$

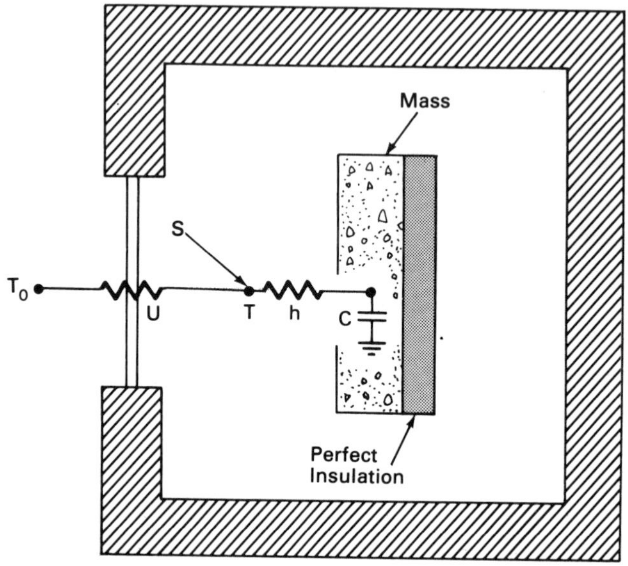

Figure 3.13
Schematic of direct-gain building and its lumped network representation. Source: Niles 1979.

$(hY)/(h + Y))$ and $B = U/A$, where $Y = j\omega C$ is the complex admittance of the lumped mass heat capacity, C.

Given the inputs in phasor form, equations (3.57) and (3.58) are combined to determine the amplitude of room air temperature swing to be:

$$|\tilde{T}| = \left| \frac{((\bar{T} - \bar{T}_0)/k)^2 + |\tilde{T}_o|^2 + \sqrt{2}|\tilde{T}_o|(\bar{T} - \bar{T}_0)/k}{1 + [(2 + h/U)(h/U)/(1 + (h/\omega c))]^2} \right|^{1/2}, \tag{3.59}$$

where $|\tilde{T}|$ and $|\tilde{T}_o|$ are the amplitudes of the room and outdoor air temperature phasors, and k is the ratio of the total daily insolation admitted into the building to the amplitude of the fundamental sinusoidal component of the insolation. The outdoor temperature was assumed to peak at 3 P.M. solar time in this analysis. The maximum and minimum temperature of the building can thus be obtained by adding and subtracting $|\tilde{T}|$ from the average indoor temperature given by equation (3.57). Niles plotted the results of this equation to include the case when the sun is all absorbed in the mass instead of the air. While not exact, these graphs afford a rapid means of determining the average temperature and temperature swing for the building on a design day. They are useful both for design purposes and for showing how the various primary building and climatic variables influence the building temperatures.

The above equations can also be obtained using the standard frequency-response methodology found in the system analysis literature (for example, see Reid 1983). In this approach, the Fourier transform is applied to the system differential equations to directly produce the frequency-domain equations. These equations contain the complex transfer functions that relate the output phasors to the input phasors; they completely characterize the building response as a function of excitation frequency. In the case of figure 3.13, these transfer functions reduce to A and B in equation (3.58), relating the room temperature dependence on the insolation and outdoor temperature.

In effect, the Fourier transform eliminates time and the time derivatives from the equations and replaces them by complex algebraic equations involving frequency. If the inputs are represented by a Fourier series (a two-term series in the above example), the building temperature response at a certain frequency can be found by forming the complex product of the transfer function and the appropriate component of the input. The total response can be found by superposition of the different frequency outputs. Kirkpatrick and Winn (1982) show how to use this frequency-response approach to study mixed direct-gain and Trombe-wall buildings.

Although the above methods use lumped masses, in general, they need not be lumped. The potential accuracy of the harmonic method results from the fact that there is no need to discretize either the time or space variables. The complex response functions for one-dimensional layers of homogeneous massive materials, with periodic inputs, are easily determined by solving the diffusion equation, as shown, for instance, by Carslaw and Jaeger (1959). They also give the method of finding the response function for multiple layers of different materials by what is commonly called the matrix method.

In the case of mass exposed to the room and adiabatic on the other side (as in figure 3.13), the distributed mass response functions can be written as complex admittances. Balcomb (1983a) gives the procedure, including a program listing, to determine the response function (he calls it the diurnal heat capacity) that gives the periodic heat flow for any number of layers of back-insulated mass. Balcomb also uses these values to determine building temperature swings. He shows how an effective total mass can be found for

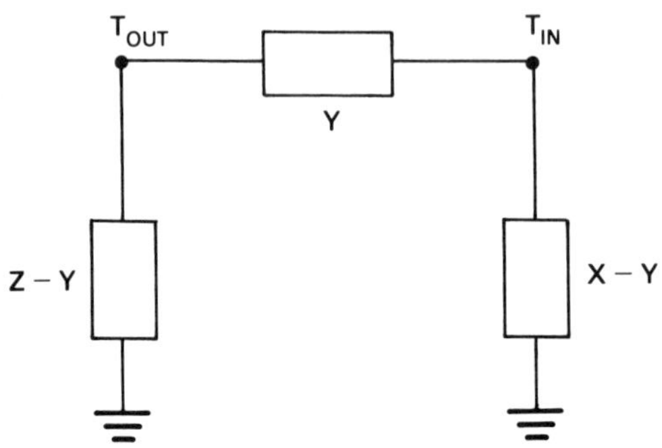

Figure 3.14
Equivalent circuit of multilayer wall. X, Y, and Z are complex admittances, T_{in} and T_{out} are surface temperatures. Source: Subbarao and Anderson 1982a.

a building by adding the magnitudes of the response functions of all the masses. He weights the terms added, somewhat arbitrarily, by factors to account for differences in exposure to the sun. This method is quite suitable as a quick design aid.

Goldstein and Lokmanhekim (1979) developed a more rigorous method intended to be used with progammable calculators or small computers. The procedure can handle a number of single-layer distributed masses, and the input will manage frequencies on either side of the diurnal frequency so that weather events longer than one day can be studied and the rapid changes in the insolation at sunrise and sunset can be accurately modeled.

Subbarao and Anderson (1982a) give a fairly comprehensive and rigorous development of the harmonic approach. Their treatment considers distributed-parameter multilayer walls exposed to the sun and air on both sides. This allows the accurate simulation of Trombe walls (noncirculating) and other external and internal mass walls, insulated or not. They show how such walls can be conveniently represented by the block diagram circuit of figure 3.14, which shows the relationship between the mass surface temperatures and the frequency-dependent complex admittances X, Y, and Z, representing the multilayer wall. These admittances can be obtained by the matrix method in Carslaw and Jaeger (1959).

For convenient use, Subbarao et al. (1982) give polar (Nyquist) plots of these admittances for a series of frequencies and materials. They show that a building with any number of layered walls and miscellaneous solar gains can be represented essentially in the form of equation (3.57), where A and B

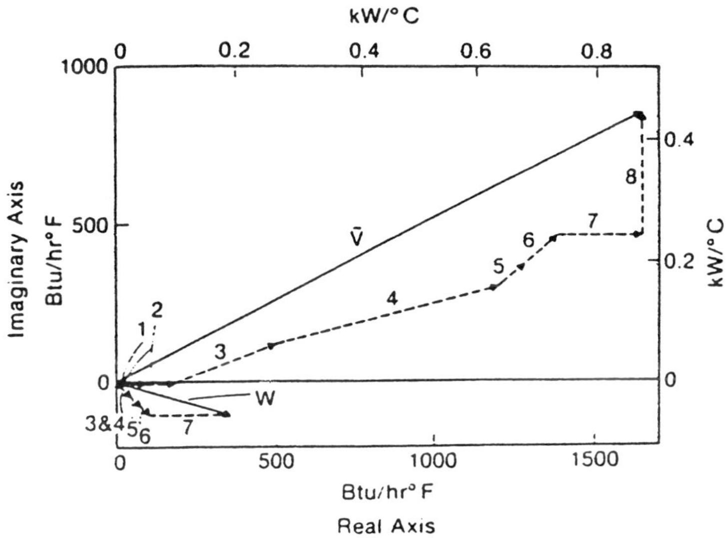

Figure 3.15
Addition of admittance vector representing (1) Trombe wall, (2) direct-gain glazing, (3) illuminated floor slab, (4) nonilluminated floor slab, (5) frame walls, (6) frame ceiling and roof, (7) infiltration, and (8) internal capacitance. Source: Subbarao and Anderson 1982a.

are determined in part by the value of V (the sum of all the X admittances) and W (the sum of all the Y admittances). The addition of the X and Y admittances is illustrated graphically in figure 3.15 for a typical mixed-gain building. Given familiarity with the relationship between admittance V and W, and the transfer functions A and B, figure 3.15 allows the assessment of the impact of each of the various building components on the final building performance.

The harmonic analysis procedures discussed so far have been developed using the combined surface coefficients, whereby the convection and linearized radiation coefficients are lumped together, as in figure 3.16b, where the single-subscripted h's are convection coefficients and those doubly subscripted are the radiation coefficients. Goldstein (1978) and Subbarao and Anderson (1982a) show how their simulation procedures can be modified to handle the radiation component in the more correct manner, as shown in figure 3.16a, where, presumably, the radiation coefficients are equal to the linearized values given, for example, by equation (3.37). Both references show that there is little change in the form of their heat balance

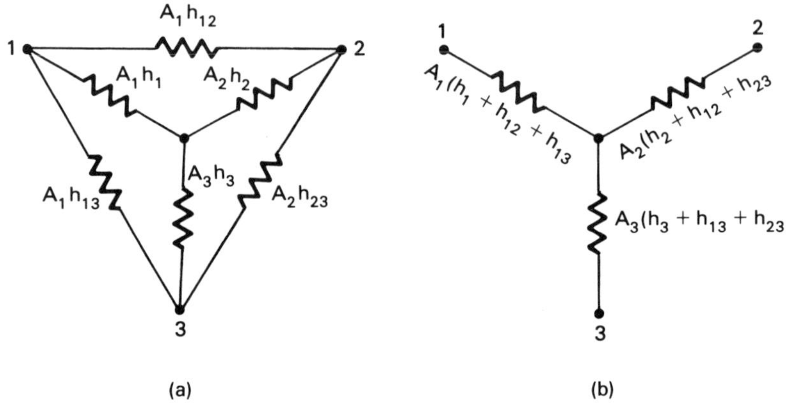

Figure 3.16
(a) Convection-radiation network for three surfaces, and (b) modified network with radiation lumped with convection. h_i is the convection coefficient for surface i. h_{ij} is the linearized radiation coefficient between surfaces i and j. Source: Subbarao 1982.

equations (such as equations [3.57] and [3.58]) when using the more correct radiation treatment.

As mentioned at the beginning of this section, the harmonic method lends itself to determining building floating temperatures and not backup use. For the linear systems considered, the essence of the harmonic method is that the contribution of the various frequency components of the driving functions and the building response can be decoupled and solved independently. However, a thermostat (and, therefore, the backup heat) does not respond to different frequencies independently but only to the sum of all the different frequency components. Thus, the decoupling of different frequency components is not possible in the presence of a thermostat. However, Subbarao et al. (1982) show that climate-specific correlations can be made between the swing in the building floating temperature on an average day of the month and the monthly backup energy that would be required if the building were constrained to be kept above a prescribed backup setpoint. More remarkable, Subbarao and Anderson (1982b) show that the backup can be calculated directly, without a correlation. This major barrier in fully using the frequency-domain approach has been overcome by Subbarao and Anderson by the technique described below.

Suppose one used the harmonic method (without backup) to simulate the performance of a building over a period of days of fluctuating weather.

The method used so far for design-day simulation can be used over extended periods if the last day of the period is wrapped onto the first day to make it steady periodic as a whole. Now, of course, more frequencies than zero and dirunal must be represented in the driving conditions. With the aid of fast Fourier transforms, Subbarao and Anderson are able to simulate building performance over multiday weather periods. Now suppose the house is constrained by a backup heater to stay above some fixed setpoint temperature. If the setpoint were high enough, such that the room temperature stayed at the setpoint, then the heater output could be determined explicitly by solving the energy balance equations for heater output in the same way they had been solved for room temperature before—indeed, this is how Mackey (1944), and Nottage (1954) first used the harmonic method to determine peak cooling loads when the inside temperature was assumed to be fixed.

Of course, if one could guess the correct time-dependent heater output profile, it could be added to the building's energy inputs, and the room temperature (not necessarily constrained to be always at the setpoint) could be determined. If the resulting inside temperature does not conform with the required setpoint constraint, one could guess again, and so on. This hints at the integral equation approach taken by Subbarao and Anderson, which uses an iterative procedure to determine the values of an ingenious single-valued function of time that simultaneously gives the time-varying values of backup heat and room temperature.

Despite the considerable developments in the harmonic methods, it is doubtful whether the method will compete as a design tool with the finite-difference approach. For most design work, hour-by-hour finite-difference simulations appear to be just too fast, flexible, and easy to understand to have any serious competition from the harmonic methods. However, the harmonic method is likely to remain the analytical, as opposed to numerical, approach of choice. Besides its potential accuracy, it produces a rich explanation of why buildings perform the way they do.

Notes

1. As given in Carslaw and Jaeger (1959), and exact solution for the temperature distribution in a continuous medium, such as the mass in figure 3.1, can be obtained from the heat conduction equation: $v\partial^2 T/\partial x^2 = \partial T/\partial t$, where $v = k/p.c$, k is the thermal conductivity, p is the mass density, and c is the specific heat capacity of the medium. See Carslaw and Jaeger 1959.

2. The convection coefficient, h, is defined as the ratio of the rate of heat convected per unit area of a surface, Q/A, to the temperature difference, ΔT, between the surface and the adjacent air: $h = (Q/A)/\Delta T$.

References

Akridge, J. M., and J. F. J. Poulos. 1983. The decremented average ground-temperature method for predicting the thermal performance of underground walls. *ASHRAE Transactions* 89(2A, B).

Altmayer, E., A. Gadgil, F. Bauman, and R. Kammerud. 1982. *Correlations for Convective Heat Transfer From Room Surfaces.* LBL-14893. Berkeley, CA: Lawrence Berkeley Laboratory.

Ambrose, C. W. 1981. Modeling losses from slab floors. *Building and Environment* 16(4).

American Society of Heating, Refrigeration, and Air-Conditioning Engineers. 1972. *ASHRAE Handbook: 1972, Fundamentals.* Atlanta, GA: ASHRAE.

ASHRAE. 1975. *Procedure for Determining Heating and Cooling Loads for Computerizing Energy Calculations; Algorithms for Building Heat Transfer Subroutines.* New York: ASHRAE.

ASHRAE. 1977. *ASHRAE Handbook: 1977, Fundamentals.* Atlanta, GA: ASHRAE.

ASHRAE. 1981. *ASHRAE Handbook: 1981, Fundamentals,* Atlanta GA: ASHRAE.

Anderson, J. V., and K. Subbarao. 1981. Spectral analysis of ambient weather patterns. Paper presented at the American Society of Mechanical Engineers' Annual System Simulation Conference, Reno, Nevada.

Andersson, B., F. Bauman, and R. Kammerud. 1980. *Verification of Blast by Comparison with Direct Gain Test Cell Measurements.* LBL-20629. Berkeley, CA: Lawrence Berkeley Laboratory.

Andrews, J. 1979. *A TRANSYS-Compatible Model of Ground Coupled Storage.* Upton, NY: Brookhaven National Laboratories.

Arumi-Noe, F., and D. O. Northrup. 1979. A field validation of the thermal performance of a passively heated building as simulated by the DEROB system. *Energy and Buildings* 2(1).

Aynsley, R. M., W. Melbourne, and B. J. Vickery. 1977. *Architectural Aerodynamics.* London: Applied Science Publishers.

Balcomb, J. D., J. C. Hedstrom, S. W. Moore, and R. D. McFarland. 1979. *Results from a Passive Thermal Storage Roof on a Mobile/Modular Home in Los Alamos.* LA-UR-79-2672. Los Alamos, NM: Los Alamos Scientific Laboratory.

Balcomb, J. D. 1983a. *Heat Storage and Distribution Inside Passive Solar Buildings.* LA-9694-MS. Los Alamos, NM: Los Alamos National Laboratory.

Balcomb, J. D., 1983b. Thermal network reduction. *Proc. 1983 Annual Meeting, American Solar Energy Society, Progress in Solar Energy,* vol. 6. Boulder, Colorado: American Solar Energy Society.

Bauman, F., A. Gadgil, R. Kammerud, and R. Grief. 1980. Buoyancy-driven convection in a rectangular enclosure: Experimental results and numerical calculations. Paper presented at the ASME Conference on Heat Transfer in Passive Solar Systems, Orlando, FL, July 27–30.

Bauman, F., A. Gadgil, R. Kammerud, E. Altmayer, and M. Nansteel. 1983. Convective heat transfer in buildings: Recent research results. *ASHRAE Transactions* 89(1A):215–233.

Berdahl, P., and R. Fromberg. 1982. The thermal radiance of clear skies. *Solar Energy* 29(4):299–314.

Berdahl, P., and M. Martin. 1982. *Emissivity of Clear Skies*. LBL-15367. Berkeley, CA: Lawrence Berkeley Laboratory.

Blanpied, M., J. B. Cummings, G. Clark, and B. Schutt. 1982a. Models and measurements of radiative cooling. *Proc. 7th National Passive Solar Conference*, Knoxville, TN, August 30—September 1, 1982. Newark, DE: American Section of the International Solar Energy Society.

Blanpied, M., G. Clark, and J. B. Cummings. 1982b. Effect of tilt angle on infrared radiation received on tilted surfaces. *Proc. 1982 Annual Meeting, American Solar Energy Society*, Houston, TX. Boulder, CO: American Solar Energy Society.

Bourdeau, L. E. 1980. Study of two passive solar systems containing phase change material for thermal storage. *Proc. 5th National Passive Solar Conference*, Amherst, MA, October 19–26, 1980. Newark, DE: American Section of the International Solar Energy Society.

Buchberg, H. 1971. Sensitivity of room thermal response to inside radiation exchange and surface conductance. *Building Science* 6:133–149.

Burns, P. J., J. Nobe, and C. B. Winn. 1979. Thermal radiation calculations for a vertical cylindrical water wall. *Proc. 4th National Passive Solar Conference*, Kansas City, MO, October 3–5, 1979. Newark, DE: American Section of the International Solar Energy Society.

Butler, P., and K. M. Letherman. 1980. A criterion for the accuracy of modelling of transient heat conduction in plane slabs. *Building and Environment* 15:143–149.

Carroll, J. A., J. R. Clinton. 1980. A thermal network model of a passive solar house. *Proc. 5th National Passive Solar Conference*, Amherst, MA, October 19–26, 1980. Newark, DE: American Section of the International Solar Energy Society.

Carroll, J. A. 1980. An MRT method of computing radiant energy exchange in rooms. *Proc. Systems Simulation and Ecomonic Analysis*, San Diego, CA, January 23–25, 1980. Golden, CO: Solar Energy Research Institute, pp. 343–348; *see also* Appendix G of Sebald and Clinton (1980).

Carroll, J.A. 1981. A comparison of radiant interchange algorithms. *Proc. 3d Annual Systems Simulation and Economic Analysis/Solar Heating and Cooling Operational Results Conference.* New York: The American Society of Mechanical Engineers, pp. 399–407.

Carslaw, H. S., and J. C. Jaeger. 1959. *Conduction of Heat in Solids*. New York: Oxford University Press-Clarendon Press.

Chapman, J., P. J. Burns, and C. B. Winn, 1980. *Users Guide for the FREHEAT Program; The Passive Solar Heating and Cooling Program.* C00-30122-20. Ft. Collins, CO: Solar Energy Applications Laboratory, Colorado State University.

Chirlian, P. M. 1973. *Signals, Systems, and the Computer*. New York: Intext Educational Publishers.

Clark, G., and C. Allen. 1978. The estimation of atmospheric radiation for clear and cloudy skies. *Proc. 3d National Passive Solar Conference*, San Jose, CA, January 11–13, 1979. Newark, DE: American Section of the International Solar Energy Society.

Clinton, J. R. 1979. The SEA-PAS passive simulation program. *Proc. 4th National Passive Solar Conference*, Kansas City, MO, October 3–5, 1979. Newark, DE: American Section of the International Solar Energy Society.

Cole, R. J., and N. S. Sturrock. 1977. The convection heat exchange at the external surface of buildings. *Building and Environment* 12.

Cooper, K. W., and D. R. Tree. 1973. A re-evaluation of the average convection coefficient for flow past a wall. Paper presented at the ASHRAE Semiannual Meeting, Chicago.

Coutier, J. P., and E. A. Farber. 1982. Two applications of a numerical approach of heat transfer processes within rock beds. *Solar Energy* 29(6):451–462.

Croft, D. R., and D. G. Lilley. 1977. *Heat Transfer Calculations Using Finite Difference Equations.* London: Applied Science Publishers.

Delsante, A. E., A. N. Stokes, and P. J. Walsh. 1983. Application of Fourier transforms to periodic heat flow into the ground under a building. *Int. J. Heat Mass Transfer* 26:121.

Dietz, P. W. 1979. Effective thermal conductivity of packed beds. *Ind. Eng. Chem. Fundam.* 18(3).

Duffie, J. A., and W. A. Beckman. 1974. *Solar Energy Thermal Processes.* New York: Wiley.

Duffie, J. A., and W. A. Beckman. 1980. *Solar Engineering of Thermal Processes.* New York: Wiley.

Eckert, E. G. R., and R. M. Drake. 1959. *Heat and Mass Transfer.* New York: McGraw-Hill.

Emery, A. F., D. R. Heerwagon, C. J. Kippenhan, S. V. Stoltz, and G. B. Varey. 1978. The optimal energy design of structures by using the numerical simulation of the thermal response—with emphasis on the passive collection of solar energy. *Energy and Buildings* 1(4):367.

Emery, A. F., et al. 1981. Computation of radiation view factors for surfaces with obstructed views of each other. ASME, 81-HT-57. Paper presented at the National Heat Transfer Conference, Milwaukee, WI, 2–5 August, 1981.

Faultersack, J., and F. Loxsom. 1982. Heat transfer inside a passively cooled building. *Proc. 1982 Annual Meeting, American Solar Energy Society*, Houston, TX. Boulder, CO: American Solar Energy Society.

Fuchs, R. and J. F. McClelland. 1979. Passive solar heating of buildings using a transwall structure. *Solar Energy* 23(2).

Gadgil, A., F. Bauman, and R. Kammerud. 1982. Natural convection in passive solar buildings: experiments, analysis, and results. *Passive Solar Journal* 1(1).

Gerald, C. F., and P. O. Wheatley. 1984. *Applied Numerical Analysis.* 3d ed. Reading, MA: Addison-Wesley.

Goldstein, D. B. 1978. *Some Analytic Models of Passive Solar Building Performance: A Theoretical Approach to The Design of Energy Conserving Buildings.* LBL-7811. Berkeley, CA: Lawrence Berkeley Laboratory.

Goldstein, D. B., and M. Lokmanhekim. 1979. *A Simple Method for Computing the Dynamic Response of Passive Solar Buildings to Design Weather Conditions.* LBL-9787. Berkeley, CA: Lawrence Berkeley Laboratory.

Grimmer, D. P., J. D. Balcomb, and R. D. McFarland. 1977. The use of small passive-solar test boxes to model the thermal performance of passively solar-heated building designs. *Proc. Annual Meeting, American Section of the International Solar Energy Society*, Orlando, FL, June 6–10, 1977. Cape Canaveral, FL: American Section of the International Solar Energy Society.

Hayt, W. H., and J. E. Kemmerly. 1962. *Engineering Circuit Analysis.* New York: McGraw-Hill.

Hittle, D. C. 1977. *BLAST—The Building Loads Analysis and System Thermodynamics Program.* Vol. 1, *User's Manual.* Report CERL-TR-E-119. U.S. Army Construction Engineering Research Laboratory.

Holman, J. P., 1981. *Heat Transfer.* New York: McGraw-Hill.

Hughes, P. J., S. A. Klein, and D. J. Close. 1976. Packed bed thermal storage models for solar air heating and cooling systems. *J. Heat Transfer*, May 1976, pp. 336–338.

Hull, J. R., J. F. McClelland, L. Hodges, J. L. Huang, R. Fuchs, and D. A. Block. 1980. Effect of design parameter changes on the thermal performance of a transwall passive solar heating system. *Proc. 5th National Passive Solar Conference*, Amherst, MA, October 19–26, 1980. Newark, DE: American Section of the International Solar Energy Society.

Hutchinson, F. W. 1964. A rational re-evaluation of surface conductances for still air. Paper presented at the ASHRAE Semiannual Meeting, New Orleans, LA, January 27–29, 1964.

International Mathematical and Statistical Libraries Incorporated. 1984. *The IMSL Library*. Houston, TX: ISML.

Jeffreson, C. P. 1972. Prediction of breakthrough curves in packed beds. *AIChE Journal* 18(2).

Jirka, G. H., M. Watanabe, K. H. Octavio, C. F. Cerco, and D. R. F. Harleman. 1978. *Mathematical Predictive Models for Cooling Ponds and Lakes*. Part A: *Model Development and Design Considerations*. Report no. 238. Cambridge, MA: Ralph Parsons Laboratory for Water Resources and Hydrodynamics, MIT.

Jones, R. W. 1982. *Monitored Passive Solar Buildings*. LA-9098-MS. Los Alamos, NM: Los Alamos Scientific Laboratory.

Kamada, R. F., and R. G. Flocchini. 1984. Gaussian thermal flux model, part I: theory, part II: validation. *Solar Energy* 32(4).

Kammerud, R., E. Altmeyer, F. Bauman, A. Gadgil. 1982. Convection coefficients at building surfaces. *Proc. Passive and Hybrid Solar Energy Update*, Washington, DC, September 15–17, 1982. Washington, DC: U.S. Department of Energy, p. 106.

Katto, Y., and T. Masuoka. 1967. Criterion for the onset of convective flow in a fluid in a porous medium. *Int. J. Heat Mass Transfer* 10.

Kimura, Ken-Ichi. 1977. *Scientific Basis of Air-Conditioning*. London: Applied Science Publishers.

Kind, R. J., D. H. Gladstone, and A. D. Moizer. 1983. Convective heat losses from flat-plate solar collectors in turbulent winds. *J. of Solar Energy Engineering* 105: 80–85.

Kirkpatrick, A. T., and C. B. Winn. 1982. A frequency response technique for the analysis of the enclosure temperature in passive solar buildings. *Proc. 1982 Annual Meeting, American Solar Energy Society*, Houston, TX. Boulder, CO: American Solar Energy Society.

Klotz, P. S. 1977. Heating requirements of swimming pools. Engineering thesis, Department of Aeronautics and Astronautics, Stanford University, Palo Alto, CA.

Kohler, J. T., P. W. Sullivan. 1979. *TEANET User's Manual: A Numerical Thermal Network Algorithm for Simulating the Performance of Passive Systems on a TI-59 Programmable Calculator*. Harrisville, NH: Total Environmental Action.

Kohler, J. T., P. W. Sullivan, and C. J. Michal. 1980. Simulation of direct gain buildings with active rockbeds on a TI-59 programmable calculator using TEANET. *Proc. 5th National Passive Solar Conference*, Amherst, MA, October 19–26, 1980. Newark, DE: American Section of the International Solar Energy Society.

Kusuda, T. 1976. *NBSLD, the Computer Program for Heating and Cooling Loads in Buildings*. NBS Building Science Series, no. 69. Washington, DC: National Bureau of Standards.

Kusuda, T. 1980. *Review of Current Calculation Procedures for Building Energy Analysis*. NBSIR 80-2068. Washington, DC: National Engineering Laboratory, National Bureau of Standards.

Kusuda, T., and J. W. Bean. 1984. Simplified methods for determining seasonal heat loss from uninsulated slab-on-grade floors. *ASHRAE Transactions* 90(1).

Lipps, F. W. 1983. Geometric configuration factors for polygonal zones using Nusselt's unit sphere. *Solar Energy* 30(5).

Löf, G. O. G., and R. W. Hawley, 1940. Unsteady state heat transfer between air and loose solids. *Ind. Eng. Chem. Fundam.* 40.

Loxsom, F., J. Mosley, B. Kelly, and G. Clark. 1981. Measured ceiling heat transfer coefficients for a roof pond cooling system. *Proc. International Passive and Hybrid Cooling Conference*, Miami Beach, FL, November 11–13, 1981. Newark, DE: American Section of the International Solar Energy Society.

Mackey, C. O., and L. T. Wright. 1944. Periodic heat flow—homogeneous walls or roofs. *ASHVE Transactions* 50:293.

MacDonald, G. R., D. E. Claridge, and P. A. Oatman. 1985. A comparison of seven basement heat loss calculation methods suitable for variable-base degree-day calculations. *ASHRAE Transactions* 91.

Marlatt, W. P., K. A. Murray, and S. E. Squier. 1984. *Roof Pond Systems.* Report ETEC-83-6. Canoga Park, CA: Engineering Technology Engineering Center, Rockwell International. Available from NTIS.

Martin, M., and P. Berdahl. 1982. *Summary of Results of Spectral and Angular Sky Radiation Measurement Program.* LBL-14913. Berkeley, CA: Lawrence Berkeley Laboratory.

McAdams, W. A. 1954. *Heat Transmission.* New York: McGraw-Hill.

McFarland, R. D. 1978. *PASOLE: A General Simulation Program for Passive Solar Energy.* LA-743-MS. Los Alamos, NM: Los Alamos National Laboratory.

Metz, P. D. 1983. A simple computer program to model three-dimensional underground heat flow with realistic boundary conditions. *Solar Energy Engineering* 105.

Meixel, G. D., and T. P. Bligh. 1983. *Earth Contact Systems.* Final Report DOE Contract DE-AC03-80SF11508. Minneapolis, MN: Underground Space Center, University of Minnesota.

Miller, W. C., and T. R. Mancini. 1977. Numerical simulation of a solar heated and cooled house using roof ponds and movable insulation. *Proc. 1977 Annual Meeting, American Section of the International Solar Energy Society*, Orlando, FL, June 6–10, 1977. Cape Canaveral, FL: American Section of the International Solar Energy Society, pp. 11–21.

Miller, W. C. 1977. *Numerical Modeling of a Passive Solar House*, Master's Thesis, Department of Mechanical Engineering, New Mexico State University, Las Cruces.

Mitalis, G. P. 1965. An assessment of common assumptions in estimating cooling loads and space temperatures. *ASHRAE Transactions* 71(II).

Mitalis, G. P. 1982. *Basement Heat Loss Studies at DBR/NRC.* Ottawa: Division of Building Research, National Research Council, Canada.

Mitalis, G. P. 1983. Calculation of basement heat loss. *ASHRAE Transactions* 89(1B).

Mitalis, G. P. 1987. Calculation of below-grade residential heat loss: low-rise residential buildings. *ASHRAE Transactions* 93(1).

Mitchell, J. W. 1976. Heat transfer from spheres and other animal forms. *Biophysical Journal* 16:561.

Mumma, S. A., and W. C. Marvin. 1976. A method of simulating the performance of a pebble bed thermal energy storage and recovery system. ASME 76-HT-73. Paper presented at ASME-AICHE Heat Transfer Conference, St. Louis, August 1976.

Muncy, R. W. R. 1979. *Heat Transfer Calculations for Buildings.* London: Applied Science Publishers.

Monsen, W. A., S. A. Klein, W. A Beckman, and D. M. Utzinger. 1979. The resistance network design method for passive solar systems. *Proc. 4th National Passive Solar Conference,* Kansas City, MO, October 3–5, 1979. Newark, DE: American Section of the International Solar Energy Society.

Muncey, R. W. R., and J. W. Spencer. 1977. Heat flow into the ground under a house. Paper presented at the International Center for Heat and Mass Transfer, International Seminar on Heat Transfer in Buildings, Dubrovnik, Yugoslavia, August-September 1977.

Nansteel, M. W. 1982. *Natural Convection in Enclosures.* Ph.D. diss., University of California, Berkeley.

National Climatic Center. 1977. *Climatic Atlas of the United States.* Asheville, NC: NCC.

Niles, P. W. 1975. *Thermal Evaluation in Research Evaluation of a System of Natural Air Conditioning.* PB 243- 498/LK. Springfield, VA: National Technical Information Service.

Niles, P. W., K. L. Haggard, and H. R. Hay. 1976. Nocturnal cooling and solar heating with water ponds and movable insulation. *ASHRAE Transactions* 82(I).

Niles, P. W., C. H. Treat, and C. Rhombs. 1978. *Cooling Rate Equations. An Assessment of Passive Cooling Processes and Systems in the United States.* First Semi-Annual Technical Progress Report Contract EG-77-C-03-1600. Edited by E. Clark. Washington, DC: U.S. Department of Energy.

Niles, P. W. B. 1979. Graphs for direct gain house performance prediction. In *Passive Systems '78.* Edited by J. Cook and D. Prowler. Newark, DE: American Section of the International Solar Energy Society, pp. 76–81.

Niles, P. W., and K. Haggard. 1980. *Passive Solar Handbook.* Sacramento, CA: California Energy Commission.

Niles, P. W. 1981. *CALPAS-1.* Sacramento, CA: California Energy Commission.

Nottage, H. B., and G. V. Parmelee. 1854. Circuit analysis applied to load estimation. *ASHVE Transactions* 90, pp. 59–102.

Oppenheim, A. K. 1956. Radiation analysis by the network method. *ASME Transactions, Heat Transfer Division,* pp. 725–735.

Ozisik, M. N. 1968. *Boundary Value Problems of Heat Conduction.* Scranton, PA: International Textbook.

Palmiter, L., and T. Wheeling. 1983. *SERI-RES: Solar Energy Research Institute Residential Energy Simulator, Version 1.0.* Golden, CO: Solar Energy Research Institute.

Peikari, B. 1974. *Fundamentals of Network Analysis and Synthesis.* Englewood Cliffs, NJ: Prentice-Hall.

Perry, E. H., G. T. Cunningham, and S. Scesa. 1985. An analysis of heat losses through residential floor slabs. *ASHRAE Transactions* 91(2).

Persons, R. W., J. A. Duffie, and J.W. Mitchell. 1980. Comparison of measured and predicted rock bed thermal performance. *Solar Energy* 24(2): 199–201.

Reid, J. G. 1983. *Linear System Fundamentals.* New York: McGraw-Hill.

Riaz, K. 1978. Analytical analysis of packed-bed thermal storage systems. *Solar Energy* 21: 123–128.

Ruberg, K. 1979. Heat distribution by natural convection: modeling procedure for enclosed spaces. *Proc. 3d National Passive Solar Conference,* San Jose, CA, January 11–13, 1979. Newark, DE: American Section of the International Solar Energy Society.

Ryan, P. J., and D. R. F. Harleman. 1973. *An Analytical and Experimental Study of Transient Cooling Pond Behavior*. Report no. 161. Cambridge, MA: Ralph Parsons Laboratory for Water Resources and Hydrodynamics, MIT.

Sandia Laboratories. 1979. *Passive Solar Buildings*. SAND 79-0824. Albuquerque, NM: Sandia Laboratories.

Schumann, T. E. W. 1929. Heat transfer: a liquid flowing through a porous prism. *J. Franklin Institute* 208.

Sebald, A. V., J. R. Clinton, and F. Langenbacker. 1979. Performance effects of Trombe wall control strategies. *Solar Energy* 23(6):479–487.

Sebald, A. V., and J. R. Clinton. 1980. *Impact of Controls in Passive Solar Heating and Cooling of Buildings*. Final Report US DOE Contract DE AC04-79AL10891. San Diego, CA: Energy Center, University of California at San Diego.

Sebald, A. V. 1981. Efficient simulation of large, controlled passive systems: forward differencing in thermal networks. *Solar Energy* 34(3):221–230.

Shah, M. M., 1981. Estimation of evaporation from horizontal surfaces. *ASHRAE Transactions* 87.

Shen, L. S., and J. W. Ramsey. 1983. A simplified thermal analysis of earth-sheltered buildings using a Fourier-series boundary method. *ASHRAE Transactions* 89(1B).

Shipp, P. H., and T. B. Brodernick. 1984. *Comparison of Annual Heating Loads for Various Basement Wall Insulation Strategies Using Transient and Steady State Models*. Granville, OH: Owens Corning Fiberglass Corporation.

Solar Energy Laboratory, University of Wisconsin. 1981. *TRNSYS—A Transient Simulation Program*. Engineering Exp. Station Rep. 38. Madison, WI: SEL.

Solomon, A. D. 1979. Melt time and heat flux for a simple PCM body. *Solar Energy* 22(3):251–257.

Sowell, E. F., and R. L. Curry. 1980. A convolution model of rock bed thermal storage units. *Solar Energy* 24(5).

Sparrow, E. M., and R. D. Cess. 1970. *Radiation Heat Transfer*. Belmont, CA: Brooks/Cole Publishing Co.

Sparrow, E. M., J. W. Nelson, and S. C. Lau. 1981. Wind related heat transfer coefficients for leeward-facing solar collectors. *ASHRAE Transactions* 87:70–70.

Sparrow, E. M., J. S. Nelson, and W. Q. Tao. 1982. Effect of leeward orientation, adiabatic framing surfaces, and eaves on solar-collector-related heat transfer coefficients. *Solar Energy* 29(1).

Subbarao, K., and J. V. Anderson. 1982a. *A Frequency-Domain Approach to Passive Building Energy Analysis*. SERI/TR-254-1544. Golden, CO: Solar Energy Research Institute.

Subbarao, K., and J. V. Anderson. 1982b. A Fourier transform approach to building simulation with thermostatic controls. *Solar Energy Engineering* 107:58.

Subbarao, K., J. V. Anderson, and J. M. Connolly. 1982. *Prediction of Thermal Performance of Passive Buildings Using Frequency-Domain Methods*. SERI/TR.-RR-254-1767. Golden, CO: Solar Energy Research Institute.

Subbarao, K. 1982. *The Dynamic Response of Thermal Massses and Their Interactions Using Vector Diagrams*. SERI/TP-254-1632. Golden, CO: Solar Energy Research Institute.

Sullivan, R., J. Bull, P. Davis, S. Nozaki, Z. Cumali, and G. Meixel. 1985. Description of an earth-contact modeling capability in the DOE 2.1B energy analysis program. *ASHRAE Transactions* 91(1A).

Sweat, M. E., and J. J. Carroll. 1983. On the use of spectral radiance models to obtain irradiances on surfaces of arbitrary orientations. *Solar Energy* 30(4).

Tavana, M., H. Akbari, T. Borgers, and R. Kammerud. 1980. A simulation model for the performance analysis of roof pond systems for heating and cooling. *Proc. Annual Meeting of the American Section of the International Solar Energy Society*, Phoenix, AZ, June 2–6, 1980. Newark, DE: American Section of the International Solar Energy Society.

Van der Mersch, P. L., P. J. Burns, and C. B. Winn. 1980. Solar radiation pattern analysis and effects for performance prediction of cylinder water walls. *Proc. 5th National Passive Solar Conference.* Edited by J. Hayes. Newark, DE: American Section of the International Solar Energy Society, pp. 224–228.

Vemuri, V. 1981. *Digital Computer Treatment of Partial Differential Equations.* Englewood Cliffs, NJ: Prentice-Hall.

von Fuchs, G. F. 1981. Validation of a rock bed thermal energy storage model. Paper Presented at the American Society of Mechanical Engineers' Annual System Simulation Conference, Reno, NV.

Walton, G. N., 1983. *Thermal Analysis Research Program Reference Manual.* NBSIR 83-2655. Washington, DC: National Bureau of Standards.

Weber, D. D., and R. J. Kearney, 1980. Natural convection heat transfer through an aperture in passive solar heated buildings. *Proc. 5th National Passive Solar Conference*, Amherst, MA, October 19–26, 1980. Newark, DE: American Section of the International Solar Energy Society, pp. 1037–1041.

Wei, J., H. Sigworth, C. Lederer, and A. H. Rosenfeld. 1979. *A Computer Program for Evaluating Swimming Pool Heat Conservation.* LBL 9388. Berkeley, CA: Lawrence Berkeley Laboratory.

White, M. D., P. J. Burns, and C. B. Winn. 1980. Optimization of numerical parameters and solution procedures for a mass wall glazing system. *Proc. 5th National Passive Solar Conference*, Amherst, MA, October 19–26, 1980. Newark, DE: American Section of the International Solar Energy Society, p. 146.

White, H. C., and S. A. Korpela. 1979. On the calculaton of the temperature distribution in a packed bed for solar energy applications. *Solar Energy* 23(2).

Willcox, T. N., et al. 1954. Analogue computer analysis of residential cooling loads. *ASHRAE Transactions*, pp. 505–524.

Yard, D. C., M. Morton-Gibson, and J. W. Mitchell, 1984. Simplified dimensionless relations for heat loss from basements. *ASHRAE Transactions* 90(1A).

4 Simplified Methods

G. F. Jones and William O. Wray

4.1 Scope of Chapter

4.1.1 Definition of Simplified Method

Generally, the performance evaluation of a passive heating system is based on a validated computer simulation (see Niles, chapter 3 of this volume) of the heat flows among the various parts of a building and its external environment. Because of the complexity of simulation and the level of technical support needed for it, simulations are not always practical. Simplified methods of analysis are developed to provide predictions of passive building performance without recourse to computer simulation. A simplified method is often, though not always, obtained from the results of a computer simulation by a correlation analysis of the kind described in section 4.2.1. The simplified method thus derived will have a restricted precision or limited range of applicability relative to the original simulation.

4.1.2 Distinction among Simplified Methods, Guidelines, and Design Tools

Simulation and other evaluation methods, such as simplified methods derived from simulation, produce building performance results based on detailed information about the building prescribed at the beginning of the method. Because the required building data is not available until the design is well advanced, these evaluation methods may be thought of as "backward-looking" methods.

Conversely, guidelines are design aids that are available at the beginning of the design procedure, and for this reason they may be thought of in a "forward-looking" manner. Because guidelines tend to be extracted from other more general methods (such as computer simulation) for specific cases, they are usually more restrictive in their range of applicability, and care should be taken when applying them.

The term design tools generally encompasses all design methods and guidelines. When it is thought to be important, the design tool described should be classified according to whether it is a backward- or forward-looking procedure. (See also Reynolds, chapter 10 of this volume.) A

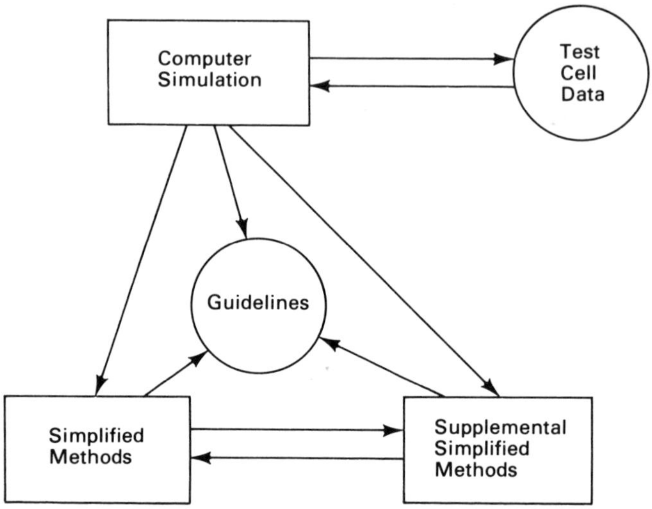

Figure 4.1
Relationship between analysis techniques.

schematic diagram depicting the relationship between analysis techniques is shown in figure 4.1.

4.2 Correlation Methods

4.2.1 Definition of Correlation Analysis

A correlation analysis is a procedure whereby a relationship among the variables for the problem at hand is determined numerically rather than from first principles. A correlation analysis is performed when this relationship is needed and when no simple analytical forms of the relationship are already available. For example, suppose we wished to estimate the inside surface temperature of a Trombe wall subject to a given time-dependent solar flux on its outside surface. This problem may be solved analytically (Arpaci 1966), but the series solution is cumbersome and may not be desirable. Instead, a computer simulation of the problem is performed and the inside temperature of the Trombe wall determined. We next assume a functional form for a correlated inside wall temperature, which contains specific independent variables describing external conditions (such as absorbed solar flux) and wall characteristics (such as thickness and density). At this point, the functional form also contains unknown coefficients. Finally, by comparing the calculated with the correlated temperatures, we choose such values of the coefficients that the difference between the two is minimized.

Specifically, in correlation analysis for passive systems, we relate a useful performance variable (indoor temperature or backup heat, for example) to building and system characteristics, and weather parameters.

4.2.2 Degree-Day Method for Heat Loss Estimation

Means of estimating heat loss are central to simplified methods of estimating room temperature and backup heat. The degree-day method, though a simplified method itself, is presented here becasue it is essential to the other methods.

The number of degree days DD_b is the difference between the base temperature T_b of the building and the mean ambient temperature T_m for that day. The base temperature is a fixed temperature, traditionally 65°F (18°C), although other base temperatures may be used. The mean temperature is traditionally defined as

$$T_m = (T_{max} + T_{min})/2, \tag{4.1}$$

where T_{min} and T_{max} are the minimum and maximum temperatures that occur during the day. If the mean ambient temperature is above the base temperature, no degree-days for heating are accrued for the day. Otherwise, degree-days are accrued for each day, i.e., the number of degree-days for a month is the sum of degree-days for each day of the month.

The building heat load over a prescribed period is the product of the building total load coefficient (TLC) and the number of degree-days for that time period. TLC is defined as the rate of heat loss through the entire building envelope, including the solar glazing, per day per unit of temperature difference between the building and the outside. Thus, in the absence of solar gains,

$$Q_{aux} + Q_{int} = \mathrm{TLC}(T_{set} - T_m), \tag{4.2}$$

where Q_{aux} is the daily auxiliary heat required to maintain the heating thermostat setpoint, T_{set}, and Q_{int} is the daily internal heat gain in the building due to appliances, lights, and people. A typical residential value of Q_{int} is 20,000 Btu/day (21 MJ/day) per occupant.

The base temperature of the building may be modified to account for daily internal heat gains by setting

$$T_b = T_{set} - Q_{int}/\mathrm{TLC}, \tag{4.3}$$

so that equation (4.2) becomes

$$Q_{aux} = \mathrm{TLC}(T_b - T_m). \tag{4.4}$$

The base temperature T_b is thus the thermostat setpoint modified by subtracting the daily internal heat gains divided by the building total load coefficient. By modifying the base temperature in this way, credit is taken for a reduced building load by the partially offsetting effect of internal heat gains.

Equation (4.4) may be written in terms of degree-days as

$$Q_{aux} = \mathrm{TLC}\ DD_b. \tag{4.5}$$

Equation (4.5) applies to a building where there is no solar gain. For a building with solar gain,

$$Q_{aux} = Q_{net} - Q_{sol}, \tag{4.6}$$

where

$$Q_{net} = \mathrm{NLC}\ DD_b. \tag{4.7}$$

The net reference load, Q_{net}, is the heat loss from the nonsolar parts of the building and Q_{sol} is the solar savings, the useful heat provided by the solar wall in excess of that required to offset the solar wall losses. In equation (4.7), NLC is the building net load coefficient based on the rate of heat loss from the nonsolar parts of the building.

By definition, the solar savings fraction, SSF, is the ratio of the solar savings to the net reference load:

$$\mathrm{SSF} = Q_{sol}/Q_{net}. \tag{4.8}$$

It follows that

$$Q_{aux} = (1 - \mathrm{SSF})\mathrm{NLC}\ DD_b. \tag{4.9}$$

Thom (1954a, 1954b) developed a method for evaluating the degree-days below a prescribed base temperature based on the monthly mean temperature and the standard deviation in monthly mean temperature. The expression for degree days is

$$DD_b = M(T_b - T_A + L\sigma_m\sqrt{M}), \tag{4.10}$$

where M is the number of days in a month, T_A is the monthly mean ambient temperature, σ_m is the standard deviation in monthly mean temperature, and

$$L = \begin{cases} 0.34 \exp(-4.7\ \mathrm{h}) - 0.15 \exp(-7.8\ \mathrm{h})(h \geq 0) \\ 0.34 \exp(4.7\ \mathrm{h}) - 0.15 \exp(7.8\ \mathrm{h}) - h(h < 0) \end{cases}, \tag{4.11}$$

where

$$h = (T_b - T_A)/(\sigma_m\sqrt{2}). \tag{4.12}$$

The quantities T_A and σ_m have been tabulated for 209 U.S. and 14 Canadian cities in appendix 3 of Balcomb et al. (1984). For those cases where the standard deviation is unknown, Erbs, Beckman, and Klein (1981) provide a simple formula to estimate it:

$$\sigma_m = \sigma_o - 0.0337\ T_A, \tag{4.13}$$

where σ_o is 4.79°F (2.66°C).

4.2.3 Solar Load Ratio Method

The solar load ratio (SLR) method, developed by Balcomb and McFarland (1978), provides month-by-month estimates of building backup heat. In the SLR method, the monthly solar savings fraction (SSF) is correlated with the monthly solar load ratio (SLR), which is generically the ratio of the monthly solar gain to the monthly building load. The exact definition of SLR has varied somewhat since it was introduced, but the current definition is (Balcomb et al. 1984, Balcomb and Wray 1988):

$$SLR = (S/DD_b - LCR_s H)/LCR, \tag{4.14}$$

where S is the monthly rate of solar radiation absorbed in the solar space per unit of solar window area projected on a vertical plane (A_p) and H is a correlation parameter. The load collector ratio (LCR) is the ratio NLC/A_p. LCR_s is the load collector ratio of the solar space (see Balcomb et al. 1984).

The correlation for solar savings fraction may be written in the form

$$SSF = 1 - C \exp(-D\,SLR). \tag{4.15}$$

The correlation parameters C, D, and H are system dependent, and extensive tables of these parameters have been compiled for specific reference system designs (Balcomb et al. 1984, Balcomb and Wray 1988). The auxiliary energy required for the building that is the object of the method is calculated from SSF as determined above, using equation (4.9).

Extensive comparisons of SLR method predictions and measured energy consumption have been reported (see Frey, chapter 9 of this volume).

4.2.4 Fast Solar Load Ratio Method

Based on ideas of Subbarao (1982), the fast solar load ratio (FSLR) method was developed by Wray et al. (1982) for the U.S. Navy (Wray, Biehl, and Kosiewicz 1983, Wray and Peck 1987, 1988). This method combines two simplifications relative to the original SLR method: (1) the monthly solar heating fraction (SHF) is correlated with a scaled monthly solar load ratio (SLR*), where the scaling factor rather than the correlation form contains the system dependence, and (2) the annual solar heating fraction is correlated with the smallest monthly solar heating fraction so that annual auxiliary heat may be estimated from just a single month's calculation. Thus, the annual auxiliary heat requirements are obtained directly from a simple expression involving the solar load ratio for a single month.

Instead of a correlation for SSF, the method uses a correlation for the annual solar heating fraction:

$$SHF_a = [1 - e^{-SLR_M^*}][1 + ae^{-SLR_M^*}], \tag{4.16}$$

where SHF_a is defined by

$$SHF_a = 1 - Q_{aux,a}/\text{TLC}\ DD_{b,a}, \tag{4.17}$$

and SLR_M^* is a scaled solar load ratio

$$SLR_M^* = F\ SLR_M, \tag{4.18}$$

where SLR_M is the minimum monthly solar load ratio for the year and F is a system-dependent scale factor. In equation (4.16), a is a location-dependent correlation parameter. The subscript a in equation (4.17) refers to annual values. Values of the correlation parameter, a, have been tabulated for 209 U.S. cities, and the scale factor, F, for 109 reference designs (Wray, Biehl, and Kosiewicz 1983, Wray and Peck 1987, 1988).

A graph of SHF as a function of the scaled monthly solar load ratio (SLR*) is shown in figure 4.2.

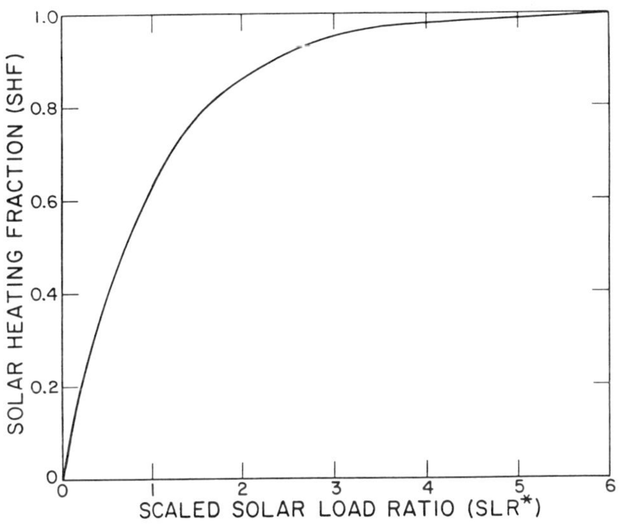

Figure 4.2
Solar heating fraction vs. scaled solar load ratio. Source: Wray et al. 1982.

4.2.5 Unutilizability Method

The unutilizability method has its roots in the utilizability method for active solar heating systems developed by Whillier (1953) and later generalized by Liu and Jordan (1963). Improvements to the method were made by Klein (1978) and Theilacker and Klein (1980). The concept of unutilizability was introduced by Monsen and Klein (1980) and applied to direct gain systems and later used by Monsen, Klein, and Beckman (1981) for thermal storage walls. Klein, Monsen, and Beckman (1981) presented a simplified version of this procedure using tabulated weather data and values of the correlation parameter, and Theilacker, Klein, and Beckman (1981) extended the method to apply to south-facing vertical surfaces shaded by an overhang.

The method makes use of a correlation between the monthly SSF and three parameters: the fraction of solar radiation incident on the solar aperture surface that has an intensity greater than that required to offset the solar window heat loss to the outside; the solar load ratio; and the ratio of the maximum monthly energy storage within the building to the maximum monthly solar heat transferred from the building.

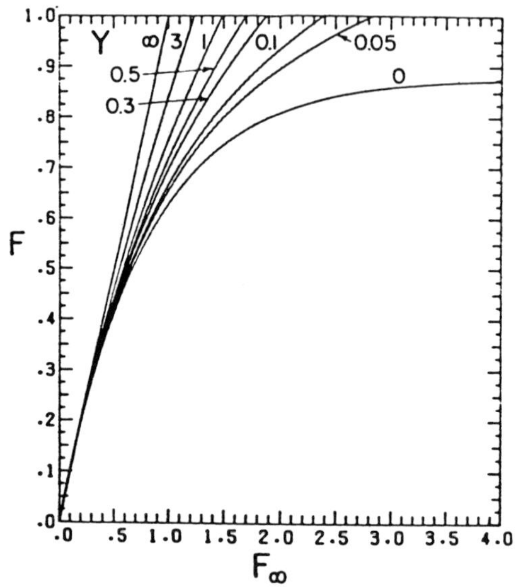

Figure 4.3
Graph of unutilizability correlation for thermal storage walls. Source: Monsen et al. 1981.

A graph of the unutilizability correlation for thermal storage walls is shown in figure 4.3, where SHF is the monthly solar heating fraction (defined as in equation (4.17) written for a month) and Y is a correlating parameter. Y is the ratio of the heat storage capacity of the building and solar wall to the energy that is rejected from a building having zero storage capacity. The subscripts ∞ and o refer to the cases of infinite and zero building heat storage capacity, respectively.

Figure 4.3 is used to estimate the monthly solar heating fraction SSF by first estimating monthly F from conventional unutilizability considerations (Monsen, Klein, and Beckman 1981). The correlating parameter Y is next entered on the graph and the value of SSF determined.

A nomograph for the unutilizability correlation for direct-gain systems is given in figure 4.4. In this figure, z, v, and x are the unutilizability, the storage-ventilation parameter, and the solar load ratio, respectively, expressions for which are given in Klein, Monsen, and Beckman (1981). Once values for these parameters are calculated, the monthly solar savings fraction, SSF, is determined directly from figure 4.4.

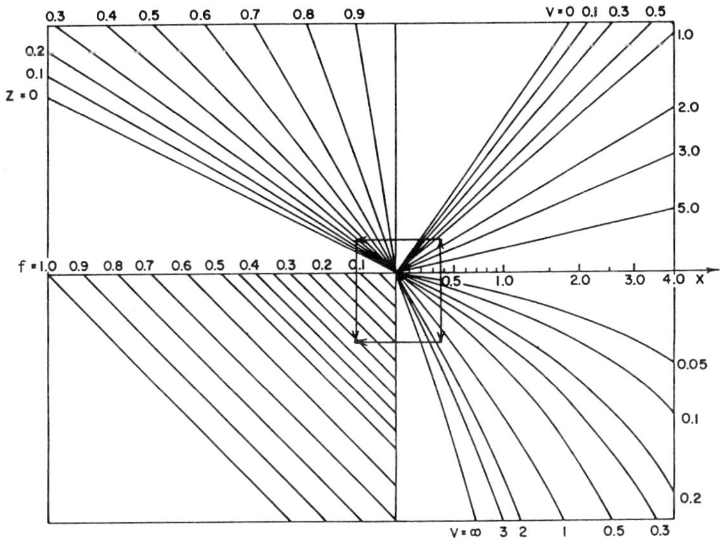

Figure 4.4
Nomograph for the unutilizability correlation for direct-gain systems. Source: Klein et al. 1981.

4.3 Supplemental Simplified Methods

4.3.1 Definition of Supplemental Simplified Methods

A supplemental simplified method is a procedure that is either derived from or directly supports the use of other simplified methods. In the cases selected below, the supplementary methods are related to the SLR method. The load collector ratio (LCR) method is derived from the SLR method and, thus, lacks the original generality but is much quicker in application. The equivalent constant thermostat setpoint method allows a night thermostat setback to be accounted for in the SLR method.

4.3.2 Load Collector Ratio Method

The load collector ratio (LCR) method is based on the use of LCR tables calculated from the SLR correlation equations. The most extensive tables presently available are in appendix 1 of *Passive Solar Heating Analysis* (Balcomb et al. 1984). The tables contain the values of annual SSF associated with a particular set of eight LCR values that span the range of practical interest and for each of 94 reference designs operating at base temperatures of either 55° or 65°F (13° or 18°C). There is a table for each of the listed 223 locations in the United States and Canada. By use of the LCR tables, the annual SSF is easily found, and the annual auxiliary heat requirement of the building can be calculated from equation (4.9).

Because the LCR method yields annual performance results directly from tabulated values, it is much quicker in application than the monthly SLR method. However, the speed of application is obtained at the expense of some loss of generality. In particular, the system definitions and the locations are limited to those in the tables.

4.3.3 Equivalent Constant Thermostat Setpoint

The rate at which a building responds to a change in thermostat setting depends on the amount of the thermal storage mass in the building and on the rate of heat loss (Wray and Kosiewicz 1984). In order to quantify these relationships, a time constant, τ, is defined as

$$\tau = \frac{24\,DHC}{TLC}, \tag{4.19}$$

where DHC is the diurnal heat capacity (see section 4.4.2) and TLC is the building total load coefficient. Note that large DHCs imply large time constants, whereas large TLCs imply small time constants. Buildings with small time constants respond to changes in thermostat setting more rapidly than buildings with large time constants.

If a nighttime setback strategy is employed in a building, the average thermostat setting over a period, P, of 24 hours is given by

$$T_{ave} = T_1(h_1/P + T_2(h_2/P), \tag{4.20}$$

where T_1 and T_2 are the daytime and nighttime settings, respectively. The duration of the daytime setting is h_1 (hours) and the setback period is h_2 (hours). To account for the effect of the setback on auxiliary heat consumption, one can perform an SLR analysis, using an equivalent constant thermostat setpoint given by the following equation:

$$T_e = T_1 - e^{-0.1\,\tau/P}(T_1 - T_{ave}). \tag{4.21}$$

To illustrate the usefulness of equation (4.21), consider the consequences of ignoring the nighttime setback in an FSLR analysis. The setback is

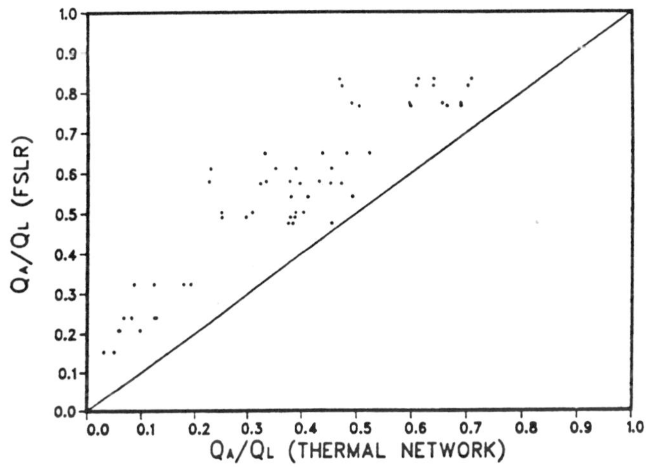

Figure 4.5
Heat to load ratio from correlation vs. heat to load ratio from thermal network calculation—no correction for thermostat setpoint. Compare with figure 4.7. Source: Wray and Kosiewicz 1984.

ignored by taking T_1 to be the constant thermostat setting, which is equivalent to employing an infinite time constant in equation (4.21). The consequences of such an approach are shown in figure 4.5, where the ratio of auxiliary heat to building load is plotted against results obtained from thermal-network calculations that modeled a variety of building load and control-strategy parameters. All buildings considered were direct gain and the following parameter values were employed:

Cities: Albuquerque, Madison

LCR: 24, 72 Btu/°F day ft^2 (136, 408 Wh/°C day m^2)

Thermal storage-mass thickness: 4 in. (10 cm)

Mass-area-to-glazing-area ratio: 3, 9

Number of glazings: 2

Night insulation thermal resistance: 0, 9 ft^2 °F h/Btu (1.6 m^2 °C/W)

T_1: 65°F (18.3°C)

Setback: 10, 20°F (5.6, 11.1°C)

h_2: 6, 13 h

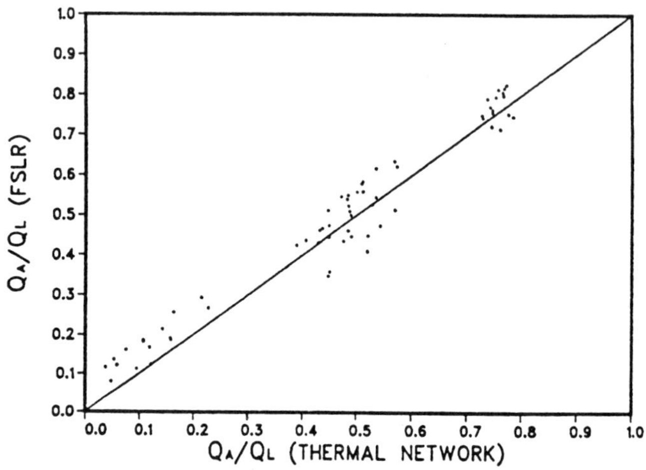

Figure 4.6
Heat to load ratio from correlation vs. heat to load ratio from thermal network calculation—thermostat setting corrected for setback. Source: Wray and Kosiewicz 1984.

The variations listed above yield a total of 64 combinations. Clearly, failing to account for nighttime setback leads to a large systematic error in which the auxiliary heat requirements are overestimated. Note that the data also exhibit considerable scatter.

The effect of using T_e in the same set of calculations is shown in figure 4.6; the systematic error and the scatter or random error are both greatly reduced.

4.4 Other Methods

4.4.1 Optimum Mix of Conservation and Solar

4.4.1.1 General Results Balcomb (1979, 1980) has developed a methodology for determining the optimum allocation of resources for conservation and passive solar strategies, assuming the initial investment is limited. The general results are

$$NLC_o = \sqrt{b'LCR/(a'R)}, \tag{4.22}$$

$$A_o = NLC_o/LCR, \tag{4.23}$$

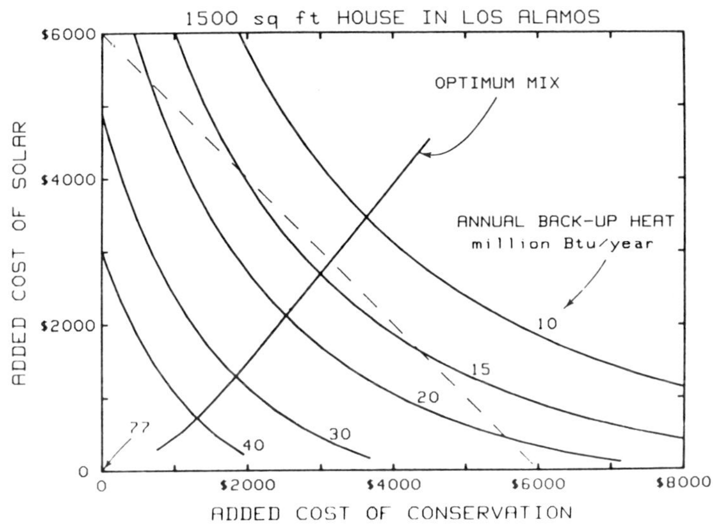

Figure 4.7
Performance map showing the energy savings expected for different initial expenditures. Source: Balcomb et al. 1984.

where

$$R = 1 + LCR(1 - SSF)/(d(SSF)/d(LCR^{-1})). \tag{4.24}$$

In equations (4.22) and (4.14), NLC_o and A_o are the optimized building net load coefficient and solar window projected area, respectively, and a' and b' are cost parameters associated with the solar system and conservation, respectively (Balcomb 1980). Equations (4.22) and (4.23) define the locus of points that represent an optimum mix between conservation and solar strategies. Results from these equations are presented in terms of performance maps showing the energy savings expected for different initial expenditures. An example of such a performance map for a hypothetical building in Los Alamos, New Mexico, is shown in figure 4.7.

4.4.1.2 Passive Solar and Conservation Guidelines Balcomb (1983a, 1983b, 1986) and Balcomb et al. (1984) present a set of guidelines for selecting passive solar aperture area and conservation levels (building infiltration levels and insulation R-values). The guidelines are based on the particular optimum mix of conservation and passive solar strategies that

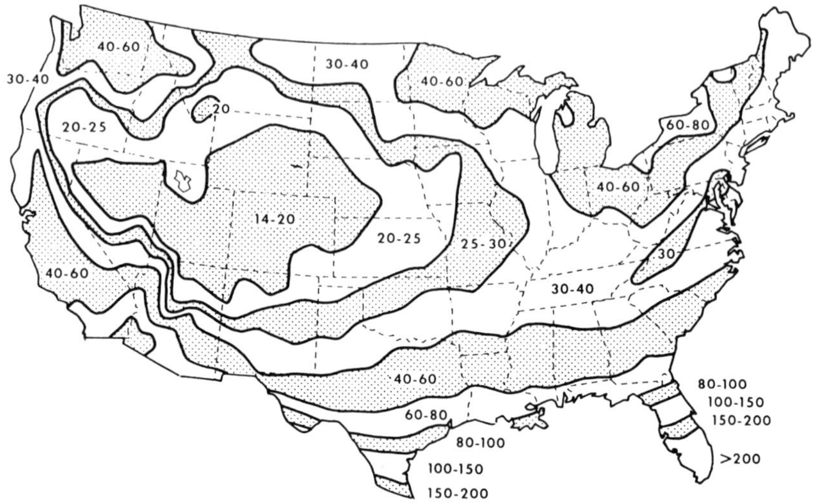

Figure 4.8
Contour map showing regional variations in optimum mix. Source: Balcomb et al. 1984.

also entails a minimum life-cycle cost of the building and heating fuel. A contour map showing regional variations in the optimum mix is shown in figure 4.8.

4.4.2 Diurnal Heat Capacity (DHC) Method

The diurnal heat capacity (*dhc*) of a material is the daily amount of heat, per unit of surface area and per degree of surface temperature swing, that is stored and then given back to the room air during a 24-hour period (Balcomb et al. 1980). The total diurnal heat capacity (DHC) in a passive solar building, $\sum dhc_i A_i$, provides a convenient measure of the daily temperature swings that might be expected in a direct-gain zone. Indeed,

$$\Delta T(\text{swing}) = 0.61\ Q_s/DHC, \tag{4.25}$$

where Q_s is the daily total solar energy absorbed in the zone (Balcomb and Wray 1988).

For a homogeneous layer,

$$dhc = F_1 s, \tag{4.26}$$

where

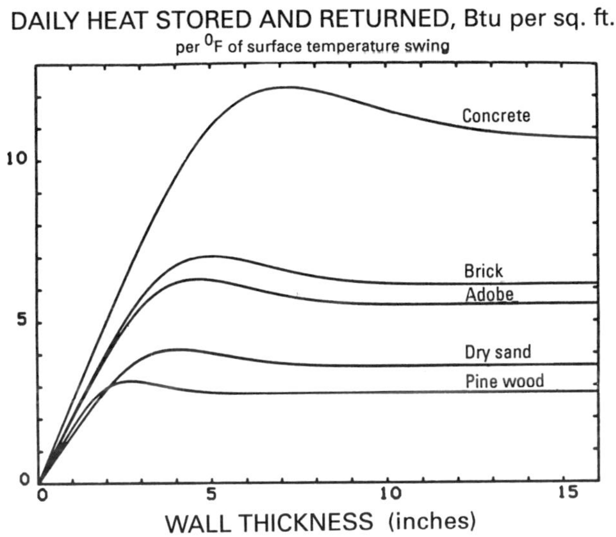

Figure 4.9
Diurnal heat capacity vs. thickness for various storage media. Source: Balcomb et al. 1980.

$$s = \sqrt{Pk\rho c/2\pi}, \tag{4.27}$$

$$F_1 = \sqrt{(\cosh 2x - \cos 2x)/(\cosh 2x + \cos 2x)}, \tag{4.28}$$

$$x = L\sqrt{\pi\rho c/Pk}, \tag{4.29}$$

and k is the thermal conductivity (Btu/ft h °F, W/m °C), ρ is the density (lbm/ft^3, kg/m^3), c is the heat capacity (Btu/lbm °F, Wh/kg °C), P is the periodicity (24 h), and L is the material thickness (ft, m).

The *dhc* for several building materials is plotted as a function of wall thickness in figure 4.9. Note that the higher density materials have larger values for *dhc* for comparable thickness. Also note that for each material, *dhc* reaches a maximum at some thickness and then asymptotically approaches a reduced value; this behavior reflects the fact that if a material layer becomes too thick, some of the heat transferred to the surface will be lost to the interior rather than returned to the room during a 24-hour period.

A method for calculating *dhc* for multi layer materials is given in Balcomb and Wray (1988).

References

Arpaci, V. S. 1966. *Conduction Heat Transfer.* Menlo Park, CA: Addison-Wesley, chap. 6.

Balcomb, J. D. 1980. Conservation and solar: working together. *Proc. 5th National Passive Solar Conference,* Amherst, MA, October 19–26, 1980. Newark, DE: American Section of the International Solar Energy Society, pp. 44–50.

Balcomb, J. D. 1980. Optimum mix of conservation and solar energy in building design. *Proc. 1980 Annual Meeting of the American Section of the International Solar Energy Society,* Phoenix, AZ, June 2–6, 1980. Newark, DE: American Section of the International Solar Energy Society, pp. 1202–1206.

Balcomb, J. D. 1983a. *Guidelines for Conservation Levels and for Sizing Passive Solar Collection Area.* LA-UR-83-359. Los Alamos, NM: Los Alamos National Laboratory. Paper presented at the Milford Event, Milford, PA, February 19–21, 1983.)

Balcomb, J. D. 1983b. Conservation and solar guidelines. *Proc. 8th National Passive Solar Conference,* Santa Fe, NM, September 7–9, 1983. Boulder, CO: American Solar Energy Society, pp. 117–122.

Balcomb, J. D. 1986. Conservation and solar guidelines. *Passive Solar Journal* 3(3): 221–248.

Balcomb, J. D., and R. D. McFarland. 1978. A simple empirical method for estimating the performance of a passive solar heated building of the thermal storage wall type. *Proc. 2d National Passive Solar Conference,* Philadelphia, PA, March 16–18, 1978. Newark, DE: American Section of the International Solar Energy Society, pp. 377–389.

Balcomb, J. D., D. Barley, R. McFarland, J. Perry, W. Wray, and S. Noll. 1980. *Passive Solar Design Handbook.* Vol. 2: *Passive Solar Design Analysis.* DOE/CS-0127/2. Washington, DC: U.S. Department of Energy, pp. G1–G6.

Balcomb, J. D., R. W. Jones, R. D. McFarland, and W. O. Wray. 1984. *Passive Solar Heating Analysis: A Design Manual.* Atlanta, GA: ASHRAE.

Balcomb, J. D., and W. O. Wray, 1988. *Passive Solar Heating Analysis. Supplement One: Thermal Mass Effects and Additional SLR Correlations.* Atlanta, GA: ASHRAE.

Erbs, D. G., W. A. Beckman, and S. A. Klein. 1981. Degree-days for variable base temperatures. *Proc. 6th National Passive Solar Conference,* Portland, OR, September 8–12, 1981. Newark, DE: American Section of the International Solar Energy Society, pp. 387–391.

Klein, S. A. 1978. Calculation of flat plate collector utilizability. *Solar Energy* 21: 393–402.

Klein, S. A., W. A. Monsen, and W.A. Beckman. 1981. Tabular data for the unutilizability passive solar design method. *Proc. 6th National Passive Solar Conference,* Portland, OR, September 8–12, 1981. Newark, DE: American Section of the International Solar Energy Society, pp. 382–332.

Liu, B. Y. H., and R. C. Jordan. 1963. The long-term average performance of flat-plate solar energy collectors. *Solar Energy* 7: 53.

Monsen, W. A. and S. A. Klein, 1980. Prediction of direct gain solar heating system performance. *Proc. Annual Meeting, American Section of the International Solar Energy Society,* Phoenix, AZ, June 2–6, 1980. Newark, DE: American Section of the International Solar Energy Society, pp. 865–869.

Monsen, W. A., S. A. Klein, and W. A. Beckman, 1981. The unutilizability design method for collector-storage walls. *Proc. Annual Meeting, American Section of the International Solar Energy Society,* Philadelphia, PA, May 27–30, 1981, Newark, DE: American Section of the International Solar Energy Society, pp. 862–866.

Subbarao, K. 1982. *Scaling Relations for the Performance of Passive Systems.* SERI-TP-254-1436. Golden, CO: Solar Energy Research Institute.

Theilacker, J. C., and S. A. Klein. 1980. Improvements in the utilizability relationships. *Proc. Annual Meeting, American Section of the International Solar Energy Society*, Phoenix, AZ, June 2–6, 1980. Newark, DE: American Section of the International Solar Energy Society, pp. 271–275.

Theilacker, J. C., S. A. Klein, and W. A. Beckman. 1981. Solar radiation utilizability for south-facing vertical surfaces shaded by an overhang. *Proc. Annual Meeting, American Section of the International Solar Energy Society*, Philadelphia, PA, May 27–30, 1981, Newark, DE: American Section of the International Solar Energy Society, pp. 853–856.

Thom, H. C. S. 1954a. The rational relationship between heating degree days and temperature. *Monthly Weather Review* 82:1–6.

Thom, H. C. S. 1954b. Normal degree days below any base. *Monthly Weather Review* 82:111–115.

Whillier, A. 1953. Solar energy collection and its utilization for house heating. Ph.D. diss., Massachusetts Institute of Technology.

Wray, W. O., and C. E. Kosiewicz, 1984. Including the effect of control strategy in solar load ratio calculations. *Proc. Annual Meeting, American Solar Energy Society*, Anaheim, CA, June 5–9, 1984. Boulder, CO: American Solar Energy Society, pp. 395–399.

Wray, W. O., F. A. Biehl, and C. E. Kosiewicz. 1983. *Passive Solar Design Manual for Naval Installations.* LA-UR-83-2236. Los Alamos, NM: Los Alamos National Laboratory.

Wray, W. O., F. A. Biehl, C. E. Kosiewicz, C. E. Miles, and E. R. Durlak, 1982. A passive solar design manual for the United States Navy. *Proc. 7th National Passive Solar Conference*, Knoxville, TN, August 30–September 1, 1982. Newark, DE: American Section of the International Solar Energy Society, pp. 183–188.

Wray, W. O., and C. E. Peck. 1987. *Military Handbook: Design Procedures for Passive Solar Buildings.* MIL-HDBK-1003/19. Washington, DC: U.S. Department of Defence.

Wray, W.O. and C.E. Peck, 1988. *Passive Solar Design Procedures for Naval Installations.* LA-11250-M. Los Alamos, NM: Los Alamos National Laboratory.

5 Materials and Components

Timothy E. Johnson

5.1 Introduction

Energy-conscious buildings conserve energy because they are designed well. Although architecture plays a pivotal role in how well buildings perform, the designer's choice of components and materials also affects occupant comfort and building operating costs. Most of the building's thermal loads can be suitably controlled at or near the outside wall by using the new glazing and thermal storage materials covered in this chapter. These materials are not only passive in nature but are also classified as architectural finish materials—materials that serve as the interior surfaces and therefore partially pay for themselves as architectural elements.

The number of energy conserving choices for glazings and window treatments has grown exponentially over the last ten years, particularly in the residential market (most of these products are the result of research programs that began as D.O.E. funded projects). Glazing choices for residences grew from three choices (single, double, or triple glazing) to dozens once inexpensive low emissivity coatings were commercialized. The already large number of glazings for commercial buildings has also been multiplied by the new coating choices for minimizing cooling loads without sacrificing daylight.

Thermal storage materials have not advanced as rapidly during this same ten-year period. Although the storage materials covered at the end of this chapter perform admirally, their continued high packaging costs have kept most of them off the market.

5.2 Residential Glazings

Materials research has focused on glazings, since windows usually create an adverse impact on building operating costs. Energy penalties caused by low insulation values and uncontrollable solar gains often outweigh energy benefits, such as daylighting. Nevertheless, windows provide essential benefits, such as view, ventilation, and a psychological (and sometimes physical) connection to the outdoors. Perhaps more important, windows are architectural elements that can set the style of a building. Any new glazing alternatives must provide all these benefits to be accepted in the marketplace.

Table 5.1
Performance of improved window glazings

City	Orientation	R-value (ft^2 °F h/Btu)	Trans-mittance	Energy balance, Nov–Mar ($kBtu/ft^2$)[a] Transmitted solar	Degree-day load	Net
Albuquerque, New Mexico	south	2.2	0.7	156	41	115
	north			24	41	−17
	south	12	0.49	104	7	97
	north			15	7	8
					R-30 wall:	−3
Columbia, Missouri	south	2.2	0.7	91	47	44
	north			21	47	−26
	south	12	0.49	61	9	52
	north			14	9	5
					R-30 wall:	−3
Caribou, Maine	south	2.2	0.7	78	76	2
	north			19	76	−57
	south	12	0.49	52	14	38
	north			12	14	−2
					R-30 wall:	−5

Source: Neeper and McFarland 1982.
a. The transmitted insolution, the thermal load, and the net gain or loss of energy from November through March are shown. For comparison, the loss (calculated on a degree-day basis) through an *R*-30 (5.3 m^2 °C/W) wall is also shown. 1 $kBbu/ft^2$ = 3.15 kWh/m^2.

There are two approaches for improving energy performance in residences that require winter-time heating during some part of the year: improve the solar transmission while maintaining the thermal insulation level, or improve the thermal resistance while maintaining the solar transmission. Either approach greatly enhances the window as a south-wall collector, and if the improvements are large enough, as an effective cloudy-day or north-facing solar collector, since the diffuse solar radiation can be effectively trapped (Neeper and McFarland 1982). Table 5.1 presents the simulated performance of improved windows in three different heating climates. Today's commercially available argon-filled low-emissivity (low-e) windows can attain performance about midway between the extremes shown in this table. The transmitted insolation, the thermal load, and the net gain (or loss) of energy from November through March are shown. For comparison, the loss through an R-30 (5.3 m^2 °C/W) wall is also shown.

The improved windows show a net energy gain for the north orientation for all but the most severe climates.

Windows for offices do not require high solar transmission, and only moderately increased thermal resistance is desired since office machines, lights, and occupants generate an adequate amount of heat during operating hours. Advances in office glazings are covered later in this chapter.

5.2.1 High Transmission Approaches

High solar transmission can be achieved without reducing the thermal resistance by reducing solar absorption and by decreasing solar reflection. Today's glazings lose so much energy by these two mechanisms, that the use of more than three layers of glass in south-facing direct-gain windows is not advantageous in most climates since the solar transmission falls off faster than the insulation builds up with additional layers.

5.2.1.1 Low Absorption Glazing Today's commercial 1/8-in. (3-mm)-thick architectural glass absorbs from 6 to 8% of the incident solar radiation due to its iron content. Most of this energy is lost to the ambient in the outside pane of a multiple glazed unit, while nearly half of the solar energy absorbed in the inner panes is also lost to the outside. Water-white, low-iron glass significantly reduces absorption to about 1% per layer (Rubin 1984). Low-iron glass is currently offered at no extra charge over ordinary float glass. However, the glass is produced by a molten glass draw process that adversely affects the flatness of the surfaces. Recent refinements have minimized the difficulty, and reflection distortions become apparent only when the viewer is told where to look for the problem.

Absorption losses can also be reduced by decreasing the glazing thickness. Plastic glazing films can reduce the absorption losses to less than 1%.

5.2.1.2 Antireflection Treatment Normal reflection losses off the two surfaces of a single layer of glass total 8%, a figure comparable to absorption losses. This figure can be lowered to 2% by altering the index of refraction at the glass surfaces with either mechanical treatments or coatings.

5.2.1.2.1 MECHANICAL TREATMENTS Graded-index treatments give a glass or plastic surface a lower index of refraction than normal. Dipping glass in a fluorosilicic-acid bath saturated with silica roughens the surface

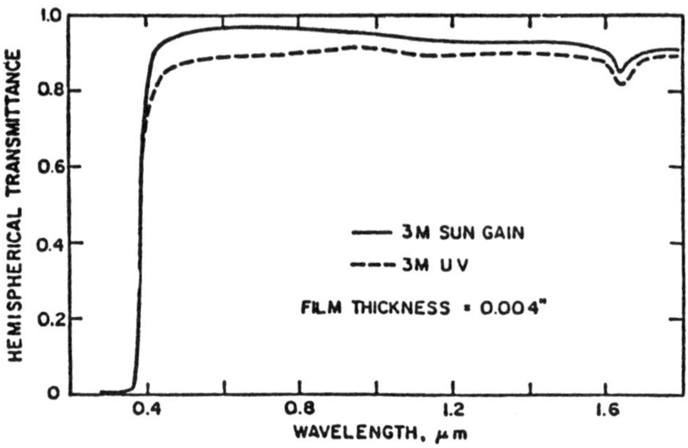

Figure 5.1
Hemispherical transmission of antireflected 3M Sungain polyester film compared to the uncoated substrate. Source: Rubin and Selbowitz 1981.

by etching out small pores in the nonsilica regions (Jurison et al. 1975). Graded-index treatments have also been used for glass solar collector tubes (Miska 1983). The Lawrence Berkeley Laboratory has shown that titanium oxynitride can be used to produce graded-index coatings by plasma-assisted chemical vapor deposition. Care must be taken to avoid fouling the surface formed by these various treatments with greases or waxes.

Production costs have been lowered by generating similar dendritic surfaces on plastic films, such as polyester (Lee and Debe 1980, Rubin and Selkowitz 1981). The process forms needlelike structures of aluminum hydroxide by steam-oxidizing aluminum film (figure 5.1).

Alternately, polymers with a low refractive index, such as Teflon® F.E.P, can be adhered to glass as films or as colloidal dispersions to increase light transmission (Lampert 1982).

5.2.1.2.2 COATINGS Thin film coatings are used to increase the solar transmission of bulk glazing materials (figure 5.2) or other thin films, such as low-emissivity (low-e) surfaces. The inorganic thin films, such as MgF_2, SiO, SiO_2, and TiO_2 are used for single or multiple interference applications. The films are usually fabricated by vapor-deposition techniques. Highly durable diamond-like transparent coatings have also been used for antireflection films (Vora and Moravec 1981).

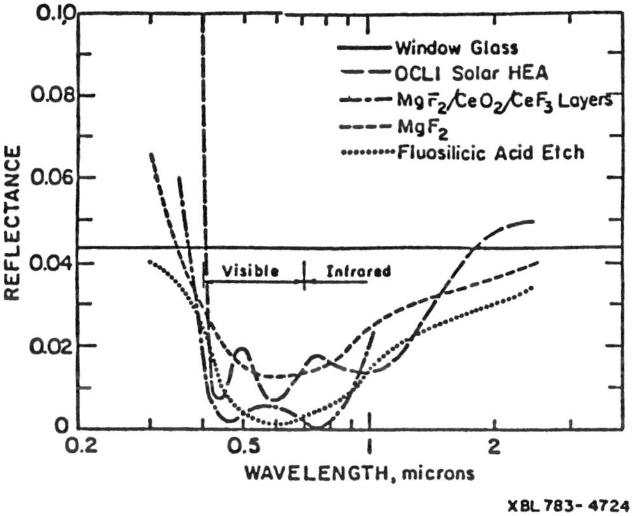

Figure 5.2
The effect of various antireflection treatments on glass. Source: Lampert 1979.

5.2.2 Increased Insulation Approaches

The other approach to improving window performance increases the thermal resistance while maintaining the solar transmission. It is possible to build R-20 (3.5 m^2 °C/W) windows, but most of these demonstrations have lacked adequate daylight transmission. Increasing the thermal resistance of windows holds the most promise because this approach also increases thermal comfort levels near the window.

5.2.2.1 Low-Emissivity Coatings Over half the thermal losses through ordinary windows are due to far-IR heat exchanges at the window surfaces (Rubin et al. 1980). Glass absorbs nearly 90% of incident room temperature thermal radiation and easily conducts this energy to the other side for reemission to the ambient. Low-emissivity (low-e) coatings on glass or plastic films are used to nearly eliminate this radiative path and essentially double the thermal resistance. The coatings are wavelength-selective, so shortwave solar gain can be maintained for residential heating purposes, while the IR is reflected back into the room (Lampert, 1980).

The selective transmitters can be classified (Vossen 1977) into three film categories: a thin multilayer stack of metal film sandwiched between anti-

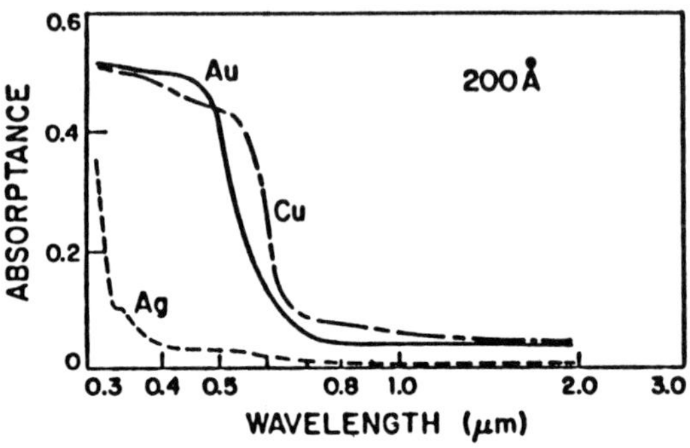

Figure 5.3
Absorbtivity of Ag, Au, and Cu films (each 200 Å thick) vs. wavelength. Source: Fan 1976a.

reflection layers, a relatively thick single-layer film of a highly doped semiconductor, or a thin film conducting microgrid. Chopra et al. (1983) gives a thorough status report on these approaches.

5.2.2.1.1 THIN FILMS The metal layer in a multilayer system is used to reflect the IR. Figure 5.3 shows the solar absorption of several 200-Å-thick metal layers. Gold and copper have too high an absorption at 0.5 μm (the middle of the solar spectrum). Silver shows significantly better performance, but the solar reflection is too high. IR-transparent coatings such as TiO_2, Al_2O_3, SiO_2, and ZnO are used to antireflect one or both sides of the metal layer (Lampert 1981b). Figure 5.4 shows the typical performance of the common $TiO_2/Ag/TiO_2$ system used for most residential applications. Far IR emissivities range from 0.10 to a low of 0.02 when a double silver stack ($TiO_2/Ag/TiO_2/Ag/TiO_2$) is used. The film can withstand short-term exposures to water and is stable up to 200°C (392°F) but must be sealed in double-glazed units to prevent abrasion damage and corrosion over the long term.

The stacked system is formed in a multistage vacuum chamber using the sputtering process (Dolenga 1986), which is favored because the layer thickness, and hence the visual appearance and color, can be uniform with good quality-control practices. Others (Haacke 1982) have listed additional materials that are potentially commercially viable thin film selective transmitters, particularly the nitride films (Karlsson and Ribbing 1982, Howson and Ridge 1982), which are more abrasion resistant (Lampert 1981b).

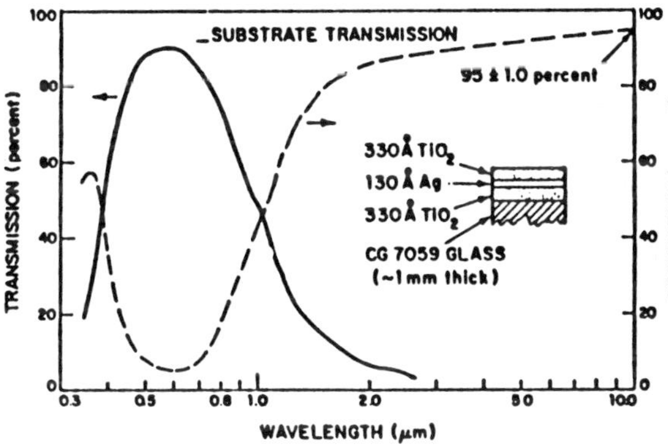

Figure 5.4
Measured optical transmission and reflectivitity of a 180 Å TiO_2/180 Å Ag/180 Å TiO_2 on Corning 7059 glass. Source: Fan 1976a.

Thin stacked coatings have also been vacuum sputtered on plastic films (Lampert 1983), such as polyester (Kiyoshi et al. 1982, Howson and Ridge 1982). Deposition on plastic is more difficult than on glass due to plastic's poor strength at elevated temperatures. Roll coating on plastic is very cost-effective since the roll can be unwound past the metal targets and rewound in a single vacuum chamber. Currently, the cost advantage is canceled by the glazing fabrication step where the plastic is mounted in a frame and protected from the weather by outboard sheets of glass (Hodge 1981). More durable coatings for plastic should negate the need for this labor-intensive framing practice.

5.2.2.1.2 THICK FILMS Doped semiconductors can be used to avoid the quality-control issues of thin films. Certain highly doped members of this class exhibit high IR-reflectance due to their controlled concentration of charge carriers. Some of these transparent conductors are SnO_2, SnO_2:F, In_2O_3, $CdSnO_4$, and In_2O_3:Sn. Typical optical characteristics are shown in figure 5.5 (Hamberg and Graqvist 1983). Emissivities below 0.15 are achieved with the thicker layers (0.3 microns), but fabrication time is extended and the product is expensive because of the high cost of indium. There has been some success with thinner coatings (Frank et al. 1982, Hamberg et al. 1982).

Sputter-coated indium-tin-oxide is particularly attractive because of its abrasion and corrosion resistance. The less expensive tin-oxide can be

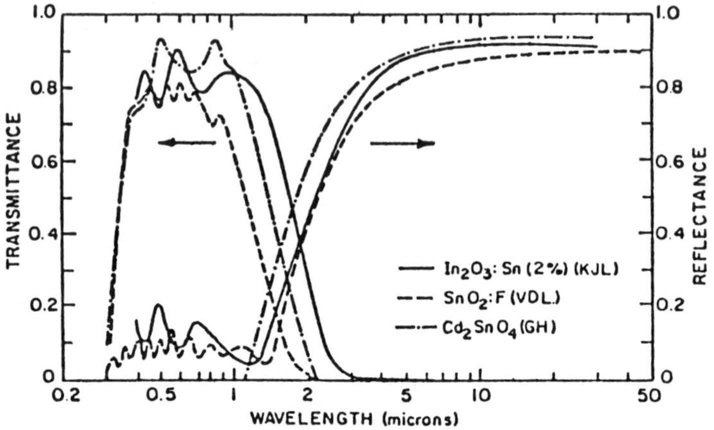

Figure 5.5
Spectral normal transmittance and reflectance of research grade thick low-e films on glass. Source: Lampert 1982.

atmospherically sprayed on molten glass (Gralenski 1982), resulting in a molecular bonding that is also abrasion and corrosion resistant. This process, known as pyrolytic coating, is faster than sputter coating, but color uniformity is still hard to achieve. The advantage of the process is that it produces glass that can be used monolithically (as a storm window, for example) at a low price. The emissivity, at 0.4, is moderate though, which noticeably reduces the window's insulation level. Recently Pilkingtion in England began producing pyrolytic coatings with emissivities below 0.19 with good color rendition, but the coating produces a slight haze on the glass.

5.2.2.1.3 MICROGRIDS Metal sheets can be etched to create openings of approximately 2.5 μm to allow solar radiation to pass through but not the IR, assuming the widths of the lines are not too narrow to prevent reflection—around 0.6 μm (Fan 1976a). The technique can also be used to improve the performance of thick films such as indium-tin-oxide.

Recent research is focusing on making the fragile sputter-coated thin films more durable, and lowering the pyrolytic coating's emissivities. The trend is toward a common performance where cost will be the deciding factor.

5.2.2.1.4 GAS FILLS Heat loss by convection becomes the major loss mechanism in multiple-glazed windows when radiation transfer is minimized with low-e coatings. Gas fills with higher viscosity and thermal resistance than air can be used to minimize convective heat transfer in sealed units. Currently, argon and krypton (which are inert) or sulphur hexafloride (which is poisonous in large amounts) are used as fills. The R-value is raised from 3.2 to 4.3°F ft^2 h/Btu (0.56 to 0.75 m^2/°C/W) when argon is used in a double-glazed, low-e unit with a 0.5-in. (1.3-cm) gap (Johnson 1982). Unfortunately, many of the metal edge spacers and organic seals in use today are not sufficiently impermeable to the gas, and thermal performance starts to deteriorate after seven to nine years. Most European building authorities recognize this and specify that only the aged R-value (which assumes an air fill) can be used in qualifying for an energy audit. Less-permeable seals are starting to emerge, however, making the gas fills viable. But the thermal losses through the metal edge spacers (or even glass welds) become a significant fraction of the window's total losses when the thermal resistance of the bulk glass approaches 5 ft^2 °F h/Btu (0.9 m^2 °C/W). Edge-spacer thermal resistance must be raised without sacrificing permeability or resiliency. Candidate materials include polymer ceramics or convoluted (lower thermal conductance) metal edge spacers adhered with low-permeable adhesives as commonly used in vacuum technology.

5.2.2.1.5 EVACUATED WINDOWS Convective heat losses in double-glazed windows can be completely eliminated if the air space is evacuated. Current research (Benson and Tracy 1985b) shows that R-values in excess of 10 ft^2 °F h/Btu (1.8 m^2 °C/W) can be achieved in evacuated double-glazed units with a low-e coating. A vacuum around 1.3×10^{-3} Pa must be maintained over thirty years if the window is to become practical. Only all-glass seals can currently do this job. A laser has been used (Benson 1984) to adequately seal borosilicate glass edges. Borosilicate glass also minimizes diffusion of atmospheric gases to acceptable time constants of over fifty years (Benson 1984). Support is provided by 1/8-in. (3-mm) glass beads arrayed on a 2-in. (5-cm) grid. A very low emissivity surface is required since the radiation path is the only principal path for heat loss. Figure 5.6 shows the effect of each internal surface's emissivity on thermal performance.

Many low-e coatings cannot withstand the high heat of laser welding. High-temperature candidates are SnO_2:F and ITO films . Large temperature differences build up across the glass weld in practice since the unit

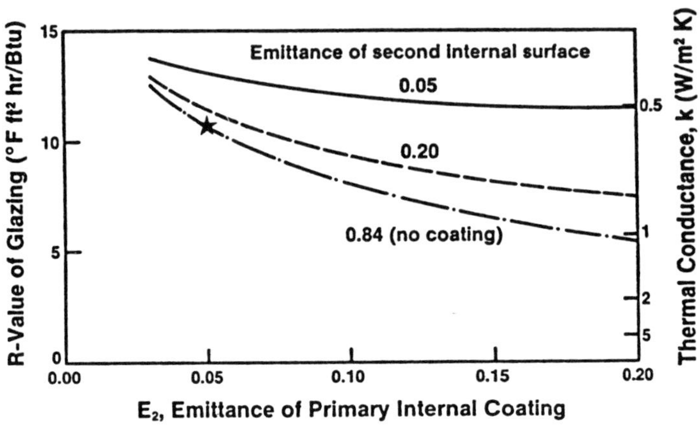

Figure 5.6
The effect of primary and secondary coating emittance on the thermal performance of an evacuated window. Source: Benson and Tracy 1985b.

insulates so well. The thermal stresses become high enough to break ordinary glass, so high-strength glass must be used. Also, the glass deflects enough over the glass beads to cause reflection distortions. Closer bead spacing minimizes the adverse reflections but increases thermal bridging and view interference. Transparent aerogels (see next section) may be used to provide uniform support (the vacuum greatly enhances the aerogel R-value). Uses may be restricted to skylights where visual appearance is not so critical.

5.2.2.1.6 WINDOWS MADE WITH BULK TRANSPARENT INSULATION Conventional bulk insulating materials are opaque because of their many internal reflecting surfaces, even though the matrix material may be transparent. It is possible to produce (Hunt 1982) a porous material, called aerogel, whose particle size is much smaller than a wavelength of light, thus rendering the bulk material transparent.

Supercritical drying of a colloidal gel of silica produces a bonded network of silica particles having a mean diameter of 100 Å (Rubin and Lampert 1983). The light transmission of the material is greater than glass of equal thickness (figure 5.7), and images are sharp. The index of refraction is between 1.01 and 1.1, depending on the material density. The insulating values are slightly superior to extruded polystyrene in air (figure

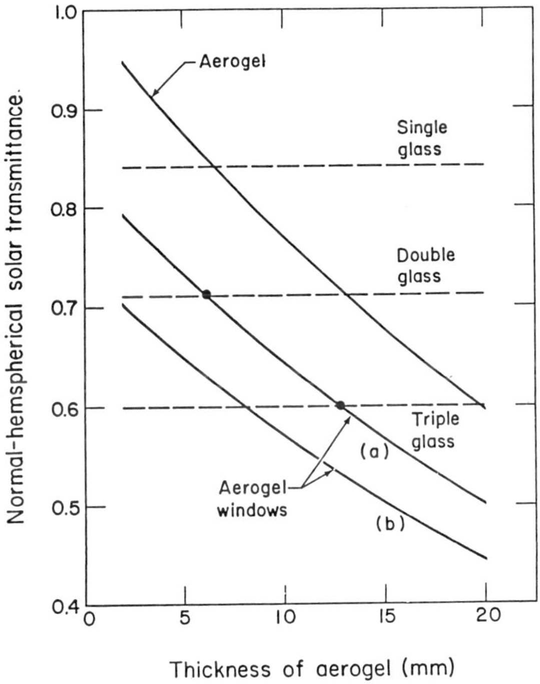

Figure 5.7
Calculated solar transmittance of aerogel windows vs. aerogel thickness compared to solar transmittance of conventional glass windows. All glass is 3-mm clear float glass except for the 3-mm low-iron glass of aerogel window (a). Source: Rubin and Lampert 1983.

5.8). Even higher insulation values could be achieved with other gases, such as Freon® or argon. Any desired insulation level could be theoretically achieved by adjusting the thickness, but the solar transmission and optical quality are currently unacceptable for thickness greater than 2 cm (Hunt 1983).

Aerogel has a slight yellow cast when viewed against a bright background, due to the scattering of blue light (Hunt 1983). The material appears milky blue when viewed against a dark background because of back scattering. Smaller partical sizes could reduce this problem.

The material must be hermetically sealed since it absorbs water vapor when exposed to the air. Usually the material is protected in sealed double glazing. This also mechanically protects the material which can fracture easily in flexure but which can readily take compression loads. Further

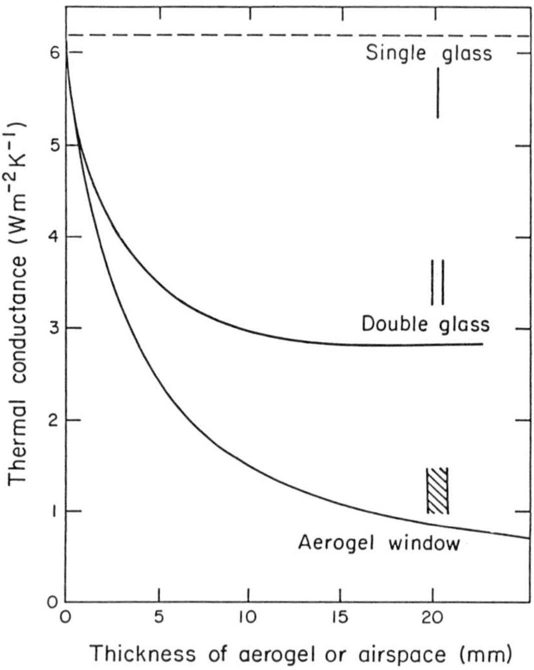

Figure 5.8
Calculated thermal conductance of an aerogel window and of conventional single and double glass windows vs. spacing between glass panes. 1 W/°C m^2 = 0.176 Btu/h °F ft^2. Source: Rubin and Lampert 1983.

difficulties are the long processing time and problems in producing uniform slabs at window size.

5.3 Office and Commercial Glazings

The previous sections have been concerned with windows for residential applications where solar gain and insulation levels are maximized for heating purposes. While offices can use better insulating windows to create better thermal comfort, they normally do not need more solar gain because of excess daytime internal gains from people, lights, and equipment (especially computers). Evenly distributed daylighting is desirable, though, because of its color and slightly better efficiency when compared to the new luminaires.

5.3.1 Low-Emissivity Coatings

Air conditioning loads can be lowered even further by using windows that admit most of the daylight while minimizing the solar gain. This is accomplished by using heavier low-emissivity coatings on specialized glazings (Johnson 1984). Figure 5.9 shows how an ideal selective transmitter for offices differs from one designed for residences where solar gain is desirable.

Both selective transmitters reflect nearly all the far-IR solar spectrum because of their equally low emissivity, so room radiant heat cannot get out, and outdoor temperature radiant heat cannot get in. But where the residential selective transmitter admits a high proportion of the entire solar spectrum, the commercial selective transmitter reflects a high proportion of the near-IR solar spectrum (where half the solar heat in located) while maintaining high daylight transmission. Typical performance for a triple-glazed unit with a thick low-e coating on the middle polyester glazing (Johnson 1984) is 50% daylight transmission, a shading coefficient of 0.42, and a U-value of 0.25 Btu/h °F ft^2 (1.4 W/°C m^2). (The shading coefficient is the ratio of solar heat gain through a window to the solar heat gain through a single layer of 1/8-in. (3-mm) clear glass under the same set of conditions.) This glazing outperforms ordinary tinted double-glazed windows with a shading coefficient of 0.60 and a daylight transmission of 48%.

Light green glass (with a higher iron content) has the unusual property of being a selective absorber. Using the low-e coating on light green glass for the cover plate in a double-glazed unit gives a shading coefficient of 0.38, a daylight transmission of 62%, and a U-value of 0.32 Btu/h °F ft^2 (1.8 W/°C m^2). Most of the daylight gets through, but a high proportion of the near-IR spectrum is absorbed. The thin low-e coating nearly doubles the thermal resistance while lowering the shading coefficient and daylight transmission slightly.

Computer models (Bartovicks 1984) have been used to compare these two approaches with ordinary clear and reflective glazings in a five-foot-strip window for an all electric, lightweight, 192 ft^2 (59 m^2) office. Figure 5.10 shows the annual energy impact of each glazing option on heating and cooling for the model office. U-values proved to be the most important parameter for reducing energy consumption in cold climates. The clear double glazing outperforms reflective single glazing, even though the re-

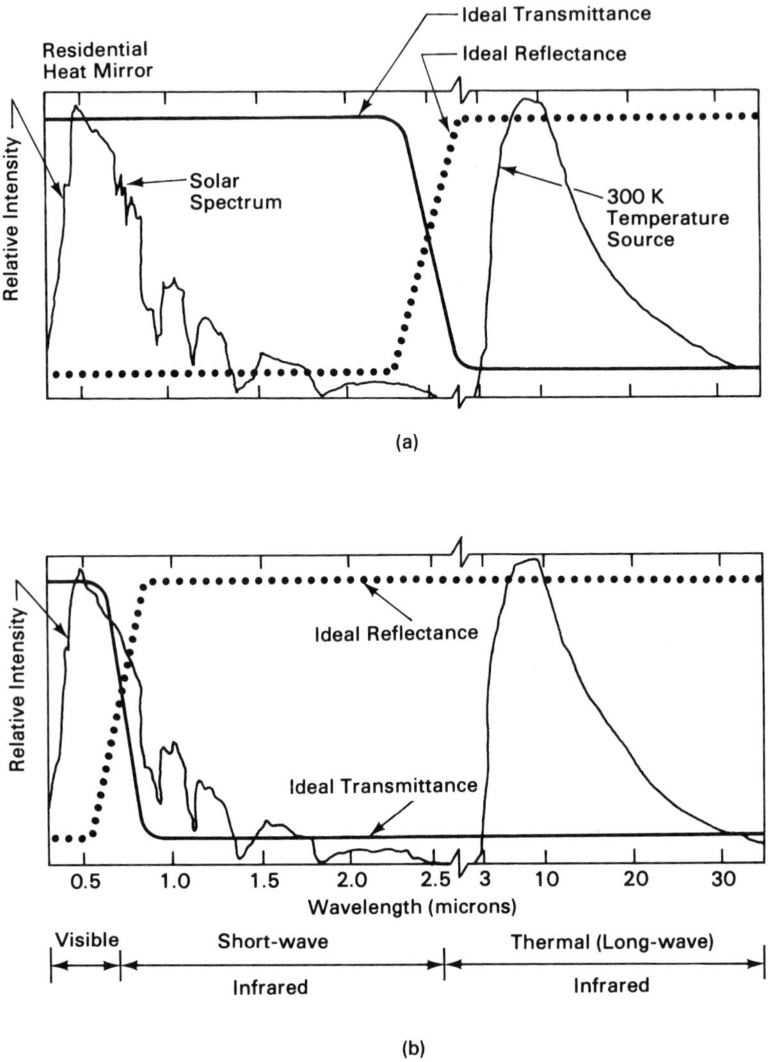

Figure 5.9
Relative solar gain vs. wavelength for (a) residential and (b) commercial selective transmitters. Source: Bartovicks 1984.

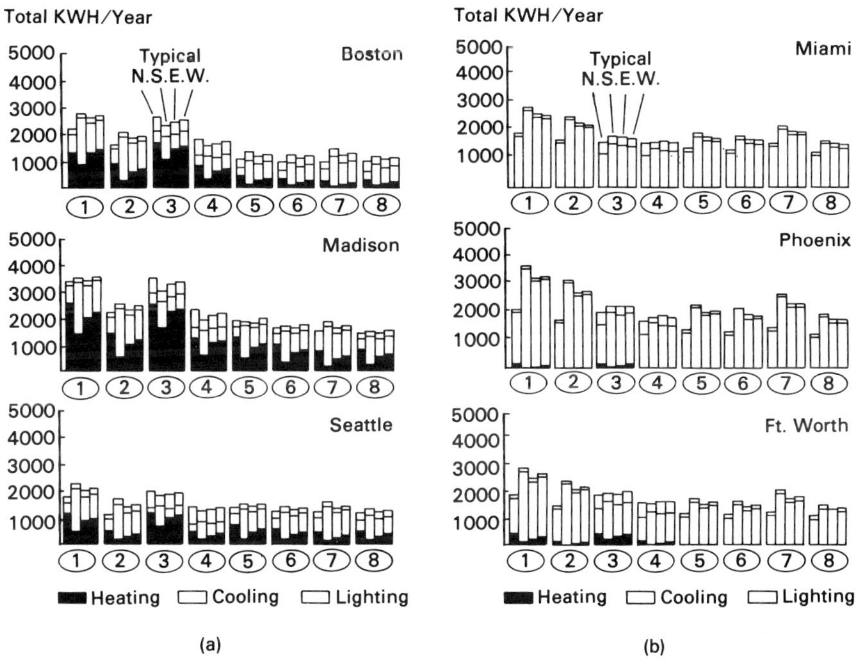

Figure 5.10
Annual load comparisons for (a) three cold-climate cities, and (b) three warm-climate cities, for low-e and ordinary office glazings. Glazing 1 = clear single-glazed, 2 = clear double-glazed, 3 = reflective single-glazed, 4 = reflective double-glazed, 5 = single-glazed, low-e coated green glass, 6 = double-glazed, low-e coated green glass, 7 = thin low-e on an interior polyester film (triple glazed), 8 = thick low-e on an interior polyester film (triple-glazed). Source: Bartovicks 1984.

flective coating lowers cooling energy consumption. Reflective double glazing saves more energy than clear double glazing, except on the north where lighting loads increase. Except for Seattle, the low-e group increases annual savings over reflective double glazing, mostly because of the lower U-values.

Figure 5.10(b) gives the annual performance for typical cooling climates. Since there is no heating load, lower U-values are no longer important. Savings as large as 25% were found for the low-e windows on the north when compared to ordinary reflective double glazing.

5.3.2 Optically Switching Windows

Offices do not always require solar heat rejection. In underheated climates, morning solar gain is desirable, particularly if the office thermostat is set back at night. One the other hand, the glare may become intolerable during certain periods of the day. The optically switching materials covered below can be used to dynamically regulate the optical and thermal properties of windows through electrochromic, thermochromic, or photochromic means. The optical change can be accomplished by transforming the material from a highly transmitting to a highly reflecting or highly absorbing material over part or all of the solar spectrum.

Additional research is needed to identifiy the most efficient applications for windows that switch to a nearly opaque state. Some researchers (Bryan 1984) have suggested that switchable glazings should be used for overhangs or sunshades.

5.3.2.1 Photochromic Materials Photochromic materials change color with light intensity. A popular application is the photochromic glass used in sunglasses. Metal halide coatings are usually used for this purpose (Smith 1966), but many other organic and inorganic compounds exhibit photochromic behavior (Lampert 1984). The organics include stereoisomers, dyes, and polynuclear aromatic hydrocarbons. The inorganics include ZnS, TiO_2, Li_3N, HgS, HgI_2, HgCNS, and alkaline earth sulfides and titanates. Many of these compounds must be doped with traces of heavy metal or a halogen to become photochromic. The slow switching speeds are not a problem with architectural applications. The major difficulty with the photochromic approach is control—it is difficult to override the change of state.

5.3.2.2 Thermochromic Materials Thermochromic materials also change color, but in response to heat. There are more than two thousand known thermochromic compounds. Gel or liquid polymer (methanol with calcium chloride in polyvinyl acetal resins, for example) and water groups can undergo a collodial change of state within a few degrees Fahrenheit. The material usually turns from transparent below the critical temperature to a milky, reflecting white above the critical temperature. The water percentage cannot vary with age, or the material will not remain reversible. Inorganic materials are more stable, such as AgI, Ag_2, HgI_4, $SrTiO_3$, and Cd_3P_3Cl. The material is usually mounted over the room side of the window, so an elevated room temperature will switch the material and reject the extra solar gain. Control can be achieved by heating the material artificially, usually with transparent electrical conductors deposited on glass. Figure 5.11 shows the optical properties vs. temperature of a Ger-

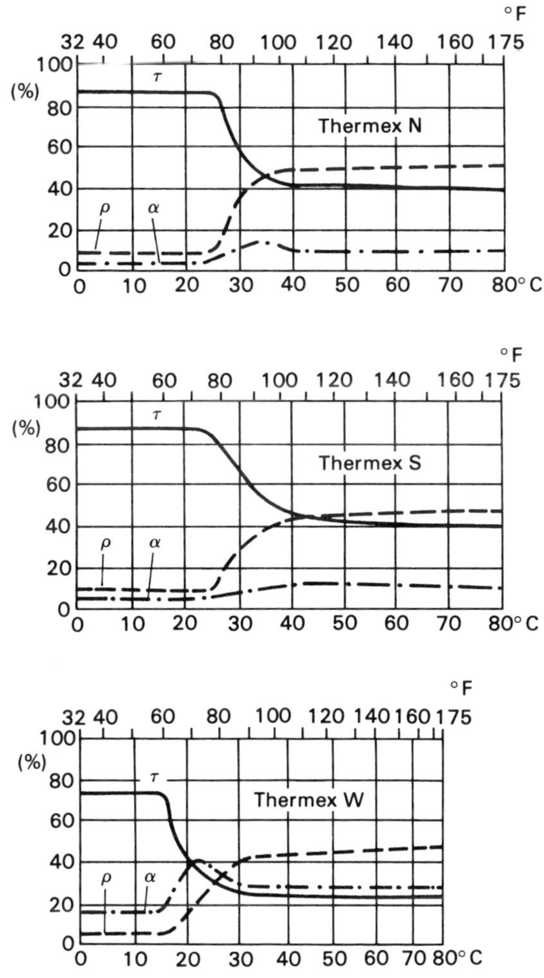

Figure 5.11
The solar transmittance (τ), reflectance (ρ), and absorptance (α) vs. temperature for a thermochromic glazing. Source: Reusch 1967.

man product (Reusch 1967) that has subsequently been removed from the market. The change from transparent to opaque is spread over at least 15°C (27°F). The solar transmission in the opaque state is 40% to 25%, depending on the product. The critical, or transition, temperature and the switching range can also be varied by varying the material's composition.

5.3.2.3 Electrochromic Materials Electrochromic materials either change their color (and sometimes also become reflective) or change their near-IR absorption without greatly changing color. Electrochromic behavior is found in large groups of both inorganic and organic materials. The two categories for windows are transistion metal oxides and organic compounds.

Some metal oxides that change color when switched are amorphous tungsten oxide a-WO_3, MoO_3, $Ni(OH_2)$, and IrO_x. The organic electrochromics are either liquid viologens, anthraquinodes, dipthalocyanides, or tetrathiafulvalens. Inclusive lists of electrochromic materials have been published (Lampert 1980 and 1984, Dautremont-Smith 1982, Rauh and Cogan 1984).

Multilayer solid-state electrochromic coatings on glass using a-WO_3 have been tested by many researchers (Deneuville et al. 1980, Benson 1984, Benson and Tracy 1985a) because the coatings' two states offer a high ratio of solar admission to rejection. The material's reflectivity in the switched state can be enhanced (Goldner et al. 1983, Goldner 1983, Goldner et al. 1984). A typical metal stack on glass (Benson et al. 1984) consists of vacuum deposited indium-tin oxide (520 nm), WO_3, (410 nm), MgF_2, (170 nm), and gold (15 nm). Figure 5.12 shows the optical spectra of a similar stack. The thin gold electrode greatly reduces the solar transmission. Subsituting ITO for the gold electrode can double the transmission (Benson et al. 1984), but the ITO coating is not conductive enough to give efficient operation.

This approach, however, has a limited lifetime in service. Stack degradation usually occurs during bleaching (switching to the clear state) in humid air, but some reseachers (Deneuville et al. 1980, Sato et al. 1982) state that a properly fabricated device need not deteriorate with time.

Liquid crystals are also candidates for window applications, since the materials can be electrically switched from a normally translucent state to a reflective state. Liquid crystals are substances that first pass through a paracrystalline stage in which the molecules are partially ordered before melting to the liquid phase. The material becomes turbulent and reflective

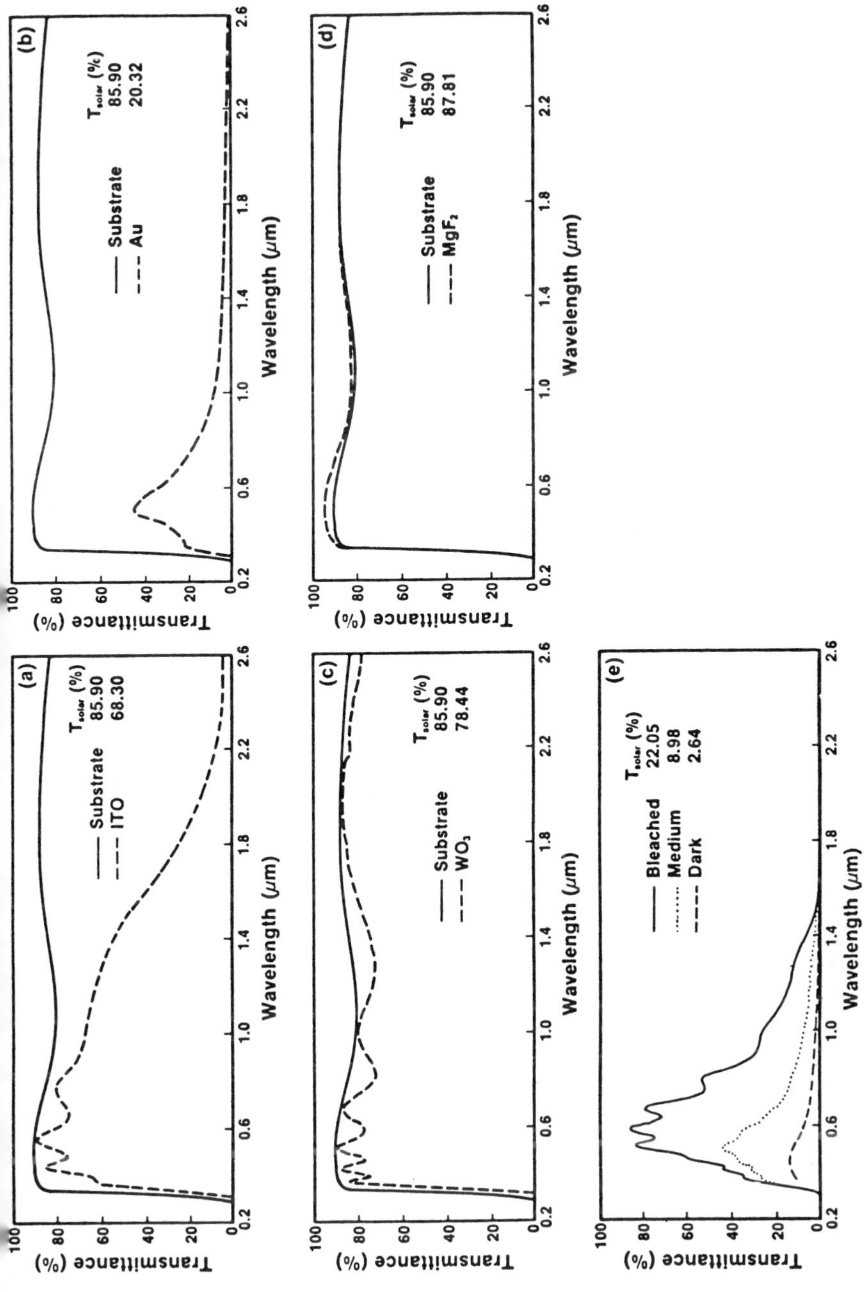

Figure 5.12
Optical spectra of component films and a completed electrochromic device. Source: Benson et al. 1984.

to light when an electric field is applied. Liquid crystals fall into three categories of structural organic mesophases. Smectic liquid crystals are composed of molecules parallel to one another, forming a layer, but no periodic pattern exists within the layer. Nematic types have the rodlike molecules oriented parallel to one another but have no layer structure. The cholesteric types have parallel molecules, but the layers are arranged in a helical, or spiral, fashion. The molecular structure of liquid crystals can be easily altered by electric and magnetic fields, mechanical stress and pressure, and temperature. The last two alteration methods become problematic when liquid crystals are used for large-scale windows. Also, where most electrochromic materials require energy only during switching, liquid crystals require continuous power to maintain the reflective state.

Computer simulations based on the same assumptions as those used in figure 5.10 were run for the same climates using a prototype electrochromic material that changed its near-IR absorption levels (Bartovicks 1984). In the clear state, the organic material has a daylight transmission of 73%, and a shading coefficient of 68%. The switched material has a daylight transmission of 43% and a shading coefficient gain of 29%. The simulations showed annual operating savings of about 20% in heating and cooling climates when compared to the commercial windows with low-e coatings as shown in figure 5.10. Savings for the electrochromic materials that underwent large changes in color were much smaller since artificial lighting was required when the material was switched to the dark state.

5.4 Movable Window Insulation

Insulation may be placed over residential windows at night to conserve heating energy and to increase thermal comfort in underheated climates. Most offices do not require such window management techniques since waste heat from interior spaces can be used to heat the windows for comfort reasons during operating hours. The usefulness of movable window insulation has declined as window R-values have increased. Movable window insulation can be classified into three groups: outdoor, between-the-glazing, and indoor.

5.4.1 Outdoor Approaches

The most popular type of outdoor window insulation is concealable, usually as a roll inside an overhead wall recess, as seen throughout Europe.

These blinds use wooden or hollow-plastic horizontal slats to provide a small but important additional insulation level. Graded-density, reaction-injection-molded polyurethane is now considered a better slat material since its hard surface can resist the weather and the center is porous for good insulation. The roll up blinds are also popular because they offer security when closed.

The other major type of outdoor window insulation is a hinged or sliding shutter that architecturally resembles the more fashionable wind shutters (Dike and Kinney 1981). Usually, remote methods of closing the shutter are available (Shurcliff 1980).

An important advantage of external insulation over indoor movable insulation is that the window surface is kept near room temperature, so condensation is avoided when the insulation is in place. Thermal shock caused by a sudden removal of the insulation can be a problem in either case. The sudden change in temperature can break the windows in severe climates.

5.4.2 Between-the-Glazing Approach

Once an insulating device is inside double glazing, it stays clean and away from mechanical abuse. Summer solar gains are still minimized since absorbed energy is not directly deposited in the room. The major disadvantage is that the device is exposed to trapped moisture that infiltrates from the room side of the window. Usually, this problem is minimized by venting the space to the outdoors. This problem can also be solved by completely occupying the space between the double glazing with the movable insulation (Harris 1972). Styrofoam® beads or, more recently, hollow beads of foam glass are blown into the cavity at night, giving an R-5 insulation level (0.88 m^2 °C/W) for an 1.25-in. (3.2-cm) air space. The beads are blown into a remote storage tank in the morning. Some difficulties with the approach (Kjelshus 1979) include static cling from accumulating electrical charges, which can be temporarily solved by spraying the beads with anti-static agents.

Simple modifications of old devices also give better performance. Metalized venetian blinds have been placed between glazings (Johnson 1980). The metal finish on one side lowers the emissivity, which lifts the double-glazed R-value to 3.5 ft^2 °F h/Btu (0.62 m^2 °C/W) when the louvers are closed at night. Others (Shurcliff 1980) have used aluminized films that can be rolled up.

Usually, high R-values are difficult to attain with the between-the-glazing approach because of the limited amount of space between the glazings, especially for operable windows.

5.4.3 Indoor Approaches

Indoor window insulation is the least effective thermally, particularly for blocking solar gains. Also, condensation can occur if well-designed edge seals are not included or if the shade itself does not act as a vapor barrier. It is not necessary to have tight edge seals if thermal insulation is the only criterion. Experiments (Shurcliff 1980) show that perimeter gaps of 0.25 in. (6 mm) only decrease the insulation value to 75% of a tightly sealed 1-in. (2.5-cm)-thick Styrofoam® shutter. The indoor insulation approach remains the most popular since it is the easiest to implement.

Storing the open shade becomes a problem when hinged, rigid insulation is used. The bifold and trifold shutters that recessed in the wide window reveal, found in American and European prerevolutionary buildings, solved the storage problem quite well. Of course, these early shutters were used only for privacy and glare control, but the same storage solution can be used for insulating shutters now that thick walls are back in the form of superinsulated houses. The edges and hinges must be sealed to prevent room air from condensing on the glass when the shade is closed. Figure 5.13 shows a plastic hinge detail for sealing the joint.

Standard indoor roll-up shades do not insulate particulary well, mostly due to the lack of edge seals. Do-it-yourself edge seals were well documented (Flower 1980), as were methods for fabricating R-5 (0.88 m^2 °C/W). thermal fabrics (Schnebly 1980, Flower 1980) that used internal aluminized films for increasing the resistance to IR heat flow. Even higher R-values were obtained by insuring that a substantial air gap was produced between the metalized surfaces when the shade was deployed—figure 5.14 (Shore 1978).

The four adjacent aluminized mylar films are automatically spaced apart when unrolled, as soon as the trapped air heats and rises internally, drawing in inflation air through the bottom vents. The shade expands against the edge tracks, forming an air-tight seal. The inflated shade has an insulation level of R-7 (1.2 m^2 °C/W). Rolling the shade up pushes the air out the bottom slots, so the layers store in a minimum volume. Less elegant solutions use two rollers to unwind two separate film/fabric combinations (Shurcliff 1980).

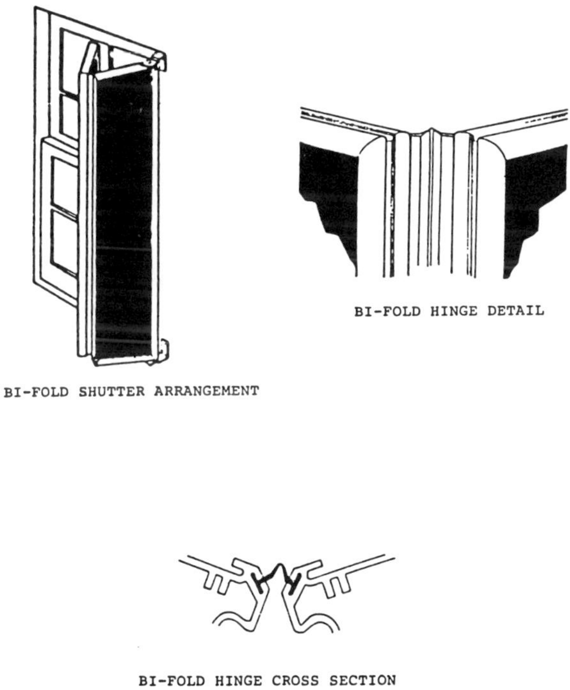

Figure 5.13
Bifold hinge details.

Perhaps the simple solution to indoor movable window insulation problems are indoor curtains (when the architecture permits). Edges are sealed to the walls with Velcro® and weighted hems seal the floor edge. Fabric, Fiber-Fil®, and aluminized film combinations give good insulation levels and vapor resistance while still remaining flexible. These fabrics, however, remain expensive.

Movable window insulation can be readily built on site and lends itself to custom design. The reader can find many published plans (Niles and Haggard 1980) detailing how to build some of the window insulation designs covered in the above categories. However, the storage problem and high price of movable window insulation (when compared with the emerging low-e windows used in new construction) will mostly likely relegate the movable insulation approach to retrofit applications where the cost of replacing the windows is not justified.

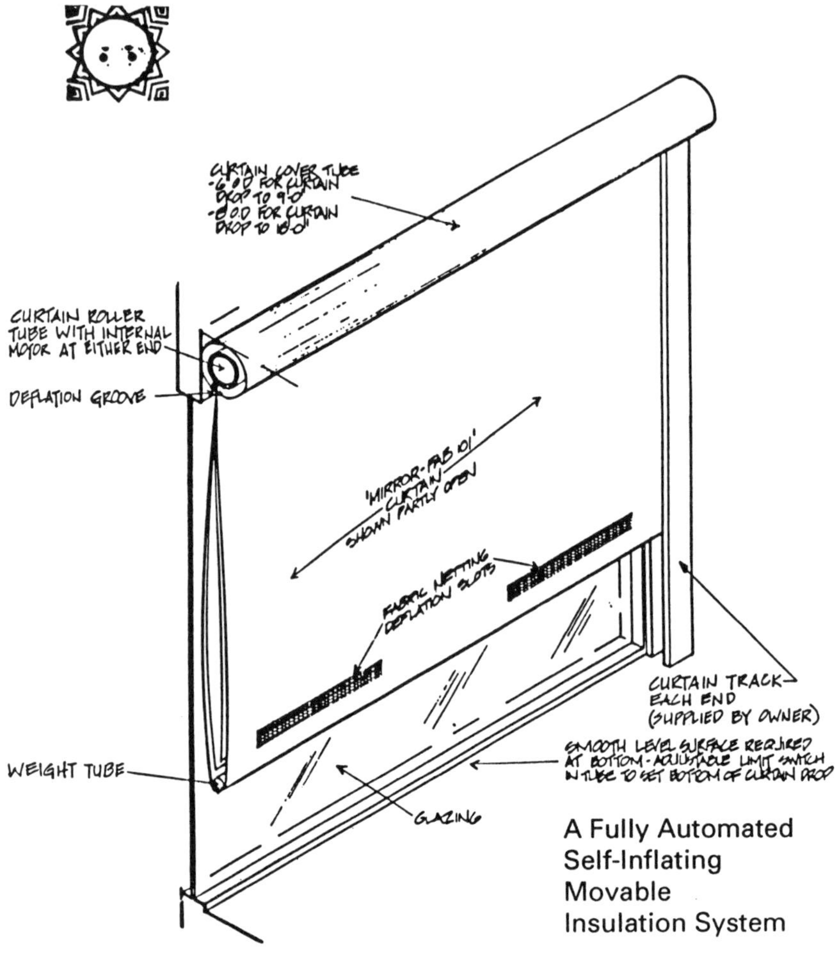

Figure 5.14
A self-inflating movable insualtion shade. Source: Shore 1978.

5.5 Thermal Storage

Passive thermal storage becomes beneficial whenever interior air or surface temperatures vary widely above and below the mean room temperature on a cyclic basis, usually in a diurnal cycle. The same passive storage techniques that are used in residences have been used successfully in offices, but office use of thermal storage is still rare because of costs and the requirement for lightweight office finish materials to overcome acoustical problems.

Residential thermal storage is usually concerned with accumulating solar energy that is delivered radiantly (as opposed to convectively).

5.5.1 Surface Treatment

If the storage material is limited in area and can only be struck by direct radiation rather than reflected or diffuse radiation, then maximum solar absorption with little back radiation is required. Ordinary surface treatments are sufficient if the sunlight can be diffused to larger areas of thermal storage.

Selective absorbers are used to capture the maximum amount of solar radiation by maximizing absorption while minimizing IR back losses. This can be accomplished in several ways: (1) a low-emissivity (low-e) metal base is coated with a far-IR transparent film that is highly absorptive to solar radiation; (2) an opaque metal or metal oxide with low IR-emissivity and low solar reflectivity may have its solar absorption increased by antireflection coatings; (3) an IR-transparent organic binder can be used to adhere inorganic semiconductor pigments to a low-e substrate (selective paint).

The most popular approach for category (1) is "black-chrome," where Cr-black is electrodeposited on copper or nickel, although many other IR-transparent materials with high solar absorption exist, such as MgO/Au, as shown in figure 5.15 (see also Angnihortri and Gupta 1981). Although the copper base gives the best performance (solar absorptivity of 0.94, IR-emissivity of 0.04), the material can oxidize in stagnation-temperature situations. Nickel substrates overcome this problem, but the emissivity rises to 0.12 (solar absorption still equals 0.94).

The Cr-black coating appears porous under a scanning electron micrograph. The porosity is such that the shortwave solar energy intercepts the particles, while the same particles become transparent to the longer-

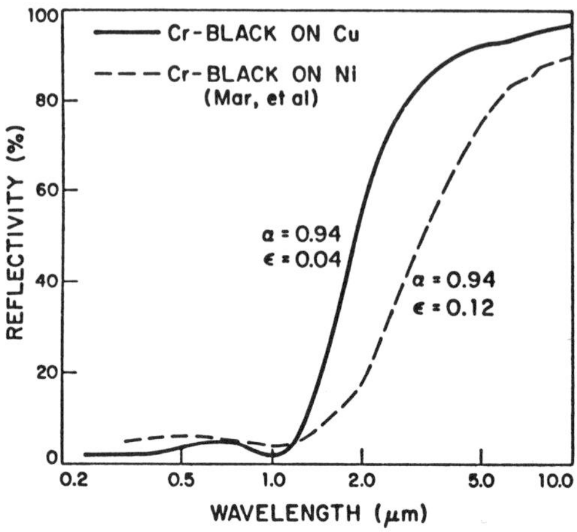

Figure 5.15
Measured optical reflectivity of a 1500 Å-thick MgO/Au film coated on Mo. Source: Fan 1976a.

wavelength IR. Molybdenum-black on nickel and variations on this material (Mason 1982, Gupta 1983) have proved to be suitably rugged for commercial use. Usually, these coatings are electroplated on a nickel or copper foil that can, in turn, be adhered to a storage medium.

An even less expensive alternative uses a fluoropolymer binder to hold solar absorbing particles to a metal plate (Moore 1982). Inherently selective inorganic pigments exhibiting semiconductor properties are adhered to metal substrates with a fluropolymer binder that must be less than 2 μm thick. The coating resists cracking, peeling, and erosion and resists outgassing better than conventional black paints. The solar absorption of the paint is 0.94, and the emissivity is 0.45. Although the performance is not as good as in the electroplated approaches, the cost benefit is currently better for the selective paint.

Ordinary off-white paint is the ideal surface treatment for thermal storage when sunlight is diffused so an entire room becomes evenly daylit. Although the surfaces only absorb 20% of the solar energy, multiple reflections and absorptions occur because the room behaves like an optical cavity. Typically, only 12% to 18% of the solar energy will escape from an off-white room with a window area at 25% of the floor area (Burkhart and Jones 1979). The light color insures that all the thermal storage surfaces participate by distributing the light evenly.

5.5.2 Bulk Materials

Thermal storage in passive designs is usually part of the architecture and acts as a finish material since the space itself is the collector and storage device. Although some energy is saved by properly applying thermal mass, thermal storage affects comfort most directly because it determines the temperature swings of the occupied space. Heat can be stored either sensibly or latently.

5.5.2.1 Sensible Storage Sensible heat storage is by far the most common and the least expensive way of storing heat in passive solar buildings. Although water is sometimes used, concrete and masonry are the most commonly used materials since they double as structural and finish materials.

The specific heat, thermal conductivity, and mass of most cementitious building materials are about the same. Some researchers have tried to improve the specific heat by imbedding foreign materials, such as magnesium (Mazria 1979), but the high cost of this approach has prohibited developement. The usual design parameter is the thickness of the material. Thick sections—8–12 in. (20–30 cm)—are used when the sun shines directly on the material, and thinner larger-area sections—4–6 in. (10–15 cm)—are used when the sunlight is diffused within a space where storage, and release occurs from the same surface (direct gain). Rules of thumb for mass placement in a space have been published (Johnson 1981, Mazria 1979).

The thermal behavior and storage capacity of these materials under the influence of a driving thermal source, such as the sun, can be accurately determined by analytical or simulation methods (see Niles, chapter 3 of this volume). The most common analytic method for thin sections is the admittance, or harmonic, method, which uses a frequency-domain response function relating the heat transferred into the wall to the surface temperature swing of the wall (Balcomb 1983). The method is limited to a building that can be represented by temperature- and time-independent parameters, and is driven by sinusoidal environmental conditions. The sun does not follow sinusoidal behavior, but the errors introduced by assuming harmonic conditions are small.

The diurnal heat capacities of materials can be given in closed form, using harmonic analysis. For thick walls, the capacity is $(k\rho cP/2\pi)^{1/2}$, where k is the thermal conductivity, ρ is the density, c is the specific heat, and P is the period, usually taken at 24 hours (Balcomb and Hedstrom 1980). Different forms must be used for thin walls. Generally, the diurnal heat capacity for cementitious materials peaks at a thickness of approximately 6–9 in. (15–23 cm)—Balcomb and Hedstrom (1980).

Graphical methods exist for modeling dynamic behavior of thermal mass. The Schmidt plots (Kreith 1958) are used to diagram temperature profiles through homogeneous materials. The approximate temperature plots are the graphical equivalent of simple averages over neighboring sections in the material if the time period chosen is equal to one-tenth of the thickness divided by the thermal diffusivity ($k/\rho c$). It always takes five of these time periods for a heat wave to travel the entire thickness of the section.

Water has been used for sensible storage because of its high specific heat and because in most cases, the entire mass will participate due to the fact that natural convection overcomes any temperature distribution within the liquid. Architecturally suitable containerization of the water has proved to be difficult, with leakage occurring in early approaches when corrosion or thermal cycling stressed the material. Modular containers (Maloney 1978) have minimized these problems. Cladding the container with more attractive architectural finishes, such as drywall, concrete, or plaster has to be done carefully to insure an intimate thermal bond (Ackridge 1985).

5.5.2.2 Latent Storage Phase-change materials are used for thermal storage when high heat capacity is required in a small lightweight space, such as a sunspace over a crawl space. The material becomes particularly useful when the building daily thermal load is lower than a clear day's solar intake. Phase-change materials can be divided into two classes depending on their transition states: solid-liquid, and solid-solid. The phase change of interest ranges from around room temperature to 40°C (104°F).

The most common solid-liquid phase-change materials are inorganic salt hydrates, such as sodium sulphate decahydrate and calcium chloride hexahydrate. These materials are preferred because of their low cost, nontoxicity, and noncombustibility. These materials melt incongruently as opposed to the organic phase-change materials (wax) that melt congruently, so that only the solid and liquid phases exist in the melt. There are more than two phases in equilibrium during melting in an incongruent phase-change material. The liquid is saturated, and the heavier anhydrous material settles to the bottom in the presence of the unmelted crystals. The anhydrous material is unable to recombine during the next change of state, so the heat storage capacity reduces with thermal cycling.

Many attempts have been made to stabilize these materials (Telkes 1949, 1964), usually by adding thickeners to retard separation. Most microencapsulation approaches (for decreasing the diffusion distance during solidification) have not been successful. Attempts to encapsulate the material in cementitious material (Chahroudi 1978) have not worked well because of chemical interactions with the matrix and because the water component permeates through the container. Recent attempts using different approaches to the separation problem have been more successful. In one, sodium sulphate decahydrate is suspended in a closed-cell matrix of polymerized latex that holds the material in a collection of permanent micropools where crystallization occurs by reversible diffusion. Stable storage capacities of over 160 J/g (69 Btu/lbm) are reported by independent tests run by the National Association of Homebuilders in 1983. The material is attractive because volume changes with change of state are small enough to prevent the stressing of any container (the container still must be impermeable to water vapor). A slightly higher stable storage capacity of 190 J/g (82 Btu/lbm) is reported by Dow Chemical Company for a previously available commerical product called Enerphase® using calcium chloride hexahydrate that was stablized with $Ca(OH_2)$ and $SrCl_2$. Additives were used to increase the solubility of the incongruent element, calcium chloride tetrahydrate. However, the material undergoes a 10% change in volume with a change in state, which presents serious long term packaging problems.

Many other inorganic candidates exist, some of which melt congruently. One is zinc nitrate hexahydrate, which melts congruently when magnesium nitrate hexahydrate is used as a nucleating agent. The heat content is 134 J/g (58 Btu/lbm). Another candidate is potassium fluoride tetrahydrate, which has a heat of fusion of 330J/g (142 Btu/lbm). The prices of these congruently melting phase-change materials are relatively high. There are many other technically feasible candidates, but their toxicity rules out their use in buildings.

Solid-solid phase-change material is intriguing: packing becomes easier since the material is self-supporting in all phases, and the material cannot leak through punctures. Although there are many candidates (Benson 1983), most have phase-change temperatures above 40°C (104°F). Several exceptions exist, including neopentyl glycol, pentaerythritol, pentaglycerine,

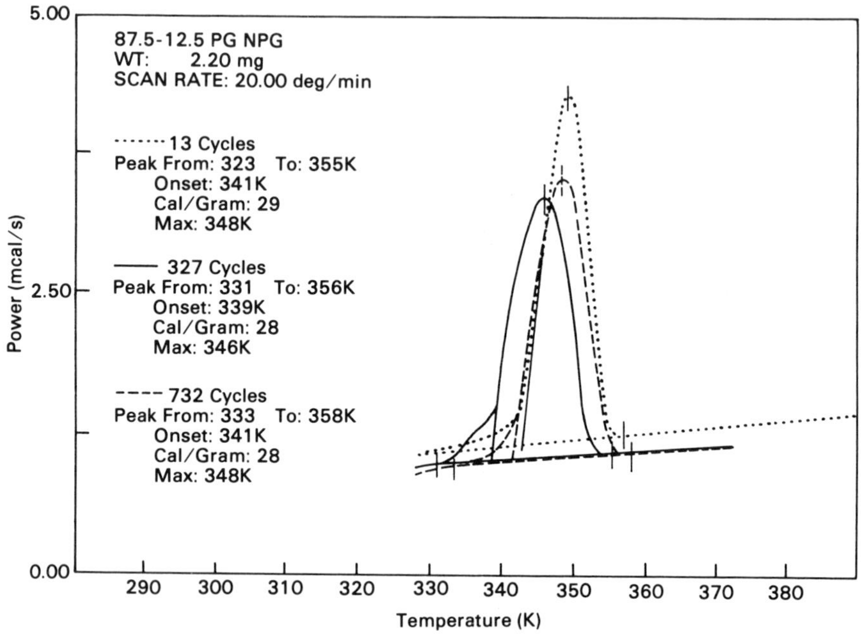

Figure 5.16
Differential thermal analysis of a mixture of neopentyl glycol and pentaglycerine solid-to-solid phase-change materials. Source: Benson 1983.

and trimethylol ethane, which melt around 40°C (104°F) and have heat of fusion around 115 J/g (49 Btu/lbm)—see figure 5.16.

Lower transition temperatures can be reached by mixing the three materials together or mixing one with a solid solution-forming compound, such as trimethylol propane (Benson and Christensen 1984). Thermal conductivity (and therefore the thermal diffusivity of the material) has been increased by adding graphite or aluminum to the molten material before casting takes place. Many drawbacks exist for the material—its cost is high, over ten times the cost of the salt hydrates, and it is flammable. But since the container price is usually the predominant factor in a phase-change product, the simplified container system for this material might give a lower overall cost (the material has been fabricated in gypsum and concrete panels).

Supercooling is also exhibited by the solid-solid phase-change materials. Powered graphite has been used as a nucleating agent (Benson and Christensen 1984), but the undercooling cannot be entirely eliminated. The cast-in-place nucleator cannot migrate as is sometimes the case with solid-liquid phase-change material, so the material remains stable. Researchers are just learning the basic mechanisms that generate the unusally high latent heat content. The current hypothesis is that the phase-change material undergoes crystallographic phase changes in which heat-storing hydrogen bonds between adjacent molecules in the crystal are broken during heating.

5.5.2.3 Containers for Solid-to-Liquid Phase Change Materials Whether the phase-change material is homogeneous or microencapsulated in a matrix to reduce separation by minimizing diffusion distances, the material still must be packaged. The container must be: (1) semirigid, (2) a heat exchanger, (3) corrosion resistant or immune to hydrocarbon attack in the case of organic phase-change materials, (4) tolerant to changes in phase-change material volume, (5) inexpensive, (6) lightweight, (7) resistant to crystal puncture, and (8) impervious to water vapor in the case of salts, since any change in water content grossly affects performance.

The last point means that plastic containers must use an additional laminate (usually aluminum foil) or thick plastic sections to act as a vapor barrier. Coating a volume of phase-change material with epoxy or other thin polymer sheet material is not sufficient. Inexpensive plastics, such as polyethylene or ABS, are sufficiently impermeable when the thickness exceeds 1.5–3 mm (0.006–0.012 in.). Plastic containers housing phase-change materials that undergo volumetric changes will experience fatique stressing that leads to cracking after several years of service. A large surface area to contained volume is desired for good heat transfer, so many packages are formed into flat, thin pouches or trays. Circular cylinders larger than 2 inches in diameter prohibit good heat transfer since the outer annular section of low heat conducting phase-change material becomes too thick to allow the core to participate. Early containers were almost exclusively large-diameter packages in an effort to minimize packaging costs.

The most effective way of driving down packaging cost is to design the container as an architectural finish material. One example of this approach (Johnson 1982) is a 2 × 2-ft (60 × 60-cm), 1 in. (2.5 cm)-thick polymer-concrete floor tile filled with two adjacent 3/8-in.(10-mm)-thick pouches of modified sodium sulphate decahydrate. Here the phase-change material is contained in a foil-laminated plastic pounch and cased over with polymer concrete, colored and patterned to represent quarry tile or slate. The polymer concrete is exceptionally strong, so thin sections can be used to promote heat transfer.

5.6 Conclusion

Passive solar heating and cooling materials research is an ongoing process. This chapter covered recent advancements in glazing materials and window treatments, thermal storage materials, and surface treatments for maximizing thermal comfort while saving energy passively. This approach to passive solar heating and cooling has become the most sensible way economically and aesthetically to achieve superior performance because the materials double as architectural finishes. As in any research endeavor, many of the potential application problems noted in this chapter have been left unresolved. On the other hand, many of the listed materials have made it to the marketplace and have performed successfully. Research will undoubtedly continue as the home owner and office worker become aware of these advances in the field and demand even higher performance.

References

Ackridge, M. 1985. *Passive Cooling Using Radiant Walls.* Department of Architecture report. Atlanta, GA: Georgia Institute of Technology.

Angnihortri, O. P., and B. K. Gupta. 1981. *Solar Selective Surfaces.* New York: Wiley-Interscience.

Balcomb, J. D., D. Barley, R. McFarland, J. Perry, W. Wray, and S. Noll. 1980. *Passive Solar Design Handbook.* Vol. 2: *Passive Solar Design Analysis.* DOE/CS-0127/2. Washington, DC: U.S. Department of Energy.

Balcomb, J. D., and J. D. Hedstrom. 1980. Determining heat fluxes from temperature measurements in massive walls. *Proc. 5th National Passive Solar Conference*, Amherst, MA, October 19–26, 1980. Newark, DE: American Section of the International Solar Energy Society.

Balcomb, J. D. 1983. *Heat Storage and Distribution Inside Passive Solar Buildings.* LA-9694-MS. Los Alamos, NM: Los Alamos National Laboratory.

Bartovicks, W. A. 1984. *The Thermal Performance of Fixed and Variable Selective Transmitters in Commercial Architecture.* Master's thesis, Massachusetts Institute of Technology.

Benson, D. K. 1983. Polyalcohol solid phase-change material. In *Review*, vol 5., no. 3. Golden, CO: Solar Energy Research Institute.

Benson, D. K. 1984. Laser-sealed evacuated window glazings. *Proc. SPIE*, San Diego, CA, August 19–24, 1984.

Benson, D. K., C. E. Tracy, and M. R. Ruth. 1984a. *Solid State Electrochromic Switchable Window Glazings.* SERI/TP-255-2455. Golden, CO: Solar Energy Research Institute.

Benson, D. K., C. E. Tracy, and M. R. Ruth. 1984b. Solid-state electrochromic switchable window glazings. *Proc. SPIE*, vol. 502, p. 46.

Benson, D. K., and C. Christensen. 1984. *Solid State Phase Change Materials For Thermal Energy Storage in Passive Solar Heated Buildings.* SERI/TP-255-2494, Golden. CO: Solar Energy Research Institute.

Benson, D. K., and C. E. Tracy. 1985a. *Amorphous Tungsten Oxide Electrochromic Coatings for Solar Windows.* SERI/TP-255-2769. Golden, CO: Solar Energy Research Institute.

Benson, D. K., and C. E. Tracy. 1985b. *Evacuated Window Glazings for Energy Efficient Buildings.* SERI/C-255-0122. Golden, CO: Solar Energy Research Institute.

Benton, C. 1979. *Off Peak Cooling Using Phase Change Material.* Master's thesis, Massachusetts Institute of Technology.

Bryan, H. 1984. Optically switched shading. *Proc. Buildings Innovative Concepts Fair*, Arlington, VA, pp. 3.1–3.11.

Burkhart, J. F., and R. E. Jones. 1979. The effective absorbance of direct gain rooms. *Proc. 4th National Passive Solar Conference*, Kansas City, MO, October 3–5, 1979. Newark, DE: American Section of the International Solar Energy Society.

Chahroudi, D. 1978. Buildings as organisms. *The CoEvolution Quarterly*, Winter 1977/78.

Chopra, K. L., et al. 1983. Transparent conductors—a status review. *Thin Solid Films* 102:1–46.

Dautremont-Smith, W. C. 1982. Transition metal oxide electrochromic materials and displays: a review. *Display Technology and Applications.*

Deneuville, A., et al. 1980. Principles and operation of an all solid state electrochromic display based on amorphous tungsten oxide. *Thin Solid Films* 70:203–223.

Dike, G., and L. Kinney. 1981. Exterior insulating shutter design and marketing studies. *Proc. U. S. Department of Energy Passive & Hybrid Solar Energy Program Update*, Washington, DC, August 9–12, 1981. Washington, DC: U. S. Department of Energy, February 1982, pp. 3–50.

Dolenga, J. 1986. Low-e: piecing together the puzzle. *Glass Magazine*, March 1986.

Fan, J. C. 1976a. Thin-film conducting microgrids as transparent heat mirrors. *Applied Physics Letters* 28:440.

Fan, J. C. 1976b. Wavelength-selective surfaces for solar energy utilization. *Proc. SPIE*, San Diego, CA, August 1976.

Flower, R. 1980. An R-5 insulating window shade designed to be constructed by the homeowner. *Proc. 5th National Passive Solar Conference*, Amherst, MA, October 19–26, 1980. Newark, DE: American Section of the International Solar Energy Society.

Frank, G., et al. 1982. Transparent heat reflecting coatings (THRC) based on highly doped tin oxide and indium oxide. *Proc. SPIE*, Los Angeles, CA, Jan. 28–29, 1982.

Goldner, R. B., et al. 1983. High near-infrared reflectivity modulation with polycrystalline electrochromic WO3 films. *Applied Physics Letters* 43.

Goldner, R. B. 1983. Electrochromic materials for controlled radiant energy transfer in buildings. *Proc. SPIE*, San Diego, CA, August 23–25, 1983, vol. 428, pp. 38–44.

Goldner, R. B., et al. 1984. Optical mobility studies on electrochromic films for variable reflectivity windows. *Proc. SPIE*, San Diego, CA, August 19–24, 1984.

Gralenski, N. 1982. Optical coatings by conveyorized atmospheric chemical vapor deposition (CVD). *Proc. SPIE*, Los Angeles, CA, January 28–29, 1982.

Gupta, B. K. 1983. Thermal stability and performance of molybdenum black solar selective absorbers. *Proc. SPIE*, San Diego, CA, August 23–25, 1983, vol. 428, pp. 182–186.

Haacke, G. 1982. Materials for transparent heat mirror coatings. *Proc. SPIE*, Los Angeles, CA, January 28–29, 1982, pp. 10–15.

Hamberg, I., et al. 1982. High quality transparent heat reflectors of reactively evaporated indium tin oxide. *Proc. SPIE*, Los Angeles, CA, January 28–29, 1982, pp. 31–36.

Hamberg, I., and C. G. Graqvist. 1983. Optical properties of transparent and heat-reflecting indium-tin-oxide films: experimental data and theorectical analysis. *Proc. SPIE*, San Diego, CA, August 23–25, 1983, pp. 2–7.

Harig, K., et al. 1983. Industrial realization of low-emittance oxide/metal/oxide films on glass. *Proc. SPIE*, San Diego, CA, August 23–25, 1983, pp. 9–13.

Harris, D. 1972, *BeadWall®*. U.S. Patent no. 3,903,665.

Herzenberg, S. A., and R. Silberglitt. 1982. Low temperature selective absorber research. *Proc. SPIE*, Los Angeles, CA, January 28–29, 1982, pp. 92–106.

Hodge, M. H. 1981. Heat-mirror transparent insulation and the trend toward high performance glazing. *Proc. U. S. Department of Energy Passive & Hybrid Solar Energy Program Update*, Washington, DC, August 9–12, 1981. Washington, DC: U.S. Department of Energy, February 1982, pp. 3–45.

Howson, R. P., and M. I. Ridge. 1982. Heat mirrors on plastic sheet using transparent oxide conducting coatings. *Proc. SPIE*, Los Angeles, CA, January 28–29, 1982, pp. 16–22.

Howson, R. P., et al. 1983. Optimized transparent and heat reflecting oxide and nitride films. *Proc. SPIE*, San Diego, CA, August 23–25, 1983, pp. 14–21.

Hunt, A. J. 1982. *Microporous Transparent Materials for Insulating Windows and Building Applications—Assessment Report.* LBL-15306. Berkeley, CA: Lawrence Berkeley Laboratory.

Hunt, A. 1983. Light scattering studies of silica aerogels. In *Science of Ceramics.* New York: Wiley.

Iler, R. K. 1952. *The Colloid Chemistry of Silica and Silicates.* Ithaca, NY: Cornell University Press.

Jacobson, M. R. 1983. Chemical vapor deposition of samarium chalcogenides: progress on fabricating thin film phase transision materials. *Proc. SPIE*, San Diego, CA, August 23–25, 1983, pp. 57–64.

Johnson, R. 1980. *Glazing Optimization Study for Energy Efficiency in Commercial Office Buildings.* LBL-12764. Berkeley, CA: Lawrence Berkeley Laboratory.

Johnson, T. 1981. *Solar Architecture: The Direct Gain Approach.* New York: McGraw-Hill.

Johnson, T. 1982. M.I.T. solar building no. 5's third year performance. *Passive Solar Journal* 1(3): 175–184.

Johnson, T. 1982. The MIT crystal pavilion, *Proc. 7th National Passive Solar Conference*, Knoxville, TN, August 30–September 1, 1982. Newark, DE: American Section of the International Solar Energy Society, pp. 255–260.

Johnson, T. 1984. Windows. *Solar Age*, August 1984.

Jurison, J., R. E. Peterson, and H. Y. B. Mar. 1975. *J. Vacuum Science and Technolgy*, 12.

Karlsson, B., and C. G. Ribbing. 1982. Optical properties of transparent heat mirrors based on thin films of TiN, ZrN, and HfN. *Proc. SPIE*, Los Angeles, CA, January 28–29, 1982, pp. 52–57.

Kiyoshi, C., et al. 1982. Transparent heat insulating coatings on a polyester film. *Proc. SPIE*, Los Angeles, CA, January 28–29, 1982, pp. 23–30.

Kjelshus, E. 1979. The Kjelshus house, Winterview, *Proc. 3rd National Passive Solar Conference*, San Jose, CA, January 11–13, 1979. Newark, DE: American Section of the International Solar Energy Society, pp. 750–753.

Klodt, G. J. 1981. Honeycomb thermal curtain development. *Proc. U.S. Department of Energy Passive & Hybrid Solar Energy Program Update*, Washington, DC, August 9–12, 1981. Washington, DC: U.S. Department of Energy, February 1982, pp. 3–16.

Kreith, F. 1958. *Principles of Heat Transfer*. Scranton, PA: International Textbook Co.

Lampert, C. M. 1979. Coatings for enhanced photothermal energy collection. *Solar Energy Materials*, vol. 1.

Lampert, C. M. 1980. *Thin Film Electrochromic Materials for Energy-Efficient Windows*. LBL-10862. Berkeley, CA: Lawrence Berkeley Laboratory.

Lampert, C. M. 1981a. Heat mirror coatings for energy conserving windows. *Solar Energy Materials* 6:1–42.

Lampert, C. M. 1981b. Materials chemistry and optical properties of transparent conducting thin films for solar energy utilization. *The Vortex of the American Chemical Society* 42(10).

Lampert, C. M. 1982. *Solar Optical Materials for Innovative Window Design*. LBL-14694, Berkeley, CA: Lawrence Berkeley Laboratory.

Lampert, C. M. 1983. *Microstructure and Optical Properties of High Transmission Coatings for Plastics*. Berkeley, CA: Lawrence Berkeley Laboratory.

Lampert, C. M. 1984. Electrochromic materials and devices for energy efficient windows. *Solar Energy Materials* 11.

Lee, P. K., and M. K. Debe. 1980. Measurement and modeling of the reflectance-reducing properties of graded index microstructured surfaces. *Photo. Science and Engineering* 24.

Low, D. 1981. Insulating curtain wall. *Proc. U.S. Department of Energy Passive & Hybrid Solar Energy Program Update*, Washington, DC, August 9–12, 1981. Washington, DC: U.S. Department of Energy, February 1982, pp. 3–53.

Maloney, T. 1978. Four generations of water-wall design. *Proc. 2d National Passive Solar Conference*, Philadelphia, PA, March 16–18, 1978. Newark, DE: American Section of the International Solar Energy Society.

Mason, J. J. 1982. Maxorb—a new selective surface on nickel. *Proc. SPIE*, Los Angeles, CA, Jan. 28–29, 1982, pp. 139–146.

Mazria, E. 1979. *The Passive Solar Energy Book*. Emmaus, PA: Rodale Press.

Miska, H. A., 1983. The use of leached gradient index anti-reflection surfaces on borosilicate glass solar collector tubes. *Proc. SPIE*, San Diego, CA, August 23–25, 1983, pp. 94–99.

Moore, S. W. 1982. Solar absorber selective paint research. *Proc. SPIE*, Los Angeles, CA, January 28–29, 1982, pp. 148–155.

Neeper, D., and R. McFarland. 1982. *Some Potential Benefits of Fundamental Research for the Passive Solar Heating and Cooling of Buildings*. LA-9425-MS. Los Alamos, NM: Los Alamos National Laboratory.

Niles, P., and C. Haggard. 1980. *Passive Solar Handbook*. Sacramento, CA: California Energy Commission.

Rauh, D., and S. Cogan. 1984. Materials for electrochromic windows. *Proc. SPIE*, San Diego, CA, August 19–24, 1984.

Reusch, 1967. Preventing the intrusion of unwelcome sunshine by means of using glass with variable transmission. *Proc. CIE. Conference.*

Roos, A. 1983. Stability problems with oxidized copper solar absorbers. *Proc. SPIE*, San Diego, CA, August 23–25, 1983, pp. 175–181.

Rubin, M., R. Creswick, and S. Selkowitz. 1980. Transparent heat mirrors for windows-thermal performance. *Proc. 5th National Passive Solar Conference*, Amherst, MA, October 19–26, 1980. Newark, DE: American Section of the International Solar Energy Society, pp. 990–994.

Rubin, M., and S. Selkowitz. 1981. Thermal performance of windows having high solar transmittance. *Proc. 6th National Passive Solar Conference*, Portland, OR, September 8–12, 1981. Newark, DE: American Section of the International Solar Energy Society.

Rubin, M., and C. Lampert. 1983. Transparent silica aerogels for window insulation. *Solar Energy Materials* 7.

Rubin, M. 1984. *Optical Constants and Bulk Optical Properties of Soda-Lime Silica Glasses for Windows.* LBL-13572. Berkeley, CA: Lawrence Berkeley Laboratory.

Sato, Y., et al. 1982. Improvement in open-circuit memory, current efficiency, and response speed of an amorphous WO_3 solid-state electrochromic device. *Japanese J. Applied Physics* 21(11): 1642–1646.

Schlotter, P., and L. Pickelmann. 1982. The xerogel structure of thermally evaporated tungsten oxide layers. *J. Electronic Materials* 2: 207–223.

Schnebly, J. 1978. The window quilt insulating shade. *Proc. 2d National Passive Solar Conference*, Philadelphia, PA, March 16–18. 1978, Newark, DE: American Section of the International Solar Energy Society, pp. 314–316.

Selkowitz, S. E. 1979. Thermal performance of insulating window systems. *ASHRAE Transactions* 6(2).

Selkowitz, S. E., C. M. Lambert, and M. Rubin. 1982. Advanced optical and thermal technologies for aperture control. *Proc. U.S. Department of Energy Passive and Hybrid Solar Energy Program Update*, Washington, DC, September 15–17, 1982. Washington, DC: U.S. Department of Energy.

Shore, R. 1978. Self-inflating movable insulation system. *Proc. 2d National Passive Solar Conference*, Philadelphia, PA, March 16–18, 1978. Newark, DE: American Section of the International Solar Energy Society.

Shurcliff, W. A. 1980. *Thermal Shutters and Shades.* Andover, MA: Brick House Press.

Smith, G. P. 1966. Chameleon in the sun: photochromic glass. *IEEE Spectrum* 3.

Telkes, M. 1949. Storing solar heat in chemicals. *Heat and Vent*, November 1949, pp. 80–86.

Telkes, M. 1964. Solar heat storage. ASME paper no. 64-WA/SOL-9.

Vora, H., and T. J. Moravec. 1981. Structural investigation of thin films of diamond-like carbon. *Applied Physics* 52: 6151.

Vossen, J. L. 1977. Transparent conducting films. *Physics of Thin Films* 9.

Wittwer, V., et al. 1983. Heat loss mechanisms in transparent insulation with honeycomb structures. *Proc. SPIE*, San Diego, CA, August 23–25, 1983, pp. 100–104.

6 Analytical Results for Specific Systems

Robert W. Jones

6.1 Scope of Chapter

This chapter describes calculated thermal performance results based on mathematical descriptions of the behavior of passive solar heating systems. The calculations were made using actual historic weather data from locations in the United States and Canada. Results describe performance features, such as expected energy savings or the degree of thermal comfort achieved by the system.

The passive solar heating systems that are covered in greatest detail are direct-gain, thermal-storage-wall, and sunspace systems. In direct-gain systems, thermal storage mass in the occupied space is directly illuminated by sunshine transmitted through windows. In thermal-storage-wall systems, thermal storage mass is placed between the window and the occupied space. The thermal-storage-wall systems include Trombe walls, in which the thermal storage mass is concrete masonry or other similar heavy materials, and water walls, in which the thermal storage mass is water in containers. The Trombe walls may be either vented or unvented. In sunspace systems, solar gains are first produced in a direct-gain space that is largely uncontrolled in temperature and that is occupied only occasionally. Solar heat produced in the sunspace is passed to the occupied space through a common wall that divides the sunspace from the adjoining room. A few other passive systems are included, but the results are less general. Such systems include roof ponds and thermosiphoning air panels.

The results included are those of sufficient generality, useful either for design or for policy making. Results that are primarily for the purpose of comparing one calculation method with another, results that are produced primarily to validate the accuracy of the method against measured data, results in the form of economic or policy generalizations, as well as results that serve the purpose of optimizing the design of a particular building in one location are not included.

Both residential and nonresidential applications of passive solar heating are included, although the majority of the results are oriented to residential applications because they emphasize the envelope-generated gains and the envelope-induced losses. The interaction of passive systems with complex mechanical equipment for backup heating and ventilation are not described.

The analysis methods represented include two types, computer simulations (see Niles, chapter 3 of this volume) and simplified methods (see Jones and Wray, chapter 4 of this volume) usually based on computer simulations. The simplified methods are predominantly correlations of computer simulation results; for example, the solar load ratio method (SLR) (see Jones and Wray).

6.2 Building Variables

Calculated results are expressed in terms of certain independent variables that characterize the building. The following are the important building variables.

For a given building type, a key variable that determines performance is the collector area (A_c). Because most passive collectors are vertical or attached to vertical walls, it is convenient for uniformity to define the *projected area* (A_p), which is the projection of the collector area on a vertical plane.

A second key variable is the building load. The building load is conveniently expressed as the *net load coefficient* (NLC), which is the heat loss from the nonsolar portions of the building envelope per day per degree of indoor-outdoor temperature difference (Btu/°F day or Wh/°C day). The building *total load coefficient* (TLC) (Btu/°F day or Wh/°C day) is the building net load coefficient (NLC) plus the load coefficient for the solar aperture (SLC), $TLC = NLC + SLC$. Because most aspects of performance scale are similarly relative to NLC and $1/A_p$, it is possible to lump the effects of these two variables into a single variable that is the product of the two. This is sometimes done as the ratio A_p/NLC, which is called a *normalized collector area*. The other possibility is

$$LCR = NLC/A_p, \tag{6.1}$$

the *load collector ratio*.

The thermostat setpoint (T_{set}) for auxiliary heat has an important effect on auxiliary heat consumption. The rate of internal heat generation (by people, lights, etc.) also has a significant effect on auxiliary heat consumption. The combined effect of these two variables may be represented by the building balance-point temperature (T_b), as in

$$T_b = T_{set} - Q_{int}/TLC, \tag{6.2}$$

where Q_{int} is the internal heat generation rate (Btu/day or Wh/day). The balance-point temperature figures in energy calculations as the degree-days base temperature. For example, $Q_{net} = NLC\ DD_b$, where DD_b is the heating degree-days calculated to the base temperature T_b (see equation [4]).

6.3 Types of Results

6.3.1 Auxiliary Heat

Performance results are expressed in terms of various performance measures. The most fundamental and frequently calculated performance measure is the *auxiliary heat*, Q_{aux}, which is the heat required by the building in a given period of time, usually a year. Q_{aux} is the actual heat delivered to the conditioned space. The fuel consumed to produce this heat would involve the efficiency of the auxiliary heating system.

The following performance measures are derived from the auxiliary heat.

6.3.1.1 Solar Savings The *solar savings*, Q_{sav}, is the auxiliary heat saved by a solar building relative to a similar nonsolar building. This quantity is, of course, ambiguous in the absence of a definition of a "similar nonsolar building." The most precise approach to such a definition is to define a conventional building that satisfies the same architectural program as does the solar building but that has no solar collection area beyond the normal complement of windows. Short of this rather elaborate procedure, a reasonable definition of a "similar nonsolar building" is widely accepted as one in which the passive solar heating apertures are replaced by perfectly insulated walls. In this case,

$$Q_{sav} = Q_{net} - Q_{aux}, \tag{6.3}$$

where Q_{net}, called the *net reference load*, is the heat loss from the nonsolar portions of the building envelope, that is, all of the building envelope except the passive solar heating apertures. If Q_{net} is calculated using a degree-day model of heat loss,

$$Q_{net} = NLC\, DD_b, \tag{6.4}$$

where DD_b is the heating degree-days relative to an appropriate base temperature T_b.

6.3.1.2 Solar Savings Fraction. Q_{sav} is often expressed in terms of dimensionless fractions. The most common one is the *solar savings fraction* (SSF), defined by

$$SSF = Q_{sav}/Q_{net}. \tag{6.5}$$

Because the quantity SSF is roughly the fraction of energy saved by the solar building compared with that of a similar but nonsolar building, it is a rough figure of merit for the passive solar system. It is, however, more precise to say that SSF is simply the dimensionless parameter that determines the auxiliary heat through the equation

$$Q_{aux} = (1 - SSF)Q_{net}. \tag{6.6}$$

6.3.1.3 Solar Heating Fraction Another dimensionless measure of solar savings is the *solar heating fraction* (SHF), although the solar savings in question is not that defined by equation (3); it is rather the heat saved by the solar building relative to a similar building in which heat is lost through the solar heating apertures but through which there are no solar gains. In other words, the solar savings in this case is $Q_{tot} - Q_{aux}$, where Q_{tot} is the total heat loss through the envelope of the solar building including the solar heating apertures. If Q_{tot} is determined by a degree-day model of heat loss,

$$Q_{tot} = TLC\,DD_b. \tag{6.7}$$

SHF is, therefore,

$$SHF = (Q_{tot} - Q_{aux})/Q_{tot}. \tag{6.8}$$

The quantity SHF is not a very useful figure of merit for a passive heating system, but it may be used as well as SSF to determine the auxiliary heat, as in

$$Q_{aux} = (1 - SHF)Q_{tot}. \tag{6.9}$$

The majority of analytical results pertain to annual auxiliary heat. These results are reviewed in section 7.4.

6.3.1.4 Performance Curves For a given passive system configuration, the annual auxiliary heat depends upon the climate, the base temperature (building balance-point temperature), and the normalized collector area, or LCR. Therefore, for a given passive system configuration in a given

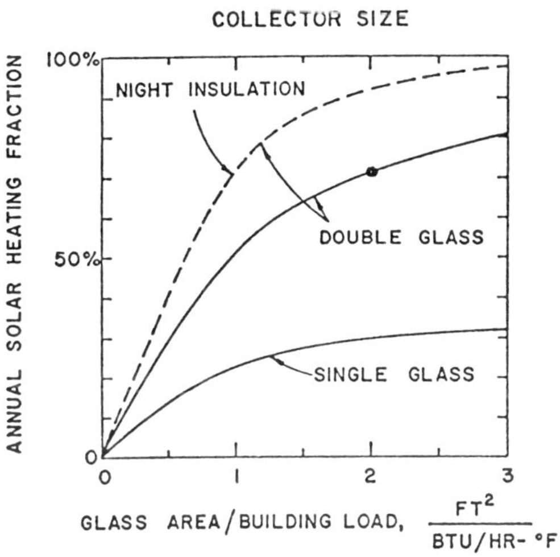

Figure 6.1
Annual solar heating fraction vs. ratio of glass area to building load, $ft^2/(Btu/h\ ^\circ F)$, for a water-wall system in Los Alamos, New Mexico. 1 $ft^2/(Btu/h\ ^\circ F) = 0.176\ m^2/(W/^\circ C)$ Source: Balcomb et al. 1977a.

location and for a given base temperature, the solar heating performance can be plotted as a function of *LCR* or normalized collector area (*1/LCR*). Such results are performance curves. For example, figure 6.1 is a plot of *SHF* versus *1/LCR* for three different water-wall systems in Los Alamos, New Mexico, assuming a base temperature of 65°F. The points on the curves were generated by computer simulation.

6.3.1.5 LCR Tables Performance results versus *LCR* presented in tabular form are called LCR tables. An LCR table applies to a single location. Several passive system configurations may be presented in the same table. The base temperature is fixed, but tables may be presented for two or more values. The most comprehensive set of LCR tables is contained in the ASHRAE manual *Passive Solar Heating Analysis* (Balcomb, et al. 1984). Each table contains 94 passive systems. There is a table for each of the listed 223 locations in the United States and Canada. Table 6.1 for St. Louis, Missouri, is an example of these tables. There are tables for two base temperatures, 55°F and 65°F. Because *SSF* is quite linear relative to base temperature in the range of 50°F to 70°F, the LCR tables can be applied with reasonable accuracy to any base temperature in this range by inter-

Table 6.1
LCR table for St. Louis, Missouri

Type	Solar savings fraction (%)																Conservation factor									
	Base temp: 55 F (2804 DD)								65 F (4754 DD)								55 F					65 F				
	LCR 100	70	50	40	30	25	20	15	100	70	50	40	30	25	20	15	100	50	30	20	15	100	50	30	20	15
WW A1	20	24	28	32	37	40	45	52	16	18	21	23	27	29	32	37	1.5	1.7	1.9	2.0	2.2	1.9	2.1	2.4	2.6	2.8
WW A2	22	28	34	39	47	52	59	67	15	20	24	28	34	38	43	50	1.2	1.4	1.5	1.7	1.8	1.5	1.7	1.9	2.0	2.2
WW A3	22	29	37	43	51	56	63	72	15	20	26	30	36	41	47	55	1.2	1.3	1.5	1.6	1.8	1.4	1.5	1.7	1.9	2.0
WW A4	22	30	38	44	53	59	66	75	14	20	26	31	38	43	49	57	1.1	1.3	1.4	1.6	1.7	1.3	1.5	1.7	1.8	2.0
WW A5	22	30	39	46	55	61	69	77	14	20	27	32	40	45	51	60	1.1	1.2	1.4	1.5	1.7	1.3	1.4	1.6	1.8	1.9
WW A6	23	31	40	47	57	63	70	79	14	20	28	33	41	46	53	61	1.1	1.2	1.4	1.5	1.7	1.2	1.4	1.6	1.7	1.9
WW B1	17	22	27	31	36	40	46	53	12	14	17	20	23	25	28	32	1.4	1.6	1.8	2.0	2.1	1.8	2.1	2.4	2.7	3.0
WW B2	23	31	40	46	56	62	69	78	15	22	29	34	41	47	53	62	1.1	1.2	1.4	1.5	1.7	1.2	1.4	1.6	1.7	1.9
WW B3	27	37	47	54	64	70	77	85	19	26	34	41	49	55	62	71	1.0	1.1	1.3	1.5	1.6	1.1	1.3	1.5	1.6	1.8
WW B4	27	38	49	56	67	73	80	87	19	27	37	44	53	59	67	76	1.0	1.1	1.3	1.4	1.6	1.1	1.2	1.4	1.5	1.7
WW B5	26	37	48	56	66	72	79	87	18	27	36	43	53	59	67	77	1.0	1.1	1.3	1.4	1.6	1.1	1.2	1.4	1.5	1.7
WW C1	29	39	49	56	66	72	79	87	20	28	37	43	52	58	65	74	1.0	1.1	1.3	1.4	1.6	1.1	1.3	1.4	1.6	1.7
WW C2	27	37	47	55	65	71	78	86	19	27	36	42	51	57	65	74	1.0	1.1	1.3	1.4	1.6	1.1	1.3	1.4	1.6	1.7
WW C3	32	44	56	64	75	80	87	93	23	33	44	52	62	69	76	85	0.9	1.0	1.2	1.4	1.5	1.0	1.1	1.3	1.4	1.6
WW C4	29	41	53	61	71	77	84	91	20	30	41	49	59	66	73	82	0.9	1.0	1.2	1.4	1.5	1.0	1.1	1.3	1.5	1.6
TW A1	20	24	28	31	36	40	45	52	16	18	21	23	27	29	32	37	1.5	1.7	1.9	2.0	2.2	1.9	2.1	2.3	2.5	2.7
TW A2	20	25	31	36	43	48	54	62	15	19	23	26	31	35	39	46	1.3	1.5	1.6	1.8	1.9	1.6	1.8	1.9	2.1	2.3
TW A3	20	26	33	38	46	51	58	66	14	18	23	27	33	37	42	50	1.2	1.4	1.5	1.7	1.8	1.5	1.7	1.8	2.0	2.2
TW A4	19	26	33	38	46	52	59	68	13	18	23	27	33	38	43	51	1.2	1.4	1.5	1.7	1.8	1.4	1.6	1.8	1.9	2.1
TW B1	19	23	28	32	37	41	46	54	15	18	21	24	27	30	34	39	1.5	1.6	1.8	1.9	2.1	1.8	2.0	2.2	2.4	2.6
TW B2	19	24	30	35	42	47	53	62	14	17	22	25	31	34	39	46	1.3	1.5	1.6	1.8	1.9	1.6	1.7	1.9	2.1	2.2
TW B3	18	24	30	35	43	48	54	63	13	17	22	25	31	35	40	47	1.3	1.4	1.6	1.7	1.9	1.5	1.7	1.9	2.0	2.2
TW B4	17	23	29	34	41	45	52	61	12	16	21	24	30	33	39	46	1.3	1.5	1.6	1.8	1.9	1.6	1.7	1.9	2.1	2.2
TW C1	18	22	27	31	37	41	47	55	14	17	20	23	28	31	35	41	1.4	1.6	1.8	1.9	2.0	1.7	1.9	2.1	2.3	2.4
TW C2	17	22	27	32	38	43	49	57	13	16	20	23	28	32	36	43	1.4	1.5	1.7	1.8	2.0	1.6	1.8	2.0	2.2	2.3
TW C3	17	21	26	30	37	41	47	55	12	15	19	23	27	31	35	42	1.4	1.6	1.7	1.9	2.0	1.7	1.9	2.0	2.2	2.3
TW C4	16	20	24	28	33	37	42	50	12	15	18	21	25	28	32	38	1.5	1.7	1.8	2.0	2.1	1.8	2.0	2.2	2.3	2.5
TW D1	15	18	23	27	32	36	41	48	10	12	15	17	20	23	26	30	1.5	1.7	1.9	2.0	2.2	1.9	2.2	2.5	2.8	3.0
TW D2	22	29	36	42	51	57	64	73	15	21	27	31	38	43	49	57	1.2	1.3	1.4	1.6	1.7	1.3	1.5	1.7	1.8	2.0
TW D3	23	31	40	46	55	61	68	77	16	22	29	34	42	47	53	62	1.1	1.2	1.4	1.5	1.7	1.3	1.4	1.6	1.8	1.9

TW D4	25	34	44	51	61	67	74	83	18	25	33	39	48	53	61	70	1.0	1.2	1.3	1.5	1.6	1.2	1.3	1.5	1.6	1.8
TW D5	25	34	44	51	61	67	75	83	17	25	33	40	49	55	62	72	1.0	1.2	1.3	1.5	1.6	1.1	1.3	1.4	1.6	1.7
TW E1	27	35	44	51	61	66	74	82	19	26	33	39	47	52	59	58	1.0	1.2	1.3	1.5	1.6	1.2	1.3	1.5	1.7	1.8
TW E2	26	35	44	51	61	67	74	83	18	26	33	39	48	53	61	70	1.0	1.2	1.3	1.5	1.6	1.2	1.3	1.5	1.6	1.8
TW E3	31	42	53	61	71	77	83	90	23	32	41	49	58	65	72	81	0.9	1.1	1.2	1.4	1.5	1.0	1.2	1.3	1.5	1.6
TW E4	29	39	50	58	68	75	81	89	21	29	39	46	56	62	70	79	0.9	1.1	1.2	1.4	1.5	1.0	1.2	1.3	1.5	1.6
TW F1	18	22	27	30	36	39	45	52	14	16	20	22	26	28	32	37	1.5	1.7	1.8	2.0	2.1	1.9	2.1	2.3	2.5	2.7
TW F2	18	23	29	34	40	45	51	60	12	16	20	24	29	32	37	43	1.3	1.5	1.6	1.8	1.9	1.6	1.8	2.0	2.2	2.3
TW F3	17	23	29	34	42	47	54	62	11	15	20	24	29	33	38	46	1.3	1.4	1.6	1.7	1.9	1.5	1.7	1.9	2.1	2.2
TW F4	15	21	27	33	40	46	52	62	9	13	18	22	28	32	38	45	1.3	1.4	1.6	1.7	1.9	1.5	1.7	1.9	2.0	2.2
TW G1	16	20	25	29	35	39	44	52	12	15	18	21	25	28	31	37	1.5	1.7	1.8	2.0	2.1	1.8	2.0	2.2	2.4	2.6
TW G2	15	20	26	30	37	42	48	57	10	13	18	21	26	29	34	41	1.4	1.5	1.7	1.8	2.0	1.6	1.8	2.0	2.2	2.3
TW G3	13	18	24	29	36	41	47	56	8	12	16	20	25	29	34	41	1.4	1.5	1.7	1.8	2.0	1.6	1.8	2.0	2.2	2.3
TW G4	11	15	21	25	32	36	42	51	6	10	14	17	22	25	30	37	1.4	1.6	1.7	1.9	2.0	1.7	1.9	2.1	2.2	2.4
TW H1	13	17	22	26	32	36	42	50	9	12	15	18	22	25	29	35	1.5	1.7	1.8	2.0	2.1	1.8	2.0	2.2	2.4	2.5
TW H2	11	15	20	24	30	35	41	49	6	10	13	16	21	24	28	35	1.5	1.7	1.8	2.0	2.1	1.8	2.0	2.1	2.3	2.5
TW H3	9	13	18	21	27	31	37	45	5	8	11	14	18	21	26	32	1.6	1.7	1.9	2.0	2.2	1.9	2.0	2.2	2.4	2.5
TW H4	7	10	14	17	22	25	30	37	4	6	9	11	15	17	21	26	1.7	1.9	2.1	2.2	2.3	2.1	2.3	2.4	2.6	2.7
TW I1	12	16	20	24	29	32	37	44	8	10	13	15	18	20	23	27	1.6	1.8	2.0	2.1	2.3	2.0	2.3	2.6	2.9	3.1
TW I2	19	25	33	39	47	53	60	70	12	17	23	28	35	39	46	54	1.2	1.3	1.5	1.6	1.8	1.4	1.5	1.7	1.9	2.0
TW I3	20	27	36	42	51	57	64	74	13	19	26	31	38	43	50	59	1.1	1.3	1.4	1.6	1.7	1.3	1.5	1.6	1.8	1.9
TW I4	22	31	40	47	57	63	71	80	15	22	30	36	44	50	58	67	1.0	1.2	1.3	1.5	1.6	1.2	1.3	1.5	1.7	1.8
TW I5	22	31	41	49	58	65	73	81	15	22	31	37	46	52	60	69	1.0	1.2	1.3	1.5	1.6	1.2	1.3	1.5	1.6	1.8
TW J1	24	33	42	49	58	64	72	80	17	23	31	37	45	50	57	66	1.0	1.2	1.3	1.5	1.7	1.2	1.4	1.5	1.7	1.8
TW J2	23	32	41	48	58	64	72	81	16	23	31	37	45	51	58	67	1.0	1.2	1.3	1.5	1.6	1.2	1.3	1.5	1.7	1.8
TW J3	28	38	50	58	68	74	81	89	20	29	39	46	56	62	70	79	0.9	1.1	1.2	1.4	1.5	1.0	1.2	1.3	1.5	1.6
TW J4	26	36	47	55	66	72	79	87	18	27	36	44	53	60	68	77	1.0	1.1	1.3	1.4	1.6	1.1	1.2	1.4	1.5	1.7
DG A1	15	20	25	28	33	36	39	43	9	12	15	18	20	22	24	26	1.4	1.7	2.0	2.5	3.0	1.8	2.2	2.7	3.7	—
DG A2	17	23	30	35	42	46	52	59	12	16	21	25	30	34	38	44	1.3	1.4	1.6	1.9	2.1	1.5	1.7	1.9	2.2	2.5
DG A3	21	28	37	44	53	59	66	74	16	21	28	33	41	46	53	62	1.1	1.2	1.4	1.6	1.8	1.3	1.4	1.6	1.7	1.9
DG B1	15	20	26	30	36	40	45	51	10	13	16	18	22	24	27	31	1.4	1.5	1.8	2.0	2.4	1.8	2.1	2.4	2.8	3.4
DG B2	18	24	31	37	45	51	58	66	12	17	22	26	32	36	41	49	1.3	1.4	1.5	1.7	1.9	1.5	1.7	1.8	2.0	2.2
DG B3	22	29	38	45	56	63	71	80	16	22	29	34	43	48	56	66	1.1	1.2	1.3	1.5	1.7	1.3	1.4	1.5	1.6	1.8
DG C1	18	24	31	36	44	49	55	63	12	16	20	23	28	31	35	41	1.3	1.4	1.6	1.8	2.1	1.6	1.8	2.0	2.3	2.6
DG C2	21	27	36	42	52	58	66	74	15	20	26	30	37	42	49	57	1.2	1.3	1.4	1.6	1.8	1.4	1.5	1.6	1.8	2.0
DG C3	24	33	43	51	62	69	77	86	19	25	33	39	48	54	63	73	1.1	1.1	1.3	1.4	1.6	1.2	1.3	1.4	1.5	1.7

Table 6.1 (cont.)

Type	Solar savings fraction (%)																Conservation factor									
	Base temp: 55 F (2804 DD)								65 F (4754 DD)								55 F					65 F				
	LCR 100	70	50	40	30	25	20	15	100	70	50	40	30	25	20	15	100	50	30	20	15	100	50	30	20	15
SS A1	24	30	36	42	49	54	60	68	19	24	29	32	38	42	47	53	1.2	1.4	1.5	1.7	1.9	1.4	1.6	1.8	2.0	2.2
SS A2	30	38	47	54	63	68	75	83	24	31	38	44	51	56	63	71	1.0	1.2	1.3	1.5	1.6	1.1	1.3	1.5	1.6	1.8
SS A3	22	27	33	38	44	49	55	62	18	21	26	29	34	37	41	47	1.3	1.5	1.6	1.8	2.0	1.5	1.8	2.0	2.2	2.4
SS A4	29	37	46	52	61	67	73	81	23	30	37	42	50	54	61	69	1.0	1.2	1.4	1.5	1.7	1.2	1.3	1.5	1.7	1.8
SS A5	24	28	34	38	44	49	54	62	19	23	27	30	35	38	42	48	1.3	1.5	1.7	1.8	2.0	1.5	1.8	2.0	2.2	2.4
SS A6	30	38	47	53	62	68	74	82	24	30	37	43	50	55	62	70	1.0	1.2	1.3	1.5	1.7	1.2	1.3	1.5	1.7	1.8
SS A7	21	25	30	33	38	42	47	54	17	20	23	26	29	32	35	39	1.5	1.7	1.8	2.0	2.2	1.7	2.0	2.3	2.5	2.8
SS A8	29	36	45	51	60	65	72	80	23	29	36	41	48	53	59	67	1.1	1.2	1.4	1.5	1.7	1.2	1.4	1.6	1.7	1.9
SS B1	21	26	32	36	43	48	54	62	16	20	24	28	33	36	41	47	1.3	1.5	1.6	1.8	1.9	1.5	1.7	1.9	2.1	2.3
SS B2	26	34	42	49	57	63	70	78	21	27	34	39	46	52	58	67	1.1	1.2	1.4	1.5	1.7	1.2	1.4	1.5	1.7	1.8
SS B3	19	24	29	34	40	44	50	58	15	18	22	25	30	33	37	43	1.4	1.5	1.7	1.9	2.0	1.6	1.8	2.1	2.3	2.5
SS B4	26	33	41	47	56	61	68	77	20	26	33	38	45	50	56	64	1.1	1.2	1.4	1.6	1.7	1.2	1.4	1.6	1.7	1.9
SS B5	19	23	28	32	37	41	46	54	16	19	22	25	29	31	35	40	1.5	1.6	1.8	2.0	2.1	1.7	2.0	2.2	2.4	2.6
SS B6	26	33	41	48	56	62	69	77	20	26	33	38	45	50	56	65	1.1	1.2	1.4	1.6	1.7	1.2	1.4	1.6	1.7	1.9
SS B7	18	21	25	28	33	36	41	48	14	17	20	22	25	27	30	34	1.6	1.8	2.0	2.1	2.3	1.9	2.2	2.4	2.7	2.9
SS B8	24	31	39	45	54	59	66	75	19	25	31	36	43	48	54	62	1.1	1.3	1.4	1.6	1.7	1.3	1.4	1.6	1.8	1.9
SS C1	19	25	32	38	45	50	57	66	13	17	22	26	31	35	40	47	1.2	1.4	1.5	1.7	1.8	1.5	1.7	1.9	2.1	2.2
SS C2	23	31	40	47	56	62	69	78	16	22	29	34	42	47	54	63	1.1	1.2	1.4	1.5	1.7	1.2	1.4	1.6	1.7	1.9
SS C3	17	22	27	31	37	41	46	54	12	15	19	22	26	29	33	39	1.4	1.6	1.8	1.9	2.0	1.7	1.9	2.1	2.3	2.5
SS C4	21	28	36	42	50	55	63	72	15	20	26	31	38	42	48	57	1.2	1.3	1.5	1.6	1.8	1.3	1.5	1.7	1.8	2.0
SS D1	28	35	43	49	58	63	69	77	22	27	33	38	44	49	55	62	1.1	1.2	1.4	1.6	1.7	1.2	1.4	1.6	1.8	2.0
SS D2	34	44	55	62	71	77	83	89	27	35	44	50	59	65	72	80	0.9	1.1	1.2	1.4	1.6	1.0	1.2	1.4	1.5	1.7
SS D3	26	32	39	44	52	57	63	71	21	25	30	34	40	44	49	56	1.2	1.3	1.5	1.7	1.8	1.4	1.6	1.8	2.0	2.1
SS D4	33	43	53	60	69	75	81	88	26	34	42	49	57	63	69	78	0.9	1.1	1.3	1.4	1.6	1.1	1.2	1.4	1.6	1.7
SS E1	24	30	38	43	51	56	63	71	18	22	28	32	38	42	47	54	1.2	1.3	1.5	1.7	1.8	1.4	1.6	1.8	2.0	2.1
SS E2	29	38	48	55	65	70	77	85	22	29	37	43	52	57	64	73	1.0	1.1	1.3	1.5	1.6	1.1	1.3	1.5	1.6	1.8
SS E3	21	26	32	36	43	47	53	61	17	20	24	28	32	35	40	46	1.3	1.5	1.7	1.8	2.0	1.6	1.8	2.0	2.2	2.4
SS E4	27	35	44	51	60	66	73	81	21	27	35	40	48	53	59	68	1.0	1.2	1.3	1.5	1.7	1.2	1.4	1.5	1.7	1.8

Source: Balcomb et al. 1984.

polation or extrapolation. Once *LCR* is known, determining *SSF* is very easy with an LCR lookup table. The disadvantage is that a given LCR table pertains to just one location and to a limited number of system configurations.

6.3.1.6 Response Surfaces Building variables in addition to *LCR* may be accounted for in tabular data the way it was done for base temperature in table 6.1, but the size and number of tables quickly get out of hand. A way to present the information in a more compact form is to fit the data to a simple algebraic form. When there are two or more variables, the algebraic form is called a *response surface.* A response surface is a very convenient result because of its ease of use in a variety of design and optimization studies and because it provides assistance in choosing the minimum number of analysis cases needed to adequately define the results. Phillips and Sebald (n.d.) describe the response surface approach.

6.3.1.7 Guideline Results In the analytical results described above, *LCR* and other system characteristics are variables. It is, however, of great interest to specify the particular system characteristics, especially *LCR*, that are appropriate for a given set of circumstances. Although many criteria are possible, one widely used approach is to design to the life-cycle-cost minimum, considering cost of fuel, cost of passive solar, and cost of conservation. This approach has been used to establish passive solar and conservation guidelines (Balcomb 1983b, Balcomb et al. 1984, Balcomb 1986). Analytical results are available in the form of tables and maps in which the passive solar heating performance is given, assuming that guideline levels of passive solar heating and conservation are employed (Balcomb et al. 1984, Balcomb 1986). An example of this approach is figure 6.2, a map of the annual solar savings produced by a direct-gain system using the guideline *LCR*.

6.3.1.8 Sensitivity Data Analytical results pertain to specific passive systems. Occasionally, principal parameters, such as *LCR* and base temperature, are assigned several discrete values so that results for any value may be estimated. Most parameters, however, are fixed. Sensitivity data relieve this situation somewhat by showing the dependence of performance on a continuous range of selected design parameters, at least for a few cities and base-case designs. Figures 6.3 and 6.4 are examples of sensitivity data, showing the sensitivity of performance to thermal storage mass for two different passive system configurations in Los Alamos, New Mexico.

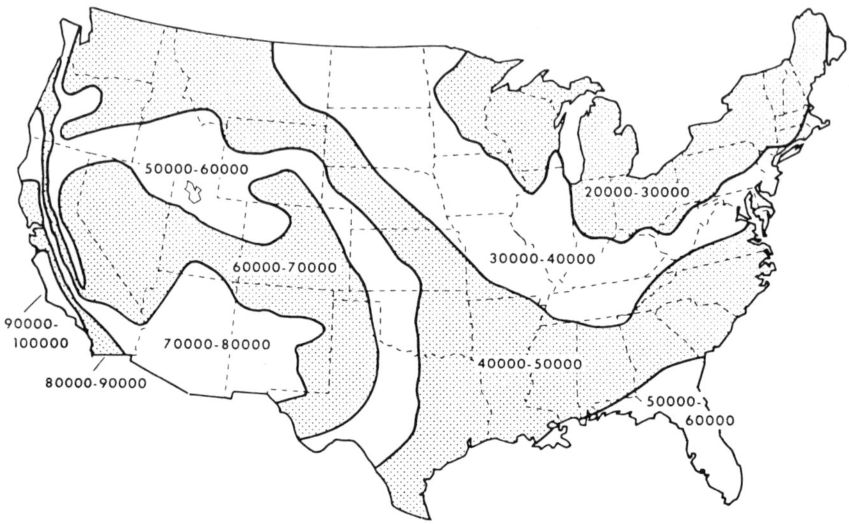

Figure 6.2
Annual solar savings per unit of south glazing area, Btu/ft^2, for a double-glazed direct-gain system. Heat storage is 4-in. (10-cm)-thick concrete with a surface area six times the south glazing area. 1 Btu/ft^2 = 3.15 Wh/m^2. Source: Balcomb 1986.

6.3.2 Backup Heating Demand

Auxiliary heat requirement relates to the total heat requirement over an extended period, such as a year; backup heating demand relates to the instantaneous rate of auxiliary heat delivery required during one hour. Of particular interest is the peak demand, which is the maximum heat delivery rate required during some worst-case "design" hour. Backup heating equipment is sized to deliver heat at this rate so the thermostat setting can be maintained even in worse-case conditions. An issue related to the back-up heating demand during a design hour is the typical backup heating demand profile by time of day. This demand profile is important when backup heat is electric, because it relates to the electric utility peaking characteristics. Results are reviewed in section 6.5.

6.3.3 Thermal Comfort Criteria

Energy performance goals for a building must not be achieved at the expense of acceptable thermal comfort conditions. Therefore, thermal comfort criteria and associated analytical results are essential to evaluate

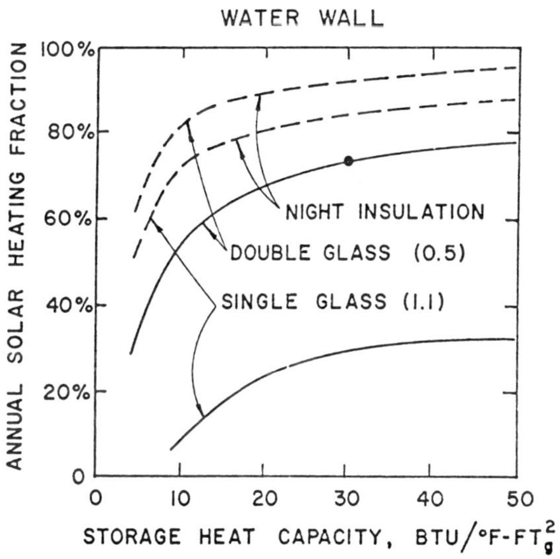

Figure 6.3
Annual solar heating fraction vs. storage heat capacity per unit of south glazing area, Btu/°F ft^2, for a water-wall system with a load collector ratio (LCR) of 12 Btu/°F day ft^2 (68 Wh/°C day m^2) in Los Alamos, New Mexico. 1 Btu/°F ft^2 = 5.67 Wh/°C m^2. Source: Balcomb et al. 1977a.

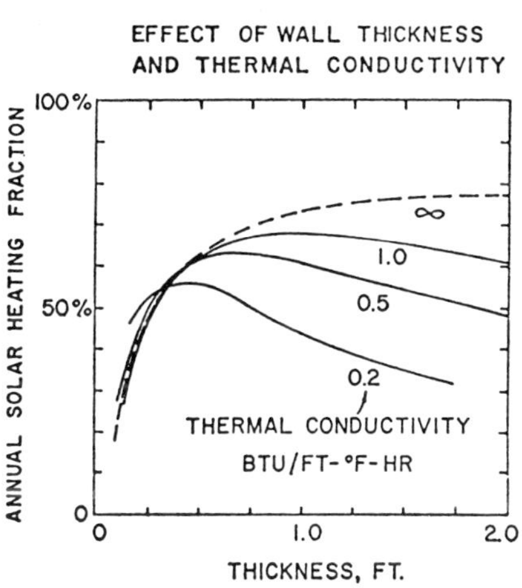

Figure 6.4
Annual solar heating fraction vs. thickness, ft, of an unvented Trombe wall system with a load collector ratio (LCR) of 12 Btu/°F day ft^2 (68 Wh/°C day m^2) in Los Alamos, New Mexico. 1 ft = 30.5 cm. Source: Balcomb et al. 1977a.

the potential success of a design. Temperature of room air is only one of several variables related to comfort, but it is of major importance and it is readily handled by computer simulation analysis. Room air temperature arises as a thermal comfort issue primarily for direct-gain systems for which daytime overheating is a common problem.

Two indicators of room temperature effects are used to evaluate direct-gain thermal comfort: maximum room temperature observed in January and diurnal temperature swing (ΔT) during clear days in January. For the latter case, some results are expressed in terms of a dimensionless comfort index (CI), which is proportional to the diurnal temperature swing (Clausing and Drolen 1979):

$$CI = \Delta T / \Delta T_a, \tag{6.10}$$

where ΔT_a is the mean indoor-outdoor temperature difference. Note that a different comfort index, also related to the diurnal temperature swing, was defined earlier by the same authors (Clausing and Drolen 1978). (Refer to section 6.6: Performance Results—Thermal Comfort.)

The effects of radiant exchanges, humidity, and air motion are, together with room air temperature, a more comprehensive indicator of thermal comfort than is room air temperature alone. An equivalent uniform temperature (T_{eu}) is defined to account for these effects. For 50% relative humidity and air motion typical of indoor conditions, the conclusion (Wray 1979b) is

$$T_{eu} = 0.55\ T_r + 0.45\ T_{mr}, \tag{6.11}$$

where T_{mr} is the mean radiant temperature and T_r is the room air temperature. An equivalent temperature is defined to include the effect of direct solar radiation (Sebald, Langenbacher, and Clinton 1979). An effective temperature (T_e) is defined to include the effect of humidity. Using a simple correlation for the ratio of indoor humidity to outdoor ambient temperature (T_a), the result (Carroll and Clinton n.d.) is

$$T_e = 0.51\ T_r + 0.42\ T_{mr} + 0.04\ T_a + 1.1°\text{C}. \tag{6.12}$$

An indicator of thermal discomfort is the error in the preferred temperature relative to a temperature reflecting the actual conditions, such as the above effective temperature. A discomfort index (DI) is defined that equals the mean square error (MSE) in the temperature plus a term proportional to the mean square ramp (MSR) in the effective temperature (Carroll and

Clinton n.d.), as in

$$DI = MSE + 5\ MSR. \tag{6.13}$$

The quantity MSR accounts for the discomfort effect of a changing temperature; it is the mean square of the rate of change of the effective temperature in degrees per hour.

6.4 Performance Results—Auxiliary Heat

6.4.1 Early Results

Various early computer simulation results apply to a very limited range of building configurations and locations. These results tend to have relatively little current practical interest because they have been largely superseded by results of greater generality. But these results were very significant in establishing the credibility of passive solar heating, and they provided a few early guidelines for the design of such systems. There is no attempt here to compile a comprehensive list, but the following are descriptions of a few representative early results.

Balcomb, Hedstrom, and McFarland (1977a) simulated a building in Los Alamos, New Mexico, with various configurations of thermal storage mass corresponding to direct-gain, water-wall, and unvented Trombe-wall systems. The most noteworthy finding at the time was simply that such systems can perform well: the performance of a passive system is comparable to that of an active system with the same collector area. The more detailed results established several now-familiar ideas. Figure 6.1 shows the dependence of the solar heating performance on a normalized collector area. The curves reveal the diminishing return of performance with added collector area. They also show the dramatic improvements obtained with double glass and night insulation. Figures 6.3 and 6.4 show the effect of thermal storage mass on performance. Figure 6.3 is for a water wall, showing that performance increases indefinitely with added thermal storage but that returns sharply diminish above about 30 Btu/°F ft^2 (170 Wh/°C m^2) (the area unit denotes a unit of passive solar aperture area). Figure 6.4 is for an unvented Trombe wall, showing that performance reaches a maximum for a particular mass thickness. The optimal thickness depends upon the thermal conductivity of the material, but it is about 10–12 in. (25–30 cm) for high-density concrete. The unvented Trombe-wall case was subsequently run for six cities besides Los Alamos and showed substantial solar heating benefit in every climate (Balcomb, Hedstrom, and McFarland 1977b).

Balcomb and McFarland (1977) simulated a Trombe wall with and without thermocirculation vents in several climates and compared the results. Unless the vents are controlled to prevent reverse flow at night, vents in a Trombe wall are worse than no vents at all. If dampers prevent reverse flow, vents improve the performance by 10–20% in severe climates, e.g., in Boston, Massachusetts, and Madison, Wisconsin, but they are not a significant advantage in mild, sunny climates, e.g., in Santa Maria and Fresno, California, and Albuquerque, New Mexico. The Trombe-wall performance was also compared with a water wall of an equal thermal capacity of 45 Btu/°F ft^2 (255 Wh/°C m^2). A water wall outperforms a vented Trombe wall, with no reverse flow, in all climates, but by only about 5–10%.

6.4.2 Direct Gain

Early analytical results for the auxiliary heat requirement of passive systems were obtained by computer simulation, using weather and solar radiation data from Los Alamos, New Mexico, and other locations, as mentioned in the preceding section (Balcomb, Hedstrom, and McFarland 1977a, 1977b). Of the cases studied by these authors that correspond to direct gain, thermal storage is represented by a single mass element (node) in thermal network representations of the systems. Three cases could correspond approximately to particular direct-gain configurations: (1) storage coupled only to room air (such as internal mass walls not in direct sunlight), (2) storage coupled to room air and to ambient (such as massive exterior walls not in direct sunlight and insulated on the outside), and (3) storage in sunlight coupled only to room air. For these three cases, there are sensitivity curves for the storage-to-room coupling coefficient. For case (3), there are sensitivity curves for the allowed room-temperature swing (assuming thermostatic controls) for two different amounts of storage.

More detailed thermal models of direct-gain systems in which the storage is represented by several mass elements (nodes) to permit an accurate calculation of the diffusion of heat into and out of storage were subsequently developed (Wray, Balcomb, and McFarland 1979). The optical relationships were later also modeled in more detail to permit a more accurate accounting of the radiative coupling between storage surfaces and between storage and glazing (Wray, Schnurr, and Moore 1980). The significant results are described below.

6.4.2.1 Performance Curves and Tables One comprehensive set of results is presented in the Los Alamos/DOE *Passive Solar Design Handbook* as solar load ratio (SLR) correlations and load collector ratio (LCR) tables for 219 locations in the United States and Canada (Balcomb, et al. 1980). The direct-gain system represented by these results is double glazed and has thermal storage in the form of heavyweight concrete with a thickness of 6 in. (15 cm) and an area of three times the glazing area. Two cases are given: one without movable insulation on the south glazing and one with movable insulation of added thermal resistance 9 ft^2 °F h/Btu (1.6 m^2 °C/W).

Another set of results is presented as performance curves (annual auxiliary heat versus south window area) in the California *Passive Solar Handbook* for fourteen locations in California (Niles and Haggard 1980). The direct-gain configuration represented in these results is double glazed and has thermal storage in the form of heavyweight concrete with a thickness of 4 in. (10 cm) and areas of 0, 1000, 2000, and 4000 ft^2 (0, 93, 186, and 372 m^2).

Volume 3 of the *Passive Solar Design Handbook* (Balcomb et. al. 1982) contains LCR tables for nine direct-gain configurations. The direct-gain configurations include various combinations of mass thickness, mass area, number of glazings, and night insulation. These tables are in terms of *SSF* and for a base temperature of 65°F (18.3°C). The ASHRAE manual *Passive Solar Heating Analysis* (Balcomb et al. 1984) contains LCR tables for the same nine direct-gain configurations as volume 3 of Balcomb et al. (1982). There is a table for each of the listed 223 locations in the United States and Canada. Table 6.1 for St. Louis, Missouri, is an example of these tables. The major improvement over previous LCR tables is that there are tables for two base temperatures, 55 and 65°F (12.8 and 18.3°C).

Performance curves (*SSF* versus *LCR*) generated by the use of the monthly SLR correlations are given for eleven locations in Nebraska for a direct-gain system both with and without movable nighttime insulation (Chen et al. 1982).

Solar load ratio (SLR) correlations for eighty-one direct-gain systems are given in *Passive Solar Heating Analysis*, supplement 1 (Balcomb and

Wray 1988). The configurations include one, two, and three glazing layers, movable night insulation added thermal resistances of 0, 4, and 9 ft^2 °F h/Btu (0, 0.7, and 1.6 m^2 °C/W), mass thicknesses of 2, 4, and 6 in. (5, 10, and 15 cm), and mass areas of three, six, and nine times the south glazing area. There are no load collector ratio (LCR) tables for these systems.

6.4.2.2 Guideline Results Annual solar savings throughout the United States are given in the form of a contour map for a direct-gain system, assuming that guideline levels of passive solar heating and conservation are employed (Balcomb 1986). The map is reproduced in figure 6.2. The direct-gain system is double-glazed, non-night-insulated, with thermal storage in the form of 4-in. (10-cm)-thick concrete with a surface area six times the south glazing area. The results range from 20,000 Btu/ft^2 (60 kWh/m^2) in the northeast to 100,000 Btu/ft^2 (300 kWh/m^2) on the central California coast. The unit of area refers to the glazing area.

6.4.2.3 Thermostat Settings Assuming a constant thermostat setting (no night setback), sensitivity curves for thermostat setting are given in the *Passive Solar Design Handbook* and in *Passive Solar Heating Analysis* for Albuquerque, New Mexico, Santa Maria, California, and Madison, Wisconsin, (Balcomb et al. 1980, 1982, 1984). These sources also contain sensitivity curves for the nighttime thermostat setback. Sensitivity curves for internal heat generation and for an allowable temperature swing of about 72.5°F (22.5°C) are given in the California *Passive Solar Handbook* (Niles and Haggard 1980) for Alturas, Fresno, and El Centro. The LCR tables in *Passive Solar Heating Analysis* give results for two base temperatures, 55 and 65°F (12.8 and 18.3°C) (Balcomb et al. 1984). (Refer to the definition of base temperature and the equivalency between thermostat setting and internal heat generation as expressed by equation [2].)

Wray and Kosiewicz (1984) calculate the effect of thermostat setback for thiry-two different direct-gain configurations in Albuquerque, New Mexico, and Madison, Wisconsin. These results are summarized in terms of an equivalent constant thermostat setpoint that produces the same annual auxiliary heat. A simple formula for the equivalent constant thermostat setting is given in terms of the building load coefficient and diurnal heat capacity.

6.4.2.4 Shading and Ventilation Daytime overheating may be controlled with shading and ventilation, but these control strategies increase auxiliary heat requirements. Sebald, Langenbacher, and Clinton (1979) investigate control strategies for shading and ventilating a direct-gain house in Albuquerque, New Mexico, Santa Maria, California, and Madison, Wisconsin. Sensitivity data for an upper thermostat setpoint of 85°F (29.4°C) compared with the nominal 75°F (23.9°C) to control ventilation or other cooling techniques are given in the *Passive Solar Design Handbook* (Balcomb et al. 1980, 1982) and in *Passive Solar Heating Analysis* (Balcomb et al. 1984) for Albuquerque, New Mexico, Santa Maria, California, and Madison, Wisconsin. Sensitivity curves for an allowable temperature swing about 72.5°F are given in the California *Passive Solar Handbook* for Alturas, Fresno, and El Centro (Niles and Haggard 1980).

The only really satisfactory solution is the use of permanent control features implicit in the building design, principally ample mass surface area in the direct-gain zones. The effect of mass on direct-gain performance is discussed in sections 6.4.2.6–6.4.2.10.

6.4.2.5 Movable Night Insulation The performance of direct-gain systems is seriously impaired, especially in cold climates, by heat loss through the glazing to ambient. This heat loss may be partially controlled by the use of movable insulation that is placed on the glazing at night and removed in the daytime.

The SLR correlations and LCR tables for double-glazed direct-gain systems in the *Passive Solar Design Handbook* and in *Passive Solar Heating Analysis* are given for no night insulation and for night insulation with added thermal resistance of 9 ft^2 °F h/Btu (1.6 m^2 °C/W) in place from 5:30 P.M. to 7:30 A.M. (Balcomb et al. 1980, 1982, 1984). The SLR correlations in *Passive Solar Heating Analysis*, supplement 1, are for no night insulation and for movable night insulation with added thermal resistance of 4 and 9 ft^2 °F h/Btu (0.7 and 1.6 m^2 °C/W) on single, double, and triple south glazing.

The performance benefits of night insulation may be determined from the difference in performance of two systems, one with and one without night insulation. This difference has been calculated throughout the United States, and the results are presented in the form of a map showing contours of energy savings per unit of glazing area (Balcomb et al. 1984, Balcomb 1986); see figure 6.5. As much as 50,000 Btu/ft^2 (160 kWh/m^2) can be saved in the extreme north-central and northeast regions by placing night insulation with added thermal resistance of 9 ft^2 °F h/Btu (1.6 m^2 °C/W) on double glazing.

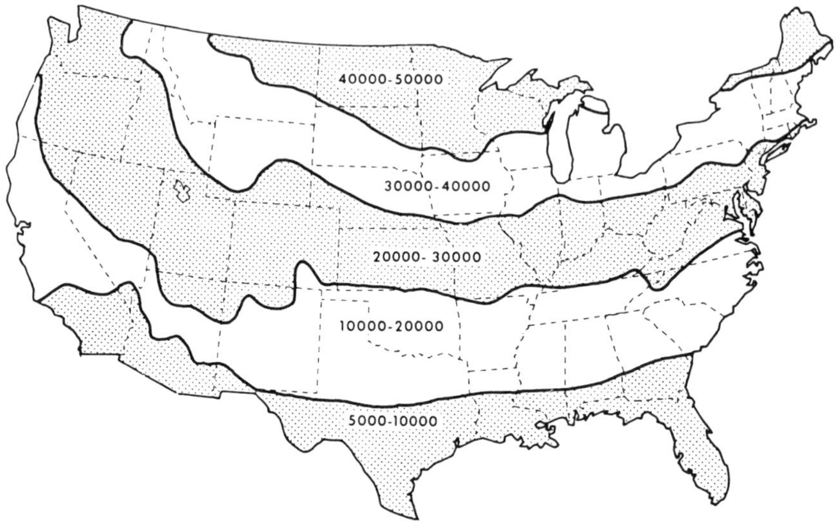

Figure 6.5
Added annual energy savings per unit of south glazing area, Btu/ft^2, due to night insulation of thermal resistance of 9 ft^2 °F h/Btu (1.6 m^2 °C/W) on a double-glazed direct-gain system. 1 Btu/ft^2 = 3.15 Wh/m^2. Source: Balcomb et al. 1984.

A universal sensitivity curve (applicable to all locations) gives the effect of the thermal resistance of night insulation on direct-gain performance (Balcomb et al. 1980, 1982, 1984). Specific sensitivity curves are given in the California *Passive Solar Handbook* for Alturas, Fresno, and El Centro (Niles and Haggard 1980).

6.4.2.6 Mass Area The most important parameter of thermal storage is mass surface area in the direct-gain zones. In some early simulation studies, Mazria, Baker, and Wessling (1977, 1978) calculate the temperature in a direct-gain room on clear January days in Portland, Oregon. Results are given for three cases in which solar radiation from a south glazing illuminates mass areas of 1.5, 3, and 9 times the glazing area. The mass area greatly affects the room temperature stability, with the room diurnal temperature swing ranging from about 42°F (23°C) for mass area 1.5 to 14°F (8°C) for mass area 9. These authors recommend diffusing the sunlight over a large area of the mass by the use of translucent glazing material, reflective blinds that redirect the light, and a light color on the first surface struck by incident sunlight, as discussed in sections 6.4.2.9 and 6.4.2.11.

Wray and Balcomb (1979) describe the effect of mass characteristics on solar heating fraction based on annual simulations using insulation and weather data for Albuquerque, New Mexico, and Madison, Wisconsin. The results, in the form of sensitivity curves for the thermal storage mass per unit of glazing area, show that the performance of direct gain increases indefinitely as the amount of thermal storage mass is increased, unlike Trombe walls, for which there is a performance maximum in the vicinity of 180–220 lbm/ft^2 (900–1100 kg/m^2), which corresponds to a wall thickness of about 14–18 in. (36–46 cm) for a wall-material density of 150 lb/ft^3 (2400 kg/m^3)—see section 6.4.3. The data are compared with similar data for Trombe walls, showing that for a given *LCR* and a given amount of mass, direct gain can outperform a Trombe wall. Curves are given for three different mass thicknesses, 4, 6, and 8 in. (10, 15 and 20 cm), showing that a given amount of direct-gain mass is best arranged in thinner layers and spread over a larger area. Similar curves for Albuquerque, New Mexico, Madison, Wisconsin, and Santa Maria, California are in the *Passive Solar Design Handbook*, volume 2 (Balcomb et al. 1980).

Studies by Arumi-Noe and Kim (1980), using the DEROB simulation program with a week of data for Madison, Wisconsin, tend to confirm the above conclusion; they show that the performance is improved by spreading available thermal storage material over a large area.

The performance curves in the California *Passive Solar Handbook* (Niles and Haggard 1980) are plotted for four different mass surface areas, all for 4-in. (10-cm)-thick heavyweight concrete.

There are SLR correlations and LCR tables in the *Passive Solar Design Handbook*, volume 3 and in *Passive Solar Heating Analysis* for two different mass-to-glazing-area ratios, three with a 6-in. (15-cm) thickness, and six with 2- and 4-in. (5- and 10-cm) thicknesses, all for heavyweight concrete (Balcomb et al. 1982, 1984). An expanded set of SLR correlations includes mass-to-glazing-area ratios of 3, 6, and 9 (Balcomb and Wray 1988).

The sensitivity curves for mass thickness in the *Passive Solar Design Handbook*, volume 3 (Balcomb et al. 1982) and *Passive Solar Heating Analysis* (Balcomb et al. 1984) are given for four different mass-to-glazing-area ratios: 2, 3, 6, and 10. These curves show that the performance increases steadily with area ratio in every case, but that the rate of performance increase diminishes. The performance increase is about the same when the area ratio increases from 6 to 10 as when the area ratio increases from 2 to 3.

A comprehensive evaluation of the effect of mass surface area and other key mass characteristics on auxiliary heat requirement is contained in the recent concept of *effective heat capacity* (Wray and Best 1987b).

6.4.2.7 Mass Thickness Wray (1980) uses annual simulations to compute the effect of mass thickness on solar savings fraction for Santa Maria, California, Albuquerque, New Mexico, and Madison, Wisconsin. The thermal storage material is heavyweight concrete with a density of 150 lbm/ft^3 (2400 kg/m^3). The results are in the form of sensitivity curves for the mass thickness for given mass areas.

A comprehensive set of sensitivity curves for mass thickness is in the *Passive Solar Design Handbook*, volume 3 (Balcomb et al. 1982) and *Passive Solar Heating Analysis* (Balcomb et al. 1984); there are curves for six U.S. cities and mass-to-glazing-area ratios of 2 through 10.

Sensitivity curves for mass thickness are contained in the California *Passive Solar Handbook* (Niles and Haggard 1980) for Alturas, Fresno, and El Centro; there are curves for heavyweight concrete with a density of 150 lbm/ft^3 (2400 kg/m^3), adobe, and carpeted concrete.

There are SLR correlations and LCR tables in the *Passive Solar Design Handbook*, volume 3 and in *Passive Solar Heating Analysis* for mass thicknesses of 2 and 4 in. (5 and 10 cm) for mass-to-glazing-area ratio of 6, and of 6 in. (15 cm) for mass-to-glazing-area ratio of 3 (Balcomb et al. 1982, 1984). An expanded set of SLR correlations includes thicknesses of 2, 4, and 6 in. (5, 10, and 15 cm), each for mass-to-glazing-area ratios of 3, 6, and 9 (Balcomb and Wray 1988).

A comprehensive evaluation of the effect of the mass thickness and other key mass characteristics is contained in the recent concept of effective capacity (Wray and Best 1987). One interesting conclusion from this work is that there is an optimum thickness for a given amount (volume) of thermal storage material, contrary to earlier conclusions that mass should be spread out in as thin a layer as possible. For heavyweight concrete, for example, the optimum thickness is in the range of 2–5 in. (5–13 cm), depending upon the total volume.

Emery et al. (1982) give curves of the sensitivity of auxiliary heat to the mass in a concrete floor slab for four values of *LCR*. The auxiliary heat is expressed as the excess over that for a superinsulated base case. The independent variable is the ratio of mass to glazing area (lbm/ft^2); because the mass area is constant, this variable is proportional to the mass thickness for a given *LCR*.

The conclusions from all of these results is that most of the benefit of mass thickness is obtained in the first 2 in. (5 cm) and that thicknesses above about 4 in. (10 cm) provide little additional benefit, assuming that the material has a density comparable to that of heavyweight concrete, i.e., about 140–150 lbm/ft^3 (2200–2400 kg/m^3).

6.4.2.8 Mass Material The sensitivity curves for thickness in the California *Passive Solar Handbook* (Niles and Haggard 1980) are given for two different materials: heavyweight concrete with a density of 140 lbm/ft^3 (2200 kg/m^3) and a less dense material, adobe. The performance of concrete is shown to be slightly greater than that of the adobe: the same performance as that for adobe may be achieved with thinner concrete, and the performance for materials thicker than 4 in. (10 cm) levels off at a higher level for concrete.

The material characteristic that affects performance is the product ρck, where ρ is the density, c is the specific heat, and k is the thermal conductivity. There are sensitivity curves for ρck, assuming a thickness of 6 in. (15 cm) and a mass-to-glazing-area ratio of 3, in the *Passive Solar Design Handbook*, volume 3 (Balcomb et al. 1982) and in *Passive Solar Heating Analysis* (Balcomb et al. 1984). The curves show that the performance increases indefinitely as ρck increases, but the rate of increase diminishes. Between ρck of 0 and 10 $Btu^2/°F^2\ h\ ft^4$ (320 $W^2h/°C^2\ m^4$), the performance increase is dramatic, but beyond 20 $Btu^2/°F^2\ h\ ft^4$ (640 $W^2h/°C^2\ m^4$), little further increase occurs. Adobe, for example, has a ρck in the range of 5–10 $Btu^2/°F^2\ h\ ft^4$ (160–320 $W^2h/°C^2\ m^4$), building brick, 10 $Btu^2/°F^2\ h\ ft^4$ (320 $W^2h/°C^2\ m^4$), and heavyweight (144 lbm/ft^3) concrete, 27 $Btu^2/°F^2\ h\ ft^4$ (860 $W^2h/°C^2\ m^4$) (Balcomb et al. 1984).

6.4.2.9 Mass Surface Characteristics The surface treatment of thermal storage materials affects their performance on direct-gain buildings. The use of a light color on the first surface struck by incident sunlight helps to diffuse the sunlight over a large area of mass (Mazria 1978).

Sensitivity curves for the solar absorptance of mass surfaces are given in the *Passive Solar Design Handbook* and in *Passive Solar Heating Analysis* (Balcomb et al. 1980, 1982, 1984). These curves show that the performance increases with increasing mass absorptance, but that the sensitivity is not very great, which suggests that light colors are acceptable. In fact, these curves were computed for a mass surface area three times the glazing area, which is a poor design. For a better design, with a ratio of 6 or 7, the sensitivity to mass absorptance should be very weak.

Studies by Arumi-Noe and Kim (1980), using the DEROB simulation program with a week of data for Madison, Wisconsin, tend to confirm this conclusion; they show that the performance is not very sensitive to the solar absorptance of mass surfaces.

The presence of a lightweight, insulating cover on mass surfaces can seriously degrade the thermal storage effectiveness. The sensitivity curves for thickness in the California *Passive Solar Handbook* (Niles and Haggard 1980) are given for carpeted and uncarpeted concrete, showing that carpeting reduces the thermal storage effectiveness by about half. The *Passive Solar Design handbook*, volume 3 (Balcomb et al. 1982) and *Passive Solar Heating Analysis* (Balcomb et al. 1984) contain sensitivity curves for the thermal resistance of mass coverings; there are curves for Albuquerque, New Mexico, Madison, Wisconsin, and Santa Maria, California.

6.4.2.10 Mass Location The effectiveness of thermal storage surfaces in direct-gain rooms depends somewhat on their location in the room. Surfaces in direct sunlight are most effective. Otherwise, vertical walls, compared with floors and ceilings, are the best locations because the heat transfer coefficients for convective exchanges to and from the surfaces are greater (Wray, Schnurr, and Moore 1980, Balcomb et al. 1982, 1984).

6.4.2.11 Glazing Properties Thermal storage mass in direct-gain rooms is most effective if solar radiation is distributed over the entire surface. Diffusive glazing can accomplish this distribution more readily than clear glazing (although diffusive glazing might be a poor choice relative to other design issues). Wray used annual simulations to compare the solar savings fractions in Albuquerque, New Mexico, for diffusive and clear glazings (Wray, Schnurr, and Moore 1980, Balcomb et al. 1982, 1984). The fractional improvement ranges from about 1% for thermal storage mass only on the floor to about 10% for mass only on the ceiling.

6.4.2.12 Number of Glazing Layers The performance of direct-gain systems depends strongly upon the insulating quality of the direct-gain glazing. Increasing the number of glazing layers increases the thermal resistance but decreases the transmittance. The net effect on direct-gain performance is indicated by sensitivity curves for the number of glazing layers, both without and with night insulation of added thermal resistance of 9 ft^2 °F h/Btu (1.6 m^2 °C/W), in the *Passive Solar Design Handbook*, volume 2 (Balcomb et al. 1980) for Santa Maria, California, Albuquerque, New Mexico, and Madison, Wisconsin. In addition, curves for Boston, Massachusetts, Medford, Oregon, and Nashville, Tennessee are included in the *Passive Solar Design Handbook*, volume 3 (Balcomb et al. 1982) and in *Passive Solar Heating Analysis* (Balcomb et al. 1984). Sensitivity data for one, two, and three glazing layers, without night insulation, are given in the California *Passive Solar Handbook* for Alturas, Fresno, and El Centro (Niles and Haggard 1980). If there is no night insulation, these data show an improvement in performance in all climates as the number of glazing layers is increased from one to four; but for mild sunny climates, such as Santa Maria and Albuquerque, there is very little improvement beyond double glazing. For cold cloudy climates, such as Boston and Madison, triple glazing is a substantial improvement over double glazing if the LCR is smaller than about 30. If there is night insulation, triple glazing is never a substantial improvement over double glazing, and, in mild sunny climates, double glazing is not an improvement over single glazing.

The air gap between the panes of double glazing is a significant parameter affecting direct-gain performance. Analytical studies have shown that increasing the air gap from 0.25 to 0.5 in. (0.6 to 1.3 cm) can increase the solar savings fraction by 10 to 15 percentage points (Wray 1980, Balcomb et al. 1980).

6.4.2.13 Glazing Orientation Direct-gain glazing is most effective when oriented due south. Sensitivity curves for off-south orientations are contained in *Passive Solar Heating Analysis* (Balcomb et al. 1984).

6.4.3 Thermal Storage Walls

Early analytical results for the auxiliary heat requirement of passive systems were obtained by computer simulation, using weather and solar radiation data from Los Alamos, New Mexico, and other locations, as described in section 6.4.1 (Balcomb, Hedstrom, and McFarland 1977a, 1977b). Of the cases studied, one corresponds to a thermal storage wall: storage is placed directly in front of the glass, sun is absorbed by storage, and storage is thermally coupled through the glass to the environment and also to the room. When thermal storage is represented by a single mass element (node) in the thermal-network representation of the system, the physical system most closely represented is a water wall. When the mass is represented by several nodes joined by appropriate conductances, the model represents an unvented Trombe wall. Balcomb and McFarland (1977) simulated a Trombe wall with and without thermocirculation vents in several climates and compared the results. Some of the results of these studies are illustrated in figures 6.1, 6.3, and 6.4. See section 6.4.1 for more discussion of the results of these early studies.

6.4.3.1 Performance Curves and Tables One comprehensive set of results is presented in the *Passive Solar Design Handbook*, volume 2, as solar load ratio (SLR) correlations and load collector ratio (LCR) tables (Balcomb et al. 1980). The thermal-storage-wall systems represented by these results are double glazed and have thermal storage capacities per unit of glazing area of 45 Btu/°F ft^2 (255 Wh/°C m^2). Two system types are included, a water wall and a vented Trombe wall. The Trombe wall is 18-in. (46-cm)-thick heavyweight concrete with backdraft-dampered thermocirculation vents totaling 6% of the wall area. Two cases are given: one without movable night insulation on the south glazing and one with movable insulation of added thermal resistance of 9 ft^2 °F h/Btu (1.6 m^2 °C/W). There are LCR tables for 219 locations in the United States and Canada.

Another set of results is presented as performance curves (annual auxiliary heat versus south window area) in the California *Passive Solar Handbook* for fourteen locations in California (Niles and Haggard 1980). The Trombe wall configuration represented in these results is double glazed and has thermal storage in the form of 12-in. (30-cm)-thick heavyweight concrete. There are no thermocirculation vents.

Volume 3 of the *Passive Solar Design Handbook* (Balcomb et al. 1982) contains SLR correlations and LCR tables for fifty-seven thermal-storage-wall configurations. The fiftween water-wall configurations include various combinations of mass thickness, number of glazings, wall surface type (normal or selective), and night insulation. The forty-two Trombe-wall configurations include various combinations of wall thickness, material density, number of glazings, wall surface type (normal or selective), night insulation, and vent type (no vents or backdraft-dampered vents). There are LCR tables for 219 locations in the United States and Canada.

Passive Solar Heating Analysis (Balcomb et al. 1984) contains SLR correlations and LCR tables for the same fifty-seven thermal-storage-wall configurations as in volume 3 of Balcomb et al. (1982). There is an LCR table for each of the listed 223 locations in the United States and Canada. Table 6.1 for St. Louis, Missouri, is an example of these tables. The major improvement over previous LCR tables is that there are tables for two base temperatures, 55 and 65°F (12.8 and 18.3°C).

Performance curves (*SSF* versus *LCR*) were generated for eleven locations in Nebraska, using the monthly SLR correlations, for the same thermal-storage-wall configurations treated in the *Passive Solar Design Handbook*, volume 2 (Chen et al. 1982).

There are SLR correlations for various concrete-block Trombe walls (Wray and Peck 1987, 1988, Balcomb and Wray 1988). The system configurations include filled and unfilled blocks. No LCR tables are given in these sources.

6.4.3.2 Guideline Results Annual solar savings per unit of glazing area throughout the United States are given in the form of a contour map for two thermal-storage-wall systems, assuming that guideline levels of passive solar heating and conservation are employed (Balcomb 1986). The thermal-storage-wall systems are an unvented, double-glazed, flat-black Trombe wall and a single-glazed, selective-surface water wall. For the Trombe wall, the results range from 20,000 Btu/ft^2 (60 kWh/m^2) in the northeast and Great Lakes region to 90,000 Btu/ft^2 (280 kWh/m^2) on the central California coast. For the water wall, they range from 50,000 to 110,000 Btu/ft^2 (160–350 kWh/m^2).

6.4.3.3 Thermostat Settings and Internal Heat Sensitivity curves for the auxiliary heat thermostat setting in a water-wall system are given in the *Passive Solar Design Handbook*, volume 2 for Albuquerque, New Mexico, and for Madison, Wisconsin (Balcomb et al. 1980). Sensitivity curves for internal heat generation and an allowable temperature swing about 72.5°F (22.5°C) in an unvented-Trombe-wall house are given in the California *Passive Solar Handbook* for Alturas, Fresno, and El Centro (Niles and Haggard 1980). The LCR tables in *Passive Solar Heating Analysis* (Balcomb et al. 1984) give results for two base temperatures, 55 and 65°F (12.8 and 18.3°C). (For the definition of base temperature and the equivalency between thermostat setting and internal heat generation, refer to equation [2].)

6.4.3.4 Shading and Ventilation Sebald and Phillips (n.d.) give the effects of various control strategies, including shading and ventilation, in controlling overheating in a Trombe-wall house in Albuquerque, New Mexico, Santa Maria, California, and Madison, Wisconsin. The result is that shading and ventilation have no effect on auxiliary heat consumption because a Trombe wall has very little tendency to overheat.

The sensitivity curves for minimum temperature setpoint in the *Passive Solar Design Handbook*, volume 2 (Balcomb et al. 1980) are plotted for two assumptions concerning the maximum temperature setpoint: (1) maximum temperature setpoint equals minimum temperature setpoint (constant room temperature), and (2) no maximum temperature setpoint (no control of overheating). The two sets of curves are not significantly different, which suggests that overheating controls are seldom necessary for a water wall.

A fixed overhang, if sized correctly, can partially control solar gains in summer while only slightly reducing the passive solar heating performance. McFarland plotted the annual cooling energy requirement and solar savings fraction (*SSF*) for a water wall in Albuquerque, New Mexico, as a function of the overhang geometry in the *Passive Solar Design Handbook*, volume 2 (Balcomb et al. 1980). The plot shows that only modest reductions of cooling load can be accomplished with overhangs, regardless of geometry, and that overhangs that do not reduce *SSF* significantly, also do not reduce cooling load very much. For example, an overhang of 0.25 times the window height located 0.25 times the window height above the top of the glass reduces *SSF* by only 2% and reduces annual cooling energy consumption by only 13%. Jones (1981) presents a series of similar plots for a water wall in four U.S. cities in which the net annual energy savings achieved by the overhang, accounting for both heating and cooling energy, is plotted relative to the overhang geometry.

6.4.3.5 Night Insulation Sebald and Phillips (n.d.) give the effect of various control strategies, including movable night insulation of added thermal resistance of 10 ft^2 °F h/Btu (1.8 m^2 °C/W), on improving the performance of a double-glazed Trombe wall in Albuquerque, New Mexico, Santa Maria, California, and Madison, Wisconsin. Night insulation has a major effect on performance in all cases.

The SLR correlations and LCR tables for thermal-storage-wall systems are given for no night insulation and for night insulation of added thermal resistance of 9 ft^2 °F h/Btu (1.6 m^2 °C/W) in place from 5:30 P.M. to 7:30

A.M. (Balcomb et al. 1980, 1982, 1984). The performance benefit of night insulation may be determined from the difference in performance of two systems, one with and one without night insulation.

A universal sensitivity formula (applicable to all locations) gives the effect of the thermal resistance of night insulation on Trombe-wall performance (Balcomb et al. 1980, 1982, 1984). Specific sensitivity curves are given in the California *Passive Solar Handbook* for Alturas, Fresno, and El Centro (Niles and Haggard 1980).

6.4.3.6 Vents Trombe walls sometimes have thermocirculation vents (openings at the top and bottom of the wall) to permit the flow of warm air to the room by natural convection from the air space between the wall and the glazing.

Balcomb and McFarland (1977) simulated a Trombe wall with and without thermocirculation vents in nine U.S. climates and compared the results. Unless the vents are controlled to prevent reverse flow at night, vents in a Trombe wall are worse than no vents at all. If dampers prevent reverse flow, vents improve the performance over a solid wall by 10–20% in severe climates, e.g., in Boston, Massachusetts, and Madison, Wisconsin, but they are not a significant advantage in mild, sunny climates, e.g., in Santa Maria and Fresno, California and Albuquerque, New Mexico. A thermostatic control was also simulated, in which dampers close when room temperature exceeds 75°F (24°C); this strategy prevents overheating and has no significant effect on auxiliary heat consumption.

McFarland used computer simulation to study the effect of vent area on passive heating performance for the *Passive Solar Design Handbook*, volume 2 (Balcomb et al. 1980) and found that there is normally an optimum vent size that depends primarily on solar savings fraction (*SSF*). The results are summarized as follows: for an *SSF* of 0.25, the performance levels off above a vent area fraction of 0.06 (the combined area of upper and lower vents as a fraction of the Trombe-wall area); for an *SSF* of 0.5, the performance levels off above a vent area fraction of 0.02; and, for an *SSF* of 0.75, the performance decreases above a vent area fraction of 0.02.

Sensitivity curves for vent area fraction are given in *Passive Solar Heating Analysis* (Balcomb et al. 1984) for six U.S. cities. These curves all resemble one another, with the performance leveling off at a vent area fraction of about 0.05, irrespective of climate and *SSF*. The performance improvement obtainable with vents depends upon *SSF*, with the fractional improvement ranging from about 3% at *SSF* of 0.8 to about 12% at *SSF* of 0.2. I have no explanation of the differences between these results and the earlier ones (Balcomb et al. 1980).

6.4.3.7 Fan It is possible to move warm air from the space between the glazing and the exterior surface of the Trombe wall by use of a thermostatically controlled fan. Sebald, Clinton, and Langenbacher (1979) simulated a thermostatically controlled fan: the fan is turned on when the outside surface temperature at the wall exceeds 85°F (29°C). There are annual auxiliary results for various wall areas and thicknesses. Comparing 400-ft^2 (37 m^2), non-night-insulated cases, the fan provides a 10-MBtu (2.9-MWh) (22%) improvement in Albuquerque, New Mexico; 5-MBtu (1.5-MWh) (20%) improvement in Santa Maria, California; and 6-MBtu (1.8-MWh) (7%) improvement in Madison, Wisconsin. Although the authors are not explicit as to whether these savings are relative to natural convection through vents or to no vents, I assume that the savings are relative to no vents. Other control strategies are investigated relative to the delivery of heat from the Trombe wall to the room. These strategies include: manual free convection vents operated on a variety of schedules, automatic free convection vents operated by a room thermostat, and insulation on the interior surface of the wall operated by a room thermostat. None of these strategies produces significant reduction of auxiliary heat relative to an uncontrolled, unvented wall with night insulation.

Sebald and Phillips (n.d.) simulated the effect of a Trombe-wall fan controlled by a thermostat; the fan is turned on if the room needs heat and when the air in the wall/glazing interface exceeds room air temperature by at least 20°F (11°C). For non-night-insulated, 400-ft^2 (37-m^2) Trombe walls, there is no significant improvement in Santa Maria, California; a 1.4-MBtu (410-kWh) (8%) improvement in Albuquerque, New Mexico; and a 1.2-MBtu (350-kWh) (3%) improvement in Madison, Wisconsin. If the comparison is made between night-insulated cases, the energy savings are similar, but the percentage savings are greater. I have no explanation of the differences between these results and those of Sebald et al. (1979).

6.4.3.8 Mass Thickness Balcomb, Hedstrom, and McFarland (1977a, 1977b) give sensitivity curves for thermal-storage-wall heat capacity per unit of glazing area, based on simulation results for Los Alamos, New Mexico. For a water wall, the performance increases indefinitely with added heat capacity, but improvement sharply diminishes above about

30 Btu/°F ft^2 (170 Wh/°C m^2), which corresponds to a water-wall thickness of about 6 in. (15 cm)—(see figure 6.3). For an unvented Trombe wall, the performance reaches a broad maximum at a thickness of about 12 in. (30 cm) for heavyweight concrete of density 150 lbm/ft^3 (2400 kg/m^3)—(see figure 6.4).

Balcomb and McFarland (1977) simulated a Trombe wall with thermocirculation vents and plotted sensitivity curves for wall thickness (both vented and unvented walls) for Madison, Wisconsin. Although the performance of the vented Trombe wall is superior to that of the unvented wall in Madison, the maximum occurs at about the same thickness, 12 in. (30 cm).

Sensitivity curves for the thickness of water walls and heavyweight concrete Trombe walls are given in the *Passive Solar Design Handbook*, volume 2 (Balcomb et al. 1980) for six U.S. cities. Sensitivity curves for the thickness of water walls and for both heavyweight concrete and adobe Trombe walls are given in the California *Passive Solar Handbook* (Niles and Haggard 1980) for Alturas, Fresno, and El Centro. A comprehensive set of sensitivity curves for thermal-storage-wall thickness is contained in *Passive Solar Heating Analysis* (Balcomb et al. 1984) for six U.S. cities; both night-insulated and non-night-insulated cases are included for the water wall, and Trombe-wall curves are given for both vented and unvented cases as well as for four different values of wall-material density.

There are SLR correlations and LCR tables in the *Passive Solar Design Handbook*, volume 3 (Balcomb et al. 1982) and *Passive Solar Heating Analysis* (Balcomb et al. 1984) for several different water-wall and Trombe-wall thickness.

6.4.3.9 Mass Material The sensitivity curves for thermal-storage-wall thickness in the California *Passive Solar Handbook* (Niles and Haggard 1980) are given for three different materials: water, heavyweight concrete, and adobe. For a given thickness of material, water is the superior performer, followed by concrete, then by adobe.

For solid materials, the characteristic that affects performance is the product ρck, where ρ is the density, c is the specific heat, and k is the thermal conductivity. There are sensitivity curves for ρck (holding the total heat capacity constant) for three U.S. cities in the *Passive Solar Design Handbook*, volume 2 (Balcomb et al. 1980). For walls with heat capacities greater than about 30 Btu/°F ft^2, the curves show that the performance increases indefinitely as ρck increases, but that the rate of increase diminishes. Between ρck of 0 and 10 Btu^2/°F^2 h ft^4 (320 W2h/°C^2 m^4), the performance increase is dramatic, but beyond 20 Btu^2/°F^2 h ft^4 (640 W^{2}h/°C^2 m^4), little further increase occurs. Adobe, for example, has a ρck in the range of 5–10 Btu^2/°F^2 h ft^4 (160–320 W^{2}h/°C^2 m^4), building brick, 10 Btu^2/°F^2 h ft^4 (320 W^{2}h/°C^2 m^4), and heavyweight (144 lbm/ft^3) concrete, 27 Btu^2/°F^2 h ft^4 (860 W^{2}h/°C^2 m^4) (Balcomb et al. 1984).

The sensitivity curves for Trombe-wall thickness contained in *Passive Solar Heating Analysis* (Balcomb et al. 1984) are given for various densities of masonry material.

6.4.3.10 Mass Surface Characteristics The treatment of the exterior surface of a thermal-storage wall has a strong effect on passive solar heating performance. Two characteristics are important: the solar absorptance and the longwave emittance. There are curves plotted for Los Alamos, New Mexico, in the *Passive Solar Design Handbook*, volume 2 (Balcomb et al. 1980) that show the sensitivity of water-wall performance to both solar absorptance (holding longwave emittance constant at 0.9) and longwave emittance (holding shortwave absorptance constant at 0.95). Curves are given for single, double, and triple glazing. A comprehensive set of similar curves is given in *Passive Solar Heating Analysis* (Balcomb et al. 1984) for six U.S. cities. Included are single, double, and triple glazing, and both vented and unvented Trombe walls. All cases are for no night insulation.

The SLR correlations and LCR tables for thermal-storage-wall systems are given for normal flat-black and selective wall surfaces (Balcomb et al. 1982, 1984). The performance benefit of a selective surface may be determined from the difference in performance of two systems, one with a normal surface and one with a selective surface. The difference in performance between a single-glazed, selective surface and a double-glazed, flat-black Trombe wall has been calculated throughout the United States, and the results are presented in the form of a contour map of energy savings per unit of glazing area (Balcomb et al. 1984, Balcomb 1986). This map is reproduced in figure 6.6. As much as 30,000 Btu/ft^2 (90 kWh/m^2) of glazing can be saved per year in the extreme north-central and northeast regions.

The heat transfer coefficient between the interior surface of a thermal-storage wall and the room air affects the wall's performance. McFarland plots sensitivity curves for the wall-to-room heat transfer coefficient for a water wall and a Trombe wall in Boston, Massachusetts, and Los Alamos,

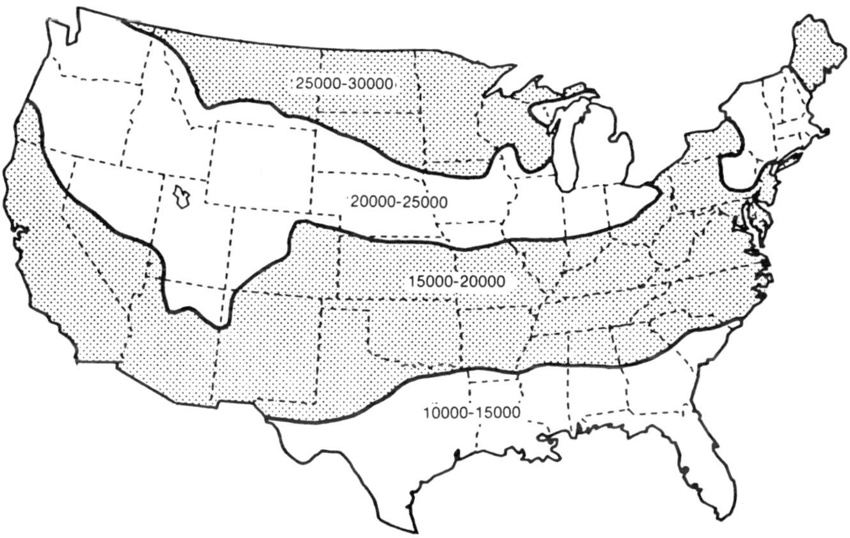

Figure 6.6
Added annual energy savings per unit of glazing area, Btu/ft^2, due to replacing one glazing in a double-glazed, flat-black Trombe wall with a selective surface on the Trombe wall. 1 Btu/ft^2 = 3.15 Wh/m^2. Source: Balcomb et al. 1984.

New Mexico in the *Passive Solar Design Handbook*, volume 2 (Balcomb et al. 1980). Fortunately for passive design, the standard value of about 1.5 Btu/h °F ft^2 (8.5 W °C m^2) is nearly optimum. Thus, special surface treatments to enhance convection are unnecessary.

6.4.3.11 Internal Mass Internal building mass, in addition to the mass in the thermal-storage wall, increases performance by storing heat that would otherwise be vented. Sensitivity curves for the area of internal mass are given in the *Passive Solar Design Handbook*, volume 2 (Balcomb et al. 1980) for a water wall in Los Alamos and Albuquerque, New Mexico, and in Madison, Wisconsin. Similar curves are given in the California *Passive Solar Handbook* (Niles and Haggard 1980) for Alturas, Fresno, and El Centro.

6.4.3.12 Glazing Properties The performance of a thermal-storage wall depends strongly upon the solar transmittance and thermal resistance of the glazing. The transmittance can be increased by the use of low-iron glass or antireflective coatings. Low-iron glass improves the transmittance by reducing the amount of energy absorbed in the glass (see Johnson, chapter 5 of this volume). McFarland computed the maximum possible benefit of low-iron glass for the *Passive Solar Design Handbook*, volume 2 (Balcomb et al. 1980); there is a table of the increases in solar savings fraction (*SSF*) due to the use of perfectly clear glass rather than of standard 1/8-in. (3-mm) glass in both double- and quadruple-layer glazing units for seven U.S. cities.

6.4.3.13 Number of Glazing Layers Increasing the number of glazing layers increases the thermal resistance but decreases the transmittance. The net effect on the performance of a thermal-storage wall is indicated by sensitivity curves for one through four glazing layers plotted by McFarland for the *Passive Solar Design Handbook*, volume 2 (Balcomb et al. 1980); there are curves for three water-wall configurations—flat-black surface and no night insulation, selective surface and no night insulation, and flat-black surface and night insulation of thermal resistance of 9 ft^2 °F h/Btu (1.6 m^2 °C/W)—for six U.S. cities. Sensitivities for one, two, and three glazing layers are given in the California *Passive Solar Handbook* (Niles and Haggard 1980) for a flat-black, non-night-insulated Trombe wall in Alturas, Fresno, and El Centro. A comprehensive set of sensitivity curves for one through four glazing layers is given in *Passive Solar Heating Analysis* (Balcomb et al. 1984); there are curves for vented and unvented Trombe walls (with and without night insulation), for a selective surface (without night insulation), and for a flat-black water wall (with and without night insulation).

The data show that increasing the number of glazing layers, up to at least four, improves the performance of a flat-black, non-night-insulated wall in all but the most mild (Santa Maria, California) climates, in which case a fourth layer fails to improve the performance. If the wall is night-insulated, there is no further improvement beyond two layers. If the wall has a selective surface, there is no further improvement beyond one layer in all but the most severe climates (Bismarck, North Dakota), and, even then, double glazing is only a slight improvement over single glazing.

The thermal resistance of multilayer glazing units has an optimum relative to the air gap between the layers. McFarland computed the effect of increasing the air gap from 0.25 to 0.75 in. (6.4 to 19 mm) for the *Passive Solar Design Handbook*, volume 2 (Balcomb et al. 1980); there is a table of the increases in *SSF* for both double- and quadruple-layer glazing units in seven U.S. cities.

6.4.3.14 Glazing Orientation Thermal-storage-wall glazing is most effective when it is oriented due south. Sensitivity curves for off-south orientation of water walls and vented Trombe walls were plotted by McFarland for the *Passive Solar Design Handbook*, volume 2 (Balcomb et al. 1980); there are curves for non-night-insulated and night-insulated cases in six U.S. cities. There are sensitivity curves for off-south orientation of a vented Trombe wall with and without night insulation in *Passive Solar Heating Analysis* (Balcomb et al. 1984) for six U.S. cities.

6.4.3.15 Glazing Tilt McFarland plotted sensitivity curves for the off-vertical tilt of a thermal-storage-wall glazing for the *Passive Solar Design Handbook*, volume 2 (Balcomb et al. 1980) for Boston, Massachusetts, and for Albuquerque, New Mexico.

6.4.4 Sunspaces

Attached sunspaces are more complex thermally than either direct-gain or thermal-storage walls; attached sunspaces combine some elements of the behavior of both of these other system types. Like direct-gain systems, sunspaces involve a rather complex process of radiative and convective exchanges between the interior sunspace surfaces and the sunspace air. Also, like direct gain, some of the surfaces are mass surfaces at which heat is stored and released on a diurnal cycle; other surfaces are lightweight surfaces at which heat is promptly transferred to sunspace air by convection. Like thermal-storage walls, sunspaces are indirect passive systems transferring their heat to occupied spaces by conduction through an intervening wall and by convection through openings in that wall.

Sunspaces are usually assumed to be relatively uncontrolled spaces with heating setpoints of 40–50°F (5–10 °C) and cooling setpoints of 90–100°F (30–40°C). Thus, unlike direct-gain rooms, sunspaces are intended for occasional rather than routine occupancy. Nevertheless, a correctly designed sunspace is comfortable most of the time and is useful for a variety of functions, such as entry, reception, gathering, and circulation.

Because of the system complexities, analytical results for sunspaces were generally delayed by a year or two while experience was developed with direct-gain and thermal-storage-wall calculations. Early results in the form of performance and sensitivity curves were obtained for Bismarck, North

Dakota (Jones 1980), and for Bradley Field, Connecticut (Direnzo and McGowan 1980).

One early comprehensive set of results is a set of solar load ratio (SLR) correlations and load collector ratio (LCR) tables for sixteen sunspace configurations (McFarland and Jones 1980b). The sunspace configurations include eight combinations of glazing geometries, common wall type, and east and west end-wall treatment (glazed or not glazed), and each combination is considered with and without movable night insulation on the sunspace glazing. There are LCR tables for 219 locations in the United States and Canada.

6.4.4.1 Performance Curves and Tables A comprehensive set of results is presented in the *Passive Solar Design Handbook*, volume 3 (Balcomb et al. 1982) as solar load ratio (SLR) correlations and load collector ratio (LCR) tables for twenty-eight sunspace configurations. The sunspace configurations include five different south glazing geometries, three different common wall types, and both night-insulated and non-night-insulated glazings. The LCR tables are available for 219 locations in the United States and Canada.

Passive Solar Heating Analysis (Balcomb et al. 1984) contains SLR correlations and LCR tables for the same twenty-eight sunspace configurations as volume 3 of Balcomb et al. (1982). There is an LCR table for each of the listed 223 locations in the United States and Canada. Table 6.1 for St. Louis, Missouri, is an example of these tables. The major improvement over previous LCR tables is that there are tables for two base temperatures, 55 and 65°F (12.8 and 18.3°C).

Weidt, Saxler, and Gerard (1983) calculate the auxiliary heat requirement of a house with what they call an attached sunspace, although the stated control assumptions raise the question of whether the system is more appropriately classed as direct gain: the heating setpoint is 65°F (18.3°C) with a 10°F (5.6°C) night setback, and the cooling setpoint is 80°F (26.7°C). The authors allude to results for combined heating and cooling energy savings relative to a base-case house for five sunspace configurations in twelve cities, although very few specifics are cited. Results of this type were later quoted for four sunspace configurations in sixteen cities in a promotional publication (Andersen Corporation, 1985).

The Sunspace Primer (Jones and McFarland 1984) contains performance maps that show the annual performance derived from SLR correlations of twelve sunspace configurations throughout the United States.

6.4.4.2 Guideline Results Balcomb et al. (1984) and Balcomb (1986) present the performance of a sunspace assuming that guideline levels of passive solar heating and conservation are employed. The results are in the form of a table of annual solar savings fractions (SSFs) throughout the United States and a contour map of solar savings and auxiliary heat per unit of glazing projected area. The sunspace system has a double glazing with a 50-degree tilt. Thermal storage is in the form of a concrete floor slab and uninsulated, 12-in. (30-cm)-thick concrete common wall separating the sunspace from the adjoining rooms. The solar savings range from 50,000 Btu/ft^2 (160 kWh/m^2) in the Northeast and Midwest to 110,000 Btu/ft^2 (350 kWh/m^2) on the central California coast.

6.4.4.3 Building Thermostat Settings Jones (1980) gives a sensitivity curve for the building lower setpoint (the auxiliary heat thermostat setting in adjoining rooms) for a sunspace in Bismarck, North Dakota. A comprehensive set of sensitivity curves for the building lower setpoint is given in the *Passive Solar Design Handbook*, volume 3 (Balcomb et al. 1982) and *Passive Solar Heating Analysis* (Balcomb et al. 1984) for six U.S. cities.

The LCR tables in *Passive Solar Heating Analysis* (Balcomb et al. 1984) give results for two base temperatures, 55 and 65°F (12.8 and 18.3°C).

Weidt, Saxler, and Gerard (1983) quote results on the effect of the duration of the thermostat night setback in Atlanta, Georgia, and in Minneapolis, Minnesota.

6.4.4.4 Sunspace Thermostat Settings The sunspace itself is considered to be a relatively uncontrolled space. The performance curves and tables described above (section 6.4.4.1) assume a lower setpoint of 45°F (7°C) and an upper setpoint of 95°F (35°C). It is assumed that auxiliary heat is used to maintain the lower setpoint, and ventilation is used to maintain the upper setpoint if the outdoor temperature permits. There are early sensitivity curves for the sunspace minimum setpoint for Bismarck, North Dakota (Jones 1980), and for Bradley Field, Connecticut (Direnzo and McGowan 1980). The latter curves are plotted for two different values of sunspace heat capacity to show that this parameter has a significant effect on the sunspace minimum temperature. A comprehensive set of sensitivity curves for the sunspace minimum setpoint is given in the *Passive Solar Design Handbook*, volume 3 (Balcomb et al. 1982) and *Passive Solar Heating Analysis* (Balcomb et al. 1984) for six U.S. cities. The curves show that there is little sensitivity to the sunspace minimum setpoint in the range below

50°F (10°C) except in extreme climates, e.g., in Bismarck, North Dakota, if the sunspace has good thermal integrity and if there is abundant heat capacity in the sunspace per unit of glazing projected area, at least 60 Btu/°F ft^2 (340 Wh/°C m^2).

Parsons (1983) and Beckman, Klein, and Parsons (1983) use hourly simulations for Madison, Wisconsin, to calculate the effect of sunspace minimum thermostat setpoint. There are plots of monthly auxiliary energy use versus month for three different minimum setpoints: 40, 52, and 65°F (4, 11, and 18°C). The cases given are double glazing with no night insulation (Beckman, Klein, and Parsons 1983), double glazing with night insulation, single glazing with no night insulation, and single glazing with night insulation (Parson 1983). Night insulation, when present, has added thermal resistance of 9 ft^2 °F h/Btu (1.6 m^2 °C/W).

6.4.4.5 Shading and Ventilation Shading, particularly of sloped glazing, and ventilation are important in maintaining summer comfort in a sunspace. Jones and McFarland (1984) give the amount of ventilation, both natural and fan-forced, required to maintain various sunspace temperatures relative to outdoor ambient.

6.4.4.6 Night Insulation Jones (1980) gives a sensitivity curve for the added thermal resistance of movable night insulation on the double glazing of a sunspace in Bismarck, North Dakota. The SLR correlations and LCR tables for double-glazed sunspace systems are given for no night insulation and for night insulation of added thermal resistance of 9 ft^2 °F h/Btu (1.6 m^2 °C/W) in place from 5:30 P.M. to 7:30 A.M. (Balcomb et al. 1982, 1984).

The performance benefits of night insulation may be determined from the difference in performance of two systems, one with and one without night insulation. This difference has been calculated throughout the United States for a sunspace with 50-degree-tilted double glazing and thermal storage in the form of a concrete floor slab and an uninsulated 12-in. (30-cm)-thick concrete wall separating the sunspace form the adjoining room (Balcomb et al. 1984 and Balcomb 1986).

A universal sensitivity formula (applicable to all locations) gives the effect of the added thermal resistance of night insulation on sunspace performance (Balcomb et al. 1982, 1984).

6.4.4.7 Common Wall Vents Solar heat is normally delivered to adjoining rooms primarily by natural convection through openings (vents) in the sunspace/room common wall. There are sensitivity curves for the common wall vent area for Bismarck, North Dakota (Jones 1980, Jones and McFarland 1980). The curves are plotted versus a "vent area fraction," which is the area of one row of thermocirculation vents (half of the total vent area) as a fraction of the total glazing projected area. The two rows of vents have a vertical separation of 8 ft (2.4 m). Curves are given for both a lightweight, insulated common wall and a heavyweight concrete, uninsulated common wall.

There is a comprehensive set of similar sensitivity curves for the common wall vent area for a variety of sunspace configurations in the *Passive Solar Design Handbook*, volume 3 (Balcomb et al. 1982) and in *Passive Solar Heating Analysis* (Balcomb et al. 1984) for six U.S. cities. The curves are plotted versus a "vent area fraction," which is the combined area of upper and lower thermocirculation vents (the total vent area) as a fraction of the total glazing projected area. The curves show that the maximum performance increases steeply up to a vent area fraction of 0.02, and the performance is level after a vent area fraction of 0.06. The area fraction equivalent to 0.06 for an 80-in. (200-cm)-door opening is 0.10.

Jones, McFarland, and Lazarus (1983) give sensitivity curves for the vent area fraction for various values of the interior room heat capacity for Boston, Massachusetts. The curves show that generous vent areas are more important if there is substantial heat capacity in the adjoining room.

6.4.4.8 Fan Heat Delivery A fan is sometimes used to supplement or supplant natural convection for the delivery of heat to adjacent rooms. In the case of supplementing natural convection, Jones, McFarland, and Lazarus (1983) give sensitivity curves for fan capacity for various values of the natural convection area for Boston, Massachusetts. The curves show that a fan makes no contribution to passive solar heating performance if there is a natural convection vent area of at least 1% of the glazing projected area. In the case of supplanting natural convection, there are sensitivity curves for fan capacity per unit of glazing projected area and fan thermostat setpoint for various values of interior room heat capacity.

Assuming a simple control strategy in which a fixed-volume fan is turned on when the sunspace temperature exceeds a thermostat setpoint, the curves show that a fan capacity in the range 2–3 cfm/ft^2 (35–50 m^3/s m^2) is near optimum (the area unit here refers to the sunspace glazing projected area), but, even under optimum circumstances, the passive solar heating performance of a sunspace with fan-forced heat delivery is less than that of heat delivery by natural convection. This difference occurs because natural convection depends upon the temperature difference between the spaces and is, therefore, more responsive to the availability of heat in the sunspace. A more elaborate strategy, based on both the sunspace and building temperatures, and on a variable-volume fan, could probably be made to mimic natural convection closely enough to reproduce comparable performance.

Parsons (1983) and Beckman, Klein, and Parsons (1983) use hourly simulations for the month of March to calculate the effect of fan delivery of heat. There are sensitivity curves for Madison, Wisconsin, plotted versus air flow rate. Several cases are examined, including both uninsulated masonry and insulated lightweight common wall, high and low heat capacity in adjoining room, high and low convective heat transfer coefficient, and high and low added heat capacity in the sunspace. Some of these cases are also calculated for Albuquerque, New Mexico, and for Nashville, Tennessee (Parsons 1983).

6.4.4.9 Mass Area The effect of the area of thermal-storage mass in a sunspace depends upon the location of the mass. There are no explicit analytical studies known by this author in which the effect of mass area is calculated if the mass is located on an exterior surface, such as a floor, which functions by the storage and release of heat at the same surface. However, the situation is very similar to the case of mass in direct-gain rooms (see section 6.4.2.6). We may assume, therefore, that the surface area of sunspace exterior mass is the most important parameter affecting performance if the material is at least 2-in. (5-cm)-thick and if it is comparable in density to heavyweight concrete.

If the mass consists of an uninsulated common wall between the sunspace and adjoining rooms, in which both sides of the wall may store and release heat, the mass area is not ordinarily a design variable independent of the glazing area. Some information on the effect of area may be extracted from the SLR correlations and LCR tables in the *Passive Solar Design Handbook*, volume 3 (Balcomb et al. 1982) and in *Passive Solar Heating Analysis* (Balcomb et al. 1984). Among the reference designs presented in these sources are cases in which the wall area equals the glazing projected area (the attached sunspaces) and cases in which the wall area is larger than the glazing area (the semienclosed sunspaces).

6.4.4.10 Mass Thickness The effect of the thickness of thermal-storage mass in a sunspace depends upon the location of the mass. If the mass is located on an exterior surface, such as a floor, there are no explicit analytical results of which the author is aware relative to the thickness. However, the situation is very similar to the case of mass in direct-gain rooms (see section 6.4.2.7). We may assume, therefore, that most of the benefit of thickness is obtained in the first 2 in. (5 cm), and thicknesses above about 4 in. (10 cm) provide little additional benefit if the material has a density comparable to that of heavyweight concrete.

If the mass consists of an uninsulated common wall between the sunspace and adjoining rooms, in which heat is stored at one surface and released at both surfaces, there are sensitivity curves for the thickness of an uninsulated water wall dividing the sunspace from adjoining rooms (Jones 1980). In this case, there is also a concrete floor for thermal storage. The performance increases indefinitely with wall thickness, but improvement sharply diminishes above about 4 in. (10 cm).

There are sensitivity curves for the thickness of an uninsulated, heavyweight concrete common wall in Bismarck, North Dakota (Jones 1980, Jones and McFarland 1980). There is a comprehensive set of sensitivity curves for the thickness of an uninsulated masonry common wall of various densities in the *Passive Solar Design Handbook*, volume 3 (Balcomb et al. 1982) and in *Passive Solar Heating Analysis* (Balcomb et al. 1984) for six U.S. cities. For a wall of heavyweight concrete, the performance reaches a broad maximum at a thickness of about 12 in. (30 cm). This situation resembles the case of a vented Trombe wall.

6.4.4.11 Water Volume Another case is thermal-storage mass in the form of water in containers in the sunspace, with the sunspace/room common wall lightweight and insulated, and with additional thermal storage in the form of a concrete floor slab. In this case, sensitivity curves show that the shape of the water containers has only a slight effect on performance (Jones 1980). Therefore, water volume rather than thickness or area is the relevant parameter. There are sensitivity curves for water volume in a sunspace in Bismarck, North Dakota (Jones 1980, Jones and McFarland 1980).

Jones (1980) compares the performance of sunspaces with equal volumes of water storage—one in which the water is in an uninsulated water wall that divides the sunspace from adjoining rooms and one in which the water is in containers on the sunspace side of a lightweight, insulated common wall. If there is no night insulation on the sunspace glazing, the water-wall case outperforms the insulated-wall case for water-storage volumes above 0.18 ft^3/ft^2 (0.05 m^3/m^2); the storage volume is expressed here as the volume of water per unit of glazing projected area. If there is night insulation of thermal resistance of 9 ft^2 °F h/Btu (1.6 m^2 °C/W) on the sunspace glazing, the water-wall case outperforms the insulated-wall case if the water volume exceeds 0.08 ft^3/ft^2 0.02 m^3/m^2). The more direct coupling of storage to room in the water-wall case is a more important design consideration than is the common-wall insulation in the insulated-wall case, except for very small heat capacities.

There is a comprehensive set of sensitivity curves for water volume in the *Passive Solar Design Handbook*, volume 3 (Balcomb et al. 1982) and in *Passive Solar Heating Analysis* (Balcomb et al. 1984) for six U.S. cities. These curves show that the performance increases indefinitely with thermal-storage volume, but most abruptly up to about 0.25 ft^3/ft^2 (0.08 m^3/m^2). At storage volumes in the range of 1.0–1.5 ft^3/ft^2 (0.30–0.45 m^3/m^2), the performances of the two thermal storage types (water in containers and masonry wall) are roughly equal. (This means, incidentally, that the heat capacity of the masonry is twice as effective as that of the water because masonry has half the volumetric heat capacity of water. The reason is, presumably, that the coupling between the storage and the occupied space is more direct for the uninsulated masonry wall than it is for the water containers behind an insulated wall.)

Jones, McFarland, and Lazarus (1983) give sensitivity curves for sunspace thermal-storage mass for various amounts of interior room heat capacity for Boston, Massachusetts. The curves are plotted versus the heat capacity of water in containers per unit of glazing projected area. It is assumed that, in addition, there is thermal storage in the sunspace in the form of a concrete floor slab. These curves show that the sensitivity to sunspace mass depends upon the amount of room mass. If the room mass is large enough, 60 Btu/°F ft^2 (340 Wh/°C m^2) or more, the sunspace passive solar heating performance is insensitive to added mass in the sunspace. Weidt, Saxler, and Gerard (1983) indicate similar results in which the sunspace heating performance is insensitive to sunspace mass, although they do not give the details of their assumptions concerning the amount of room mass.

6.4.4.12 Mass Material The sensitivity curves for thermal-storage mass in an uninsulated masonry common wall for various densities of masonry material are given in the *Passive Solar Design Handbook*, volume 3 (Balcomb et al. 1982) and in *Passive Solar Heating Analysis* (Balcomb et al. 1984).

6.4.4.13 Mass Surface Characteristics There are sensitivity curves for the solar absorptance of massive surfaces in a sunspace in Bismarck, North Dakota (Jones 1980). A comprehensive set of similar curves is given in the *Passive Solar Design Handbook*, volume 3 (Balcomb et al. 1982) and in *Passive Solar Heating Analysis* (Balcomb et al. 1984).

Sensitivity curves for the infrared emittance of water-storage containers in a sunspace in Bismarck, North Dakota (Jones 1980) show that there is no advantage to a selective surface in this application.

6.4.4.14 Internal Mass Internal building mass, in addition to mass in the sunspace or in the sunspace/room common wall, increases performance by storing heat that would otherwise be vented. Sensitivity curves for the heat capacity of internal building mass are given in Jones, McFarland, and Lazarus (1983) for a sunspace in Boston, Massachusetts.

6.4.4.15 Lightweight Surfaces Jones (1980) gives sensitivity curves for the solar absorptance of the lightweight ceiling and end walls of a sunspace in Bismarck, North Dakota. These curves show almost no dependence of passive solar heating performance on the absorptance of lightweight surfaces. However, the extensive presence of lightweight objects, such as furniture and plants, that absorb solar radiation diminishes the passive solar performance. The effect of lightweight objects is simulated by assuming that a certain fraction, the "lightweight absorption fraction," of solar radiation directly heats sunspace air after it is transmitted through the glazing and each time it is reflected from interior surfaces. Initially, sensitivity curves for the lightweight absorption fraction were given for Bismarck, North Dakota (Jones 1980, Jones and McFarland 1980), then a comprehensive set was published in the *Passive Solar Design Handbook*, volume 3 (Balcomb et al. 1982) and in *Passive Solar Heating Analysis* (Balcomb et al. 1984) for six U.S. cities.

6.4.4.16 Glazing Properties The performance of a sunspace depends strongly upon the solar transmittance and the thermal resistance of the glazing. The glazing transmittance can be increased by using a glazing material with a higher intrinsic transparency (lower extinction coefficient), thinner layers, or antireflective coatings. There are sensitivity curves for glazing thickness, assuming double-layer glass and an extinction coefficient of 0.5/in. (0.2/cm), in the *Passive Solar Design Handbook*, volume 3 (Balcomb et al. 1982) and in *Passive Solar Heating Analysis* (Balcomb et al. 1984) for six U.S. cities. Note that the effective parameter is the product of the thickness and the extinction coefficient. Consequently, the abscissa of the sensitivity curves for thickness may be converted to the product of thickness and extinction coefficient.

6.4.4.17 Number of Glazing Layers Increasing the number of glazing layers increases the thermal resistance but decreases the transmittance. The net effect on sunspace performance is indicated by sensitivity curves for one through four glazing layers in the *Passive Solar Design Handbook*, volume 3 (Balcomb et al. 1982) and in *Passive Solar Heating Analysis* (Balcomb et al. 1984) for six U.S. cities. The curves show that increasing the number of glazing layers, up to at least four, improves the performance of a non-night-insulated sunspace, but the fourth layer produces only a very small improvement. If the sunspace glazing is night-insulated, there is no significant improvement beyond two layers.

The thermal resistance of multilayer glazing units depends upon the air gap between the layers. There are sensitivity curves for air gap, assuming double-layer glass, in the *Passive Solar Design Handbook*, volume 3 (Balcomb et al. 1982) and in *Passive Solar Heating Analysis* (Balcomb et al. 1984) for six U.S. cities. The curves show a small improvement in performance up to about 0.5 in. (1.3 cm) and very little thereafter.

6.4.4.18 Glazing Orientation Sunspace glazing is most effective when it is oriented due south. There are sensitivity curves for off-south orientation of sunspace glazing for six U.S. cities in the *Passive Solar Design Handbook*, volume 3 (Balcomb et al. 1982) and in *Passive Solar Heating Analysis* (Balcomb et al. 1984). The orientation may be up to 15 degrees off south with a performance penalty of no more than 5%. For orientations of 45 degrees off south, the performance penalty ranges from 10 to 30%, depending upon the location.

Weidt, Saxler, and Gerard (1983) calculate the effect of off-south orientation of sunspace glazing in the form of combined heating and cooling energy savings relative to a base-case house for twelve U.S. cities. These authors do not give their individual results but the range of energy penalties for 45 degrees off south as 1–23%, depending upon the climate. More specific results are quoted for eight orientations in sixteen cities in a promotional publication (Andersen Corporation 1985).

The effect of east and west glazings on sunspace end walls may be assessed from the SLR correlations and LCR tables (Balcomb et al. 1982, 1984), for which there are sunspace configurations both with and without end-wall glazings. It is evident from the LCR tables that east and west sunspace glazings harm the passive solar heating performance.

6.4.4.19 Glazing Tilt Jones (1980) gives sensitivity curves for the tilt (angle relative to horizontal) of sunspace glazing for Bismarck, North Dakota. A comprehensive set of sensitivity curves for sunspace glazing tilt is given in the *Passive Solar Design Handbook*, volume 3 (Balcomb et al. 1982) and in *Passive Solar Heating Analysis* (Balcomb et al. 1984). In these curves, the glazing area is held constant, while the tilt is varied. The curves show that the optimum glazing tilt depends not only upon the latitude but also upon the climate, the building *LCR*, and other details. (For example, the presence of movable insulation of thermal resistance of 9 ft^2 °F h/Btu [1.6 m^2 °C/W] on the sunspace glazing typically increases the optimum tilt by about 5 degrees). Therefore, there is no simple rule for tilt that depends upon latitude alone. But the performance is very insensitive to tilt within several degrees of the optimum. As a general rule, a tilt in the range of 50–65 degrees is very close to the optimum. Within this range, for every case for which there is a sensitivity curve, the performance is at least 95% of the optimum.

There are also sensitivity curves for glazing tilt holding glazing projected area (rather than actual glazing area) constant. These curves represent a variation in both glazing tilt and area, and so they are hard to interpret; they correspond, however, to a realistic situation: often the building south wall area to be enclosed by a sunspace (hence, the projected area) is fixed by architectural constraints. Sensitivity curves for glazing tilt with projected area constant are available for Bismarck, North Dakota, and Madison, Wisconsin (Jones 1980), for Bismarck, North Dakota, and Albuquerque, New Mexico (Jones and McFarland 1980), and for four other U.S. cities in the *Passive Solar Design Handbook*, volume 3 (Balcomb et al. 1982) and in *Passive Solar Heating Analysis* (Balcomb et al. 1984).

6.4.4.20 Two-Glazing-Tilt Geometries Sunspace designs often have two different glazing tilts, especially when one of the glazed surfaces is vertical.

There are sensitivity curves for such cases relating to the height of vertical glazing and to the tilt of sloped glazing for Bismarck, North Dakota (Jones 1980, Jones and McFarland 1980) and for five other U.S. cities (Balcomb et al. 1982, 1984).

6.4.5 Other Systems

Other systems were studied soon after the early calculations on the standard systems. The analytical results are of limited generality, but they indicate comparative performance.

6.4.5.1 Roof Ponds Houses with four different roof-pond configurations in six U.S. cities were simulated by researchers at Trinity University (Clements 1980). Initially, the simulations were performed for one typical January day. Subsequently, four cities and the optimum configuration for each city were chosen for month-long January simulations. The results, expressed in terms of solar heating fractions for the month, are: Albuquerque, New Mexico, 98%; Bismarck, North Dakota, 22%; Boston, Massachusetts, 78%; Fresno, California, 100%; and Omaha, Nebraska, 76%. Miami, Florida, was also simulated with a resulting 100% solar heating fraction, but Miami is not an appropriate location for roof-pond passive solar heating due to the negligible heating load.

There are performance curves (annual auxiliary heat versus pond area) generated by annual simulations for several roof-pond configurations in the California *Passive Solar Handbook* (Niles and Haggard 1980) for fourteen locations in California. There are also sensitivity curves for three locations for R-value of movable night insulation, water depth, allowed temperature swing, internal heat generation, window area, reflector proportions, roof parapet height, absorptance of building exterior, and added slab area.

The roof-pond results mentioned here are well summarized in Marlatt, Murray, and Squier (1984).

6.4.5.2 Transwall Fuchs and McClelland (1979) calculate the performance of a partially transparent water-wall (called "transwall" by the authors) using just one day of "synthetic weather" data. The results have no general interest other than in comparison with the other systems calculated with the same data. The results show that the partially transparent water wall performed better than the unvented or direct-gain unvented Trombe wall. Unfortunately, no comparisons were made with a standard opaque water wall or with a standard water wall combined with direct gain, either of which might be considered to be the competitor of such a system. Hull et al. (1980) give performance curves for Boston Massachusetts, Madison, Wisconsin, and Albuquerque, New Mexico. There are also sensitivity curves for Boston for wall thickness, plate absorptance, and internal mass. The maximum performance is for a plate absorptance of 1.0, meaning that a standard, opaque water wall outperforms a partially transparent water wall, but that the performance is nearly flat for absorptances above 0.8. Thus, daylighting and view may be achieved with practically no compromise in solar heating performance.

6.4.5.3 Thermosiphoning Air Panels Thermal network models validated with test room data are the basis of hour-by-hour simulation results for the performance of thermosiphoning air panels. The sensitivity of auxiliary heat requirement in Albuquerque, New Mexico, to various design parameters is described (Schnurr and Wray 1984). Simulation results for ten U.S. cities are summarized by solar load ratio correlations (Wray, Schnurr, and Kosiewicz 1984). These correlations have also been published in Wray and Peck (1987, 1988) and in Balcomb and Wray (1988).

6.4.5.4 Radiant Panels A solar radiant panel is simply the metal absorber plate of a south-wall solar collector directly exposed to the room. The room is heated by direct thermal radiation. Massive room surfaces function as thermal storage. Hour-by-hour simulation results for ten U.S. cities are summarized by solar load ratio correlations (Wray, Schnurr, and Kosiewicz 1984). These correlations have also been published in the *Military Handbook: Design Procedures for Passive Solar Buildings* (Wray and Peck 1987) and in *Passive Solar Heating Analysis*, supplement 1 (Balcomb and Wray 1988).

6.4.6 Mixed Systems

The combination of two or more passive solar systems delivering heat to the same thermal zone is called a mixed system. Passive systems in practice are almost always mixed systems for a variety of architectural and thermal reasons. The most common mixed system is direct gain plus one of the indirect systems, such as an unvented Trombe wall or a sunspace.

6.4.6.1 Direct-Gain and Thermal-Storage Wall Sebald and Phillips (n.d.) use computer simulation to calculate the auxiliary heat requirement of a passive building with a combination of Trombe-wall and direct-gain glazing areas for three cities: Albuquerque, New Mexico, Madison, Wisconsin, and Santa Maria, California. The direct-gain glazing areas of south, east, and west, and south Trombe wall are variables. The results are expressed in terms of response surfaces for Santa Maria, Albuquerque, and Madison, of the form

$$Q_{aux} = a_0 + a_1 S + a_2 E + a_3 W + a_4 T + a_5 E^2 + a_6 W^2 + a_7 S^2 + a_8 T^2 + a_9 EW + a_{10} SW + a_{11} ET + a_{12} WT + a_{13} ST, \tag{6.14}$$

where S, E, and W are the south, east, and west direct-gain glazing areas, respectively, and T is the south Trombe-wall glazing area. The coefficients a_0, a_1, etc., depend upon the city and the control strategy.

There are performance curves for a mixed direct-gain/water-wall system in the California *Passive Solar Handbook* (Niles and Haggard 1980) for fourteen locations in California. There are also sensitivity curves for Alturas, Fresno, and El Centro for the thermal resistance of movable night insulation, mass thickness and material, allowed temperature swing, internal heat generation, non-south window area, shading strategy, number of glazing layers, added slab area, and fraction of south window covered by storage.

6.5 Performance Results—Backup Heating Demand

Wray (1979a) gives simulation results on backup heating demand for the case of a Trombe wall in Albuquerque, New Mexico. The auxiliary heat demand peaks at about 8 A.M. and is very low between about 11 A.M. and 7 P.M., the winter peak demand hours for the electric utility. Some coincident peaking occurs during early-morning and late-evening hours. The use of night insulation on the Trombe wall substantially reduces the evening demand. Wray investigated the possibility of eliminating the evening coincident peak by the use of controls and an auxiliary heater coupled to storage. An underslab heater is effective it suitably controlled: for December, 95% of the time the room temperature slips no more than 2°F (1°C) below the normal 65°F (18°C) thermostat setting. During 4% of the time, or about one hour per day, the temperature is between 2 and 4°F (1 and 2°C) below the thermostat setting; this occurs primarily between 10 P.M.

and midnight. There is some penalty in total auxiliary heat consumption due to overheating and ventilation in the daytime.

Another study of minimizing demand with underslab electric backup is of an actual house in Santa Fe, New Mexico (Murray, Melsa, and Balcomb 1980). The study concludes that about 65% of the backup heating occurs off peak and that assuming an idealized smart controller, it is possible, in principal, that all of the backup heating could be off peak with small temperature errors relative to the thermostat setting. A subsequent study focuses on control strategies to minimize cost and discomfort (Winn and Winn 1982).

Niles and Haggard (1980) calculate the peak backup heating demand for each of the California locations and system types for which they provide annual auxiliary heat performance curves. The results are expressed as curves plotted versus the passive solar aperture area. The peak heating demand depends primarily upon the location and upon the assumed heat loss coefficient for the building. The peak heating demand also depends, although weakly, upon the passive solar aperture area. Depending upon the location and system type, the peak heating demand may increase or decrease with aperture area. For direct-gain systems in relatively cold climates, the peak heating demand increases with south window area, indicating that the added heat loss from the window is the predominant effect. For direct-gain systems in mild climates (Santa Maria, Los Angeles, San Diego), the peak heating demand decreases with south window area, indicating that stored solar energy more than offsets window loss during peak heating conditions, up to a certain window area that depends upon location and the area of thermal storage. For Trombe walls in all locations, the peak heating demand decreases with south window area.

Wray and Best (1987a) calculate the backup heating demand for a large number of passive systems in several climates; their results are expressed in terms of a generalized design method for sizing backup heating equipment.

6.6 Performance Results—Thermal Comfort

6.6.1 Direct Gain

6.6.1.1 Room Temperature Mazria, Baker, and Wessling (1977, 1978) calculate the temperature in a direct-gain room for a series of identical, consecutive clear January days in Portland, Oregon. Heat capacity is in the form of concrete of density of 140 lbm/ft^3 (2200 kg/m^3). The cases examined include mass areas of 1.5, 3, and 9 times the glazing area, each for mass thicknesses of 4, 8, and 16 in. (10, 20, and 40 cm). The results are presented as plots of room temperature versus time of day. The mass area has a pronounced effect on room temperature, with the room temperature ranging diurnally approximately from 46 to 90°F (8 to 32°C) for the smaller mass area (1.5 times the glazing area) and from 62 to 75°F (17 to 24°C) for the larger mass area (nine times the glazing area). The mass thickness, in contrast, has a very small effect on room temperature, the minimum nighttime room temperature being 2–4°F (1–2°C) higher for the 8- and 16-in. (20- and 40-cm) thicknesses than for the 4-in. (10-cm) thickness; otherwise, there are no significant differences between the performances of the three thickness cases. The absence of a significant effect in these results due to mass thickness is caused by the fact that the maximum potential benefit is largely reached by the 4-in. (10-cm) thickness.

The case of 8-in. (20-cm) mass thickness and area three times the glazing area is analyzed for a variety of mass materials, including concrete, brick, magnesite brick (with magnesium added to increase the thermal conductivity), adobe, and water (modeled as isothermal) in containers. Among the solid materials, the largest temperature fluctuations occurred with adobe, and the smallest with magnesite brick. The least fluctuations of all were for water.

6.6.1.2 Maximum Room Temperature Wray (in Balcomb et al. 1982, 1984) computes the maximum room temperature in January for direct-gain systems in Albuquerque, New Mexico, Madison, Wisconsin, and Santa Maria, California. The results are expressed as functions of the ratio of the thermal storage mass area to the glazing area for various values of *LCR* and the mass thickness. A typical example of these results is shown in figure 6.7. It is assumed that the air temperature is not limited by any form of shading or ventilation, so the extremely high temperatures calculated for some configurations are not representative of the actual temperatures that would be experienced. They do indicate, however, as the author puts it, "the extent to which the occupant may be kept busy adjusting ventilation rates and shading in order to keep the structure comfortable." Clearly, severe problems can arise for small *LCR*s and small mass areas. There is also a set of results expressed as functions of the fraction of solar

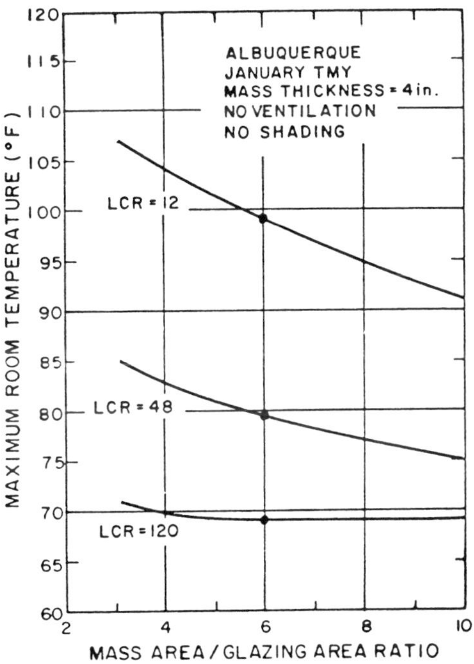

Figure 6.7
Maximum room temperature, °F, vs. ratio of mass area to glazing area. Source: Balcomb et al. 1982.

gains absorbed by lightweight surfaces (as opposed to thermal-storage surfaces). All of the indicated factors, *LCR*, mass area, and lightweight absorption are important determinants of maximum January temperature.

6.6.1.3 Diurnal Temperature Swing Clausing and Drolen (1978, 1979) present computer simulation results for the diurnal temperature swing in terms of the dimensionless ratio $\Delta T/\Delta T_a$—see equation (10). Historical weather data are not used, but rather a sequence of identical days. Results for direct-gain systems are presented as a function of the building time-constant in which the building mass is represented by a single heat-capacity node. Results were also generated for vented and unvented Trombe walls and water walls (Drolen 1979)

Frequency-domain analysis methods are well suited to calculations of diurnal temperature swing (see also Niles, chapter 3 of this volume). Several authors have developed simple, analytical expressions suitable for

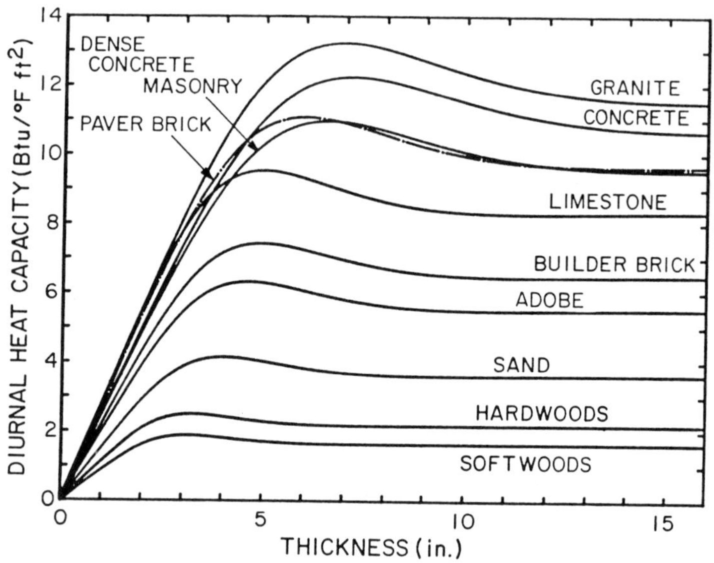

Figure 6.8
Diurnal heat capacities, Btu/°F ft^2, of various materials as a function of material thickness, in. 1 Btu/°F ft^2 = 5.67 Wh/°C m^2. 1 in. = 2.54 cm. Source: Balcomb 1983a.

estimating mean temperature and diurnal temperature swings in passive buildings. A few numerical results for direct gain are given primarily as examples of the methods (Niles 1978, Balcomb et al. 1980, Kirkpatrick and Winn 1982). Balcomb presents an extensive development of this approach applied to estimating the diurnal temperature swing in a direct-gain building on a clear winter day (Balcomb et al. 1980, Balcomb, 1983a, Balcomb and Wray 1988). The method shows that the fundamental building variable that determines the diurnal temperature swing is the diurnal heat capacity of the building relative to the direct-gain glazing area. The diurnal heat capacity is directly proportional to the surface area of the mass and depends upon the thickness, as shown in figure 6.8. It is clear that the diurnal heat capacity depends strongly upon thickness up to about 4–6 in. (10–15 cm) for dense materials, such as concrete, but a thickness greater than about 4–6 in. (10–15 cm) does not contribute significantly to the diurnal heat capacity.

6.6.1.4 Hours of Discomfort Sebald, Langenbacher, and Clinton (1979) give the relative discomfort of a variety of control strategies for direct-gain systems in Albuquerque, New Mexico, Santa Maria, California, and Madison, Wisconsin. They express the discomfort in terms of the number of hours in which the equivalent temperature (see section 6.3.3) exceeds 80°F (27°C).

6.6.2 Thermal-Storage Walls

Sebald, Clinton, and Langenbacher (1979) give histograms of room air temperature in the absence of backup heating for vented Trombe walls with and without thermostatic fan-forced air circulation in Albuquerque, New Mexico. The results show that fan control produces greater comfort than does natural convection, even for 6-in. (15-cm) walls, compared with walls of any thickness without fans.

Sebald and Phillips (n.d.) give the effects of various control strategies, including shading and ventilation, to control overheating in a Trombe-wall house in Albuquerque, New Mexico, Santa Maria, California, and Madison, Wisconsin. The result is that shading and ventilation have very little or no effect on thermal comfort (the effects of shading and ventilation on thermal comfort are not quoted separately) because a Trombe wall has very little tendency to overheat.

A frequency-domain analysis method is applied by Sebald and Vered (1981) to a Trombe wall in Albuquerque, New Mexico. The result is expressed in the from of contributions to the room temperature by both outdoor ambient temperature and the solar energy absorbed by the Trombe wall. There are several significant frequency components with periods ranging from 4 to 365 days.

6.6.3 Sunspaces

6.6.3.1 Minimum and Maximum Sunspace Temperature Jones, McFarland, and Lazarus (1983) express thermal comfort in the sunspace itself in terms of January temperature extremes, the minimum and maximum temperatures experienced in the sunspace in the absence of auxiliary heat or ventilation. Results are given for Boston, Massachusetts, in the form of plots of the January temperature extremes versus several design variables.

In the case of heat delivery by natural convection, the January temperature extremes are plotted versus the area of the natural convection vents, the heat capacity in the sunspace, and the heat capacity in the adjoining room. The maximum sunspace temperature is quite sensitive to the vent area, up to an area of about 2% of the glazing area, after which little further

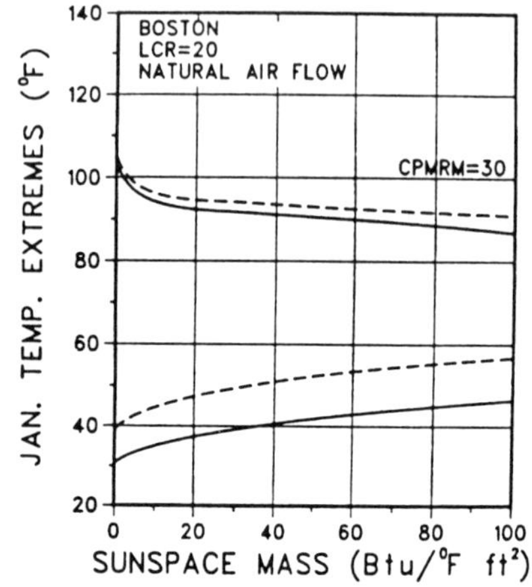

Figure 6.9
January temperature extremes, °F, in the sunspace as a function of heat capacity added to the sunspace per unit of glazing projected area, Btu/°F ft^2. The load collector ratio of the sunspace-building system is 20 Btu/°F day ft^2 (113 Wh/°C day m^2). The heat capacity of the adjoining building per unit of sunspace glazing projected area (the parameter CPMRM) is 30 Btu/°F ft^2 (170 Wh/°C m^2). The solid curves represent no night insulation on the sunspace glazing. The dashed curves are for night insulation of thermal resistance of 5 ft^2 °F h/Btu (0.9 m^2 °C/W). Source: Jones et al. 1983.

reduction of the maximum temperature occurs; the minimum sunspace temperature is insensitive to the vent area.

Both the maximum and minimum temperatures are sensitive to the sunspace heat capacity. In the case studied, the sunspace has an exposed concrete floor slab and additional heat capacity in the form of water in containers. The temperature extremes are plotted versus the amount of additional heat capacity per unit of glazing projected area—see figure 6.9. The curves indicate that thermal comfort is improved indefinitely by added mass, but the greatest improvement is achieved by the first 20 Btu/°F ft^2 (100 Wh/°C m^2). The sunspace temperature extremes are insensitive to the heat capacity in the adjoining room.

In the case of heat delivery by fan-forced convection, the fan is assumed to have a fixed volume flow capacity and is switched on and off with a single thermostat setpoint in the sunspace. The January temperature extremes are plotted versus fan capacity, fan thermostat setpoint, heat capacity in the sunspace, and heat capacity in the adjoining room. Qualitatively, the sensitivity of temperature extremes to fan capacity is very similar to the sensitivity to vent area in the natural convection case. The maximum sunspace temperature is very sensitive to the fan capacity, expressed as volume flow rate per unit of glazing projected area, especially below 2 cfm/ft^2 (35 m^3/s m^2). The minimum sunspace temperature is insensitive to fan capacity.

If the fan capacity were large enough, the maximum sunspace temperature would be limited to the fan thermostat setpoint; but, in typical cases, the fan is much too small for the setpoint to have much effect on the maximum temperature. The maximum fan capacity (per unit of glazing projected area) studied in Jones, McFarland, and Lazarus (1983) is 5 cfm/ft^2 (90 m^3/s m^2), and at this capacity, the maximum sunspace temperature is about 95°F (35°C), even though the setpoint is 80°F (27°F). There is a plot of maximum sunspace temperature versus setpoint for a fan capacity of 3 cfm/ft^2 (55 m^3/s m^2), which shows almost no sensitivity of sunspace maximum temperature to setpoint.

Both the maximum and minimum sunspace temperatures are sensitive to the sunspace heat capacity in the case of fan-forced convection. The sensitivity of the minimum sunspace temperature is almost identical to the corresponding curves for the natural convection case—see figure 6.9. The sensitivity of the maximum sunspace temperature is even more pronounced. The sunspace temperature extremes are insensitive to the heat capacity in the adjoining room.

6.6.3.2 Temperature Histograms Parsons (1983) gives histograms of sunspace air temperature in March in Madison, Wisconsin, for two different assumptions regarding sunspace heat capacity.

6.6.4 Mixed Systems

Sebald and Phillips (n.d.) calculate the discomfort index (see equation [13]) of combinations of Trombe-wall and direct-gain glazing areas, and of various control strategies for three cities. These results were obtained along with the data for auxiliary heat requirement (see section 6.4.6.1). The results show that Trombe-wall systems provide a good and relatively steady level of thermal comfort despite large variations in auxiliary energy requirement for the various configurations examined. The best control strategies provide a good level of thermal comfort as well as low auxiliary heat consumption. Controls can also be optimized relative to thermal comfort, providing low but not minimum, auxiliary energy requirement.

Winn (1981) applies a frequency-domain analysis method to a mixed direct-gain/Trombe-wall system in Albuquerque, New Mexico.

References

Andersen Corporation. 1985. *Andersen Concept IV Sunspaces.* 2d. ed. Bayport, MN: Andersen Corporation.

Arumi-Noe, F. and J. J. Kim. 1980. Parametric studies on the thermal response of a direct gain room to the distribution of massive elements on the walls. *Proc. 5th National Passive Solar Conference*, Amherst, MA, October 19–26, 1980. Newark, DE: American Section of the International Solar Energy Society, pp. 96–100.

Balcomb, J. D. 1983a. *Heat Storage and Distribution Inside Passive Solar Buildings.* LA-9694-MS. Los Alamos, New Mexico: Los Alamos National Laboratory.

Balcomb, J. D. 1983b. Conservation and solar guidelines. *Proc. 8th National Passive Solar Conference*, Santa Fe, NM, September 7–9, 1983, Boulder, CO: American Solar Energy Society, pp. 117–122.

Balcomb, J. D. 1986. Conservation and solar guidelines. *Passive Solar Journal* 3(3):221–248.

Balcomb, J. D., D. Barley, R. McFarland, J. Perry, W. Wray, and S. Noll. 1980. *Passive Solar Design Handbook.* Vol. 2: *Passive Solar Design Analysis.* DOE/CS-0127/2. Washington, DC: U.S. Department of Energy.

Balcomb, J. D., J. C. Hedstrom, and R. D. McFarland. 1977a. Simulation analysis of passive solar heated buildings—preliminary results. *Solar Energy*, 19(3):277–282.

Balcomb, J. D., J. C. Hedstrom, R. D. McFarland. 1977b. Simulation as a design tool. *Proc. Passive Solar Heating and Cooling Conference and Workshop*, Albuquerque, NM, May 18–19, 1976. LA-6637-C. Los Alamos, NM: Los Alamos Scientific Laboratory, pp. 238–246.

Balcomb J. D., R. W. Jones, C. E. Kosiewicz, G. S. Lazarus, R. D. McFarland, and W. O. Wray. 1982. *Passive Solar Design Handbook.* Vol. 3: *Passive Solar Design Analysis.* Edited by R. W. Jones. DOE/CS-0127/3. Washington, DC: U.S. Department of Energy; Boulder, CO: American Solar Energy Society, 1983.

Balcomb, J. D., R. W. Jones, R. D. McFarland, and W. O. Wray. 1984. *Passive Solar Heating Analysis: A Design Manual.* Atlanta, GA: ASHRAE.

Balcomb, J. D., and R. D. McFarland. 1977. Simulation analysis of passive solar heated buildings—the influence of climate and geometry on performance. *Proc. Annual Meeting, American Section of the International Solar Energy Society*, Orlando, FL, June 6–10, 1977. Cape Canaveral, FL: American Section of the International Solar Energy Society, pp. 11.1–11.4.

Balcomb, J. D., and R. D. McFarland. 1978a. A simple empirical method for estimating the performance of passive solar heated buildings of the thermal storage wall type. *Proc. 2d National Passive Solar Conference*, Philadelphia, PA, March 16–18, 1978. Newark, DE: American Section of the International Solar Energy Society, pp. 377–389.

Balcomb, J. D., and R. D. McFarland, 1978b. A simple empirical method for estimating the performance of a passive solar heated building of the thermal storage wall type. *Proc. Annual Meeting, American Section of the International Solar Energy Society*, Denver, CO, August 28–31, 1978. Newark, DE: American Section of the International Solar Energy Society, pp. 89–96.

Balcomb, J. D., and W. O. Wray, 1988. *Passive Solar Heating Analysis.* Supplement 1: *Thermal Mass Effects and Aditional SLR Correlations.* Atlanta, GA: ASHRAE.

Beckman, W. A., S. A. Klein, and B. K. Parsons, 1983. Performance prediction of attached sunspaces. *Proc. 1983 Annual Meeting, American Solar Energy Society.* Progress in Solar Energy, vol. 6. Boulder, CO: American Solar Energy Society, pp. 671–675.

Carroll, J. A., and J. R. Clinton. N.d. An index to quantify thermal comfort in homes. Appendix D of *Impact of Controls in Passive Solar Heating and Cooling of Buildings, Final Report.* La Jolla, CA: University of California, San Diego.

Chen, B., E. Hollingsworth, K. E. Pedersen, J. Maloney, D. Stangl, J. Thorp, and J. Rives. 1982. *Path to Passive: Nebraska's Passive Solar Primer.* Lincoln, NE: Nebraska Energy Office.

Clausing, A. M., and B. L. Drolen. 1978. Parameters for quantifying the thermal performance of passive solar heating systems. *Proc. 5 Annual Meeting, American Section of the International Solar Energy Society,* Denver, CO, August 28–31, 1978. Newark, DE: American Section of the International Solar Energy Society, pp. 78–82.

Clausing, A. M., and B. L. Drolen. 1979. The characterization of the performance of passive solar heating systems. *Sun II: Proc. International Solar Energy Society Silver Jubilee Congress,* Atlanta, GA, May 28–June 1, 1979. New York: Pergamon Press, pp. 1540–1544.

Clements, E. A. 1980. Development and validation of a computer model for predicting heating performance of a roof pond house, Master's thesis, Trinity University, San Antonio, TX.

Clinton, J. R., A. V. Sebald, F. Langenbacher, and E. Hendricks. 1978. Effects of controls on water wall performance. *Proc. 2d National Passive Solar Conference,* Philadelphia, PA, March 16–18, 1978. Newark, DE: American Section of the International Solar Energy Society, pp. 262–266.

Direnzo, R. J. 1981. Thermal performance model and design sensitivity studies for attached sunspaces. Master's thesis, University of Massachusetts, Amherst.

Direnzo R., and J. G. McGowan. 1980. Thermal performance model and design sensitivity study for attached sunspaces, *Proc. 5th National Passive Solar Conference,* Amherst, MA, October 19–26, 1980. Newark, DE: American Section of the International Solar Energy Society, pp. 262–266.

Drolen, B. L. 1979. Thermal performance of passive solar heating systems. Master's thesis, University of Illinois, Urbana.

Emery, A. F., D. R. Heerwagen, B. R. Johnson, C. J. Kippenhan, and J. E. Lakin. 1982. The development of sensitivity curves for designing passive solar houses for the Pacific Northwest. *Proc. 7th National Passive Solar Conference,* Knoxville, TN, August 30–September 1, 1982. Newark, DE: American Section of the International Solar Energy Society, pp. 93–98.

Fuchs, R., and J. F. McClelland. 1979. Passive solar heating of buildings using a Transwall structure. *Solar Energy* 23(2): 123–128.

Hull, J. R., J. F. McClelland, L. Hodges, J. L. Huaug, R. Fuchs, and D. A. Block. 1980. Effect of design parameter changes on the thermal performance of a Transwall passive solar heating system. *Proc. 5th National Passive Solar Conference,* Amherst, MA, October 19–26, 1980. Newark, DE: American Section of the International Solar Energy Society, pp. 394–398.

Jones, R. W. 1980. *Passive Solar Heating of Buildings with Attached Greenhouses, Final Report for Period August 31, 1979–August 30, 1980.* DOE/30242-4. Washington, DC: U.S. Department of Energy.

Jones, R. W. 1981. Summer heat gain control in passive solar heated buildings: fixed horizontal overhangs. *Passive Cooling: Proc. International Passive and Hybrid Cooling Conference,*

Miami Beach, FL, November 11–13, 1981. Newark, DE: American Section of the International Solar Energy Society, pp. 402–406.

Jones, R. W., and R. D. McFarland. 1980. Attached sunspace heating performance estimates. *Proc. 5th National Passive Solar Conference*, Amherst, MA, October 19–26, 1980. Newark, DE: American Section of the International Solar Energy Society, pp. 426–430.

Jones, R. W., and R. D. McFarland. 1984. *The Sunspace Primer: A Guide for Passive Solar Heating*. New York: Van Nostrand Reinhold Co.

Jones, R. W., R. D. McFarland, and G. S. Lazarus. 1983. Mass and fans in attached sunspaces. *Proc. 3d Energy Conserving Greenhouse Conference*, Hyannis, MA, November 19–21, 1982. New York: American Solar Energy Society, pp. 81–90.

Kirkpatrick, A. T., and C. B. Winn. 1982. A frequency response technique for the analysis of the enclosure temperature in passive solar buildings. *Proc. 1982 Annual Meeting, American Solar Energy Society*, Houston, TX. Boulder, CO: American Solar Energy Society, pp. 773–777.

McClelland, J. F., and R. Fuchs. 1979. A preliminary study of passive solar heating performance and visual clarity for a Transwall structure. *Proc. 3d National Passive Solar Conference*, San Jose, CA, January 11–13, 1979. Newark, DE: American Section of the International Solar Energy Society, pp. 107–113.

McFarland, R. D., and J. D. Balcomb. 1979. The effect of design parameter changes on the performance of thermal storage wall passive systems. *Proc. 3d National Passive Solar Conference*, San Jose, CA, January 11–13, 1979. Newark, DE: American Section of the International Solar Energy Society, pp. 54–60.

McFarland, R. D., and R. W. Jones. 1980a. Performance estimates for attached-sunspace passive solar heated buildings. *Proc. Annual Meeting of the American Section of the International Solar Energy Society*, Phoenix, AZ, June 2–6, 1980. Newark, DE: American Section of the International Solar Energy Society, pp. 784–788

McFarland, R. D., and R. W. Jones. 1980b. *Performance Estimates for Attached-Sunspace Passive Solar Heated Buildings. Includes LCR tables*. LA-UR-80-1482. Los Alamos, NM: Los Alamos National Laboratory.

Marlatt, W. P., K. A. Murray, and S. E. Squier. 1984. *Roof Pond Systems*. ETEC-83-6. Canoga Park, CA: Energy Technology Engineering Center.

Mazria, E. 1978. A design sizing procedure for direct gain, thermal storage wall, attached greenhouse, and roof pond systems. *Proc. Annual Meeting, American Section of the International Solar Energy Society*, Denver, CO, August 28–31, 1978. Newark, DE: American Section of the International Solar Energy Society, pp. 162–165.

Mazria, E., M. S. Baker, and F. C. Wessling. 1977. An analytical model for passive solar heated buildings. *Proc. Annual Meeting, American Section of the International Solar Energy Society*, Orlando, FL, June 6–10, 1977. Cape Canaveral, FL: American Section of the International Solar Energy Society, pp. 11.10–11.14

Mazria, E., M. S. Baker, and F. C. Wessling. 1978. Predicting the performance of passive solar heated buildings—a two-year study. *Proc. 2d National Passive Solar Conference*, Philadelphia, PA, March 16–18. 1978, Newark, DE: American Section of the International Solar Energy Society, pp. 393–397.

Murray, H. S., J. L. Melsa, and J. D. Balcomb. 1980. Control system analysis for off-peak auxiliary heating of passive solar systems. *Proc. Systems Simulation and Ecomonic Analysis*, San Diego, CA, January 23–25, 1980. Golden, CO: Solar Energy Research Institute.

Niles, P. W. B. 1978. A simple direct-gain passive house performance prediction model. *Proc. 2d National Passive Solar Conference*, Philadelphia, PA, March 16–18, 1978. Newark, DE: American Section of the International Solar Energy Society, pp. 534–537.

Niles, P. W. B., and K. L. Haggard. 1980. *Passive Solar Handbook*, Sacramento, CA: California Energy Commission.

Palmiter, L. 1979. Development of an effective solar fraction. *Sun II: Proc. International Solar Energy Society Silver Jubilee Congress*, Atlanta, GA, May 28–June 1, 1979. New York: Pergamon Press, pp. 843–846

Parsons, B. K. 1983. The simulation and design of building attached sunspaces. Master's. thesis, University of Wisconsin, Madison.

Phillips, G. M., and A. V. Sebald. N.d. Response surfaces: summarizing complex simulation results in a manner suitable for inexpensive use by architects and builders. Appendix F of *Impact of Controls in Passive Solar Heating and Cooling of Buildings, Final Report*. La Jolla, CA: University of California, San Diego.

Schnurr, N. M., and W. O. Wray, 1984. Numerical and experimental studies of thermosiphon passive heating systems. *Proc. 9th National Passive Solar Conference*, Columbus, OH, September 1984, Boulder, CO: American Solar Energy Society, pp. 580–585.

Sebald, A. V., J. R. Clinton, and F. Langenbacher. 1979. Control considerations in the Trombe wall. *Proc. 3d National Passive Solar Conference*, San Jose, CA, January 11–13, 1979. Newark, DE: American Section of the International Solar Energy Society, pp. 48–53.

Sebald, A. V., F. Langenbacher, and J. R. Clinton. 1979. On proper control of direct gain passive solar houses. *Proc. 4th National Passive Solar Conference*, Kansas City, MO, October 3–5, 1979. Newark, DE: American Section of the International Solar Energy Society, pp. 235–239.

Sebald, A. V., and D. Munoz. N.d. On eliminating peak load auxiliary energy consumption in passive solar residences during winter. Appendix C of *Impact of Controls in Passive Solar Heating and Cooling of Buildings, Final Report*. La Jolla, CA: University of California, San Diego.

Sebald, A. V., and G. M. Phillips. N.d. On the controlled tradeoffs for rockbins. Appendix B of *Impact of Controls in Passive Solar Heating and Cooling of Buildings, Final Report*. La Jolla, CA: University of California, San Diego.

Sebald, A. V., and G. Vered. 1981. On extracting useful building performance characteristics without simulation. *Proc. Annual Meeting, American Section of the International Solar Energy Society*, Philadelphia, PA, May 27–30, 1981. Newark, DE: American Section of the International Solar Energy Society, pp. 1016–1020.

Weidt, J. L., R. J. Saxler, and S. Gerard. 1983. Predicting performance of prototypical sunspaces attached to typical residences in various climatic regions. *Proc. 1983 Annual Meeting, American Solar Energy Society*. Progress in Solar Energy, vol. 6. Boulder, CO: American Solar Energy Society, pp. 81–85.

Winn, C. B. 1981. A simple method for determining the average temperature, the temperature variation and the solar fraction for passive solar buildings. *Proc. 6th National Passive Solar Conference*, Portland, OR, September 8–12, 1981. Newark, DE: American Section of the International Solar Energy Society, pp. 303–307.

Winn, R. C., and C. B. Winn. 1982. Optimal control of passive solar buildings with load managed storage. *Proc. 1982 Annual Meeting, American Solar Energy Society*, Houston, TX. Boulder, CO: American Solar Energy Society, pp. 685–690.

Wray, W. O. 1979a. Off-peak auxiliary heating of passive solar homes, *Proc. 3d National Conference and Exhibition on Technology for Energy Conservation*, Tucson, AZ, January 23–25, 1979.

Wray, W. O. 1979b. A simple procedure for assessing thermal comfort in passive solar heated buildings. *Sun II: Proc. International Solar Energy Society Silver Jubilee Congress*, Atlanta, GA, May 28—June 1, 1979. New York: Pergamon Press, pp. 1530–1534.

Wray, W. O. 1980. Additional solar load ratio correlations for direct gain buildings. *Proc. 1980 Annual Meeting, American Section of the International Solar Energy Society*, Phoenix, AZ, June 2–6, 1980. Newark, DE: American Section of the International Solar Energy Society, pp. 870–874.

Wray, W. O., and J. D. Balcomb. 1979. Trombe wall vs. direct gain: a comparative analysis of passive solar heating systems. *Proc. 3d National Passive Solar Conference*, San Jose, CA, January 11–13, 1979. Newark, DE: American Section of the International Solar Energy Society, pp. 41–47.

Wray, W. O., J. D., Balcomb, and R. D. McFarland. 1979. A semi-empirical method for estimating the performance of direct-gain passive solar heated buildings. *Proc. 3d National Passive Solar Conference*, San Jose, CA, January 11–13, 1979. Newark, DE: American Section of the International Solar Energy Society, pp. 395–402.

Wray, W. O., and E. D. Best. 1987a. Sizing backup heating equipment in passive solar heated buildings. *Proc. 12th National Passive Solar Conference*, Portland, OR, July 11–16, 1987. Boulder, CO: American Solar Energy Society, pp. 59–68.

Wray, W. O., and E. D. Best. 1987b. Thermal storage in direct gain buildings: the effective heat capacity and a generalized solar load ratio correlation. *Passive Solar Journal* 4(1):41–61.

Wray, W. O., and C. E. Kosiewicz, 1984. Including the effect of control strategy in solar load ratio calculations. *Proc. Annual Meeting, American Solar Energy Society*, Anaheim, CA, June 5–9, 1984. Boulder, CO: American Solar Energy Society, pp. 395–399.

Wray, W. O., and C. E. Peck. 1987. *Military Handbook: Design Procedures for Passive Solar Buildings*. MIL-HDBK 1003/19. Washington, DC: U.S. Department of Defense.

Wray, W. O., and C. E. Peck. 1988. *Passive Solar Design Procedures for Naval Installations*. LA-11250-M. Los Alamos, NM: Los Alamos National Laboratory.

Wray, W. O., N. M. Schnurr, and C. E. Kosiewicz. 1984. Fast solar load ratio correlations for simple radiant panels and thermosiphoning air panels. *Proc. 9th National Passive Solar Conference*, Columbus, OH, September 1984. Boulder, CO: American Solar Energy Society, pp. 101–104.

Wray, W. O., N. M. Schnurr, and J. E. Moore. 1980. Sensitivity of direct gain performance to detailed characteristics of the living space. *Proc. 5th National Passive Solar Conference*, Amherst, MA, October 19–26, 1980. Newark, DE: American Section of the International Solar Energy Society, pp. 92–95.

7 Test Modules

Fuller Moore

7.1 Types of Modules

7.1.1 Classification

Research in passive solar heating of buildings has been enhanced through the use of physical test modules. The principal advantages of test modules are that their cost is low (compared with full-scale testing of occupied buildings) and that data can be taken under carefully controlled and well-known conditions. Types of test modules include test boxes (reduced height, width, and depth), test rooms (full height, reduced width and depth), and test buildings (full size).

Small-scale insulated test homes have been successfully used for proof-of-concept demonstration as well as for quantifiable analogies of passive solar-heated buildings. Full-height test rooms permit the observation of vertical convection effects that approximate those of actual buildings. The advantage, coupled with increased interior accessibility and dimensional similarity to actual buildings, accounts for the extensive use of test rooms in passive solar research. Test buildings are similar in size and construction to actual buildings but are designed and constructed for test purposes only and not intended for occupancy. Such buildings offer the opportunity for both multizone testing and direct analogy to actual building performance.

The three classes of test modules are reviewed in detail in sections 7.3, 7.4, and 7.5.

7.1.2 Test Objectives

Test modules can be used to accomplish a variety of experimental objectives such as proof-of-concept demonstrations and operational testing ("Does this work?") and comparative testing of components or systems in matched test modules ("Which works best?"). Single test modules can be used to empirically measure heat transfer coefficients for components or an entire system ("Why does this perform as it does?") and for computer model validation ("How closely can a computer simulate the measured conditions of a test module, given identical weather data as input?"). Finally, test modules can be used as direct physical analogs of actual buildings ("How will an actual building perform in terms of comfort and auxiliary energy usage?"). The experimental objectives of a project must be clearly defined prior to the selection of a test-module type.

7.1.3 Operating Modes

7.1.3.1 Proof-of-Concept Experiments Proof-of-concept experiments are often used by investigators to determine whether further, more rigorous testing is justified. The objective is to isolate the principle under investigation and minimize unrelated variables. In the interest of expediency and economy, there is a temptation to proceed with the experiment before performing the theoretical analysis necessary to maximize the predicted effect.

Consider, for example, the early heating experiment by Harold Hay in which he used a foam box to prove the concept of his roof-pond system (Hay and Yellott 1970). With the box filled with water, Hay removed the lid to admit insolation during the day and replaced the lid at night to reduce heat loss. The measured increase in water temperature demonstrated the heating potential of the system. By covering the box during the day (and uncovering it at night), the cooling potential was similarly demonstrated.

It is easy to imagine a set of conditions (low sun angle, rapid air movement, low humidity) that actually would have resulted in a cooling effect in Hay's experiment, rather than the heating anticipated. Conversely, early (even premature) physical experimentation can be conducive to enhancing the intuitive process. Many researchers gain most of their insights while working with such physical models, whereas others favor sketching or theoretical calculations.

Another important use of proof-of-concept experiments is for educational purposes. Such single-purpose test modules are particularly valuable for demonstrating various passive solar principles. Several such experiments using various test modules are described by Benton and Akridge (1981).

7.1.3.2 Comparative Experiments Comparative experiments may be direct (where two or more systems are compared in one experiment under identical conditions) or indirect (where two or more experimental systems are compared indirectly in separate experiments using a third, "control" module). In the latter, some method of normalization is required to provide a basis of comparison between the competing experimental systems.

In direct-comparison experiments, the objective is to compare the performance of two competing components or systems through simultaneous testing. The experiment should isolate the difference(s) of the two systems,

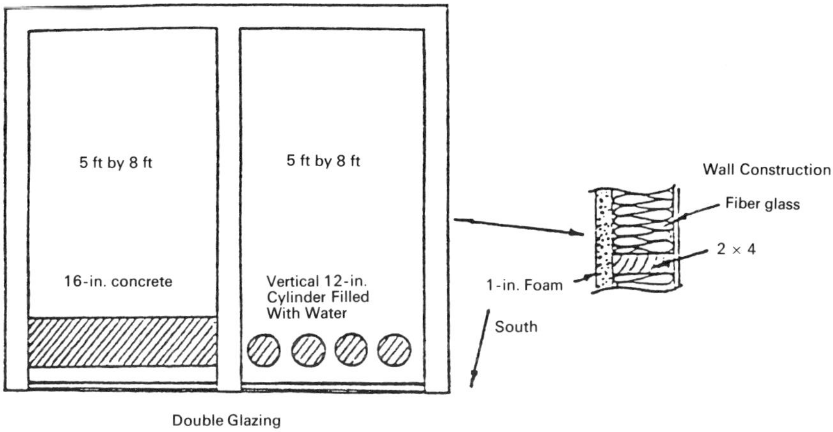

Figure 7.1
Plan for Los Alamos test room water-wall/Trombe-wall experiment.

minimizing the unrelated variables. Any performance differences can thus be assumed to be a result of the differing system characteristics

For example, an early experiment at Los Alamos (Balcomb, Hedstrom, and McFarland 1977) directly compared a masonry thermal-storage wall system and a water tube thermal-storage system in adjacent test rooms (see figure 7.1). Several comparative experiments were run while changing such parameters as the vents in the masonry wall and the blocking of the space between the water tubes. These procedures gave a direct comparison of overall system performance for the particular configurations tested

Note, however, that it is not possible to draw general conclusions from this experiment about masonry versus water as a thermal-storage medium. In this experiment, readily available standard components were used, that is, 12-in. by 8-ft (30 by 240 cm) fiber glass water tubes and 16-in. (41-cm)-thick solid masonry. The resulting thermal-storage capacities, per unit of glazing area, were similar, 35 Btu/°F ft^2 (199 Wh/°C m^2) for water versus 32.5 Btu/°F ft^2 (186 Wh/°C m^2) for masonry. However, this thermal-storage limitation resulted in spacing the water tubes 2.4 in. (6 cm) apart and leaving a 24-in. (61-cm)-gap above the tubes. Foam insulation was inserted into the spaces between and above the water tubes (to reduce the convective air transfer between the glazing and room interior). This prevented any direct solar gain between the tubes at midday and partially blocked sun from the sides of the tubes in the morning and afternoon. If

Table 7.1
Test module facilities in the United States and Canada

Test module facilities	year operational	discontinued	boxes	rooms	buildings	modules per structure	height, feet	width, feet	depth, feet	walls, R-value	floor, R-value	roof, R-value	# of attached greenhouses	solar aperture, sq. ft.	configurable mass	configurable glazing	configurable partitions	Auxiliary heat type	Auxiliary heat control	Infiltration control	reference	remarks
Skytherm Processes & Engineering	67	y	1			1	9	10	12	7	S	P		120				FC	T	M	Hag & Yellott 1970	roof pond
Environmental Research Laboratory	81				4	1	12	40	24	V	V	V		104		y			/	M	Peck & Kessler 1981	cooling experiments
Pacific Gas and Electric	77		4			2	9	10	10	20	20	20		64	y			ER	D	M	D'Albora 1981	
Farallones Institute	76	78			5	1	16	14	12	11	8	19		V						M	Calthorpe 1978	occupied during testing
Pala Passive Saar Project	81				8	1	9	16	16	11	8	19	1	V				FC	D	M	Clinton 1980	2 zone
Habitat Center	79	81	4			2	10	14	22	14	16	30	4	117	y					M	Martin et al 1981	
National Bureau of Standards	83		4			4	8+	12	27	28	9	32		106	y	y	y	ER	T	M	Solar group 1983	
Purdue University	45	y			2	1	8	?	?	12	?	?		112/4	y			ER	T	M	Hutchinson 1947	18 30 S.F. floor area
Massachusetts Institute of Technology	47	50	6			6	10	6	14	14	14	14		32	y	y	y	ER	T	M	Hollingsworth 1947	variable north insulation
National Center for Appropriate Technology	78	80	4			2	10	5	8	19	24	24	2	35	y	y		ER	T	M	Palmiter, et al 1978	plus zretrofit + 2 greechhouse
Universits of Nebraska	78		6			1	10	5	8	22	18	28	1	30	y	y	?			?	Chen et al 1979	
Florida Solar Energy Center	81				1	1	24	40	40	V	S	V		V	y	y	y	ER	D	M	Fairey 1981	cooling experiments
Los Alamos National Laboratory	80		2			2	20	40	1	40	40	4		V	y	y				M	Morris 1980	convective collector
Los Alamos National Laboratory	75		14			2	10	5	8	19	19	19	2	34	y	y	y	ER	D	F	Balcomb et al 1977a	

Los Alamos National Laboratory	75		8			1	2	2	2	32	32	32		1.7	y	y	y	ER		M	Grimmer, et al 1979	
Solar Room Company	77	y		4		4	9	8	22	40	40	40	3	96				ER	T	M	Kerin and Betz 1978	product development
Wessling Consulting	77	78		4		4	8	8	8	V	V	V	3	V						M	Wessling 1978	retrofitted to residence
Scott Morris	77	79	4			1	6	2	3	9	18	9		12	y					M	Morris 1978	convective collectors
Sundwellings Demonstration Center	76	77			4	1	10	40	17	22	25	25		V				ER	T	M	Rogers 1980	adobe construction
Ames Laboratory	79			3		1	8	12	12	28	28	24		45	y					M	Mercer et al 1981	
Waterwall Engineering	80	82		2		1	9	6	4	14	20	20		32	y	y				F		product development
Pennsylvania State University	80	81		4		4	8	12	8	10	S	2	4	V	y			ER	D	M		plus attached greenhouse
University of Pennsylvania	77	y		2		2	10	5	8	19	19	19		36	y	y				M	Prowler 1978	
Tennessee Valley Authority	81			4/2		1/2	8	5/12	8	19	19	25		35	y		y	ER	D	M	Kuberg 1981	
Trinity Universit	81				2	1	10	40	20	?	S	V		800	y	y	y	FC	D	M	Doderer 1981	roof pond; cooling
University of Alberta	78				6	1	12	24	22	20	?	40		119	y		y	ER	T	M	Dale, et al 1980	1 passive solar/5 conserv.
Lakehead University	79	80	2			2	4	4	4	20	20	32		7/10	y	y		ER	T	M	Jones, 1979	
Solar Energy Research Institute	81		2	10		2	9	6	10	25	25	25		40	y	y	y	ER	T	M	Andrews 1980	test boxes 4′/side
Zomeworks	70	y	4			1	3	2	3	15	15	15		6	y	y	y			M	Baer 1977	waterwall test boxes

notes
y—yes
g—gravel
p—pond
s—slab
v—varies
?—information unavailable

FC—Fan coil aux. heating
ER—Electric resistance aux. heating
T—Thermostat controlled
D—Data Acquisition controlled
F—Forced infiltration
M—minimized infiltration

the experimental objective had been to compare thermal-storage media, the following characteristics (unrelated to medium type) should have been identical: exposed mass area and shape, venting, thermal-storage capacity, and interior mass area and shape. This would have implied a very non-standard configuration of at least one of the systems (that is, vertical concrete cylinders or flat water wall).

The experiments at Los Alamos reported by Hyde (1980) are examples of indirect comparative testing. In this series, the performance of various experimental thermal-storage-wall systems (varying storage medium, glazing, surface selectivity, and night insulation) was compared with that of a control system (double-glazed, flat-black, unvented Trombe wall, no night insulation). The basis for comparing the systems was Q_{solar} (net energy delivered through the south wall). This value was calculated from the energy balance:

$$Q_{solar} + Q_{aux} = Q_{load}. \tag{8.1}$$

Q_{aux}, the auxiliary heat, was measured directly, and Q_{load} was calculated from the measured room heat-loss coefficient, the measured average room temperature, and the measured ambient temperature.

The control Trombe-wall system (room no. 8) was operated for the duration of all test periods, whereas the various experimental systems were tested during shorter periods (see table 7.1). Note that not all experimental systems were directly compared with each other (because of the limited number of rooms available). Instead, the various experimental systems were normalized against the control; that is, the ratio of a test room's net energy to the control room's net energy (Q_{solar}/Q_{8c}) was used as a direct basis for performance comparison.

For such indirect comparisons to be valid, the ratio Q_{solar}/Q_{8c} for a given system must remain constant over changing weather conditions. Hyde (1980) validated the normalization parameter by operating each of the experimental systems in at least two test periods. A comparison of Q_{solar}/Q_{8c} (table 7.2) for a given experimental system in two different test periods provides a measure of the validity of this normalization method for each of the various systems.

Test module indirect comparison experiments are frequently used by commercial manufacturers to compare the performance of competing systems while using a minimum number of test modules. Usually only two

Table 7.2
Performance summaries, indirect comparison test-room experiments, normalized on the basis of $Q_{solar}/Q_{solar,\ control}$

Test room No. Description	Q_{solar} kBtu/day (kWh/day)	Q_{aux}	Q_{load}	Ambient temp. °F(°C)	Insolation Btu/ft² day (Wh/m² day)	*SSF*	$Q_{solar}/Q_{s,c}$
1 TW, FB, SG	11.4 (3.3)	19.7 (5.8)	30.9 (9.1)	32.4 (0.2)	1630 (5141)	0.38	0.63
2 TW, BC, SG	19.1 (5.6)	13.1 (3.8)	31.8 (9.3)			0.59	1.05
5 TW, BC, SG	25.5 (7.5)	10.1 (3.0)	34.9 (10.2)			0.68	1.40
6 WW, FB, SG	15.2 (4.5)	20.4 (6.0)	34.8 (10.2)			0.36	0.83
7 WW, FB, DG, NI	22.1 (6.5)	10.5 (3.1)	32.1 (9.4)			0.67	1.31
8 TW, FB, SG	16.9 (5.0)	15.4 (4.5)	31.8 (9.3)			0.52	1.00
1 TW, FB, SG, NI	13.7 (4.0)	14.6 (4.3)	28.4 (8.3)	37.6 (3.1)	1214 (3829)	0.49	0.91
2 TW, BC, SG	15.1 (4.4)	12.8 (3.7)	27.9 (8.2)			0.55	1.01
5 WW, BC, SG	17.8 (5.2)	12.5 (3.7)	30.3 (8.9)			0.56	1.19
6 WW, FB, SG, NI	22.8 (6.7)	9.9 (2.9)	32.7 (9.6)			0.65	1.52
7 TW, FB, DG, NI	18.2 (5.3)	11.2 (3.3)	29.4 (8.6)			0.61	1.31
8 TW, FB, DG	13.9 (4.1)	13.6 (4.0)	27.6 (8.1)			0.52	1.00
1 TW, FB, DG	11.4 (3.3)	17.2 (5.0)	28.6 (8.4)	37.4 (3.0)	1430 (4510)	0.40	0.92
2 TW, BC, DG	14.7 (4.3)	13.1 (3.8)	27.8 (8.1)			0.54	1.19
5 WW, BC, DG	19.9 (5.8)	10.1 (3.0)	30.0 (8.8)			0.65	1.61
6 WW, FB, DG	14.4 (4.2)	15.9 (4.7)	30.3 (8.9)			0.44	1.16
7 TW, FB, DG, NI	Malfunction in bead system						
8 TW, FB, EG	11.5 (3.4)	16.2 (4.7)	27.7 (8.1)			0.43	1.00
1 TW, FB, DG	22.4 (6.6)	16.4 (4.8)	38.8 (11.4)	23.5 (−4.7)	1950 (6150)	0.57	1.00
2 TW, BC, DG	28.5 (8.3)	10.4 (3.0)	38.9 (11.4)			0.72	1.27
5 WW, BC, DG	40.0 (11.7)	3.5 (1.0)	43.5 (12.7)			0.91	1.78
6 WW, FB, DG	37.4 (11.0)	5.6 (1.6)	44.1 (12.9)			0.82	1.67
7 TW, FB, DG, NI	29.9 (8.8)	10.0 (2.9)	39.8 (11.7)			0.74	1.44
8 TW, FB, DG	20.8 (6.1)	18.0 (5.3)	38.8 (11.4)			0.53	1.00

Source: Hyde 1980.

Notes:

TW	Trombe wall	BC	black chrome	NI	night insulation
WW	water wall	SG	single glazed	SSF	solar savings fraction
FB	flat black	DG	double glazed		

modules are employed. However, the use of three units provides the opportunity to monitor continuously the validity of the normalization measure ($Q_{exper}/Q_{control}$ in the Hyde experiments). It should be noted that an accurate normalization procedure (in either direct or indirect comparisons) raises the level of results from "Which is better?" to "How much better?"

7.1.3.3 Model Validation Thermal-network models (see figure 7.2) have become widely accepted as an accurate technique for simulating the dynamic performance of passive solar buildings (Perry 1977, Wray 1980, Clinton 1979, Sebald 1981, McFarland 1978, Kohler and Sullivan 1979, Arumi-Noe 1978, Abrash et al. 1978, and Judkoff et al., 1980). (See also Niles, chapter 3 of this volume.) Before a computer model can be accepted as an accurate simulation of the thermal behavior of a building, it must be validated against the actual performance of a monitored structure. Typically, the actual building-performance data are directly compared with the simulation results obtained using the recorded test-period weather data. Statistical techniques are employed to measure the accuracy of the comparison. Such comparisons provide a basis for refining the model. Although the calculational procedure used by the model may be precise, the determination of some of the system input parameters may not be. Test

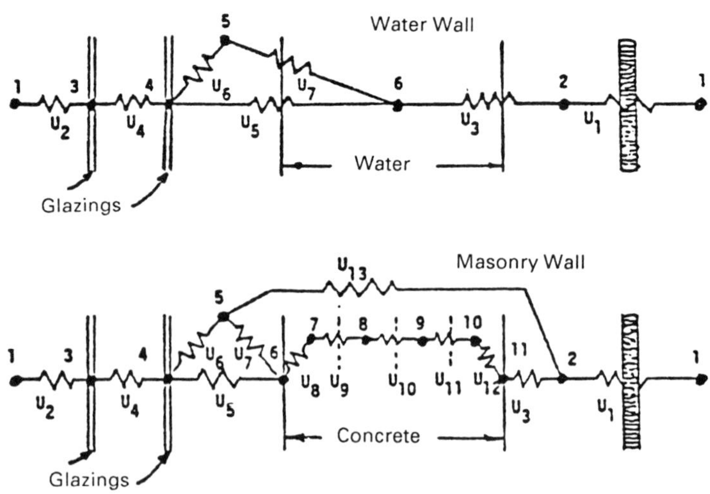

Figure 7.2
Thermal network diagrams for water-wall and Trombe-wall test rooms.

module operating data and one-time measurements may be used to determine these parameters.

Test modules usually provide the best basis for this code validation process for several reasons: they are easily reconfigurable, they facilitate sensor placement, the effects of unrelated variables can be minimized, they are unoccupied and can be used with or without auxiliary heat.

Although test boxes and test buildings have been used for code validation, full-height, small test rooms are the most widely used; in fact, in many research programs, code validation is the primary purpose of these rooms.

Because of the detailed scrutiny that code validation data will ultimately receive, the most rigorous experimental procedures must be employed. For example, sensors must be located to correspond to thermal-network nodes. Not only must sensor locations be planned to maximize their value in code validation, but these locations must be precise and accurately recorded for future reference. In addition, the experimenter must assure that no sensor significantly perturbs the experiment. For example, conduction along the stem of a thermocouple can alter the local temperature.

7.1.3.4 Component Performance Test modules can be used to obtain data on component performance under realistic operating conditions. The data might be used for evaluating the effectiveness of a night insulation system or measuring the effective heat loss coefficient of a selective surface (Balcomb 1981).

7.1.3.5 Building-Performance Experiments When test modules are to be used as a direct analog of an occupied building, the primary measures of performance are usually human thermal comfort and auxiliary heating performance over an extended period. Successful direct simulation of thermally complex, occupied buildings requires that the size and zonal configuration of the test module exceed the capabilities of a test room and approach the characteristics of an actual building. Because of these difficulties, it may be preferable to monitor an occupied building; this is the focus of the SERI Class B Passive Solar Buildings Monitoring Program (Frey and McKinstry 1980). However, it is difficult to draw general conclusions from specific experiments when monitoring occupied buildings that present a number of experimental measurement and control problems. These include:

1. Variation in infiltration rate (operation of doors and windows, and exhaust-fan operation)

2. Measurement of net contribution of auxiliary heat from heat pumps, wood stoves, fireplaces, oil or gas furnaces, etc.
3. Miscellaneous heat gain from internal sources (water heaters, cooking, bathing, laundry, lighting, occupant body heat, etc.)
4. Inaccessibility of the building for measurement and observation
5. Configuration impermanence

The experimental alternative to monitored occupied buildings is unoccupied, full-scale test buildings. In such a facility, the above-named variables can be controlled and measured, although they present inherent disadvantages in the cost and difficulty of reconfiguration (compared with smaller test rooms).

7.2 Instrumentation

To monitor relevant ambient and experimental conditions in passive solar test modules, sensors are required to measure at least some of the following conditions: solar radiation, temperature (air, globe, surface, solid, and liquid), air movement (infiltration rate, wind speed, and interior convection), auxiliary heat, and events (see section 7.2.1). The measurements are usually recorded automatically by a digital data acquisition system (DAS)—see section 7.2.2.

7.2.1 Data Acquisition Sensors

7.2.1.1 Radiation Sensors Pyranometers are used to measure total, direct (from the sun only), and diffuse (reflected/refracted by clouds, atmosphere, and ground) radiation, and are of two types: thermal and photovoltaic. At least two pyranometers are typically used for test module research. One is positioned to measure total horizontal radiation to permit comparisons with other standard radiation data. The second is positioned to measure total radiation in the plane of the test module glazing (usually vertical, south facing). Pyranometers can be equipped with a shadow band mask, permitting the approximate measurement of diffuse radiation directly (and beam radiation indirectly by subtraction from global).

The pyrheliometer measures beam radiation only. It resembles a small telescope and is aimed directly at the sun by a mechanical tracking system. While beam radiation is of primary concern in concentrating collector research, it is of minor interest in passive test module research.

7.2.1.2 Temperature Sensors Direct-reading thermometers are useful for limited measurement of temperature in certain experiments; however, sensors with automatically recordable output are generally preferable for monitoring test modules. Four types are in general use: thermocouples (widest use), platinum resistance detectors, thermistors, and transducers.

When measuring air temperature, care must be taken to avoid inaccuracies due to radiation. Radiation shielding is usually accomplished by a wood-louvered instrument shelter (Weather Bureau type for measuring ambient air temperature) or an unaspirated, tubular radiation shield for other locations.

Globe temperature is often measured in the "room" portion of a test module to assess the combined effect of interior air temperature and mean radiant temperature on occupant comfort. Precision globes are available, but all but one of the facilities surveyed (Moore and McFarland 1982) used a temperature sensor suspended within a sealed standard polyethelyne toilet float.

Surface temperature is usually measured by scoring a groove in the surface and embedding a sensor using a cement having a conductance similar to the original surface material.

Solid temperature (such as the interior of a masonry Trombe wall) is measured with a temperature sensor embedded in the material. Usually, a series of five or more is cast when the wall component is formed. Because sensor lead wires are more thermally conductive than the wall material, they are usually confined to an isothermal location for the first few inches (that is, perpendicular to the direction of heat flow).

Liquid temperatures are measured simply by immersing waterproofed sensors into the liquid.

7.2.1.3 Air Movement Infiltration is potentially a major source of heat loss from a test module. One strategy (employed by most researchers) is to minimize infiltration and determine its effect based on one-time measurements using tracer gas or a blower door. The combined effects of stack-effect infiltration and wind-induced infiltration are continuously estimated by monitoring wind speed (using an anemometer) and temperature difference, and applying these measurements to a formula using coefficients determined by measurements. A second strategy, used at Los Alamos, employs a small fan to pressurize the test module. This strategy overcomes any wind or stack effect, and results in a constant infiltration rate equal to the volume flow rate of the fan.

7.2.1.4 Auxiliary Heat Measurement If a test module receives auxiliary electric resistance heat (preferred because its 100% conversion efficiency makes measurement easy), some method of monitoring electric power consumption is required. If an automatic digital DAS is used to sample data frequently, the simplest method is to use a fixed-rate resistance heater and periodically measure its on/off status. Power consumption can also be measured using a watt-hour meter that has been equipped with a pulse initiator, and counting the pulses with the DAS. A third alternative is a clip-on Hall-effect current transducer to measure amperes in conjunction with a line voltage sensor.

Standard thermostats are not accurate enough as auxiliary heat controllers to be used in test modules. Most facilities utilize the DAS for auxiliary heat control.

7.2.1.5 Event Sensors The two types of event sensors of interest are status sensors and counting sensors. Physical status (such as the position of night insulation) and electrical status (such as the on/off status of auxiliary heat) can be monitored by the DAS using a simple microswitch on a low voltage circuit. Event counting (such as door openings or anemometer rotations) uses a similar principle but requires a special microswitch resistant to "bounce," and a signal conditioning card to accumulate the counts during the sampling interval.

7.2.2 Data Acquisition Systems

Although direct graphic recording of data (strip-chart recorder) can be used, the large amount of data generated by test module experiments, coupled with recent developments in digital system hardware, has made the digital data acquisition system (DAS) the preferred choice for this application. Digital systems can be further categorized as data loggers (which simply record raw data) and those that are programmable and thus able to store calculated results during the monitoring period. The Acurex Autodata Nine®, as an example of the former, was used extensively in the mid-1970s for monitoring passive and active solar buildings. As the cost of microcomputers dropped dramatically in the late 1970s, they were adapted by most test module researchers for the combined tasks of data acquisition, recording, simultaneous data reduction, and auxiliary heat control. Most of these systems included the following components:

1. A cold reference junction (required for thermocouple temperature sensors only)

2. A digital voltmeter (for converting analog voltage sensor signals to digital data for recording and processing)

3. A digital clock to control data sampling

4. A programmable microcomputer to control the clock and to read, process (if required), and store data

5. A data-storage device (printer, cassette, or disk drive)

6. A visual display to facilitate programming or temporarily display data

7.2.3 Instrumentation Design

Test module experiments for simple demonstrations and comparisons require minimal instrumentation (see figure 7.3). Experiments for code validation require more sophisticated instrumentation, including the many temperature sensors necessary to compare with thermal network nodal temperatures—see figure 7.4.

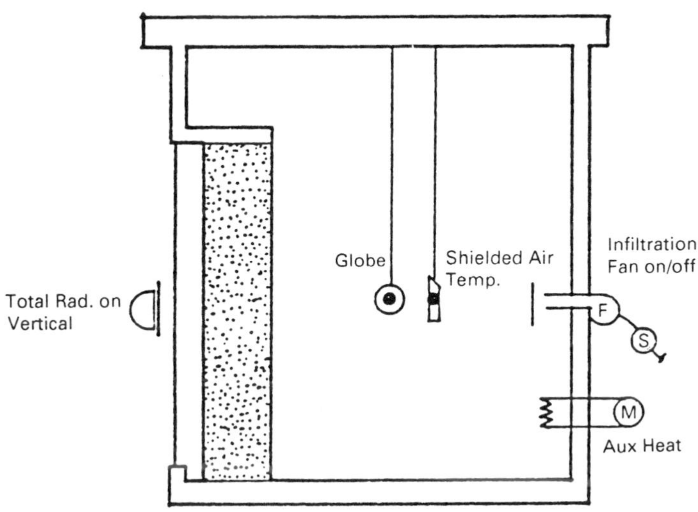

Figure 7.3
Schematic section of Trombe-wall test room showing typical instrumentation required for comparison experiments.

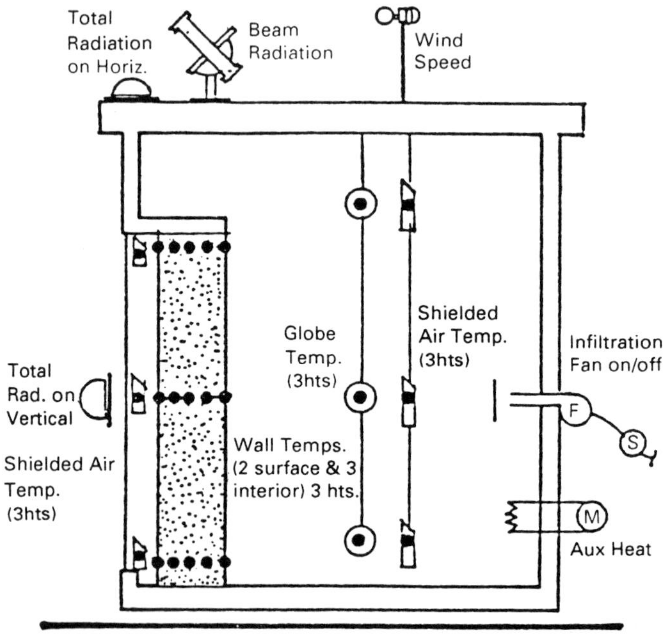

Figure 7.4
Schematic section of Trombe-wall test room showing typical instrumentation required for computer simulation code validation.

7.2.4 Test Module Calibration

If test modules are to be used only for direct comparison experiments, it is sufficient to ensure that the various parameters (heat loss, aperture, size, tilt, glazing, etc.) are identical. This condition can be achieved by configuring the modules identically and by monitoring interior temperatures after an initial stabilization period.

However, if test modules are to be used for code-validation purposes, the values of all parameters affecting performance must be known by direct measurement, and this purpose requires a precise calculation procedure.

The following one-time measurements are typical:

1. Absorber-surface characteristics (emissivity and absorptivity)
2. Glazing characteristics (transmissivity, absorptivity, reflectivity, area, tilt, and orientation)

3. Thermal-storage characteristics (surface area, thickness, density, thermal conductivity, and specific heat)

4. Heat loss coefficient (including infiltration) of entire test cell, excluding solar wall (solar wall is covered with heavily insulated panel and auxiliary heat consumption is monitored)

5. Infiltration rate

7.2.5 Auxiliary Heating

A fundamental option in passive test modules is the use of auxiliary heating. Most proof-of-concept experiments are "free running" (no auxiliary heat). The choice is less clear for comparison and code-validation experiments; studies concerning realistic building performance must eventually employ auxiliary heat. Preliminary free-running experiments may be more useful in the validation of specific algorithms or the isolation of a particular thermal behavior in a component; however, this type of investigation may distort the overall behavior of a passive system because of higher or lower operating temperatures. For example, in a Los Alamos experiment on the effect of reverse thermosiphoning in vented Trombe walls, it was found that the use of a check-valve (to prevent reverse thermosiphoning) made little difference in thermal performance when the system was free running, but significantly affected increased performance when auxiliary heat was used.

In comparative experiments (section 7.1.3.2) the amount of auxiliary heat required is the primary performance indicator of passive buildings.

7.3 Test Module Facilities

The following sections describe existing facilities in the United States and Canada that contain test modules. Table 7.3 summarizes these facilities. (The operational status of their test module programs was current when the manuscript was submitted in 1983. In the subsequent years, several programs have been discontinued and other new ones begun.)

7.3.1 Test Boxes

Test boxes are used primarily for comparative testing where convective flow is not a parameter. Because the principal concern is thermal similarity

Table 7.3
Test module facility locations and reporting author or contact person

A. Harold Hay, Skytherm Processes and Engineering. Tempe, AZ. J. Yellott, 901 W. El Caminito, Phoenix, AZ 85021.

B. Environmental Research Laboratory. Tucson International Airport, Tucson, AZ 85706. J. Peck.

C. Pacific Gas and Electric, Department of Engineering Research. San Ramon, CA 94583. H. Seielstad.

D. Farralones Institute. Sonoma County, CA. P. Calthorpe, 55c Gate 5 Rd., Sausalito, CA 94964.

E. Pala Passive Solar Project. San Diego County, CA. J. Clinton, 4325 Donald, San Diego, CA 92117.

F. Habitat Center. 2293 Olympic Blvd., Walnut Creek, CA 94595. M. Martin.

G. Solar Energy Research Institute. 15/3, 1617 Cole Blvd., Golden, CO 80401. K. Harr.

H. Solar Equipment Group, Building Equipment Division, Center for Building Technology, NEL, National Bureau of Standards. Building 226, Room B306, Washington, DC 20234. B. Mahajan.

I. Purdue University, College of Architecture. W. Lafayette, IN 47907.

J. Massachusetts Institute of Technology, Department of Building Engineering Construction. Cambridge, MA 02139. F. Hollingsworth.

K. National Center for Appropriate Technology, R/D Branch. P.O. Box 3838, Butte, MT 59701. L. Palmiter.

between competing boxes, the boxes are typically constructed of foam insulation with removable glazing. The ratio of heat loss to collector area should be similar to that of a passive building. The result is a thick-walled box with little similarity in appearance to a scale building model.

7.3.1.1 Harold Hay The foam ice chest experiments (described in section 7.1.3.1) in the mid-1960s (Hay and Yellott 1970) were apparently the first use of test boxes for demonstrating passive solar heating.

7.3.1.2 Zomeworks In 1970, Zomeworks Corporation (Baer 1977) used several monitored 15-in. (38-cm) by 30-in. (76-cm) by 22-in. (56-cm) foam boxes with 3-in. (8-cm) foam insulation sides and south-facing glass to test passive solar water-wall systems in Albuquerque, New Mexico. Based on the results of these experiments, Baer proposed a calculation procedure for determining the thermal storage capacity of passive water-wall systems. Between 1971 and 1976, he publicly demonstrated passive solar principles using similar boxes in workshops in various climatic locations around the United States. Baer recommended to his clients that they build and test

Figure 7.5
Test boxes at Los Alamos.

similar test boxes in their local climate prior to the construction of a passive home.

Baer (1977) also used air convection test boxes in 1971 to test various absorber configurations. These monitored boxes were approximately 30 in. (76 cm) wide by 60 in. (152 cm) deep by 72 in. (183 cm) high with sloped glazing. These and later experiments demonstrated the superiority of wire mesh over sheet metal absorbers/heat exchangers by measuring glazing temperatures (lower glazing temperatures accompanied more effective convection heat transport away from the absorber).

7.3.1.3 Los Alamos National Laboratory Subsequent to Baer's work in Albuquerque, Grimmer et al. (1979) used cubic test boxes, 24 in. (61 cm) on a side, at Los Alamos National Laboratory for direct comparison with each other as well as for normalized, indirect comparison with test rooms (see figure 7.5).

Grimmer (1979) theoretically considered using small test boxes to represent multiroom passive solar designs, including realistically massive walls, thermocirculation, infiltration effects, edge-effect corrections, and microclimate shading effects, in terms of how each responds to dimensional analysis and scaling.

7.3.1.4 Solar Energy Research Institute The Solar Energy Research Institute (SERI) utilized two test boxes. They are 39-in. (99-cm) cubes, with 4-in. (10-cm) expanded polystyrene foam sides, and have one side glazed and one side a black-coated, copper radiation plate with removable insulation (to allow radiative cooling for varying the thermal load). These are used for preliminary Trombe-wall, water-wall, and direct-gain configurations prior to more detailed experiments with test rooms.

7.3.1.5 Lakehead University R. E. Jones (1979) has described the use of boxes at Lakehead University similar to those in use at SERI. These were 4-ft (1.2-m) cubes of frame construction, used for experiments in a colder Canadian climate.

7.3.1.6 Scott Morris Morris (1978) used four similar convection test boxes (two tilted, two vertical) with integral water-thermal storage. Approximately 24 in. (61 cm) wide by 36 in. (91 cm) deep by 72 in. (183 cm) high, these boxes were used to directly compare the performance of four convective configurations in terms of collection efficiency: sloped U-tube (40%), sloped center-glazed U-tube (40%), vertical wall (36%), and vertical S-loop (36%).

7.3.2 Test Rooms

Test rooms usually have the following physical characteristics:

1. Full scale in height, reduced scale in width and depth
2. Permanent construction, deliberately of low mass
3. Ratio of heat loss to collector area, similar to that of actual passive buildings
4. Solar aperture size, tilt, and glazing, reconfigurable
5. Thermal mass, reconfigurable
6. Infiltration rate, forced to be constant, or continuously monitored

7.3.2.1 Massachusetts Institute of Technology One of the earliest examples of the use of test rooms for passive solar research was the Massachusetts Institute of Technology facility described by Dietz and Czapek (1950) and Hollingsworth (1947). Constructed in 1947 in Cambridge, Massachu-

Figure 7.6
Roof-pond test room, Tempe, Arizona (photo credit: J. Yellott).

setts, the single structure housed six reconfigurable test rooms, a direct-gain low-mass room, a control, and an instrument room. The vertically glazed 6-ft (1.8-m) by 14-ft (4.3-m) rooms featured a reconfigurable north wall to permit adjustment of heat loss, auxiliary heat, movable insulation, and a variety of configurable water thermal-storage walls.

7.3.2.2 Skytherm® Processes and Engineering In 1967, a test room was constructed in Tempe, Arizona, by Hay and Yellott (1970) to investigate roof-pond cooling and heating. The 10-ft (3.0-m) by 12-ft (3.7-m) structure was heavily instrumented and was monitored from 1967 to 1968 (see figure 7.6).

7.3.2.3 Los Alamos National Laboratory At Los Alamos National Laboratory, fourteen test rooms were constructed for validation of the PASOLE simulation code (Balcomb, McFarland, and Moore 1977) as well as for comparison testing (Hyde 1980) and heat transfer measurements (Balcomb 1981). Each of the seven test structures contains two south-facing, single-zone rooms separated by an insulated partition. In addition to fiberglass insulation in the wood-frame construction, each interior is completely lined with 1-in. (2.5-cm) extruded polystyrene-foam insulation to minimize the effect of the test-cell mass itself. The thermal load can be varied by changing the rate of fan-forced infiltration. The fan is controlled by a damper, and the flow rate is measured periodically using an American Society for Testing and Materials (ASTM)-type calibrated nozzle. The accuracy of this measurement has been validated using BF6 tracer-gas tests. Because the interiors are maintained under positive pressure, the effects of temperature difference and wind speed on the infiltration rate are small (see figure 7.7).

When required, auxiliary heat is added using incandescent light bulbs that are controlled by the data acquisition computer. Auxiliary power is monitored by a relay-controlled on/off status sensor and calculated from a one-time measurement of electrical consumption of typical bulbs.

Between 1980 and 1981, Los Alamos test room experiments included Trombe walls with and without selective surfaces, water wall, phase change wall, direct gain, heat pipe collectors, and sunspaces (McFarland 1982a).

Figure 7.7
Test rooms at Los Alamos (photo credit: Los Alamos National Laboratory).

The 1981–82 experiments included direct gain, unvented Trombe wall, water wall, and phase change wall. Strategies for reducing heat loss were tested during the last period, included selective surfaces, two types of improved glazing systems, a heat-pipe system, and convection suppression baffles.

Morris (1980) has described a pair of natural convection collector test modules that were constructed at Los Alamos. The unit is reconfigurable and capable of varying collector length, tilt, airflow channel depth, glazing, absorber, transfer to simulated storage, airflow resistance, pressure drop through storage, and operating temperature of the collector system. Using these modules, Biehl (1981) found that collector flow rates decrease and efficiency increases with decreasing channel depth, and that flow channel corner design had a substantial impact on efficiency.

7.3.2.4 National Center for Appropriate Technology Two-room test buildings (virtually identical to those at Los Alamos) were constructed and operated both at National Center for Appropriate Technology (NCAT), Butte, Montana (Palmiter, Wheeling, and Corbett 1980), and at the University of Pennsylvania, Philadelphia (Prowler 1978). The NCAT test rooms were used for direct-gain and Trombe-wall experiments to validate the computer simulation code SUNCAT (Palmiter, Caswell, and Corbett 1978). In addition, two retrofit, two attached-greenhouse, and two detached-greenhouse test rooms have been constructed and tested at NCAT.

7.3.2.5 Solar Energy Research Institute An extensive test room program was conducted at the Solar Energy Research Institute (SERI). The ten SERI rooms are configured in pairs (similar to the earlier Los Alamos rooms) for a total of five structures. The 10-ft (3.0-m) by 12-ft (3.7-m) units are completely reconfigurable (suitable for two zones in each module) and portable. Fixed mass and infiltration are minimized, and auxiliary heat is available in the form of monitored, portable, electric resistance heaters. They are used principally for comparison testing and code validation (see figure 7.8).

7.3.2.6 University of Nebraska Chen et al. (1979) describe the six passive solar test rooms constructed at the University of Nebraska (see figure 7.9). The modules are being used to directly compare direct-gain, Trombe-wall, water-wall, double-envelope, earth-sheltered, greenhouse, and direct-gain/ceiling storage passive strategies; they are also used for simulation

Figure 7.8
Test rooms at Solar Energy Research Institute, Golden, Colorado (used with the permission of Kelvin Harr).

code validation. It is noteworthy that this rigorous research program is being accomplished almost entirely by undergraduate students under faculty direction. The test rooms were constructed by building-construction students; the instrumentation and DAS were designed, constructed, and installed by electrical engineering and physics students; signage and demonstration graphics were done by graphic-design students; and materials were donated by local sources. The program is considered exemplary not only as an integrated test room research program but also as a successful example of the fulfillment of diverse undergraduate educational objectives through a cooperative research effort.

7.3.2.7 Wessling Consulting Wessling (1978) describes the use of four different test rooms retrofitted to an existing residence in Albuquerque, New Mexico. The modules were thermally isolated, but all supplied heat to the house. The systems tested included: 1. a perforated masonry storage wall, 2. a distributed mass direct gain (concrete wall with iron oxide

Figure 7.9
Test rooms at University of Nebraska.

additive), 3. rock storage in direct gain, and 4. direct gain with massive floor. Wessling concluded that the iron oxide additive increased conductivity and significantly reduced temperature swings compared with the control direct gain. The perforated storage wall also exhibited considerably reduced temperature swings and resulted in a superior quantity and quality of daylight illumination.

7.3.2.8 Habitat Center Four test rooms in two buildings were constructed by the Habitat Center, Walnut Creek, California, to compare the performance of retrofit sunspaces (Martin et al. 1981). Each test room is a sunspace attached to a "dwelling" room, the two comprising a test unit. Each of the two buildings were 26 ft (7.9 m) wide by 22 ft (6.7 m) deep by 9 ft (2.7 m) high and contained two adjacent test units separated by an insulated wall. One test unit served as the control (free-standing mass within the sunspace, with vented, insulated common wall), whereas the common walls (between the greenhouse and dwelling) were varied in the other three units during the various test periods (unvented concrete, vented

Figure 7.10
Attached sunspace test rooms at Habitat Center, Walnut Creek, California (used with the permission of Muscoe Martin, Habitat Center).

concrete, and vented adobe). No auxiliary heat was used during the tests (see figure 7.10). Based on measurements at this facility, it was concluded that an optimum system should include a freestanding thermal mass (adjacent to the common wall) that can be isolated from the sunspace by the use of movable insulation.

7.3.2.9 Pennsylvania State University Eleven passive greenhouse test modules (four attached to a building and seven isolated for crop growing only) have been constructed at Pennsylvania State University. Each encloses 1,400 ft^2 (130 m^2) of floor area and is reconfigurable in mass and glazing.

7.3.2.10 Ames Laboratory Three 12-ft (3.7-m) by 12-ft (3.7-m) by 8-ft (2.4-m) passive solar rooms were constructed at the Passive Technologies Test Facility, Ames Laboratory, Iowa State University in Ames (Mercer et al. 1981). The three rooms are thermally isolated and housed in a single, insulated, earth-sheltered envelope. The facility was primarily used for testing Transwall, a transparent water-wall system using various absorber positions, glazings, and night insulation. Mercer reported a significant improvement of the Transwall system over a comparable direct-gain system in both heating and cooling modes.

7.3.2.11 Solar Room Company Four attached greenhouse test rooms were constructed in Taos, New Mexico, by the Solar Room Company (Kenin and Betz 1978) for the direct comparison of various storage media in double-wall, polyethelene-film, inflated greenhouses. The four test rooms were attached to one structure, with the adjacent rooms isolated with R-40 walls.

7.3.2.12 Tennessee Valley Authority Four test rooms 5 ft (1.5 m) wide and 8 ft (2.4 m) deep, and two test rooms 12 ft (3.7 m) by 8 ft (2.4 m) were used by the Solar Applications Division, Tennessee Valley Authority in Chattanooga, Tennessee. The smaller rooms are used as single units, whereas the 12-ft (3.7-m)-wide modules can accommodate a north/south partition (Kuberg 1981).

7.3.2.13 National Bureau of Standards Four test rooms 12 ft (3.7 m) by 27 ft (8.2 m) within a single building 52 ft (15.8 m) wide and 28 ft (8.5 m) deep were constructed at the National Bureau of Standards (NBS) in Gaithersburg, Maryland (LASL 1983). In its present configuration, the rooms are being used for a component calorimeter, collector/storage wall, control, and direct gain. The rooms are being used for the evaluation of a previously proposed standardized test procedure for passive solar components (McCabe et al. 1981) as well as for Class A performance testing of passive subsystems and components.

7.3.3 Test Buildings

Existing test buildings range from small, reconfigurable, two-zone structures (only slightly larger than test rooms) to full-size, habitable (but usually uninhabited) residences. Because most were designed to meet specific research objectives, they share only a few common characteristics. In addition to being unoccupied, most have at least limited reconfigurability of aperture, mass, and zones.

7.3.3.1 Purdue University One of the earliest examples of passive solar test buildings was the installation of two test houses at Purdue University (Hutchinson 1947). Constructed in Lafayette, Indiana, in 1945, one

Figure 7.11
Two-zone test buildings at the Pala Passive Solar Project, San Diego County, California.

house was used as a control and the other for comparative direct-gain experiments.

7.3.3.2 Pala Passive Solar Project The eight two-zone test buildings of the Pala Passive Solar Project were constructed northeast of San Diego at Pala, California, as part of a passive solar research program at the University of California, San Diego (Clinton 1980). The buildings, managed by the Solar Energy Analysis Laboratory of San Diego, were constructed by a consortium of local utilities; monitoring, analysis, and simulation studies were supported by these agencies and the U.S. Department of Energy. The 16-ft (4.9-m) by 16-ft (4.9-m) buildings are permanent structures built to reflect standard construction practices (at some expense for ease of reconfigurability). These structures are among the few available for full-scale analysis of the effects of controls (fans, remote storage) on the performance of multizone passive structures. The data were used by Sebald and Vered (1981), Carroll and Clinton (1980), McFarland and Lazarus (1989), and UCSD graduate engineering students for code validation and simplified design-tool development (see figures 7.11 and 7.12).

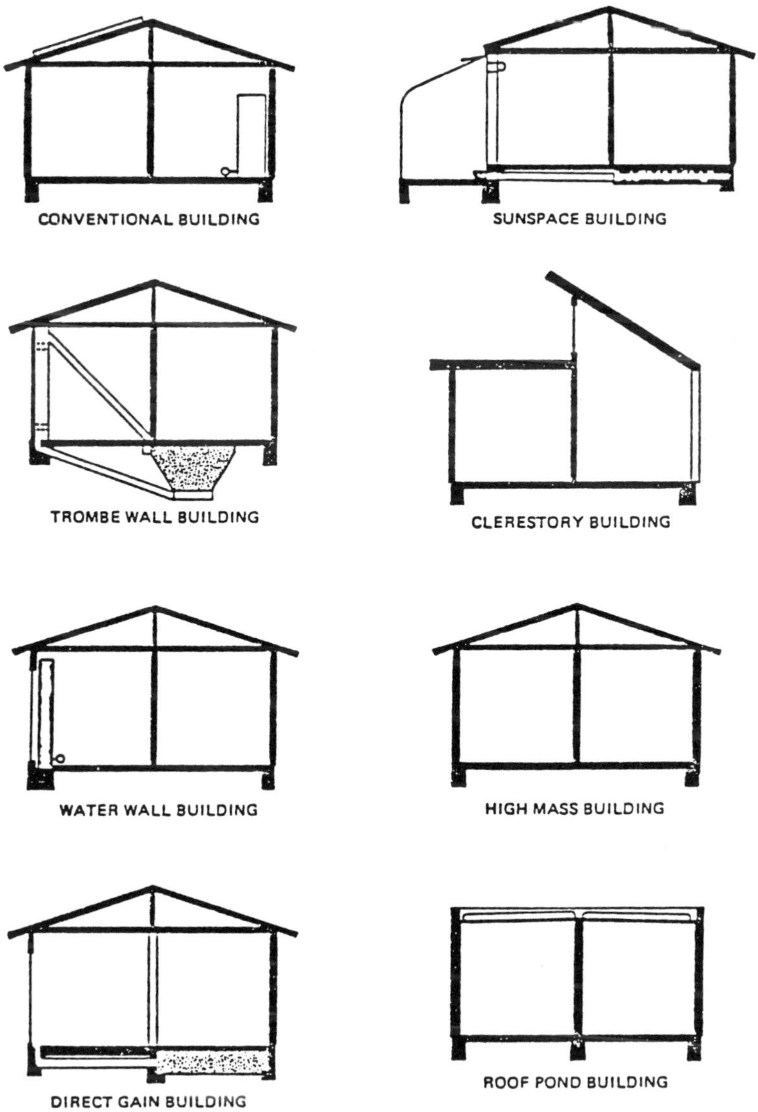

Figure 7.12
Schematic sections for test buildings, Pala Passive Solar Project (drawings after Clinton 1980).

Clinton (1983) compared gain values for sunny days for each building and found that a simple linear correlation closely fits the monitored data. With an offset of 7°F (4°C) (solar gain with zero south window area), gain values were proportional to south window area for every building. Based on comparisons of measured results with simulation analyses, he also concluded that an "inordinate degree of caution is required before either handbook values of heat transfer coefficients are believed or models are applied to situations different from those that 'validated' the method."

7.3.3.3 Trinity University Two identical reconfigurable test buildings were constructed at Trinity University, San Antonio, Texas. Doderer, Bentley, and Allen (1981) and Doderer and Bentley (1981) describe the facilities. Giolma and Clower (1981) describe the instrumentation. Each building can be configured into the following systems: roof pond with movable insulation, trickle roof, low mass roof, and plenum roof. Data from the test buildings have been used for validation of computer simulation codes. Results from these studies include the development of simple procedures for estimating temperature and nocturnal heat loss from roof-pond buildings (Clark et al. 1981), measured ceiling heat transfer coefficients for roof-pond buildings (Lowsom et al. 1981), the determination of the geographic limits of comfort in roof-pond cooled residences in the United States (Fleishhacker, Clark, and Giolma 1983), and a method for predicting energy savings in roof-pond residences (Vieira, Clark, and Faultersack 1983).

7.3.3.4 Environmental Research Laboratory Four test buildings constructed at the Environmental Research Laboratory, Tucson, Arizona, are primarily used for study of various cooling strategies in hot, arid climates; some passive heating is included (Peck and Kessler 1981). The size, mass, configuration, and thermal load are different for each structure. Reconfigurability is limited by the relatively permanent nature of the massive components used.

7.3.3.5 Florida Solar Energy Center The Florida Solar Energy Center (Cape Canaveral) constructed a unique building module for cooling experiments in which all walls (including exterior walls) are reconfigurable (Fairey 1981). The structure allows investigation of thermal stack spaces up to 24 ft (7.3 m) high. The concrete slab floor is supported by hollow concrete masonry blocks on a second slab allowing reconfigurable air

Figure 7.13
Test building (before installation of reconfigurable walls) at the Florida Solar Energy Center, Cape Canaveral (used with the permission of Philip W. Fairey, Florida Solar Energy Center).

distribution and thermal storage (see figures 7.13 and 7.14). Recent results from this program have included a design handbook for ventilated buildings (Chandra, Fairey, and Houston 1983), validation of an experimental technique using physical scale models in natural wind to predict ventilation performance (Chandra, Ruberg, and Kerestecioglu 1983), the empirical determination of heat transfer coefficients in naturally ventilated rooms using full-scale measurements (Chandra and Kerestecioglu, 1983), measurement of the thermal performance of selected building components suitable for use in warm, humid climates (Faire 1981), and a study of the effects of infrared radiation barriers on the effective thermal resistance of building envelopes (Fairey 1983).

7.3.3.6 University of Alberta Dale et al. (1980) describe a set of six test buildings constructed at the University of Alberta, Edmonton, Alberta, Canada, to compare the thermal performance of standard construction with four alternatives (superinsulated, passive solar, active air solar, active

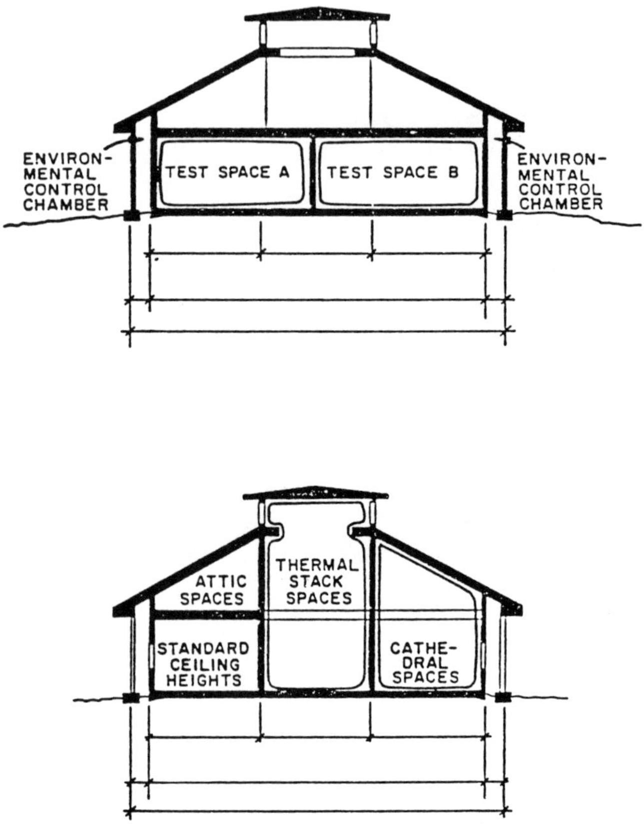

Figure 7.14
Schematic sections showing possible test configurations of the test building at the Florida Solar Energy Center (used with the permission of Philip W. Fairey, Florida Solar Energy Center).

Figure 7.15
Sundwellings Demonstration Center, Ghost Ranch, New Mexico.

liquid solar). The sixth building was used for short-term testing. The 22-ft (6.7-m) by 24-ft (7.3-m) buildings incorporated frame construction, roof trusses, one zone above grade, and basement.

7.3.3.7 Sundwellings Demonstration Center The Sundwellings, four two-room adobe test buildings (control, sunspace, direct gain, and Trombe wall) were constructed at Ghost Ranch near Abiquiu, New Mexico. Designed by Peter van Dresser and constructed by unskilled labor, these 17-ft (5.2-m) by 40-ft (12-m) buildings were not reconfigurable. They were extensively instrumented by Los Alamos National Laboratory before occupancy (see figure 7.15). Jones and McFarland (1979) used data from the sunspace unit to validate a sunpace simulation model.

7.3.3.8 Farralones Institute Five frame-construction test buildings are located at the Farralones Institute, northern California. Calthorpe (1978) has described the design and performance of these occupied test structures; experimental objectives included direct comparison and demonstration of the various passive solar heating types (see figure 7.16).

Figure 7.16
Test buildings at the Farralones Institute, Sonoma County, California (used with the permission of Peter Calthorpe, Van de Ryn; Calthorpe and Partners).

7.3.3.9 Pacific Gas and Electric Company Four test rooms arranged in two unoccupied buildings, 9 ft (2.7 m) high by 10 ft (3 m) wide by 10 ft (3 m) deep, were used by the Pacific Gas and Electric Company (D'Albora 1981) at their San Ramon, California, Passive Solar Test Facility (see figure 7.17).

7.3.3.10 Solar Energy Research Institute A single test building at the Solar Energy Research Institute (SERI) is full-size frame construction. Though not designed as such, it was tested in five configurations (control, water wall, water wall/greenhouse, low-mass green house, and air convection). The house is extensively instrumented and will be used for coheating experiments as well as for code validation.

7.4 Conclusions

The most significant contribution of test modules has been their use for computer simulation code validation. Once validated, these codes have

Figure 7.17
Test rooms at the Pacific Gas and Electric Company, San Ramon, California (used with the permission of H. Seielstad, Pacific Gas and Electric Company).

been used to generate simplified tools for analysis of passive solar heated buildings. Of all the test module programs surveyed, the Los Alamos program has made the greatest contribution to passive solar research through the development of the SLR method of building performance prediction. This method has become the standard of the field and has provided the basis for a wide variety of graphic and microcomputer design tools. The cost of the Los Alamos program may have been considerable, but the ultimate results are impressive.

Any conclusion regarding the use of test modules for passive research would be incomplete without special recognition of two pioneers in the field, Harold Hay and Steve Baer. Their early use of test modules demonstrated the potential of passive heating and cooling, and there can be little doubt that subsequent research benefited from the test-module experimental methodologies developed by these researchers.

References

Abrash, M., R. Wirtshafter, P. Sullivan, and J. Kohler. 1978. Modeling passive buildings using TRNSYS. *Proc. 2d National Passive Solar Conference*, Philadelphia, PA, March 16–18, 1978. Newark, DE: American Section of the International Solar Energy Society, pp. 398–403.

Arumi-Noe, F. 1978. A model for the DEROB/PASOLE System. *Proc. 2d National Passive Solar Conference*, Philadelphia, PA, March 16–18, 1978. Newark, DE: American Section of the International Solar Energy Society, pp. 529–533.

Baer, S. 1977. *Sunspots*. 2d ed. Albuquerque, NM: Zomeworks Corporation, pp. 118–135.

Balcomb, J. D. 1981. Dynamic measurement of nighttime heat loss coefficients through Trombe wall glazing systems. *Proc. 6th National Passive Solar Conference*, Portland, OR, September 8–12, 1981. Newark, DE: American Section of the International Solar Energy Society, pp. 84–88

Balcomb, J. D., J. C. Hedstrom, and R. D. McFarland. 1977. *Passive Solar Heating of Buildings*. LA-UR-77-1162. Los Alamos, NM: Los Alamos National Laboratory.

Balcomb, J. D., R. D. McFarland, and S. W. Moore. 1977. Simulation analysis of passive solar heating buildings—comparison with test room results. *Proc. Annual Meeting, American Section of the International Solar Energy Society*, Orlando, FL, June 6–19, 1977, vol. I, sec. II. Cape Canaveral, FL: International Solar Energy Society, pp. 1–4.

Benton, C. C., and J. M. Akridge. 1981. *Interim Report: Passive Solar Curriculum Development Project*. Atlanta, GA: Department of Architecture, Georgia Institute of Technology, pp. 7–10.

Biehl, F. A. 1981. Test results and analysis of a convective loop solar air collector. *Proc. 6th National Passive Solar Conference*, Portland, OR, September 8–12, 1981. Newark, DE: American Section of the International Solar Energy Society, pp. 79–83.

Calthorpe, P. 1978. The Farralones Institute study of five passive and hybrid space heating systems. *Proc. 2d National Passive Solar Conference*, Philadelphia, PA, March 16–18, 1978. Newark, DE: American Section of the International Solar Energy Society, pp. 146–152

Carroll, J. A., and J. R. Clinton. 1980. A thermal network model of a passive solar house. *Proc. 5th National Passive Solar Conference*, Amherst, MA, October 19–26, 1980. Newark, DE: American Section of the International Solar Energy Society, pp. 257–261.

Chandra, S., P. Fairey, and M. Houston. 1983. *A Handbook for Ventilated Buildings*. FSCE-CR-93-83. Cape Canaveral, FL: Florida Solar Energy Center.

Chandra, S., and A. A. Kerestecioglu. 1983. *Heat Transfer in Naturally Ventilated Rooms: Data from Full-Scale Measurements*. FSEC-PF-45-83. Cape Canaveral, FL: Florida Solar Energy Center.

Chandra, S., K. Ruberg, and A. A. Kerestecioglu. 1983. Outdoor testing of small scale naturally ventilated models. *Building and Environment* 18(1/2):45–53.

Chen, B., J. Maloney, J. Thorp, K. Pederson, W. Holmes, C. Sedlacek, R. Sash, and E. Hollingsworth. 1979. Preliminary winter comparison results of four passive test cells in Omaha, Nebraska. *Sun II: Proc. American Section of the International Solar Energy Society Annual Meeting*, Atlanta, GA, May 1979. Newark, DE: American Section of the International Solar Energy Society, pp. 1583–1587.

Clark, G., F. Lonsom, B. Schutt, and Faultersack. 1981. Simple estimation of temperature and nocturnal heat loss for a radiantly cooled roof mass. *Passive Cooling: Proc. International Passive and Hybrid Cooling Conference*, Miami Beach, FL, November 11–13, 1981. Newark, DE: American Section of the International Solar Energy Society, pp. 244–248.

Clinton, J. R. 1979. The Sea-Pass passive simulation program. *Proc. 4th National Passive Solar Conference*, Kansas City, MO, October 3–5, 1979. Newark, DE: American Section of the International Solar Energy Society, pp. 202–206.

Clinton, J. R. 1980. The Sea-Lab passive test building project. *Proc. Systems Simulation and Ecomonic Analysis*, San Diego, CA, January 23–25, 1980. Golden, CO: Solar Energy Research Institute, pp. 91–94.

Clinton, J. R. 1983. Results from the Pala passive solar project. *Proc. 8th National Passive Solar Conference*, Santa Fe, NM, September 7–9, 1983. Boulder, CO: American Solar Energy Society, pp. 15–20.

D'Albora, G. 1981. *Performance Results of Passive Test Rooms*. Research Report 005-81.2. San Ramon, CA: Pacific Gas and Electric Company.

Dale, J. D., M. Ackerman, R. R. Gilpin, and T. Forest. 1980. The Alberta home heating research facility update I—the 1979–1980 heating season. *Proc. of Solweet 80 Joint Solar Conference*, Vancouver, BC, Canada, August 6–10, 1980. Winnipeg, MB, Canada: The Solar Energy Society of Canada, pp. 304–307.

Dietz, A. G. H., and E. L. Czapek. 1950. Solar heating of houses by vertical south wall storage panels. *Heating, Piping and Air Conditioning*, March 1950, pp. 118–125.

Doderer, E., and D. Bentley. 1981. Design and construction of a passive test facility for warm humid climate thermal experimentation. *Passive Cooling: Proc. International Passive and Hybrid Cooling Conference*, Miami Beach, FL, November 11–13, 1981. Newark, DE: American Section of the International Solar Energy Society, pp. 446–449.

Doderer, E., D. Bentley, and C. Allen. 1981. A flexible test facility for passive cooling and solar heating. *Proc. Annual Meeting, American Section of the International Solar Energy Society*, Philadelphia, PA, May 27–30, 1981. Newark, DE: American Section of the International Solar Energy Society, pp. 941–945.

Fairey, P. W. 1981. A passive cooling experimental facility for warm humid climates. *Proc. U. S. Department of Energy Passive & Hybrid Solar Energy Program Update*, Washington, DC, August 9–12, 1981. Washington, DC: U. S. Department of Energy, February 1982, pp. 6.4 to 6.6.

Fairey, P. W. 1983. *Effects of Infrared Radiation Barriers on the Effective Thermal Resistance of Building Envelopes*. FSEC-PF-37-83. Cape Canaveral, FL: Florida Solar Energy Center.

Fleishhacker, P., G. Clark, and P. Giolma. 1983. Geographic limits for comfort in unassisted roof pond cooled residences. *Proc. 8th National Passive Solar Conference*, Santa Fe, NM, September 7–9, 1983. Boulder, CO: American Solar Energy Society, pp. 835–838.

Frey, D., and M. McKinstry. 1980. *Installation Manual: SERI Class B Passive Solar Data Acquisition System*. Golden, CO: Solar Energy Research Institute.

Giolma, J. P., and C. A. Clower. 1981. Instrumentation of a passive test facility. *Passive Cooling: Proc. International Passive and Hybrid Cooling Conference*, Miami Beach, FL, November 11–13, 1981. Newark, DE: American Section of the International Solar Energy Society, pp. 442–445.

Grimmer, D. P. 1979. Theoretical considerations in the use of small passive-solar test-boxes to model the thermal performance of passively solar-heated building designs. *Solar Energy* 22:343–350.

Grimmer, D. P., R. D. McFarland, and J. D. Balcomb. 1979. Initial experimental tests on the use of small passive-solar test-boxes to model the thermal performance of passively solar-heating building designs. *Solar Energy* 22:351–354.

Hay, H. R., and J. I. Yellott. 1970. A naturally air-conditioned building. *Mechanical Engineering* 32(1):19–23.

Hollingsworth, F. N. 1947. Solar heat test structure at MIT. *Heating and Ventilating*, May 1947.

Hutchinson, F. W. 1947. The solar house: analysis and research. *Progressive Architecture*, March 1947, pp. 90–94.

Hyde, J. C. 1980. Performance of night insulation and selective absorber coatings in LASL test cells. *Proc. 5th National Passive Solar Conference*, Amherst, MA, October 19–26, 1980. Newark, DE: American Section of the International Solar Energy Society, pp. 277–281.

Jones, R. E. 1979. Results from passive solar test room investigations in a cold climate. *Proc. 4th National Passive Solar Conference*, Kansas City, MO, October 3–5, 1979. Newark, DE: American Section of the International Solar Energy Society, pp. 53–58.

Jones, R. W. and R. D. McFarland. 1979. Simulation of the Ghost Ranch Greenhouse-Residence. *Proc. 3d National Passive Solar Conference*, San Jose, CA, January 11–13, 1979. Newark, DE: American Section of the International Solar Energy Society, pp. 35–40

Judkoff, R., D. Wortman, C. Christensen, B. O'Doherty, D. Simms, and M. Hannifan. 1980. A comparative study of four passive building energy simulations: DOE-2.1, BLAST, SUNDAT-2.4, DEROB-III. *Proc. 5th National Passive Solar Conference*, Amherst, MA, October 19–26, 1980. Newark, DE: American Section of the International Solar Energy Society, pp. 126–130.

Kenin, S. R., and W. B. Betz. 1978. The $2.50/ft^2 solar collector: a description and cost analysis. *Proc. 2d National Passive Solar Conference*, Philadelphia, PA, March 16–18, 1978. Newark, DE: American Section of the International Solar Energy Society, pp. 588–590.

Kohler, J. T., and P. W. Sullivan. 1979. *TEANET User's Manual.* Harrisville, NH: Total Environmental Action.

Kuberg, D. 1981. *Calibration of Test Cells.* Test Plan no. 804002. Chattanooga, TN: Tennessee Valley Authority.

Los Alamos Scientific Laboratory. 1983. *National Bureau of Standards Direct-Gain Test Cell Site Handbook.* LA-9786-MS, Los Alamos, NM: LASL.

Lowsom, F., J. Mosley, B. Kelly, and G. Clark. 1981. Measured ceiling heat transfer coefficients for a roof pond cooling system. *Passive Cooling: Proc. International Passive and Hybrid Cooling Conference*, Miami Beach, FL, November 11–13, 1981. Newark, DE: American Section of the International Solar Energy Society, pp. 270–273.

McCabe, M. E., W. Ducas, M. J. Orloski, and K. N. Decorte. 1981. *Passive/Hybrid Solar Components—An Approach to Standard Thermal Test Methods.* NBSIR 81-2300. Washington, DC: National Bureau of Standards.

McFarland, R. D. 1978. *PASOLE: A General Simulation Program for Passive Solar Energy.* LA-7433-MS. Los Alamos, NM: Los Alamos Scientific Laboratory.

McFarland, R. D. 1982a. *Los Alamos Passive Test Cell Results for the 1980–81 Winter.* LA-9300-MS. Los Alamos, NM: Los Alamos Scientific Laboratory.

McFarland, R. D. 1982b. *Los Alamos Passive Test Cell Results for the 1981–82 Winter.* LA-9543-MS. Los Alamos, NM: Los Alamos Scientific Laboratory.

McFarland, R. D. and G. S. Lazarus. 1989. *Monthly Auxiliary Cooling Estimation for Residential Buildings.* LA-11394-MS. Los Alamos, NM: Los Alamos National Laboratory.

Martin, M., L. Nelson, A. Wexler, and S. R. Schiller. 1981. Results of a retrofit sunspace testing program. *Proc. 6th National Passive Solar Conference*, Portland, OR, September 8–12, 1981. Newark, DE: American Section of the International Solar Energy Society, pp. 778–782.

Mercer, R., J. McClelland, L. Hodges, and R. Szydlowski. 1981. Recent developments in the transwall system. *Proc. 6th National Passive Solar Conference*, Portland, OR, September

8–12, 1981. Newark, DE: American Section of the International Solar Energy Society, pp. 178–182.

Moore, E. F., and R. D. McFarland. 1982. *Passive Solar Test Modules.* LA-9421-MS. Los Alamos, NM: Los Alamos Scientific Laboratory.

Morris, W. S. 1978. Natural convection solar collectors. *Proc. 2d National Passive Solar Conference*, Philadelphia, PA, March 16–18, 1978. Newark, DE: American Section of the International Solar Energy Society, pp. 596–601.

Morris, W. S. 1980. Development of an experimental test apparatus for natural convection solar collectors. *Proc. 5th National Passive Solar Conference*, Amherst, MA, October 19–26, 1980. Newark, DE: American Section of the International Solar Energy Society, pp. 1032–1041.

Palmiter, L., T. Wheeling, and R. Corbett. 1978. Performance of passive test units in Butte, Montana. *Proc. 2d National Passive Solar Conference*, Philadelphia, PA, March 16–18, 1978. Newark, DE: American Section of the International Solar Energy Society, pp. 581–585.

Palmiter, L., T. Wheeling, and B. Corbett. 1980. Performance of passive test units in Butte, Montana. *Proc. 5th National Passive Solar Conference*, Amherst, MA, October 19–26, 1980. Newark, DE: American Section of the International Solar Energy Society, pp. 591–595.

Palmiter, L., T. Wheeling, and B. Wadsworth. 1979. Summary of passive test unit performance. *Proc. 4th National Passive Solar Conference*, Kansas City, MO, October 3–5, 1979. Newark, DE: American Section of the International Solar Energy Society, pp. 698–699.

Peck, J. F., and H. J. Kessler. 1981. A passive cooling experimental facility for a hot/arid climate. *Passive Cooling: Proc. International Passive and Hybrid Cooling Conference*, Miami Beach, FL, November 11–13, 1981. Newark, DE: American Section of the International Solar Energy Society, pp. 437–441.

Perry, J. E. 1977. *Mathematical Modeling of the Performance of Passive Solar Heating Systems.*.LA-UR-77-2345. Los Alamos, NM: Los Alamos National Laboratory.

Prowler, D. 1978. Testing and simulation of passive solar systems. *Proc. 2d National Passive Solar Conference*, Philadelphia, PA, March 16–18, 1978. Newark, DE: American Section of the International Solar Energy Society, pp. 581–587.

Sebald, A. V. 1981. *Efficient Simulation of Large, Controlled Passive Solar Systems: Forward Differencing in Thermal Networks, Final Report.* Washington, DC: U.S. Department of Energy.

Sebald, A. V., and G. Vered. 1981. On extracting useful building performance characteristics without simulation. *Proc. Annual Meeting, American Section of the International Solar Energy Society*, Philadelphia, PA, May 27–30, 1981. Newark, DE: American Section of the International Solar Energy Society, pp. 1016–1020.

Vieira, R. K., G. Clark, and J. Faultersack. 1983. Energy savings potential of humidified roof pond residences. *Proc. 8th National Passive Solar Conference*, Santa Fe, NM, September 7–9, 1983. Boulder, CO: American Solar Energy Society, pp. 829–834.

Wessling, F. G. 1978. Solar retrofit test modules. *Proc. 2d National Passive Solar Conference*, Philadelphia, PA, March 16–18, 1978. Newark, DE: American Section of the International Solar Energy Society, pp. 445–449.

Wray, W. O. 1979. A semi-empirical method for estimating the performance of direct gain passive solar heated buildings. *Proc. 3rd National Passive Solar Conference*, San Jose, CA, January 11–13, 1979. Newark, DE: American Section of the International Solar Energy Society, pp. 395–402.

Wray, W. O. 1980. A quantitative comparison of passive solar simulation codes. *Proc. 5th National Passive Solar Conference*, Amherst, MA, October 19–26, 1980. Newark, DE: American Section of the International Solar Energy Society, pp. 121–125.

8 Building Integration

Michael J. Holtz

8.1 Scope of Chapter

Passive solar heating systems use elements of the building to collect, store, and distribute energy. Passive cooling also uses elements of the building to store and distribute energy, and, when prevailing conditions are favorable, to discharge heat to the cooler parts of the environment (sky, atmosphere, ground). In all cases, energy transfers to, from, and within the building rely primarily on natural processes—conduction, convection, and radiation—with minimum dependence on mechanical equipment, such as fans, pumps, and compressors. Mechanical equipment can be used effectively to augment natural energy flows when the capital costs and operating energy are justified by the improved system performance.

Since the collection, storage, discharge, and distribution of energy is generally accomplished by the architectural elements and features of the building, the passive system components are not easily distinguishable from the remainder of the structure. Therefore, integration of the passive systems in the building will, in large measure, determine the building's energy performance and comfort. This being the case, the way designers and builders have dealt with this integration issue is important to the continued development and application of passive systems.

This chapter reviews the evolution of research, demonstration projects, and design competitions concerned with the building integration of passive systems. Its focus includes new and existing residential and commercial buildings. Also, due to the integral nature of passive systems, the discussion addresses the heating, cooling, and daylighting design of passive buildings. Thus, this chapter will touch on areas covered in other chapters of this volume and other volumes in the series. In particular, analytical and empirical results obtained from modeling and monitoring are described elsewhere (see Jones and Frey, chapters 6 and 9 of this volume). These results indicate the performance sensitivities of various passive design parameters and the measured performance of built passive solar projects. The findings from these analytical and empirical research methods define the successfulness of the building integration process.

Volume 8 of this series, *Passive Cooling*, discusses natural cooling design alternatives and their integration in residential and commercial buildings.

Chapter 6, in particular, addresses the question of subsystem integration and measured performance of buildings utilizing natural cooling.

This chapter is organized into four topic areas. The first area defines passive *subsystems* and describes the passive subsystem integration issues. The second area discusses the building integration issues surrounding the use of passive systems in residential buildings. The third topic area concerns passive systems integration into commercial buildings, and the final area assesses the impact of previous building integration experience on current design practice and energy savings, and indicates future research direction.

8.2 Passive Subsystems: Definition and Description of Building Integration Goals

8.2.1 Passive Subsystem Definition

Numerous taxonomies have been developed for defining the elements/subsystems of passive heating and cooling systems. Some define system types such as direct gain, thermal-storage wall, sunspace, thermal-storage roof, and convective loop (Balcomb 1976). Others use the manner in which energy is delivered to the conditioned space, such as direct, indirect, and isolated (Holtz, Place, and Kammerud 1979). Still others consider the nature of the energy flow, natural or forced, between the collection aperture and storage, and from storage to the space (AIA/RC 1980). While each of these passive system taxonomies may be useful ways of categorizing different passive heating and cooling system concepts, they do little to address the question of building integration. For this, it is necessary to define the architectural elements and features that the designer is manipulating to create space and form while at the same time creating the passive system. Knowing these elements, it is possible to assess the energy performance implications of their integration into buildings.

The primary passive system design elements manipulated during the design process are as follows:

1. *Building Site* The site and the positioning and orientation of the building on the site are design elements that may have a significant influence on building energy performance and comfort. Unobstructed access to solar radiation and daylight, adjacency to surface water, and exposure to sea-

sonal winds are among the factors that will influence the energy-use characteristics of the building.

2. *Building Shape/Form* The two- and three-dimensional expression of the building on the site can have a profound influence on its energy performance.

3. *Insulation Levels* The amount of wall, floor, and ceiling or roof insulation may significantly influence the building load coefficient in some building types and the need for heating and cooling energy.

4. *Infiltration Rate* Product selection and construction detailing affect the likely rate of natural infiltration that will occur in the building. Obviously, construction quality and occupant behavior are also determinants of the actual rate. See also *Ventilation* below.

5. *Glazing Area, Location, and Type* The areas, locations, and types of windows, skylights, clerestories, and roof monitors are primary variables related to the issues of view, visual privacy, daylighting, and solar gains. Indeed, the combination of glazing considerations with construction technology and interior spatial arrangements to a large extent defines the passive system employed, whether the system consists of direct-gain windows alone or also includes other technologies, such as thermal-storage walls, sunspace or an atrium.

6. *Construction Technology* The structural system, including exterior and interior wall and floor materials, has architectural-design and passive system performance implications. The choice of construction technology determines the intrinsic heat capacity of the building and the potential to add thermal mass.

7. *Added Thermal Mass* Additional thermal mass, above the inherent heat capacity of the building, may be conventional construction materials, such as brick, concrete, or quarry tile, or materials specifically designed for thermal storage, such as phase-change materials or contained water.

8. *Interior Spatial Arrangement* The layout of interior space influences the thermal zoning and heat distribution from the passive system. Unconditioned buffer spaces, such as a sunspace, storage rooms, and a garage, will reduce building energy consumption.

9. *Shading* Window shading may be provided through architectural elements, such as fixed overhangs or side fins, and through window treatments, such as draperies, window blinds, or movable insulation.

10. *Ventilation* Ventilation may be required during both summer and winter to maintain acceptable levels of thermal comfort and indoor air quality. The potential for natural ventilation is an important aspect of passive solar design. Forced ventilation may be required, especially in commercial buildings, but also in residential buildings with very low rates of natural infiltration.

11. *Heating, Cooling, and Lighting Systems* The heating, cooling, and lighting loads of the building are established by the characteristics of the building energy regulation or the equipment selected. The passive strategies integrated into the design will influence the design of the heating, cooling, and lighting systems, and the associated controls.

8.2.2 Building Integration Goals

Within the context of the architectural design process and other project objectives and constraints, building integration involves the manipulation of the passive elements presented in Section 8.2.1 for the purposes of balance, economy, and aesthetics, as explained below.

Balance between heating and cooling performance is accomplished if the passive architectural elements are well adapted to all of the seasons: summer, autumn, winter, and spring. That is, the passive architectural elements respond to their changing performance role in the different seasons rather than create an imbalance resulting in discomfort or increased energy consumption for one or more seasons of the year. For example, large south glazing for winter heating may lead to summer overheating and high cooling costs or may lead the occupant to cover the south windows with shades or draperies, thus reducing daylight. Balancing heating and cooling performance means recognizing the year-round influence of the passive architectural elements and not allowing one season to dominate the passive design concept.

Balance between energy conservation and solar energy design is achieved by economically justified levels of conservation and solar design. That is, the design balances the economic benefits of isolating the building interior from outside climatic conditions to minimize heating and cooling loads and the benefits of selectively opening the building up to effectively utilize the natural climatic resources to meet all or part of these loads. Choosing the proper mix of energy conservation features (insulation levels, glazing type, infiltration and ventilation rates, mechanical equipment) and passive solar features (glazing area, thermal mass levels, shading, and control strategies) represents an important and complex passive building integration goal. The energy savings achieved by the various energy conservation and solar design features and their respective costs is used to establish the economically optimum mix. However, nonenergy related considerations, such as views, code compliance, and market requirements, act as constraints in determining the final mix of energy conservation and passive solar design features.

Balance between auxiliary energy and comfort is a fundamental issue in passive design integration. Is the goal to achieve the lowest possible energy consumption for the building or to provide acceptable year-round comfort? Generally, thermal comfort cannot be sacrificed to achieve low energy consumption. For example, consideration of glare may dictate a reduction of window area below the recommended levels for energy savings.

Economics often plays a limited role in passive integration of residential buildings. In most instances, no attempt is made to achieve an economic optimum in energy design. Either budget is not a primary consideration, or initial cost is the only concern. In commercial building design, economic justification of the passive architectural elements is often mandatory for their inclusion in the design. A variety of economic criteria may be used depending on the financial requirements of the client, such as a minimum life-cycle cost or a minimum rate of return on the energy investment. A detailed discussion of economic analysis methods and how they are employed in passive building integration is outside the scope of this chapter. However, economics is briefly discussed in section 8.4 on commercial buildings where it plays a significant role.

The aesthetic requirements of passive building integration are the same as in architectural design in general: to achieve a pleasing interior and exterior visual appearance in harmony with the occupants' needs and with the surrounding man-made and natural environment. Thus, discussions of passive solar building aesthetics must be held within the context of the broader and more fundamental discussions of contemporary architectural philosophy and theory. Yet, attempts to include passive solar design in the philosophical and theoretical discussion of the evolution of architecture have met with very limited success. Perhaps a major cause of this failure is the lack of a formal language of passive solar architecture and an ignorance of thermal and luminous phenomena in the mainstream of architectural practice. A thorough review of this issue, however, is outside the scope of this chapter.

The designer's ability to simultaneously achieve multiple architectural integration goals determines the ultimate success or failure of the passive solar building. Various thermal, daylighting, and economic decision tools have been developed to assist the designer during the design process (see Reynolds, chapter 10 of this volume). Through these tools and through the experience of designing, constructing, and occupying a passive solar building, intuition and confidence are developed.

8.3 Residential Building Integration

Residential buildings were the principal focus of the initial research and design concerning building integration of passive systems. This focus was certainly a natural and logical starting point because of the rich and numerous examples of indigenous, climate-adapted housing found throughout the world. Many early, innovative structures, often homes designed

Figure 8.1
Taos Pueblo, Taos, New Mexico.

and built by their owners and occupants, provided the basis for the passive system concepts that would emerge. The initial goal was to identify relevant passive systems, to understand how they worked, and to effectively integrate these systems into residential buildings.

This section begins by briefly reviewing the state of the art of residential passive system integration as of 1972. The focus is on post–World War II examples. With this historical perspective in place, the intent is to identify the key research and design activities that significantly advanced the state of knowledge on integrating passive systems in residential buildings. The section concludes with a review of the lessons learned about residential passive system integration from these research and design activities.

8.3.1 State of the Art as of 1972

In 1972, *passive* was not a term used by residential designers and builders. Excluding indigenous American architecture, few passive solar homes existed in the United States, and the ones that did were usually experimental structures built for research or occupied by their designers. An exception were the homes designed by Chicago architects George and Fred Keck for homebuilders and developers in the Midwest during the 1940s and 50s. Their designs used large expanses of south facing glass that allowed the low winter sun to heat interior masonry floors and walls of the building during the day, which in turn radiated the stored heat to the spaces during the evening. A reporter for the Chicago Tribune began to describe houses with these features as "solar houses," and other designers and builders for a time copied these design features (Holtz 1976).

Seeing the potential comfort and energy-saving advantages of these solar houses, university researchers and designers initiated projects to study the design and performance of houses incorporating passive solar design features. In 1945, F. W. Hutchinson of Purdue University, for example, designed and built two identical 1,830-ft^2 (170-m^2) homes on adjacent lots in Lafayette, Indiana. One house had 112 ft^2 (10.4 m^2) of south-facing double glazing, while the second house had only 41 ft^2 (3.8 m^2) of south-facing double glazing. Both homes were insulated to identical levels. Thermal storage was in the fabric of the buildings, especially in the concrete floor slab. Wide overhangs shaded the windows in the summer months, and large operable windows provided natural ventilation. Electric resistance heaters provided auxiliary energy and were metered to determine the

Figure 8.2
Montezuma's Castle, Arizona.

Figure 8.3
Keck prototype, speculative home, suburban, Chicago, Illinois.

Figure 8.4
Keck prototype, interior photo.

energy savings due to the larger south-facing windows. His studies substantiated the intuitive claims made by the Keck brothers (Holtz 1976).

The systematic approach employed by Hutchinson to investigate the performance of south-facing windows led to similar research projects on passive subsystems. One of the better known is the series of studies on the Massachusetts Institute of Technology (MIT) solar house conducted from 1939 to 1956. Funded by the Cabot Foundation, MIT Solar House 2 was a series of test rooms for testing different glazing and heat storage alternatives. The energy-saving advantage of added thermal mass was shown, although the savings were relatively modest (Dietz and Czapek 1950).

During this time, engineer Felix Trombe in cooperation with architect Jacques Michel designed and built a number of innovative passive solar homes in Odeillo, France. The one-story, two-bedroom structures had dark-colored solid-concrete, south-facing walls covered with two layers of glass. Solar radiation was collected on the massive wall surface and stored in the thickness of material. Heat was distributed to the interior by conduction through the wall and radiation and convection from the interior surface. High and low vents were placed in the wall to allow a portion of the

Figure 8.5
Trombe/Michel house, Odeillo, France.

solar heat to enter the house directly through thermosyphon circulation. This general concept became known as the Trombe wall.

Other passive solar pioneers designed and built passive solar homes prior to 1972. These included Peter van Dresser of Santa Fe, New Mexico, who developed a hybrid solar heating system using air-cooled active solar collectors with an underfloor rockbed storage; Norman Saunders of Weston, Massachusetts, who employed south-facing double-glazed windows and massive building construction; and Steve Baer of Corrales, New Mexico, who designed and built several innovative passive solar homes using water walls and thermosiphon air systems. Given the pioneering efforts of these innovative designers and researchers, what was the state of the art of building integration of passive systems in 1972? Using the definition of passive architectural design elements and the building integration goals discussed in the section 8.2, we can construct an overall picture.

The passive solar homes existing in 1972 were designed almost exclusively for heating season performance. The sizing of south glazing area was determined by heating season requirements and not the potential for cooling-season overheating. Fixed overhangs were typically used to shade a portion of the glazing during the summer months. Nevertheless, it is remarkable that summer comfort was generally not sacrificed due to the high levels of thermal mass and adequate window shading. We will see that this was not always the case in succeeding generations of passive solar homes.

It is apparent from the designs that thermal or economic optimization of the passive architectural design elements was not performed. Energy conservation was considered important, but the dynamic relationship between the building load and the passive glazing systems was not fully understood. Insulation levels, infiltration rates, and construction technology are typical for the year of construction. Exceptions are the massive construction used by Norman Sanders in New England and the "zome" wall-panel system used by Steve Baer in New Mexico.

The building shape and the interior spatial arrangement in most cases reflects the designer's intention to maximize solar gain and heat distribution. There is recognition that thermal mass is important for heat storage and overheating reduction, but there is little sensitivity to how the thermal mass is integrated with the conventional building features. The Keck brothers' homes appear most successful in this regards.

Economic analysis does not seem to have been an important consideration in the design decisions other than the obvious need to control the cost. Because the Keck brothers' passive solar homes were designed for the speculative home market, they were probably the most responsive to total building cost. However, in no case was a detailed economic analysis performed, partly because the primary purpose of most of these early buildings was proof of concept and partly because there were no tools to predict the energy savings that could be expected.

Aesthetically, these early passive solar homes represented a significant departure from traditional design. Large expanses of south-facing glass, large fixed overhangs, greater amounts of exposed massive materials, and nontraditional building shapes collectively defined a new aesthetic related to the passive solar approach integrated into the design. That the results were not entirely successful is evidenced by the sometimes less than flattering articles that appeared in the home press about these "odd" new buildings.

8.3.2 Major Milestones

In this section, an attempt is made to identify key events or results that represent major milestones in the development of residential passive systems integration. Of interest are the research projects, demonstration programs, competitions, prototype designs, and architectural examples that significantly advanced the understanding and knowledge of residential passive systems integration. Such a selection is not unique. However, the milestones listed are indicative of the type of activities that have given rise to the current state of the art.

The OPEC oil embargo of 1973 played a major role in igniting research, development, and design activities in solar energy. The recognition of the vulnerability of the national economy and security caused by dependence on imported oil resulted in the creation of major national research, development, and demonstration programs in solar heating and cooling. Foremost among these, concerning residential solar design, was the U.S. Department of Housing and Urban Development (HUD) Residential Solar Heating and Cooling Program (ERDA 1976). This program consisted of annual cycles of demonstration solicitations, at which time representative solar housing projects were selected for construction and evaluation, the concept being that with each subsequent cycle, new and improved solar technology would be available and ready for demonstration.

Prior to initiation of the demonstration cycles, HUD commissioned a publication on solar home design to assist home builders and designers to develop viable solar design solutions. Prepared by the American Institute of Architects Research Corporation (AIA/RC) (Holtz 1976), *Solar Dwelling Design Concepts* was the first publication to systematically study the influence of solar heating and cooling on dwelling design and site planning. Climate, comfort, building characteristics, and solar system characteristics are discussed in terms of their impact on residential architecture. Once described, numerous dwelling and site design concepts are presented for various housing types, solar systems, and climates. Much of the information in this publication was derived from the numerous subcontractors to AIA/RC who participated in the research project (Arizona State University 1975, American Society of Landscape Architects 1977, Massdesign 1975, TEA 1974, The Architects Taos 1975, Watson 1975, Giffels Associates 1974, The Continuum Team 1975, Joint Venture 1974, RTL 1975, University of Detroit 1975).

The first cycle of the HUD Residential Solar Demonstration Program resulted in fifty-three awards of which only three could be considered passive solar designs (USHUD 1976). Two of the awards were for passive system integration concepts. Communico received an award for Unit 1 of First Village, a solar subdivision in Santa Fe, New Mexico (USHUD 1976). The L-shaped, two-story design incorporated a sunspace between the legs of the *L* with thick adode walls for thermal storage between the sunspace and the primary living spaces behind. The design represented a passive solar architectural prototype that was to be repeated numerous times in the future. Self-Help Enterprises received an award for five single-story houses using the roof pond Skytherm® process invented and patented by Harold Hay (USHUD 1976). The design was a second generation of the full-scale prototype built by Hay in Atascadero, California, in 1973 (Hay 1976).

Seeing the need to determine the state of the art of passive solar research and design, the Energy Research and Development Administration's Solar Heating and Cooling Research and Development Branch sponsored the first National Passive Solar Heating and Cooling Conference in May 1976 in Albuquerque, New Mexico. The conference was significant because it brought together designers and researchers from all parts of the United States and many foreign countries to share their ideas and results on passive solar home design and research. The exposure of the conference

Figure 8.6
Unit 1 First Village, Santa Fe, New Mexico, HUD Phase 1 Demonstration award winner.

participants to the information presented at the conference would cause an explosion of activity in passive solar design and research over the next five years. The first generation of modern passive solar designs were presented by Saunders (1976), Hay (1976), Schmitt (1976), Kelbaugh (1976), Nichols (1976), Anderson (1976), Hammond (1976), Yanda (1976), Rogers (1976), Terry (1976), and Haggard (1976). The design of most all of these solar homes was based on intuition and a few simple calculations. The presumed performance requirements of the passive system dominated any concern for building integration. Many of the homes' architectural concepts looked like schematic heat transfer diagrams. Nevertheless, due to the design and technical skills of the owners/designers, many of these homes have quite a satisfactory overall appearance.

To explore the aesthetic qualities of the emerging solar designs, the AIA/RC—through a grant from Exxon Corporation—sponsored a student design competition (AIA/RC 1977). The entries would demonstrate a reduction in residential reliance on fossil fuels by integrating solar energy for space heating, cooling, and domestic hot water into prototypical hous-

Figure 8.7
Kelbaugh Residence, Princeton, N.J.

ing designs. The goal was to provide a forum for innovation in solar design while retaining traditional architectural design concerns for beauty, economy, and humanity. Regional aspects of solar design were stressed. During the fall semester of 1976, over 2,000 architectural students in the United States and Canada participated in the competition. The ten winners and four honorable mentions represented the full spectrum of housing types, climate regions, and innovativeness. Some accentuated the solar aspects of their designs, while others attempted to blend the solar features into more traditional architectural forms. The competition successfully engaged a large number of architectural students and faculty in the process of solar design integration.

From 1976 to early 1978, HUD sponsored three more cycles of its residential solar demonstration program (USHUD 1977a, 1977b, 1979). Each cycle was dominated by active solar demonstration projects, and few awards were made to passive solar homes. This was in part due to the small number of proposals submitted but also to the lack of quantitative design methods for passive solar systems. Grants in these cycles that were notable from a systems integration point of view were to South Central

Community Action Program, Inc. for a Skytherm-North project in South Dakota (USHUD 1977), Greenmoss Builders for a hybrid heating system in Vermont using a sunspace and an interior hollow concrete wall (USHUD 1977), and Living Systems for the Suncatcher direct-gain house in California. These projects proposed innovative approaches for integrating the passive system design features with the architectural concept of the building.

Despite the series of competitions described above, passive solar design remained far removed from the mainstream of residential design and construction. To overcome this lack of information and knowledge on passive solar home design, three significant projects were initiated during 1977 and 1978. The first was a Passive Solar Program Research and Development Announcement issued by the U.S. Energy Research and Development Administration, and managed by Los Alamos National Laboratory (ERDA 1976). Eight awards were made to study the performance of passive systems integrated into new and existing residential buildings. A wide range of passive concepts were studied including active solar collectors located in a clerestory of a sunspace connected to an underfloor rockbed, roof ponds, sunspaces, atria, thermal-storage walls, and roof apertures. Although not all of these projects were successful, this program was the start of a more systematic investigation of passive system integration.

The second of the projects involved documenting existing passive solar buildings (AIA/RC 1980, 1981) and developing regional guidelines for designing energy efficient passive solar homes (AIA/RC 1978). The significance of these two efforts was the presentation of energy conservation and passive solar design strategies appropriate to different climate regions, and the publication of examples of passive solar buildings as a means to encourage creative passive solar design. *A Survey of Passive Solar Buildings* (AIA/RC 1981) provided needed images of what passive solar heating and cooling meant to architectural form. The regional guidelines booklet (AIA/RC 1978) provided a starting point for determining what passive solar strategies were appropriate to particular climate conditions, although, other than a useful tabulation of regional climatic data, little quantitative energy design guidance was provided.

The two preceding passive systems integration research activities together with research conducted in the new passive solar energy program of the U.S. Department of Energy (DOE) provided the foundation for another attempt to introduce passive solar design into the homebuilding industry—the HUD Residential Passive Solar Design Competition undertaken during the summer and fall of 1978. A unique feature of the design competition was the introduction of a standard procedure, the solar load ratio (SLR) method, for calculating the energy performance of the passive solar home. Heretofore, designers used their own methods for estimating their building's thermal load, auxiliary energy needs, and solar contribution. Over 550 grant applications were submitted; 162 of these passive designs were selected for awards, 145 for new homes, and 17 for retrofit installations of passive solar elements on existing homes (Franklin Research Center 1979). In one major event, hundreds of passive system integration concepts were added to the state of the art, and the number of passive solar homes doubled. The value and impact of design competitions for supporting an emerging technology was firmly established.

The state of the art of passive systems integration was redefined at the second National Passive Solar Conference held in Philadelphia in March of 1978. Clearly evident was the progress made in designing and integrating passive solar heating and cooling systems into residential buildings. Second-generation architect-designed custom passive solar homes were typically more sensitive to the form implications of the various passive concepts. With the availability of passive solar calculation procedures, greater emphasis was placed on sizing the passive system elements—glazing area, mass area, insulation levels, and so on. Representative of this growing sophistication of residential passive systems integration were projects presented by Schiff (1978), Watson (1978), Lambeth (1978), Pfister (1978), Banwell, White and Arnold, Inc. (1978), and Garrison (1978). However, little headway was seen of the widespread use of passive systems in the homebuilding industry. Builders and developers were still not convinced that the costs and performance of passive homes were justified by market demand. Few marketable passive solar designs existed. Builders remained concerned about the overall attractiveness of passive solar homes.

The fact that passive solar design was not finding widespread acceptance in the moderately priced home market was recognized early in 1979. At the third National Passive Solar Conference held in January of 1979 in San Jose, California, Leach and DeWitt (1979) and Born and Chase (1979) presented prototype passive solar homes targeted for the moderately priced, speculative market. Direct-gain, thermal-storage-wall, and sunspace concepts were integrated into conventional architectural styles. Significant attention was paid to the energy conservation features of the designs. On the whole, the designs represented an improved balance between heating and cooling performance and between energy conservation and solar energy utilization.

Village Homes, a subdivision within the city of Davis, California, emerged as a symbol of architectural and planning integration of energy conservation and passive solar energy. Here, an entire subdivision was planned and built to conserve energy and use passive systems, representing a new paradigm for suburban housing for middle America. The planning concepts and passive solar home designs were the prototypes that homebuilders, developers, and designers were looking for to replicate for their own markets. Bainbridge, Corbett, and Horfarce (1979) packaged these concepts in an attempt to encourage their adoption and use by other developers and designers.

A critical need remained to enable effective residential passive system integration to occur—design guidance and analysis procedures. Designers needed information on the interaction of passive system components, the sizing and performance of these components, and methods to calculate anticipated auxiliary energy consumption. Few such procedures existed, and none of them found widespread use. Los Alamos National Laboratory had published papers on the solar load ratio (SLR) method of energy calculation (Balcomb and McFarland 1978, Wray, Balcomb, and McFarland 1979), but the design handbook (Balcomb et al. 1980) had not yet been published. Mazria (1979) filled this void with *The Passive Solar Energy Book*. The book was organized around a series of patterns or guidelines for designing passive solar homes. It was written in a style and at a level appropriate to designers, although builders, homeowners, and others immediately adopted its recommendations. The significance of the book lay in its provision of quantified design guidelines based on simulation of various passive design approaches. At the time, it represented the state of the art of passive design integration information and in combination with an energy calculation procedure provided a comprehensive tool for developing integrated residential energy design solutions. *The Passive Solar Energy Book* was the first serious and successful attempt to provide a link between the research and the practice of passive solar design.

In January and March of 1980, DOE published the long-awaited *Passive Solar Design Handbook* (TEA 1980, Balcomb et al. 1980). The handbook came in two volumes. Volume 1, written by Total Environmental Action (TEA), was subtitled *Passive Solar Design Concepts* and contained an introduction to passive solar design concepts and principles (TEA 1980). Volume 2, by authors at Los Alamos National Laboratory, was subtitled *Passive Solar Design Analysis* and contained simplified methods for calculating the thermal and economic performance of various passive solar design strategies (Balcomb et al. 1980).

Volume 1 of the DOE *Passive Solar Design Handbook* was essentially a compilation of information on passive solar heating design organized by five physically identifiable system types: direct gain, thermal-storage wall, thermal-storage roof, sunspace, and convective loop. The volume discussed the advantages and disadvantages of each passive heating system type and presented built examples. A small chapter was included on passive cooling concepts.

Volume 2 of the *Handbook* was a systematic presentation of results and conclusions derived from experiments and analysis at Los Alamos National Laboratory (LANL). LANL had been engaged in a systematic study of the various passive solar heating concepts using test boxes, test rooms, and simulation. Through the validation using test-room data of several simulations, known as PASOLE and SUNSPOT, and the exercise of these simulations for several locations and passive system characteristics, LANL researchers developed a monthly, correlation-based passive solar heating calculation procedure—the solar load ratio (SLR) method. Also developed from tabulated results derived from the SLR method was an annual passive solar heating calculation procedure—the load collector ratio (LCR) method. Volume 2 presented these methods as part of a procedure for performing thermal and economic analysis during the residential design process. This procedure represented a breakthrough in passive system integration. It was now possible for the home designer to evaluate the performance of passive solar heating concepts. Energy savings due to the passive system could be calculated and used to determine the cost effectiveness of the design solution and to optimize the passive design features. Volume 2 gave home builders and designers a powerful tool for investigating passive design integration options. A limitation of the method was that it addressed only passive solar heating concepts and thus was unable to effectively deal with cooling-season issues. This limitation constrained the handbook's ability to develop balanced heating and cooling design solutions. Volume 2 was revised and expanded between 1980 and 1982, resulting in Volume 3 (Balcomb et al. 1982), and a second round of revisions and expansions was eventually published by ASHRAE in 1984 as *Passive Solar Heating Analysis* (Balcomb et al. 1984).

Sandia Laboratories of Albuquerque, New Mexico (Sandia 1979) initiated a technology-transfer program to make available case studies of passive solar buildings. The case studies presented information on the architectural design of the buildings, a description of the passive and energy conservation features, and a short period of performance data. This compilation was one of the first attempts to show the design and performance of passive solar buildings as a complete, integrated system. These case studies were also important in showing the subtleties and complexities of passive solar design and the nature of the low-quality heat that was being captured, stored, and controlled in passive solar design. Also shown was the important fact that passive solar buildings work—save energy—and provide levels of comfort comparable to those in conventional buildings. More information on performance monitoring/evaluation of passive solar buildings is described elsewhere (see Frey, chapter 9 of this volume).

The continuing and growing awareness of passive systems integration was reflected in numerous papers presented at the fourth National Passive Solar Conference held in October 1979 in Kansas City, Missouri. Furthermore, in addition to concerns for technical aspects of design integration, a greater emphasis was now being placed on the cost-effectiveness of passive solar design. Davis (1979) presented the results of the California passive solar design competition. The competition design categories included four types of dwelling units, six climate regions, and specific total home costs. Seventy-five designs were submitted for review, and thirteen of these received awards. One aspect of the jury evaluation was the detailed calculation of construction costs and energy performance. On the basis of this cost and performance data, the cost-effectiveness was determined, stated as the net present value of energy savings. The values ranged from \$2,748 to \$29,298 (Davis 1979).

A similar cost-effectiveness study was reported by Taylor (1979). Here, a passive solar home's costs and energy performance was compared to a conventional home in Santa Fe, New Mexico. The projected yearly heating bill for the conventional home was \$710, and for the passive solar home, \$28. This performance improvement was achieved at a cost per unit of conditioned floor area of approximately \$4/ft^2 (\$40/m^2).

While Davis (1979) and Taylor (1979) are concerned with determining cost-effectiveness on a project-by-project basis, Noll and Thayer (1979) sought a broader, more universal approach to optimizing the design of passive solar heating, auxiliary heating, and building conservation features. They proposed a graphical technique to visually depict the optimization trade-offs. The importance of these studies is the introduction of a more rigorous economic analysis into the passive systems integration process. While previously designer experience and intuition solely determined the design and integration of passive elements, now both energy savings and cost could be considered to determine the most economical solution. Other economic studies of note reported at the Kansas City conference were by Sullivan and Malcolm (1979), Coonley (1979), and Thayer and Noll (1979).

Seeking ways to influence a large percentage of U.S. residential construction, DOE, in August 1979, announced the Passive and Hybrid Solar Manufactured Buildings Program (USDOE 1979). Manufactured housing was estimated to account for approximately 30% of annual housing starts. Unlike the HUD Passive Solar Home Design Competition, this program was structured to link building manufacturers with solar design consultants. The program represented a major challenge and opportunity for introducing passive solar concepts into mobile, modular, panelized, and precut homes. The manufacturers' concerns for assembly line production, transportability, and costs had to be integrated with solar designers' concerns for energy performance. Nineteen awards were made to home manufacturer/solar consultant teams. The types of units included mobile homes, panelized, log, and modular housing. Several innovative design concepts were developed (SERI 1983). In general, energy savings of 50–70% were achieved in all of the prototype designs at a cost per unit of conditioned floor area of \$1–4/ft^2 (\$10–40/m^2) (SERI 1983). The process of linking the building manufacturers with solar consultants proved to be an effective means of transferring solar technology. However, except for those manufacturers who participated in the program, widespread industry adoption of passive solar design did not occur.

The Denver Metropolitan Solar Homebuilders Program conducted by the Solar Energy Research Institute (SERI) utilized a similar concept to the Passive Solar Manufactured Building Program by linking homebuilders and developers with experienced solar design consultants (Baccei et al. 1980). No funding was provided for constructing the homes; however, analytical support, performance monitoring, and promotional assistance were provided by SERI. Integration of the passive systems into the design

Figure 8.8
Acorn prototype manufactured house, Boulder, Colorado.

improved dramatically over the HUD demonstration projects; however, since the builders were assuming the majority of the risk, optimal designs from an energy point of view were not developed in all cases. What this program did accomplish was to dispell the myth that passive solar design was complex, inflexible, and unattractive, and, instead, opened the homebuilders' eyes to its simplicity, flexibility, and marketing advantages. A two-day tour of these passive solar homes was attended by over 100,000 people. Providing a variety of passive architectural concepts for homebuilders and homebuyers to consider, changed the marketplace in Denver.

The success of the Denver Metropolitan Solar Homebuilders Program spawned numerous similar projects, each built around the concept of developing prototypical passive design solutions for the residential marketplace. The Mid-America Solar Energy Complex (MASEC) Solar 80 resulted in ten prototypical single-family passive solar home designs made available to homebuilders and designers through a plan book (MASEC 1980). The program stressed the integration of simple energy conservation measures with passive solar features. Typically, greater insulation levels and lower infiltration rates were combined with simple direct-gain windows

Figure 8.9
Denver Metro Solar Homebuilders Program Home, Golden, Colorado.

and some added interior thermal mass. The designs were *sun-tempered*, as opposed to high-performance, state-of-the-art systems, to respect market preferences in the region. Several of the prototype homes were constructed and tested as part of the MASEC 80 program.

The Northeast Solar Energy Center (NESEC) conducted a similar project but focused on multifamily passive solar home designs (NESEC 1980). Five prototypical design solutions were developed for townhouse, apartment, and retrofit conditions. Each design concept was presented as a case study intended to influence housing developers and designers on the potential of passive solar design. Greater insulation levels and direct-gain windows were the predominant energy-saving strategies. However, several designs employed thermal-storage walls and sunspaces. As with the MASEC designs, the approach employed sun-tempered designs that were most easily understood and adapted by homebuilders and designers.

Western Solar Utilization Network (SUN) in cooperation with the Bonneville Power Administration instituted a Solar Homebuilders Program modeled after SERI's Denver Metro Homebuilders Program (Parkin 1981). Initially planned for seven cities, the program was completed in only two—Portland, Oregon, and Spokane, Washington. Twelve passive solar homes were designed, built, and monitored in Portland, and seven passive solar homes were constructed in Spokane. Passive system integration was again the primary focus of the program with the majority of the homes utilizing enhanced energy conservation with direct-gain windows and sunspaces. Homebuilders and designers were now "getting the feel" for how to successfully integrate energy conservation and passive system concepts into attractive, marketable homes.

During the same time period, the Commonwealth of Massachusetts initiated a Multi-Family Passive Solar Program (Rouse 1980, Rouse and Noble 1981, Noble and Lofchie 1985). The program was targeted at new state-financed housing for the elderly. However, the intent was to influence other public and private multifamily housing projects. Nineteen projects comprising almost 500 apartments received awards totaling over $1.5 million. Each project received careful thermal and economic analysis by the Massachusetts Office of Energy Resources. Results of this program showed that considerable design flexibility exists when energy conservation, especially increased insulation, cross ventilation, and east/west win-

Figure 8.10
Solar Homebuilders Program Home, Spokane, Washington.

dow shading, and simple passive solar system concepts are employed in multifamily housing. The small thermal loads necessitate careful attention to solar-aperture sizing and location so as to avoid winter and summer overheating problems.

By the time the fifth National Passive Solar Conference was held in Amherst, Massachusetts, in October 1980, passive system integration for new residential buildings was becoming a formalized concept. However, the basis of this understanding was primarily experience, a little performance data, and some systems analysis results. Passive systems integration research was not advanced enough to have conclusive results on the relationship between energy conservation, passive solar, and mechanical systems. Kelbaugh (1980) enumerated seven principles of energy-efficient architecture in an attempt to codify the factors that may emerge as primary design determinants in a modern architecture responsive to energy and climate. His principles ranged from use of materials with low-energy intensity to opening the building up to receive solar gains and cooling breezes to recycling a building's materials and wastes. However, these were architectural principles, not specific design guidelines or rules of thumb that enabled one to develop an integrated residential design solution. In espousing these principles, Kelbaugh sought to bring passive solar design into the mainstream debate on architectural form and purpose.

Until this time, residential systems integration research and design had been dominated by a concern for space heating. Cooling or overheating avoidance was rarely considered by researchers developing passive heating concepts or design tools. However, in 1980 a major initiative was undertaken by DOE to introduce space cooling as an area for research and design activities. With funding provided by DOE, the University of Miami's Center for Advanced International Studies organized an International Expert Group Meeting to assess the state of the art of passive and low-energy cooling, heating, and dehumidification. Fifteen foreign and two U.S. experts attended the meeting and presented two papers each: a country or regional monograph that described climatic, cultural, and demographic conditions and the state of the passive art, and a subject monograph that described in depth one of the passive cooling technologies. This meeting set the groundwork for DOE funding of passive cooling research and an international conference on passive cooling. The International Passive and Hybrid Cooling Conference organized by the Passive Systems Division of the American Solar Energy Society was held at Miami Beach in Novem-

Figure 8.11
Sundance I Home, Reston, Virginia, Passive Design Competition award winner.

ber, 1981. The proceedings from the conference (Bowen, Clarke, and Labs 1981) represent an important contribution to understanding the impact of cooling on the design of energy efficient homes. (See also volume 8 of this series, *Passive Cooling*.)

Recognizing the growing maturity of passive solar home design and seeking to introduce energy as a formal design determinant, the Passive Systems Division of the American Solar Energy Society sponsored the first National Passive Solar Design Competition (Cook 1984a, 1984b). Announced in the spring of 1980, the awards were presented at the fifth National Passive Solar Conference. Twelve residential projects received awards, seven in the designed (but not built) category and five in the built category. Of the twelve projects, four involved the redesign of existing buildings. On the whole, the projects showed remarkable sensitivity to the integration of the energy conservation and passive solar features into the overall architectural concept. Most chose simple direct-gain or sunspace passive heating and natural and forced ventilation cooling. The influence of solar energy on architectural form is clearly evident in all of the projects, thus reinforcing the emergence of a solar aesthetic theme.

Figure 8.12
Pfister residence solar retrofit, Minneapolis, Minnesota.

Although passive system integration into existing residential buildings had always been recognized as a research and design concern, this area was not systematically addressed until the publication of the *Passive Retrofit Handbook* (Thompson Hancock Witte & Associates 1980) and the *Passive Solar Retrofit Guidebook* (Londe Parker Michels 1981). Prior to these publications, passive retrofit research and design was on a project-by-project basis.

Yanda (1976) popularized the attached sunspace as a retrofit. He attempted to reconcile the requirements of a sunspace for growing plants and for providing heat. He presented some initial design guidelines and a look at the economic viability of attached sunspace retrofits.

Michels (1979) investigated the energy and economic performance of a passive and conservation retrofit on a forty-year-old masonry home in St. Louis. Rigid insulation was applied to nonsouth exterior walls, and passive solar measures consisted of a sunspace and a thermal storage wall. Of significance was the thoughtful integration of the energy-saving elements into the overall architectural design. Through these efforts, and those of DeWitt (1980), Pfister (1978), Brant and Holtz (1981), the potential of passive solar retrofit became an increasingly viable and attractive option for energy savings in existing residential buildings.

The *Passive Retrofit Handbook* (Thompson Hancock Witte & Associates 1980) was a sourcebook of information on designing and constructing passive solar retrofits in the southeast United States. It defined a range of issues from infiltration reduction to building code compliance that must be considered in designing passive solar retrofits. It also provided design guidance for sizing the passive solar aperture and thermal storage as well as recommended construction details for building the most common passive solar retrofit solutions. This publication's significance was the systematic approach to passive solar retrofit, which enabled designers and builders to effectively integrate the passive system into the existing home. The publication's shortfall was the lack of information and design methods for passive cooling techniques.

The *Passive Solar Retrofit Guidebook* (Londe Parker Michels 1981) was prepared as part of a passive retrofit workshop program sponsored by the Mid-America Solar Energy Complex. The workbook is a thorough presentation of the concepts, calculations, and construction details needed to design and build passive solar heating retrofits. Numerous design tools are provided to assess solar access, size the solar aperture and storage mass, and calculate the economic viability of the passive retrofit.

While passive solar retrofit has remained an area of interest to designers, research in this area has diminished considerably since 1981. The seventh National Passive Solar Conference held in Knoxville, Tennessee, in August 1982 had a technical session devoted to passive solar retrofit. The papers in this session focused primarily on examples of passive retrofit projects and very little on general strategies, design methodologies, or innovative concepts for passive solar retrofits. Papers of interest regarding passive system integration include Theis (1982), Arasteh and Lindsey (1982), and Andrews and Snyder (1982).

Also at this conference, the winners of the second National Passive Solar Design Competition were announced. In the residential buildings category, four awards were made and seven projects of merit were cited. William Leddy, architect of San Francisco, received a first-place award for a single-family remodel/design in San Francisco (*Passive Solar Journal* 1983a). Faced with a narrow, steep site, adapting the current architectural design to the Victorian style of the neighborhood, developing a northeast view of the city, maintaining an outdoor space on the west side of the building, and meeting local seismic design requirements, the architect successfully integrated energy conservation and passive solar design features. The design featured a sunspace incorporated on the third floor, with excess heat removed from the sunspace and stored in a ground-floor rock bed.

Eric Meng Associates of Seattle, Washington, received a first-place award in the multifamily category for eleven two-story townhomes located in Seattle (*Passive Solar Journal* 1983b). The designers organized the townhouses around a central courtyard on the south-sloping odd-shaped lot. Direct-gain windows and a sunspace provide solar gains. The precast, hollow-core concrete floors and ceilings double as thermal-storage elements. Heat is removed from the sunspace and circulated through the cores.

Both award winners exhibited considerable architectural skill in integrating the conservation and passive solar features into the designs. The projects represented excellent architectural design and illustrated the form implications of passive solar energy utilization. Compared to the first design awards, the projects selected in the second competition were more understated in the impact solar energy had on the overall design concept.

This may have been the result of a growing awareness of the flexibility possible in passive design.

An economic optimization of the energy conservation and passive solar design of residential buildings is central to achieving cost-effective, energy-efficient residential buildings. Balcomb (1980) proposed a methodology based on a constrained optimum (initial investment in energy savings is limited) for allocating resources between energy conservation and passive solar strategies. His work built upon the efforts of Noll and Thayer (1979) to develop an economic optimization methodology for designing and sizing solar systems. Balcomb's interest was to develop a simple scaling law that would allow the designer to understand the trade-offs between energy conservation and passive solar investments.

The theme of optimizing conservation and solar design was continued at the sixth National Passive Solar Conference held in Portland, Oregon, in September 1981. Hannifan, Christensen, and Perkins (1981) investigated the energy requirements and comfort implications of various window areas and locations, thermal mass levels, and insulation levels. Their results clearly showed that year-round energy performance and comfort required a balanced design relative to these variables. The significance of this research was the addition of cooling-season energy performance and comfort to the usual emphasis on heating-season performance. This research set the stage for further investigations into balanced heating- and cooling-season design. Other papers of interest on the issue of optimizing energy conservation and passive solar design are Palmiter (1981), Duffield (1981), and McDonald and Tsongas (1981).

Research on the optimal mix of conservation and solar energy, and other observations (Derickson and Sadlon 1981) began to reveal design patterns that ran counter to the current notions of passive solar design. Energy conservation together with window redistribution to the south were seen as the most important and cost-effective energy-saving strategies. Consequently, significantly less south window area and thermal mass could be used to achieve high levels of energy savings relative to current recommendations. These conclusions were significant because they opened new flexibilities and a new area of research in integrated energy design.

Since 1981–1982, when the U.S. Department of Energy's funding for passive solar research was significantly reduced, the amount of research into residential passive systems integration has dropped dramatically. An important activity, especially in light of reduced funding in the United States, is an International Energy Agency project on Passive and Hybrid Solar Low Energy Buildings, of which the United States was a participant (Holtz 1983). This fourteen-country cooperative research project was concerned with understanding the design and performance of residential buildings using active and passive solar and conservation technologies, the interactions of these technologies, and their effective combinations in various climate regions. As one element of this large international research project, the participants undertook a comprehensive set of parametric sensitivity studies to investigate design integration issues of combining energy conservation, passive/hybrid solar, and advanced mechanical systems in new residential buildings. The results were presented as a series of Design Information booklets (1988–1991) and technical reports.

The New Mexico Showcase of Solar Homes presented a variation to previous design-build demonstration programs (Balcomb et al. 1983). The objective of the Showcase project was to demonstrate that solar homes are affordable and attractive. Unlike previous such projects, quantitative guidelines were established that had to be met by homebuilders seeking entry into the Showcase project. The guidelines were based on research performed at Los Alamos National Laboratory (Balcomb 1983) and addressed the balance of energy conservation, passive solar design, and occupant comfort. Twenty homes were selected to be built in two of the fastest-growing population areas in the state. Statewide publicity and a home tour were used to inform potential homebuyers and the homebuilding industry of these prototype passive solar homes.

The refinement of residential passive systems integration achieved by experienced designers was clearly evident in the projects presented at the eighth National Passive Solar Conference. Barrett (1983) presented a thoughtful discussion on the use of thermal mass in passive solar homes as both an architectural and an energy-saving element. Reeder and Merkezas (1983a, 1983b) discussed the use of passive design strategies in the context of restoration, renovation, and infill of historic buildings. Lee, Kelbaugh, and Holland (1983) presented a project of a twenty-unit housing for the elderly in New Jersey where numerous energy-saving concepts are fully integrated throughout the design. Each of the above projects represents a level of sophistication in residential passive systems integration that exemplifies the state of the art at that time.

8.3.3 Lessons Learned

After more than fifteen years of research and design activities in the area of residential passive systems integration, the question is what has been learned. As is common in research and design, residential passive systems integration has advanced by trial and error. One important aspect, however, is the necessity of maintaining the link between research and practice. Summarized below are a few of the most significant lessons learned.

- **Greater design flexibility exists than previously assumed.**

Early passive solar design exemplified rigid principles of glass sizing and placement, mass sizing and placement, and spatial arrangement. Plan and section drawings looked like heat transfer diagrams of the passive systems employed. Results from monitored passive solar homes and from parametric sensitivity analyses indicated that a continuum of residential energy design exists—from improved conservation to sun-tempered, passive solar, and super-insulated—and that greater design flexibility was possible while achieving substantial energy savings. Once the passive solar myths were exposed and refuted, home designers had much greater flexibility in integrating passive systems into residential designs compatible with local construction practices and market preferences. This advance enabled passive solar design to enter the mainstream of the homebuilding industry.

- **The various passive system types have about equal performance.**

Early passive solar studies did not clarify whether one passive heating system type was a better performer than another. Differences in system size, climate, and operation complicated comparisons of system types. Testing and analysis of passive solar homes in a wide range of climates did not show any pattern that one passive solar heating system type was a substantially better performer than another. Direct-gains windows, thermal-storage walls, and sunspaces, when properly designed, all seemed to achieve equivalent levels of thermal performance. This knowledge enhanced design flexibility and enabled issues of construction costs, aesthetics, view, marketplace appeal, thermal zoning, and other building integration requirements to determine the passive system type.

- **Validated numerical simulations are essential for studying passive system integration.**

The ability to identify and quantify the dynamic thermal interactions occurring in passive solar homes has been fundamental to developing procedures for their design. The development of building energy analysis simulations has greatly reduced the time, expense, and effort of understanding the sensitivity of the passive system performance to climate and design parameter variations. Without validated simulations, passive system integration research would be entirely dependent on a small number of empirical tests of a few passive system designs in a few climates. The extrapolation of these empirical results to other designs and climates would be an uncertain procedure.

- **Design guidelines are important in the early stages of design.**

Home designers and builders require guidance at the beginning of design. Rules of thumb should address solar aperture area and placement, thermal mass characteristics and placement, glazing type, insulation levels, infiltration reduction, nonsouth windows, shading, ventilation, and mechanical equipment. Such rules used early in the design process can lead efficiently to a design that is economical and balanced with respect to heating and cooling, energy conservation and passive solar, and auxiliary energy and comfort.

- **Simplified calculation procedures are necessary to assess the complex thermal interactions in passive solar homes.**

The home designer, in manipulating the many factors influencing thermal performance, needs to know the performance implications of each design decision. This knowledge can be obtained with the aid of a procedure that calculates the heating and cooling performance of the building. Such a procedure must be simple to use, include the essential energy design features, have few built-in assumptions, and be accurate. A number of passive solar calculation procedures have been developed, but only a few meet all of these criteria.

- **The additional cost of passive systems in homes is minor.**

The added cost of passive features, if they are carefully integrated into the design, can be negligible to minor, especially when passive elements perform double duty—as elements of both the passive energy system and the building system. However, where specialized elements are used, such as phase-change or containerized-water storage, there is an additional cost per unit of floor area, typically \$1–5/ft^2 (\$10–50/m^2). This additional cost is often offset by the energy savings. Some financial institutions recognize the benefits and adjust accordingly the principal limits in a long-term mortgage.

• **Design competitions and demonstrations are an effective means to promote innovation.**
Experience from over ten passive solar design competitions and four demonstration programs shows that these activities are a useful means for developing innovative passive solar design solutions, especially when the lack of rigid program requirements and constraints provide the necessary design freedom. The designs are often not optimum with respect to the passive system performance, but they provide the starting point for refinement into useful passive prototype buildings. Design competition and demonstration programs appear to work best when organized in two or more phases where review and refinement follow initial selection.

• **Careful construction detailing and execution is essential for long-term performance, durability, and reliability.**
The development of an optimum passive solar design does not guarantee that optimum performance will be realized when the building is constructed and occupied. The designer must carefully select the materials to be used, considering their performance under a range of thermal and environmental conditions. Standard building materials are usually the best choices. Also, these materials must be properly organized and assembled so that their performance is not impaired over time by thermal stress, moisture, and structural change. Standard designs are the most reliable. And lastly, construction quality control is mandatory to ensure that the materials are properly installed. Standard construction practices should be used whenever possible.

• **Passive solar design is as much an amenity as it is an energy saving strategy.**
Passive solar design can improve the quality of the living environment as well as save energy. The quality of light and thermal comfort achievable with daylighting and passive solar design can be superior to those aspects of conventional designs and is one reason why home builders, designers, and buyers want passive design features. The experience of the passive solar demonstration programs is that the market demands passive solar designs that are attractive, cost effective, and comfortable.

8.4 Commercial Building Integration

Based on the successful integration of passive systems into residential buildings, designers began to see the possibilities of passive systems in commercial buildings. The first attempts were the direct application of residential passive system concepts to commercial buildings. While this approach worked reasonably well for residential-scale commercial buildings, designers soon learned that commercial building energy use was substantially different than that of residential buildings and that the design and integration of passive systems for commercial buildings had to be substantially different, too. Despite the design complexities associated with commercial buildings, designers and researchers learned quickly, and soon a well-defined passive design process emerged. Through this process, designers were able to accurately define the energy characteristics of commercial buildings and to develop appropriate energy conservation and passive design solutions.

This section focuses on the current status of commercial passive system integration but, like the residential section, begins by briefly reviewing the state of the art of commercial passive system integration as of 1972. This historical overview is followed by a description of the key research and design activities that significantly advanced the state of knowledge on integrating passive systems in commercial buildings. The section concludes with a review of the lessons learned from these research and design activities about commercial passive system integration.

8.4.1 State of the Art as of 1972

For the past thirty years, artificial and mechanical environmental controls have dominated commercial and institutional buildings—electric illumination, fan ventilation, and thermal tempering with mechanical heating and cooling systems. However, such environmental control systems did not always exist, and earlier building designers relied on the natural environment for sources of heating, ventilation, and lighting. To illustrate the predecessors of modern passive solar designs, notable examples of this earlier design activity are presented in this section. The intent is to show that current commercial passive system integration is rooted on age-old architectural and engineering concepts.

8.4.1.1 The Wainwright Building The Wainwright Building, designed by Louis Sullivan, was constructed in St. Louis, Missouri, in 1890–91 (Ternoey et al. 1985). Mechanical cooling and humidity control did not exist. Incandescent lighting was recently available, but its high cost and large heat release relegated it to use as a supplement to daylighting. Daylight and natural ventilation requirements dictated a narrow architectural form to permit work-areas proximity to windows. The depth of the building was limited by the extent to which light and air could penetrate. Ceiling height was determined by the need to allow daylight to penetrate the interior work spaces. A rule of thumb was that useful daylight on a cloudy day will be available in a building up to 1.5–2 times the height of the window. Since the architectural form had to be narrow and extended, a U-shaped plan permitted a compact site. The environmental-control requirements were so constraining that the designer had few choices on the overall organization and concept of the building. So typical was the architectural form of the Wainwright building during this period that it was rarely mentioned in the writings of Sullivan or architectural critics of that time.

8.4.1.2 The Larkin Building The Larkin Building, designed by Frank Lloyd Wright, was constructed in Buffalo, New York, in 1904–1906 (Ternoey et al. 1985). New developments in mechanical environmental-control systems and the polluted external environment combined to cause the development of a unique approach to environmental control in commercial buildings that was to emerge as a passive prototype over 70 years later.

The Larkin Building was one of the first "hermetically sealed" buildings ever designed. All thermal and ventilation needs were met internally through mechanical equipment. Daylight was the primary light source, again because of the cost and large heat gain of incandescent lighting. Pollution was removed and air cooled during the summer months by passing outdoor air through a water-wash filtering system.

To resolve the conflicting requirements for a narrow and extended form to enhance daylighting and for a compact form to reduce the heating and cooling requirement, Wright created a glazed central atrium around which were located all the work areas. This solution provided a light source on two sides of the work area; yet reduced the total exposed exterior skin.

8.4.1.3 William W. Caudill The architectural prototypes represented by the Wainwright and Larkin Buildings were replicated during the first half of the twentieth century. At the same time, practical mechanical air-conditioning systems with temperature and humidity control were developed in the 1920s, and the fluorescent lamp became available in the 1930s. However, these technological advances did not immediately lead to commercial buildings with total internal environmental control. Narrow buildings using daylight and natural ventilation were still common until the 1950s. Improved equipment reliability, lower equipment cost, and cheap electricity finally made full internal environmental control the standard solution.

Representative of designers who still believed in climate-responsive architecture was William W. Caudill. Caudill was very interested in school architecture and the need to provide a total environment conducive to learning (Oklahoma A&M College Magazine 1951). As a research architect with the Texas Engineering Experiment Station at College Station, Texas, Caudill conducted research on daylighting, ventilation, and sound conditions in classrooms using wind tunnels, sky domes, and full-scale mock-ups. Numerous design recommendations for achieving adequate air flow for ventilation cooling, use of windows for daylighting, and solar gains for heating resulted from this research (American School & University 1953–54), and the ideas were incorporated into numerous schools designed by his architectural firm of Caudill, Rowlett and Scott during the 1950s. Many prototypical design solutions emerged from Caudill's school projects that were widely publicized and copied by other school architects.

8.4.1.4 Wallasey School The Annex of St. George's School, Wallasey, United Kingdom (near Liverpool), was designed in 1961 by architect E. A. Morgan. The school has been continuously occupied since 1962 (Perry 1976). The design incorporates an immense "solar window" on the south facade, 230 ft (70 m) long and 27 ft (8 m) high. Behind the solar window on two levels are classrooms, an assembly hall, art studios, and a gymnasium. The remainder of the structure is exposed poured-in-place reinforced concrete. The art-room floor (ceiling of classrooms) is 9 in. (23 cm) of concrete. The sloping roof is 7 in. (18 cm) of concrete. The interior partition walls and a portion of the north exterior wall are made up of 9-in. (23-cm) bricks. A 5-in. (13-cm) thickness of expanded polystyrene was placed outside the exterior walls and sloping ceiling for insulation of the envelope.

The south wall is double glazed with an air space of 24 in. (64 cm) between panes. The wide air space was for repairs and maintenance rather than for thermal reasons. Solar energy collected through the south windows represented the largest source of heat gain at approximately 50% of the total, followed by the incandescent lights at 34%, and the students providing about 16%. The auxiliary hot-water radiator heating system has been used only once during a short winter period when a power failure occurred. Incandescent lighting, unlike the previous examples, was chosen for its large heat generation as an integral part of the overall heating concept. The lights operate on a timer. The students and teachers understand the use of the lights, ventilation, and clothing for temperature control and adjust each accordingly. Problems of glare from the large solar aperture and inadequate ventilation have accompanied the otherwise excellent thermal performance achieved by the design.

While certainly not an optimized solution, the Wallasey School is a powerful example of the heating and lighting potential from passive systems even in a northern climate. The integration of the solar window, thermal mass of the building structure, the incandescent lights, and even the occupants is essential to the overall performance of the building.

8.4.2 Major Milestones Since 1972

Because of the greater complexity of the energy issues, the foundations of commercial passive system integration are less deep than those of residential passive system integration. Consequently, designers and researchers have had to undertake greater exploratory development in order to understand the commercial building energy problems and the passive system integration issues. This section identifies major milestones in this development process.

The major milestones in commercial passive system integration research and design are represented by government-initiated programs and by individual building projects. Through both types of activities, the nature of commercial building energy use was defined and appropriate passive system concepts developed. What emerged was not only a new set of architectural prototypes based on energy concerns but also a new process of designing energy-responsive commercial buildings. In this section, the major programs and projects that have advanced the state of knowledge on commercial passive system integration are discussed.

8.4.2.1 Demonstration Programs Similar to the HUD residential solar demonstration program, the U.S. Energy Research and Development Administration (ERDA) and its successor organization, the U.S. Department of Energy (DOE), planned and implemented a commercial solar heating and cooling demonstration program (ERDA 1976). Between the fall of 1975 and spring of 1979, six Program Opportunity Notices were issued to solicit projects to demonstrate the energy-saving potential of solar heating and cooling in commercial buildings. There were 291 awards during this period, of which 215 projects were actually undertaken. Only three of the projects used passive or hybrid systems. This small number speaks of the lack of awareness and understanding of passive systems in general and their application to commercial buildings in particular.

The three commercial demonstration projects funded involved the application of standard residential passive heating strategies to commercial buildings. The first passive project award was for the retrofit of a 8,640-ft^2 (803-m^2) building-materials warehouse with a direct-gain passive heating system (Keller, Johnson, and Sedrick 1978). The existing south facade of the building was simply removed and replaced with a double-glazed plastic window panel system. The existing concrete slab and stored building materials provided thermal storage. In addition to providing solar gain for heating, improved daylighting of the warehouse resulted from the retrofit.

The second project funded was the New Mexico State Office Building in Taos, New Mexico (Raskin and Lenker 1982). This 12,000-ft^2 (1100-m^2) single-story structure employed a roof aperture system that combined

Figure 8.13
Warehouse retrofit, Manchester, Vermont.

Figure 8.14
Taos state office building, Taos, New Mexico.

passive solar collection, thermal storage in water tanks, control through an operable shutter system, and daylighting in one integrated system. The roof aperture system was fully integrated with the mechanical system (including controls). The project was extremely innovative in its overall systems integration but perhaps overly complex in its mechanical operation and control requirements. Architecturally, the project was extremely successful, in large part due to the daylighting provided by the roof aperture system.

The third and last passive/hybrid project funded under the DOE Commercial Demonstration Program was a sunspace retrofit of an elementary school in Telluride, Colorado (Raskin and Lenker 1982). The sunspace attached to the south side of the school provided heat during sunny days to the north-side classrooms.

8.4.2.2 Solar in Federal Buildings Program The application of residential passive heating strategies to commercial buildings continued under the Solar in Federal Buildings Program (SFBP). The objectives of this program were to support efforts to shift from nonrenewable to renewable energy sources, stimulate the solar industry, encourage the continued development and refinement of solar technologies, and assert the leadership role of the federal government in promoting the widespread use of technically and economically feasible solar technologies (USDOE 1979). Seventeen of the SFBP projects incorporated passive and hybrid systems. In general, all these projects involved the application of well-characterized passive heating systems (direct-gain windows, thermal-storage walls, and sunspaces) to conventional building designs. Very little attention was directed to the cooling and lighting needs of the buildings. In several instances, the passive heating strategy increases the cooling load. The one building in which heating, cooling, and lighting solutions were employed is the U.S. Army Reserve Center in Suffolk, Virginia. Designed by Edward Mazria and Associates, the building utilizes clerestory windows for solar heat gain and daylighting. Ventilation cooling is employed for cooling when ambient conditions allow, with heat pumps providing cooling during extreme periods. The architect recognized the unique energy-use requirements of commercial buildings and attempted to provide a workable architectural solution that integrates passive heating, cooling, and lighting systems that function effectively together.

8.4.2.3 Ehrenkrantz Group Study The Ehrenkrantz Group (1978) studied the cost and performance impacts of passive systems applied to a conventional office building. The study was based on assessing the impact of incremental design changes on a conventional office building. A base building representing a typical energy conserving building in a New York City climate was developed and its construction cost and energy performance estimated. Thirty-three passive design options were then analyzed one at a time. The resultant energy and economic performance were determined and compared against the base building. In general, envelope-related design strategies did not show significant energy savings. Perimeter daylight, on the other hand, shows energy savings of over 20% with an internal rate of return of 67% for the building owner. The significance of this study was the systematic characterization of the energy problem in energy terms, and in cost terms, and the systematic investigation of energy-saving design alternatives. Drawbacks of the analytical method employed were the large number of computer runs required and the reliance on a conventional base building. The approach did not account for performance improvements by drastically changing the base building design.

These limitations would be overcome in future research sponsored by DOE.

8.4.2.4 Passive Solar Commercial Buildings Program Building on the experience of the DOE Solar Heating and Cooling Demonstration Program and the Solar in Federal Buildings Program, DOE, in spring of 1979, issued a Program Opportunity Notice (PON) entitled "Passive Solar Commercial Buildings Design Assistance and Demonstration" (USDOE 1979). This program represented the first organized effort to investigate and demonstrate on a national level the potential of passive systems in meeting the heating, cooling, and lighting energy needs of commercial buildings. Two of the objectives of the program as stated in the PON were to support the design and implementation of exemplary and prototypical passive solar commercial buildings, and to identify the cost and performance characteristics of passive systems in commercial buildings.

Thirty-three projects located throughout the United States received awards. The buildings ranged in size from 600 to 64,000 ft^2 (56 to 5950 m^2) and represented a variety of building types, including offices, retail stores, educational facilities, government buildings, community and visitor centers, and private, special-use facilities.

The significance of this program lay in its three-phase organizational structure. In phase I, designs were to be developed with mandatory schematic and final-design reviews by a team of experts. Phase II consisted of the construction of a selected number of promising projects. Phase III completed the program by evaluating the energy performance and the user response, and by disseminating the findings to the larger design community.

The impact of the Passive Solar Commercial Building Program (PSCP) has been so pervasive on the design and integration of passive systems in commercial buildings that it is difficult to summarize the results of this program in a few brief paragraphs. Discussed below are the general findings of the program; however, it is recommended that the extensive literature on this program be consulted for specific conclusions and lessons learned. In particular, the following reports are recommended: SERI (1980) and BHKRA (1984). BHKRA (1984) is an anthology of articles and papers on the PSCP through 1984 where the numerous facets of the program are discussed in considerable detail and references are made to the numerous reports prepared as part of the program. Figure 8.18 compares the average

Figure 8.15
Wells State Bank, Wells, Minnesota.

Figure 8.16
Johnson Controls building, Salt Lake City, Utah.

Figure 8.17
Interior, Princeton Professional Park, Princeton, New Jersey.

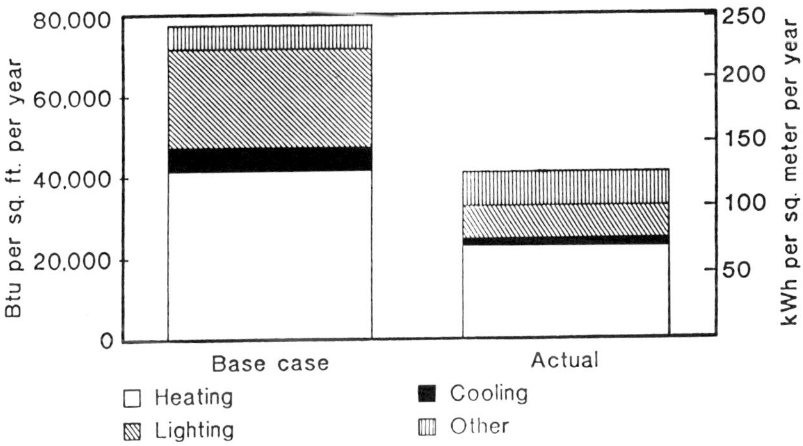

Figure 8.18
Actual versus predicted performance of twelve PSCP buildings.

measured performance of twelve PSCP buildings against the average predicted performance of the base buildings for these same twelve buildings. Energy consumption is halved by the use of passive-heating, cooling, and lighting systems.

General observations resulting from the PSCP are as follows:

• *Identify the Right Energy Problem Early.* Architects and engineers rarely identify the energy-use characteristics of a commercial building without performing some type of analysis. The energy problem in commercial buildings must be defined in terms of load, consumption, and cost. The relative importance of heating, cooling, and lighting can vary significantly from one building type to the next. Energy cost is typically more important to the building owner than energy consumption.

• *Select Energy-Saving Strategies Appropriate to the Energy Problem.* Once the energy use is defined, the appropriate energy-saving strategy must be developed. The solution must be simple, reliable, and effective. Complicated mechanical approaches or ones that involve significant occupant or owner participation should be avoided. Creativity and engineering judgment are essential to identify and evaluate the proposed solution.

• *Passive Solar Commercial Buildings Cost Only Slightly More.* The cost increase above a similar conventional building was always less than 10%, and in some instances there was no added cost. The economic viability or cost-effectiveness of the passive buildings must be determined on a case-by-case basis. In general, the energy savings from the passive buildings provide a reasonable internal rate of return to the building owner to justify the investment. However, since these were first-generation passive buildings, it can be expected that added costs will decrease as passive design becomes more commonplace.

• *Energy Analysis Expertise Is Required throughout the Design Process.* The energy performance of the building must be analyzed throughout the design process to obtain an effective design solution. This is a necessity if the balancing of heating, cooling, and lighting is to be successful. No single analytical tool is the best.

• *Daylighting is a Major Determinant of Architectural Form.* Energy consumed for lighting and the cooling loads created by lighting comprise the largest single energy cost in most commercial buildings. As building size increases, so does size and cost of these internal loads. Daylight is effective in reducing electrical lighting usage and thus the cooling loads associated with it. Consequently, designers explored numerous daylighting strategies, usually in combination with other heating or cooling strategies. Unless building occupants consistently and appropriately use manual lighting controls, automatic lighting controls are essential if energy savings from daylighting are to be achieved.

• *Occupant Behavior and Satisfaction Can Be Important to Energy Performance.* Although the PSCP projects represent a small sample, results of post-occupancy evaluation show that passive buildings provide their users with environments that are thermally comfortable.

Apart from the buildings themselves, the single most important outcome of the DOE Passive Solar Commercial Building Program was the book *The Design of Energy-Responsive Commercial Buildings* (Ternoey et al. 1985). The development of the background, case studies, concepts, and research that make up the book paralleled the design, construction, and evaluation of the exemplary passive solar commercial buildings. As a result, there was considerable interaction between the authors, the research teams, the designers, and the building owners. This interaction was responsible for the development of a unique design approach. Rather than focusing on component-level, incremental supplements to existing designs, the authors take a *whole-building* point of view.

The authors define a range of whole-building design solutions from "the climate adapted building" to "the climate rejecting building." Case studies of numerous buildings are presented to illustrate examples along this continuum of design solutions. A framework for design is presented that builds on the concept of energy-use characterization, establishing the energy end-uses and their relationships within a building. A number of analytical techniques, such as base-building analysis, elimination parametrics, rainbow plots, and economic analysis, are presented as means to understand the specific energy problems and patterns. Also presented is the means of using this information to identify, develop, and evaluate alternative schematic design solutions, illustrated by three examples.

The Design of Energy-Responsive Commercial Buildings was a tremendous intellectual breakthrough on how to think about energy use in commercial buildings and how to develop and integrate appropriate energy design solutions. As the authors state on page 169 of the book:

> The most important lesson we extracted from the study of energy-related design approaches is that process serves intuition. Process is a paradox in the design of energy-responsive commercial buildings. On the one hand, the purpose of process is to explore and test that which uninformed intuition cannot accurately appraise. Yet, regardless of what information is clarified by process, the solution generated is bound by the concepts one intuitively offers to test or study. Intuition is both the problem and the key to the design of energy-responsive commercial buildings. Design approaches are a framework to inform, stimulate, and educate intuition, not a mechanistic, computerlike series of steps that guarantees that the best possible solution can be found for any project.

8.4.2.5 Passive Solar Manufactured Buildings Program Another activity focusing on commercial passive systems integration was the nonresidential buildings portion of the DOE/SERI Passive Solar Manufactured Buildings Program (SERI 1983). Four nonresidential projects were awarded, three to steel building manufacturers and one to a precast concrete component manufacturer. Prototype designs were developed and analyzed by all four awardees, but only three prototypes were constructed and tested (SERI 1983). Two of the prototype designers, Banes and Porter (SERI 1983), incorporated primarily passive heating system strategies into their standard products. Banes, for example, proposed three options for its portable classroom: a combination water wall and direct-gain system, a greenhouse system, and a thermosyphon wall panel. Recognizing that daylighting was also important, Banes included light shelves in two of the design options. A monitored prototype with the water wall and light shelf showed a 68% energy savings over a monitored conventional structure also manufactured by Banes.

Porter, on the other hand, chose to incorporate a site-built thermal mass wall into its steel frame-sandwich panel system. Both solid masonry and water-wall options were studied. Little consideration was given to lighting or cooling. A prototype was built and tested using a solid mass wall and thermosyphon air panels (TAP). Results showed poor performance of the mass wall and good performance of the TAP in Holland, Michigan. The prefabrication of the TAP proved more economical for the manufacturer than even windows.

The most significant nonresidential passive solar manufactured building project involved the Butler Manufacturing Company with Princeton Energy Group as design consultants. The design team explored all aspects of energy demand in Butler's Landmark Building System: heating, cooling, and lighting. A variety of subsystem and component-level options were analyzed for their performance and compatibility with current construction techniques. A roof aperture system was finally selected as the most flexible for meeting the heating and lighting requirement of the most common building configurations. Numerous roof aperture systems were studied before selecting one with north- and south-facing vertical glass and a movable panel for insulation and shading. Three types of in-plenum heat storage subsystems were considered: suspended water-filled pipes, water bags on a suspended platform, and phase-change materials used as interior ceiling panels. Water bags were selected to be tested in the prototype building.

The significance of the Butler project lay in the thoroughness of the evaluation of the energy problem and the study of alternative solutions. The result was an energy-saving concept fully integrated with the requirements of a prefabricated metal building system. As a first-generation building solution, the energy systems were perhaps overly reliant on mechanical components. However, the project represented an important contribution to understanding energy use in nonresidential buildings in general and in manufactured buildings in particular.

8.4.2.6 Design & Energy Design & Energy, a nationwide student design competition conducted by the Association of Collegiate Schools of Architecture in 1980, is another element of the exploratory research and design effort undertaken to investigate the influence of energy conservation and passive solar energy on architectural form (ACSA 1980). The competition objectives were to involve students and faculty in a rigorous studio design exercise emphasizing energy considerations as a major design determinant, solicit new, imaginative energy concepts for buildings, and assess the level of interest in energy concerns in architectural schools.

The competition was conducted in two categories: an International House (a multifunction residence for American and foreign students) and any medium-sized architectural project of the students' choosing. The evaluation criteria used by the jury included integration of energy concerns and concepts with other design considerations, response to the user's specific energy requirements, adaptation to climatic conditions without excessive reliance on mechanical or electrical equipment, and application of well-understood principles of passive solar energy in innovative and appropriate ways.

Each of the ninety-two U.S. and Canadian architectural schools held internal design competitions to select the school representative to the national design competition. Over 300 entries were presented to a distinguished panel of jurors. The quality of the submissions was high, and most of the entrants attempted to integrate architectural design with innovative strategies for passive heating, passive cooling, ventilation, and daylighting. However, as the jury noted, many of the submissions exhibited an imbalance between passive technology and architectural design: "either an exciting idea dominated rather than generated form, or an exciting formal concept precluded the careful manipulation of materials and climatic factors for energy purposes" (ACSA 1980). Given the scarcity of prototypes and case studies of medium-scale passive solar commercial buildings, the results of Design & Energy were remarkable. The architectural students and faculty who participated helped advance the state of knowledge of commercial passive system integration by putting in front of the design professions a wealth of carefully reasoned, energy-responsive design solutions (Smeallie 1980).

8.4.2.7 National Passive Solar Design Awards The first National Passive Solar Design Awards Program (Cook 1984a, 1984b) was in progress at the same time as Design & Energy. Sponsored by the Passive System Division of the American Section of ISES and the New England Solar Energy Association, the Design Awards program sought exemplary passive commercial building projects in three categories: built, buildable (designed), and retrofit. A total of eight awards were made, one in the first category, six in the second, and one in the third. The award-winning projects represented a variety of scales, building types, and climate regions. The three office projects ranged from a 253,000-ft^2 (23,500-m^2), 18-story tower in Spokane, Washington, to a 64,000-ft^2 (5900-m^2), single-story professional park in Princeton, New Jersey. Other building types included a vocational-technical school in St. Paul, a nature center in Englewood, New Jersey, a pharmacy in New Mexico, a Girl Scout center near Philadelphia, and a retrofit of a historic building for apartments and shops in Providence.

In terms of commercial passive systems integration, the projects tended to fall into two categories: those applying primarily passive heating system concepts to residential-scale commercial buildings, and those applying integrated passive heating, cooling, and daylighting strategies to large internal-load-dominated commercial buildings. In both cases, the energy loads of the building were well characterized, and the choice of passive system was appropriate for meeting those loads. The passive systems were skillfully manipulated to blend with and enhance the architectural form of the building but also integrate with the mechanical and electrical systems.

The second National Passive Solar Design Awards were announced at the seventh National Passive Solar Conference held in September of 1982 in Knoxville, Tennessee. In the category of Commercial and Institutional Buildings, fourteen projects were selected for recognition, with three receiving design awards (*Passive Solar Journal* 1982, 1983c, 1983d).

8.4.2.8 State of California Under the leadership of the governor and the state architect, the State of California initiated, in early 1979, a major office-building construction program. A central element of this program was the reduction of nonrenewable energy use through the application of energy conservation and renewable energy sources. In total, seven large office buildings were designed during the period 1979–1982, four in Sacramento and one each in Long Beach, San Jose, and Santa Barbara.

The significance of this program, apart from the leadership role of the state in developing architectural innovation, lay in the exploratory research and design conducted to understand and respond to the energy situation of each building. Extensive analysis was performed to define the climatic conditions, the building energy problem, and the appropriate energy design response. The resulting architectural forms represented a total integration of heating, cooling, ventilation, and lighting needs. Together, the seven state office buildings established a new paradigm of office architecture, with energy considerations being a key determinant of architectural form. These buildings and the process by which they were designed became the prototypes that other office-building designers could learn from and copy. The process described by Ternoey et al. (1985) was by and large used by all of the design teams.

8.4.2.9 Energy Conscious Design Awards The Owens Corning Fiberglass (OCF) Energy Conscious Design Awards Program represents a unique opportunity to observe the trends of commercial passive systems integration over a ten-year period. The design awards program began in 1971 to recognize architectural excellence that was also energy conserving. Awards were made annually from 1972 until 1984, when the program was canceled.

Over a ten-year period from 1972 to 1981, sixty-five projects, representing over twenty building types, were selected for award. The projects were geographically dispersed throughout the continental United States. Over 60% of the projects exceeded 100,000 ft^2 (9000 m^2) in size, and the projects were evenly split between government and private sector ownership.

Shibley, Hartman, and Hart (1982) analyzed the OCF award winners, using the energy conservation and solar design categories defined in table 8.1. The general trends observed by the authors were the following (quoted):

1. The most commonly used energy strategies involved modification of the building envelope and improving the HVAC system.

2. Active solar has been commonly used in this (OCF) building sample and has shown a relative decline in recent years.

3. Passive solar, particularly the use of daylighting, has increased in popularity.

4. A number of strategies were intermittently reported and indicated no definitive trends.

5. Overall there appears to be some shift from mechanical system emphasis to more integrated architectural solutions.

The authors further observe that the number of energy-conserving strategies employed in a single building as well as the number of alternative,

Table 8.1
Categories of energy conservation and solar energy design strategies

General strategies	Specific energy features
Siting/berming	Orientation, siting, landscaping, berming, earth contact
Envelope	Multiple glazing, extra insulation, infiltration control, shading, reflective glass
HVAC	Heat recovery, air-to-air heat exchangers, heat pumps, variable air volume distribution, thermal storage
Lighting	Task lighting, high performance lighting
Controls	All types including time clocks, enthalpy controls, energy management systems, facility automation systems
Active solar	Space heating and/or cooling, domestic water heating
Passive solar heating	Direct gain (walls or floors), Trombe walls
Natural cooling	Natural ventilation, night flushing, evaporation, radiation, dehumidification
Daylighting	Window strategies (extra glass, placement), light shelves, clerestories, roof monitors, skylights, atria, sunspaces
Central plants/process	Multibuilding utility plants, cogeneration, industrial process heat

Source: USDOE 1982.

feasible options has increased over time (table 8.2). Also, these multiple strategies are being employed interdependently in integrated architectural solutions. The ability to design and analyze integrated energy-saving concepts improved considerably over the ten-year period of the OCF awards program.

8.4.2.10 Selected Buildings Apart from the major government-sponsored programs, design competitions, and design awards, commercial passive systems integration was advanced by a small number of individual building projects in which owners and design teams addressed energy considerations as a primary architectural design determinant. The buildings received national publicity and became prototypes of energy-responsive architectural solutions. In this section, nine buildings are selected as being pivotal in the evolution of commercial passive system integration in the United States. The buildings are presented in chronological order beginning in 1974.

1974, Pitkin County Airport Terminal, Aspen, Colorado
This was one of the first commercial buildings to fully integrate passive solar heating and daylighting concepts: roof apertures with Skylids® and

Table 8.2
General strategies: (a) early, (b) middle, and (c) recent years.

	72					73						74				75		
Project #	1	2	3	4	5	6	7	8	9	10	11	12	13	14	15	16	17	18
Siting/Beaming		•				•						•	•		•	•	•	
Envelope	•	•	•			•							•	•	•	•	•	
HVAC	•	•				•	•	•					•	•	•	•	•	
Controls	•		•			•		•						•			•	
Lighting	•		•			•										•	•	
Active Solar								•				•	•	•	•	•	•	•
Passive Heating																•		
Passing Cooling																		
Daylighting	•												•		•	•	•	
Central Plants			•				•									•		

(a)

	76						77							78									
Project #	19	20	21	22	23	24	25	26	27	28	29	30	31	32	33	34	35	36	37	38	39	40	41
Siting/Beaming		•		•		•	•	•		•							•			•		•	
Envelope	•						•			•				•	•	•	•	•	•	•	•	•	•
HVAC	•						•	•	•	•				•	•	•	•		•	•	•	•	
Controls	•								•								•					•	•
Lighting										•				•	•					•	•		
Active Solar			•		•	•		•						•	•		•				•	•	
Passive Heating																		•					
Passing Cooling		•					•			•				•	•			•					
Daylighting	•	•		•			•	•		•					•	•	•	•		•	•		•
Central Plants									•										•				

(b)

	79											80							81					
Project #	42	43	44	45	46	47	48	49	50	51	52	53	54	55	56	57	58	59	60	61	62	63	64	65
Siting/Beaming	•	•	•		•	•			•	•						•				•	•	•		•
Envelope	•	•	•	•				•	•	•	•	•		•	•	•	•		•	•	•	•	•	•
HVAC	•		•		•				•			•	•	•		•		•		•	•	•	•	
Controls		•										•	•	•	•					•	•		•	
Lighting		•			•	•	•		•			•	•		•					•			•	
Active Solar	•	•	•	•		•	•	•	•		•				•	•		•				•	•	•
Passive Heating		•								•							•			•		•	•	•
Passing Cooling	•	•	•	•		•		•		•	•			•					•			•	•	•
Daylighting	•	•	•	•	•	•	•	•		•		•	•	•	•	•		•	•	•	•	•	•	•
Central Plants																								

(c)

Figure 8.19
Pitkin County airport terminal, Aspen, Colorado.

direct-gain windows with Bead Wall® provide solar gains and daylighting to the 16,000-ft^2 (1500-m^2), single-story building. Massive interior construction was used for thermal storage. Shore 1976.

1979, Carey Arboretum Administration and Research Building, Millbrook, New York

This 28,000-ft^2 (2600-m^2) building incorporated numerous energy conservation and solar measures into the overall architectural design concept. Daylighting is used throughout as well as massive interior construction for thermal mass.

1981, Bateson State Office Building, Sacramento, California

This is the first large nonresidential building to consider passive solar design as primary design determinant. The 267,000-ft^2 (24,800-m^2) office building employed an interior courtyard (atrium), exposed structural mass, nighttime ventilation, and automatic and fixed shading to balance heating, cooling, and lighting needs year-round.

1981, Hooker Chemical and Plastics Corporation Office Building, Niagara Falls, New York
This "high-tech" 200,000-ft^2 (19,000-m^2) office building uses a double-skin wall to control heat gains and motor controlled louvers to control daylighting and shading. Photocell lighting controls maximize daylighting benefit. All building environmental and lighting systems use a computerized automatic control system.

1982, Justice Department Building, Sacramento, California
There is innovative integration of advanced mechanical, electrical, and control systems in this two-story, 300,000-ft^2 (28,000-m^2) building. Roof monitors, high windows, and exterior courtyards provide daylighting. Photocells control electric lighting. Heat from lights is extracted with return air. There are enthalpy and automatic controls for ventilating the building and cooling structural mass. Dubin 1982.

1982, Shell Oil Company Office Complex, Houston, Texas
This large, energy-efficient office complex employs many innovative concepts for integrating the energy-saving features into the overall architectural concept. Light shelves enclose the HVAC distribution system, interior windows light the central corridor, and an enclosed atrium provides light and common space for the interior offices. Ternoey et al. 1985.

1983, TVA Office Complex, Chattanooga, Tennessee
The 1979 design process included very rigorous analyses of energy conservation, passive solar, and daylighting strategies. It was one of the first projects to make extensive use of daylighting models to study the lighting performance of the central atrium and the proposed direct-beam daylighting system. This 750,000-ft^2 (70,000-m^2) building broke new ground in atrium design, daylighting concepts, lighting controls, and HVAC design. Ternoey et al. 1985.

1983, Prudential Insurance Company's Enerplex Office Buildings, Plainston, New Jersey
The complex consists of two 130,000-ft^2 (12,000-m^2) office buildings that were designed to explore numerous innovative concepts of energy efficiency. Daylighting was employed through roof monitors, atria, reflective blinds, and sawtooth windows. Nighttime ventilation of the building is used to cool structural mass. An innovative HVAC system includes ice-pond storage, earth cool tubes, and underground thermal storage.

1983, Lockheed Company Office Building, Sunnyvale, California
Like the TVA building, Lockheed's 600,000-ft^2 (56,000-m^2) office complex underwent extensive research and analysis of energy-saving concepts during the design process. The final design included a central enclosed atrium with a sawtooth roof structure, light shelves, and sloped interior ceilings to reflect daylight deep into the space. The design reconfirmed and refined many of the lessons learned from the TVA project.

8.4.3 Lessons Learned

Unlike residential passive systems integration, designers and researchers are still on the steep slope of the learning curve relative to commercial passive systems integration. Much has been learned and put into practice. However, more is still to be learned about the complex interactions that occur in commercial buildings between the occupants, the shell, the HVAC systems, and the lighting systems and controls. Based upon commercial passive systems integration research and design experience to date, the lessons learned are as follows:

• Energy loads, consumption, and costs must be characterized. Commercial building energy use is not intuitive. Energy loads, consumption, and cost must be carefully characterized to define the energy problem. Otherwise, inappropriate energy-saving concepts will likely to be chosen.

• Proven, reliable, and cost-effective conservation techniques should be used. A tight, well-insulated envelope, high-efficiency HVAC system, low-voltage lighting, and energy management and control systems are now commonplace in many new commercial building design projects.

• Daylighting is an effective energy-saving strategy. Although not as widespread as conservation techniques, daylighting (with photocell-controlled stepped lighting) is gaining acceptance as a viable, effective, and economically attractive energy-saving strategy. When the reduction in the cooling load is considered, daylighting will become the single most important energy-saving and architectural concept.

• Energy efficient, passive commercial buildings need not cost more. While many of the first-generation passive commercial buildings cost 5–10% more than similar conventional buildings, the trend is to lower added cost because of the passive systems. This is due to improved design integration of the passive systems components with the other architectural elements of the building.

• Greater potential for energy savings occurs during the early stages of the design process. Unless energy saving strategies are considered early in the design process, the potential for integrating them into the design later is reduced. Obviously, architectural approaches (such as atria and daylighting) must be considered when the design concept is in the formative stage. Mechanical concepts can be included later, but their energy-saving potential will be reduced.

• Multidisciplinary skills are required. All disciplines—architectural, mechanical, electrical, and structural—must be involved in concept development to achieve cost-effective integration of the passive systems.

8.5 Future Direction

Although passive systems integration research has made a major contribution to the advancement and use of passive solar design in the United States since 1972, considerable work remains to be done. Listed below are some of the major areas where passive system integration research is needed.

• Improved Analytical Techniques. The optimum integration of passive systems is dependent on the ability to analyze the interaction of numerous passive architectural elements. While many building energy analysis techniques exist, most can analyze only the simplest of interactions among the various passive architectural elements. Most are unable to analyze combined active, passive, and hybrid solar techniques. Research is needed to improve the ability of building energy analysis tools to analyze the dynamic interaction of energy conservation, passive/hybrid solar, daylighting, and HVAC systems. These analytical techniques must be relatively fast and easy to use. Given the growth and widespread use of computer-aided design and drafting (CADD), the integration of building energy analysis software with CADD may be an appropriate way to proceed.

• Comprehensive Design Guidelines. Despite the years of building energy research, satisfactory location-specific design guidelines do not exist, especially for nonresidential buildings. Since design guidelines have proven to be an effective means of promoting the use of passive solar design strategies, more research is needed to evaluate and establish design guidelines methodologies and to prepare residential and commercial building design guidelines for cities where significant construction activity is likely to occur. Issues of thermal analysis, modeling assumptions, economic analysis, and building integration should be considered when developing or evaluating design guideline procedures. Clear and concise design guideline booklets should be prepared for use by architects, engineers, homebuilders, realtors, and lenders.

• Diagnostic Procedures. Diagnostic procedures are needed to evaluate the in-situ performance of passive solar buildings and to isolate design or operation problems. The results from such a procedure will identify problems of passive design integration and suggest new avenues of research.

• More Sophisticated Research Facilities. More sophisticated passive system integration research facilities must be developed. This is especially the case for commercial buildings. To date we have relied primarily on analysis simulation. Only very limited experimental data has been acquired from a few existing passive solar commercial buildings, and these buildings do not permit highly controlled investigations of the dynamic interaction between a variety of passive system elements. What is needed is a reconfigurable, full-scale, multistory, multizone test facility. Such a facility would permit definitive measurements of passive component interactions and the testing of advanced component and system concepts.

• Advanced Materials Assessment. A revolution in materials and products is occurring in the building industry, for example, in high performance glazings (switchable optical properties and low-emissivity coatings) and improved thermal-storage materials (solid-solid phase-change materials and embedded-wax gypsum board). The commercial availability of new materials and products will dramatically alter previous concepts of passive systems integration. Continued research should be conducted to guide the development of new materials and products and to assess their impact on building performance, comfort, and cost, so that building designers can effectively integrate these materials and products when available.

• Zero-Purchased-Energy Buildings. Previous and ongoing research suggests that zero-purchased-energy buildings may be practical in some locations in a few years. Weekly or monthly energy-storage systems, dessicant cooling, and other advanced passive system concepts may change the cost-benefit ratio of a zero-purchased-energy building.

• Indoor Air Quality. The maintenance or improvement of indoor air quality must be part of any passive systems integration research.

• Demonstrations and Design Competitions. Demonstration projects and architectural design competitions have proven to be an effective way to develop and disseminate knowledge on innovative passive system integration concepts. Such demonstrations and competitions should occur regularly and be tied to the application and integration of the latest advances.

References

American Institute of Architects Research Corporation. 1977. *Capturing the Sun: Designs from an Architectural Student Competition.* Washington, DC: AIA/RC.

American Institute of Architects Research Corporation. 1978. *Regional Guidelines for Building Passive Energy Conserving Homes.* Washington, DC: U.S. Department of Housing and Urban Development.

American Institute of Architects Research Corporation. 1980. *A Survey of Passive Solar Homes.* Washington, DC: U.S. Department of Housing and Urban Development.

American Institute of Architects Research Corporation. 1981. *A Survey of Passive Solar Buildings.* Washington, DC: U.S. Department of Housing and Urban Development.

American Society of Landscape Architects Foundation. 1977. *Landscape Planning for Energy Conservation.* Respn, VA Environmental Design Press.

Anderson, B. 1976. The solar heating and cooled Tyrell house. *Proc. Passive Solar Heating and Cooling Conference and Workshop,* Albuquerque, NM, May 18–19, 1976. LA-6637-C. Los Alamos, NM: Los Alamos Scientific Laboratory, pp. 150–152.

Andrews, S., and R. Snyder. 1982. Passive solar retrofits in Denver: highlights of five effective and elegant solutions. *Proc. 7th National Passive Solar Conference,* Knoxville, TN, August 30–September 1, 1982. Newark, DE: American Section of the International Solar Energy Society, pp. 231–206.

Arasteh, D., and L. Lindsey. 1982. Large scale (residential) passive solar retrofits: a nationwide survey. *Proc. 7th National Passive Solar Conference,* Knoxville, TN, August 30–September 1, 1982. Newark, DE: American Section of the International Solar Energy Society, pp. 203–206.

Architects Taos, The. 1975. *Solar Energy and Housing: Three Passive Prototypes* (unpublished report).

Architectural Energy Corporation. 1986. *Design Guidelines for Energy Efficient Passive Solar Homes.* Denver, CO: AEC.

Arizona State University College of Architecture. 1975. *Solar-Oriented Architecture.* Washington, DC: American Institute of Architects Research Corporation.

Association of Collegiate Schools of Architecture. 1980. *Design & Energy: Results of a National Student Design Competition.* Washington, DC: U.S. Department of Energy and the Brick Institute of America.

Baccei, B., B. Parkin, M. McCray, D. Bates, and P. Wrenn. 1980. Denver metro home builders program: a pilot program to move passive solar into the mainstream of residential construction. *Proc. 5th National Passive Solar Conference*, Amherst, MA, October 19–26, 1980. Newark, DE: American Section of the International Solar Energy Society, pp. 583–585.

Bainbridge, D., J. Corbett, and J. Horfarce. 1979. *Village Homes' Solar House Designs: A Collection of 43 Energy-Conscious House Designs.* Emmaus, PA: Rodale Press.

Balcomb, J. D. 1976. Summary of the Passive Solar Heating and Cooling Conference. *Proc. Passive Solar Heating and Cooling Conference and Workshop.* LA-6637-C. Albuquerque, NM, May 18–19, 1976, Los Alamos, NM: Los Alamos Scientific Laboratory, pp. 1–4.

Balcomb, J. D. 1980. Conservation and solar: working together. *Proc. 5th National Passive Solar Conference*, Amherst, MA, October 19–26, 1980. Newark, DE: American Section of the International Solar Energy Society, pp. 44–50.

Balcomb, J. D. 1983. Conservation and solar guidelines. *Proc. 8th National Passive Solar Conference*, Santa Fe, NM, September 7–9, 1983. Boulder, CO: American Solar Energy Society, pp. 117–122.

Balcomb, J. D., D. Barley, R. McFarland, J. Perry, W. Wray, and S. Noll. 1980. *Passive Solar Design Handbook.* Vol. 2: *Passive Solar Design Analysis.* DOE/CS-0127/2. Washington, DC: U.S. Department of Energy.

Balcomb, J. D., R. W. Jones, R. D. McFarland, and W. O. Wray. 1984. *Passive Solar Heating Analysis: A Design Manual.* Atlanta, GA: ASHRAE.

Balcomb, J. D., C. E. Kosiewicz, G. S. Lazarus, R. D. McFarland, and W. O. Wray. 1982. *Passive Solar Design Handbook.* Vol. 3: *Passive Solar Design Analysis.* Edited by R. W. Jones. DOE/CS-0127/3. Washington, DC: U.S. Department of Energy; Boulder, CO: American Solar Energy Society, 1983.

Balcomb, J. D., and R. D. McFarland. 1978. A simple empirical method for estimating the performance of a passive solar heated building of the thermal storage wall type. *Proc. 2d National Passive Solar Conference*, Philadelphia, PA, March 16–18, 1978. Newark, DE: American Section of the International Solar Energy Society, pp. 377–389.

Balcomb, S., S. Pyde, M. Conkling, and L. Vigil. 1983. Passive solar design guidelines and evaluation criteria for buildings. *Proc. 8th National Passive Solar Conference*, Santa Fe, NM, September 7–9, 1983. Boulder, CO: American Solar Energy Society, pp. 623–627.

Banwell, White & Arnold, Inc. 1978. White Mountain School Living Center, Littleton, New Hampshire. *Proc. 2d National Passive Solar Conference*, Philadelphia, PA, March 16–18, 1978. Newark, DE: American Section of the International Solar Energy Society, pp. 173–179.

Barrett, D. 1983. Building as hearth: the articulation of building mass in Colorado architecture. *Proc. 8th National Passive Solar Conference*, Santa Fe, NM, September 7–9, 1983, Boulder, CO: American Solar Energy Society, pp. 921–924.

Booze Allen & Hamilton, Burt Hill Kosar Rittelman, and William I. Whiddon & Associates. 1981. Economic analysis of selected passive solar commercial buildings. *Proc. 6th National Passive Solar Conference*, Portland, OR, September 8–12, 1981. Newark, DE: American Section of the International Solar Energy Society, pp. 631–635.

Booze Allen & Hamilton, Burt Hill, Kosar Rittelman, F. Sizemore, W. I. Whiddon and Associates, and Hart McMurphy & Parks. 1983. *Passive Solar Commercial Building Program, Case Studies.* Washington, DC: U.S. Department of Energy.

Born, B., and E. Chase. 1979. Thomas Village three prototype passive homes for the Tennessee Valley. *Proc. 3d National Passive Solar Conference*, San Jose, CA, January 11–13, 1979. Newark, DE: American Section of the International Solar Energy Society, pp. 877–880.

Bowen, A., E. Clark, and K. Labs, eds. 1981. *Passive Cooling: Proc. International Passive and Hybrid Cooling Conference*, Miami Beach, FL, November 11–13, 1981. Newark, DE: American Section of the International Solar Energy Society.

Brant, S., and M. Holtz. 1981. Low cost passive retrofits for new and existing mobile homes. *Proc. 6th National Passive Solar Conference*, Portland, OR, September 8–12, 1981. Newark, DE: American Section of the International Solar Energy Society, pp. 788–792.

Bryan, H., and V. Bazjanac. 1983. Energy-conscious design features for new California state office building—site 1-C. *Proc. 8th National Passive Solar Conference*, Santa Fe, NM, September 7–9, 1983. Boulder, CO: American Solar Energy Society, pp. 1023–1028.

Burr, P. A., and R. J. Fowler. 1981. South Florida retrofit for semi-automatic living. *Passive Cooling: Proc. International Passive and Hybrid Cooling Conference*, Miami Beach, FL, November 11–13, 1981. Newark, DE: American Section of the International Solar Energy Society, pp. 112–117.

Burt Hill Kosar Rittelman & Associates. 1984. *Anthology of Articles and Papers: Passive Solar Commercial Buildings Program.* Washington, DC: U.S. Department of Energy.

Caudill, W. W. 1951. Architect for children. *Oklahoma A & M College Magazine*, January 1951, pp. 7–11.

Cole, W. J., and J. J. Barron. 1980. The New York state passive design/build competition and follow-on activities. *Proc. 5th National Passive Solar Conference*, Amherst, MA, October 19–26, 1980. Newark, DE: American Section of the International Solar Energy Society, pp. 599–603.

Continuum Team, The. 1975. *Design of Multi-Family Solar Dwellings for Minneapolis and Phoenix* (unpublished report).

Cook, J. 1984a. *Award Winning Passive Solar Building Designs.*

Cook, J. 1984b. *Award-Winning Passive Solar Houses.* Pownal, VT: Garden Way Publishing.

Coonley, D. R. 1979. Deadweight on toothpicks, or putting concrete & brick on wood (and living to tell about it). *Proc. 4th National Passive Solar Conference*, Kansas City, MO, October 3–5, 1979. Newark, DE: American Section of the International Solar Energy Society, pp. 666–669.

Corson, B. A. 1978. An energy efficient office building for the state of California. *Proc. 2d National Passive Solar Conference*, Philadelphia, PA, March 16–18, 1978. Newark, DE: American Section of the International Solar Energy Society, pp. 233–239.

Davis, S. 1981. *Designing for Energy Efficiency: A Study of Eight California State Office Buildings.* Berkeley, CA: Department of Architecture, University of California.

Davis, W. B. 1979. Results of the California Passive Design Competition. *Proc. 4th National Passive Solar Conference*, Kansas City, MO, October 3–5, 1979. Newark, DE: American Section of the International Solar Energy Society, pp. 83–86.

Derickson, R. G., and K. S. Sadlon. 1981. Flexibilities in passive design: examining some limiting solar myths. *Proc. 6th National Passive Solar Conference*, Portland, OR, September 8–12, 1981. Newark, DE: American Section of the International Solar Energy Society, pp. 333–337.

DeWitt, S. 1980. The Utica Street residence: a passive solar retrofit. *Proc. 5th National Passive Solar Conference*, Amherst, MA, October 19–26, 1980. Newark, DE: American Section of the International Solar Energy Society, pp. 1312–1313.

Dietz, A. H., and E. C. Czapek. 1950. *Solar Heating of Houses by Vertical South-Wall Storage Panels, Heating and Piping and Air Conditioning*, pp. 118–125.

Dubin, F. S. 1982. *Department of Justice office building, Sacramento, California. Proc. 7th National Passive Solar Conference*, Knoxville, TN, August 30–September 1, 1982. Newark, DE: American Section of the International Solar Energy Society, pp. 267–270.

Duffield, J. 1981. Joint optimization of solar and superinsulation in a cold climate. *Proc. 6th National Passive Solar Conference*, Portland, OR, September 8–12, 1981. Newark, DE: American Section of the International Solar Energy Society, pp. 228–231.

Ehrenkrantz Group, The. 1978. *Cost Benefit Analysis of Passive Solar Design Alternatives: New Office Building Temperate Climate.* Washington, DC: U.S. Department of Energy.

Emery, A. F., B. R. Johnson, D. R. Heerwagen, and C. J. Kippenhan. 1981. Assessing the benefit-costs of employing alternative shading devices to reduce cooling loads for three climates. *Passive Cooling: Proc. International Passive and Hybrid Cooling Conference*, Miami Beach, FL, November 11–13, 1981. Newark, DE: American Section of the International Solar Energy Society, pp. 417–421.

Energy Research and Development Administration. 1976. *Interim Report: National Program Plan for Research and Development in Solar Heating and Cooling.* Washington, DC: ERDA, p. 186.

Fisher, W. J., and R. C. Shibley. 1980. Passive solar opportunities in commercial buildings: technical insights and a model for professional development. *Proc. 6th National Passive Solar Conference*, Portland, OR, September 8–12, 1981. Newark, DE: American Section of the International Solar Energy Society, pp. 616–620.

Franklin Research Center. 1979. *The First Passive Solar Home Awards.* Washington, DC: U.S. Department of Housing and Urban Development.

Funaro, G., and A. Cellie. 1985. The passive solar Italian schools. *Proc. 10th National Passive Solar Conference*, Raleigh, NC, October 15–20, 1985. Boulder, CO: American Solar Energy Society, pp. 194–198.

Garrison, M. 1978. The double box. *Proc. 2d National Passive Solar Conference*, Philadelphia, PA, March 16–18, 1978. Newark, DE: American Section of the International Solar Energy Society, pp. 180–184.

Giffels Associates, Inc. 1974. *Solar Energy and Housing Design Concepts* (unpublished report).

Gordon, H. T., and W. J. Fisher. 1982. Commercial passive solar buildings—performance evaluation. *Proc. 7th National Passive Solar Conference*, Knoxville, TN, August 30–September 1, 1982. Newark, DE: American Section of the International Solar Energy Society, pp. 531–536.

Haggard, K. 1976. First cost economic evaluation of the Atascadero Skytherm House. *Proc. Passive Solar Heating and Cooling Conference and Workshop*, Albuquerque, NM, May 18–19, 1976. LA-6637-C. Los Alamos, NM: Los Alamos Scientific Laboratory, pp. 250–253.

Hagman, J. T., and D. L. Graybeal. 1980. Obermeyer headquarters and warehouse. *Proc. 6th National Passive Solar Conference*, Portland, OR, September 8–12, 1981. Newark, DE: American Section of the International Solar Energy Society, pp. 597–601.

Hammett, W. S. 1985. Decorating the NCSU Solar House—an integration of interior design with passive solar concepts. *Proc. 10th National Passive Solar Conference*, Raleigh, NC, October 15–20, 1985. Boulder, CO: American Solar Energy Society, pp. 299–302.

Hammond, J. 1976. Winters House. *Proc. Passive Solar Heating and Cooling Conference and Workshop*, Albuquerque, NM, May 18–19, 1976. LA-6637-C. Los Alamos, NM: Los Alamos Scientific Laboratory, pp. 153–156.

Hannifan, M., C. Christensen, and R. Perkins. 1981. Comparison of residential window distributions and effects of mass and insulation. *Proc. 6th National Passive Solar Conference*, Portland, OR, September 8–12, 1981. Newark, DE: American Section of the International Solar Energy Society, pp. 213–217.

Hart, G. K. 1984. *Measuring Economic Performance in Non-Residential Buildings.*

Hartkopf, V., and V. Loftness. 1985. Energy-efficient urban infill homes in Pittsburgh reveal outstanding alternative to fuel subsidies. *Proc. 10th National Passive Solar Conference*, Raleigh, NC, October 15–20, 1985. Boulder, CO: American Solar Energy Society, pp. 493–497.

Hartmann, R. J. 1985. Cedarwood Solar Park: a passive solar subdivision. *Proc. 10th National Passive Solar Conference*, Raleigh, NC, October 15–20, 1985. Boulder, CO: American Solar Energy Society, pp. 680–685.

Hay, H. 1976. Atascadero residence. *Proc. Passive Solar Heating and Cooling Conf. and Workshop*, Albuquerque, NM, May 18–19, 1976. LA-6637-C. Los Alamos, NM: Los Alamos Scientific Laboratory, pp. 101–107.

Holton, J. K., and S. H. Pansky. 1983. Energy strategy for small office design. *Proc. 8th National Passive Solar Conference*, Santa Fe, NM, September 7–9, 1983. Boulder, CO: American Solar Energy Society, pp. 413–418.

Holtz, M. J. 1976. *Solar Dwelling Design Concepts.* Washington, DC: U.S Department of Housing and Urban Development.

Holtz, M. J. 1983. International Energy Agency task VIII: description and selected results. *Proc. 8th National Passive Solar Conference*, Santa Fe, NM, September 7–9, 1983. Boulder, CO: American Solar Energy Society, pp. 889–894.

Holtz, M. J., ed. 1988–1991. Design Information Booklet Series, vols. 1–8. Washington, DC: U.S. Government Printing Office.

Holtz, M. J., W. Place, and R. Kammerud. 1979. *A Classification Scheme for the Common Passive and Hybrid Heating and Cooling Systems.* Golden, CO: Solar Energy Research Institute.

Joint Venture. 1974. Here Comes the Sun. (unpublished report).

Jones, R. W. 1981. Summer heat gain control in passive solar heated buildings: fixed horizontal overhangs. *Passive Cooling: Proc. International Passive and Hybrid Cooling Conference*, Miami Beach, FL, November 11–13, 1981. Newark, DE: American Section of the International Solar Energy Society, pp. 402–406.

Kantrowitz, M. 1983. Occupant effects and interactions in (passive) solar commercial buildings: preliminary findings from the US DOE Passive Solar Commercial Buildings Program. *Proc. 8th National Passive Solar Conference*, Santa Fe, NM, September 7–9, 1983. Boulder, CO: American Solar Energy Society, pp. 433–438.

Kelbaugh, D. 1976. Kelbaugh House. *Proc. Passive Solar Heating and Cooling Conference and Workshop*, Albuquerque, NM, May 18–19, 1976. LA-6637-C. Los Alamos, NM: Los Alamos Scientific Laboratory, pp. 119–128.

Kelbaugh, D. 1980. The seven principles of energy efficient architecture. *Proc. 5th National Passive Solar Conference*, Amherst, MA, October 19–26, 1980. Newark, DE: American Section of the International Solar Energy Society, pp. 12–16.

Keller, B. M., W. C. Johnson, and A. V. Sedrick. 1978. Passive solar heated warehouse. *Proc. 2d National Passive Solar Conference*, Philadelphia, PA, March 16–18, 1978. Newark, DE: American Section of the International Solar Energy Society, pp. 52–56.

Kroner, W. M. 1981. A passive solar commercial building: new applications of an old technology leads to architectural innovations. *Proc. 6th National Passive Solar Conference*, Portland, OR, September 8–12, 1981. Newark, DE: American Section of the International Solar Energy Society, pp. 703–706.

Kurkowski, T. L., and S. E. Ternoey. 1981. An overview of designs from the DOE Passive Commercial Buildings Program. *Proc. 6th National Passive Solar Conference*, Portland, OR, September 8–12, 1981. Newark, DE: American Section of the International Solar Energy Society, pp. 636–640.

Kurkowski, T. L., and H. T. Gordon. 1981. Cooling Strategies in the DOE Passive Solar Commercial Buildings Program. *Passive Cooling: Proc. International Passive and Hybrid Cooling Conference*, Miami Beach, FL, November 11–13, 1981. Newark, DE: American Section of the International Solar Energy Society, pp. 492–496.

Lambeth, J. 1978. Direct gain passive design Delap residence. *Proc. 2d National Passive Solar Conference*, Philadelphia, PA, March 16–18, 1978. Newark, DE: American Section of the International Solar Energy Society, pp. 43–46.

Leach, J. W., and S. C. DeWitt. 1979. A moderate cost passive solar housing design. *Proc. 3d National Passive Solar Conference*, San Jose, CA, January 11–13, 1979. Newark, DE: American Section of the International Solar Energy Society, pp. 805–808.

Lee, S. J., D. Kelbaugh, and E. Holland. 1983. Senior citizen housing project at Roosevelt, NJ. *Proc. 8th National Passive Solar Conference*, Santa Fe, NM, September 7–9, 1983. Boulder, CO: American Solar Energy Society, pp. 1029–1034.

Levin, D. 1978. A model design process for passive solar architecture. *Proc. 2d National Passive Solar Conference*, Philadelphia, PA, March 16–18, 1978. Newark, DE: American Section of the International Solar Energy Society, pp. 57–61.

Londe Parker Michels. 1981. *Passive Solar Retrofit Guidebook*. Washington, DC: U.S. Department of Energy.

Lutha, R., P. G. Rockwell, and W. J. Fisher. 1983. The DOE Passive Solar Commercial Building Program: preliminary results of performance evaluation. *Proc. 8th National Passive Solar Conference*, Santa Fe, NM, September 7–9, 1983. Boulder, CO: American Solar Energy Society, pp. 397–400.

Malt, H. L., and J. Ripoli. 1981. A contemporary naturally cooled residence in Coconut Grove, South Florida. *Passive Cooling: Proc. International Passive and Hybrid Cooling Conference*, Miami Beach, FL, November 11–13, 1981. Newark, DE: American Section of the International Solar Energy Society, pp. 105–109.

Massdesign. 1975. *Solar Heated Houses for New England* (unpublished report).

Mazria, E. 1979. *The Passive Solar Energy Book*. Emmaus, PA: Rodale Press.

McCluney, R., and S. Chandra. 1984. A comparison of window shading strategies for heat gain prevention. *Proc. 9th National Passive Solar Conference*, Columbus, OH, September 1984. Boulder, CO: American Solar Energy Society, pp. 414–419.

McDonald, C. L., and G. A. Tsongas. 1981. A comparative analysis of conservation and passive solar strategies in multi-family residences. *Proc. 6th National Passive Solar Conference*, Portland, OR, September 8–12, 1981. Newark, DE: American Section of the International Solar Energy Society, pp. 237–241.

Michels, T. I. 1979. Results: the retrofit of an existing masonry home for passive space heating. *Proc. 4th National Passive Solar Conference*, Kansas City, MO, October 3–5, 1979. Newark, DE: American Section of the International Solar Energy Society, pp. 564–565.

Nichols, W. D. 1976. Unit 1, First Village. *Proc. Passive Solar Heating and Cooling Conference and Workshop*, Albuquerque, NM, May 18–19, 1976. LA-6637-C. Los Alamos, NM: Los Alamos Scientific Laboratory, pp. 137–149.

Noble, E., and B. Lofchie. 1985. Massachusetts multi-family passive solar program: experience with performance monitoring of multi-family buildings. *Proc. 10th National Passive Solar Conference*, Raleigh, NC, October 15–20, 1985. Boulder, CO: American Solar Energy Society, pp. 315–320.

Noll, S., and M. Thayer. 1979. Passive solar, auxiliary heat, and building conservation optimization: a graphical analysis. *Proc. 4th National Passive Solar Conference*, Kansas City,

MO, October 3–5, 1979. Newark, DE: American Section of the International Solar Energy Society, pp. 128–131.

Northeast Solar Energy Center. 1980. *Five Award-Winning Solar Designs for Multi-Family Housing.* Washington, DC: U.S. Department of Energy.

Palmiter, L. 1981. Optimum conservation for northwest homes. *Proc. 6th National Passive Solar Conference*, Portland, OR, September 8–12, 1981. Newark, DE: American Section of the International Solar Energy Society, pp. 223–227.

Parkin, B. 1981. Solar Homebuilders Program—getting builders to build solar. *Proc. 6th National Passive Solar Conference*, Portland, OR, September 8–12, 1981. Newark, DE: American Section of the International Solar Energy Society, pp. 574–577.

Passive Solar Journal. 1983a. San Francisco residence and remodel. *Passive Solar Journal* 2(1) (Winter): 8–13.

Passive Solar Journal. 1983b. Wildwood Place residential townhomes. *Passive Solar Journal* 2(1) (Winter): 14–17.

Passive Solar Journal. 1983c. The Society for the Protection of New Hampshire Forests Conservation Center. *Passive Solar Journal* 2(1) (Winter): 18–23.

Passive Solar Journal. 1983d. Government Service Insurance System headquarters building. *Passive Solar Journal* 2(1) (Winter): 24–28.

Peck, J. R., and B. W. Reams. 1981. Analysis of a passive solar atrium for a high-rise office building. *Passive Cooling: Proc. International Passive and Hybrid Cooling Conference*, Miami Beach, FL, November 11–13, 1981. Newark, DE: American Section of the International Solar Energy Society, pp. 457–461.

Perry, J. E., Jr. 1976. The Wallasey School. *Proc. Passive Solar Heating and Cooling Conference and Workshop*, Albuquerque, NM, May 18–19, 1976. LA-6637-C. Los Alamos, NM: Los Alamos Scientific Laboratory, pp. 223–237.

Pfister, P. J. 1978. The Lindberg residence: a DOE funded hybrid solar house. *Proc. 2d National Passive Solar Conference*, Philadelphia, PA, March 16–18, 1978. Newark, DE: American Section of the International Solar Energy Society, pp. 122–127.

Raskin, S., and J. Lenker. 1982. *Design Assistance Package: Pacific Telephone New Headquarters Project, California Energy Commission.*

Real Estate Research Corporation. 1982a. *Passive Solar Homes in the Marketplace, Final Report Findings of the 1978–79 Passive Grant Awards.* Volume I, Discussion, Winter 1981–82. Washington, DC: U.S. Department of Housing and Urban Development, pp. 1–4.

Real Estate Research Corporation. 1982b. *Passive Solar Homes in the Marketplace, Final Report Findings of the 1978–79 Passive Grant Awards.* Volume II, Data: Winter 1981–82. Washington, DC: U.S. Department of Housing and Urban Development, pp. 1–2.

Reeder, B. C., and C. Merkezas. 1983a. Solar elements as classical counterpoint: passive solar design and the revitalization of historic buildings. *Proc. 8th National Passive Solar Conference*, Santa Fe, NM, September 7–9, 1983. Boulder, CO: American Solar Energy Society, pp. 983–988.

Reeder, B. C., and C. Merkezas. 1983b. Passive solar strategies as a logic for improved architectural design: two prototypes for modular housing. *Proc. 8th National Passive Solar Conference*, Santa Fe, NM, September 7–9, 1983. Boulder, CO: American Solar Energy Society, pp. 993–998.

Reeder, B. C., and C. Merkezas. 1985. Evolution of a passive solar housing prototype. *Proc. 10th National Passive Solar Conference*, Raleigh, NC, October 15–20, 1985. Boulder, CO: American Solar Energy Society, pp. 199–204.

Rogers, B. T. 1976. Some performance estimates for the Wright house—Santa Fe, New Mexico. *Proc. Passive Solar Heating and Cooling Conference and Workshop*, Albuquerque, NM, May 18–19, 1976. LA-6637-C. Los Alamos, NM: Los Alamos Scientific Laboratory, pp. 189–199.

Rouse, R. E. 1980. Passive solar is cost-effective—observations on financing conservation and passive solar features in public sector housing. *Proc. 5th National Passive Solar Conference*, Amherst, MA, October 19–26, 1980. Newark, DE: American Section of the International Solar Energy Society, pp. 697–701.

Rouse, R. E., and E. C. Noble. 1981. The Massachusetts multi-family passive program—recent activities and findings. *Proc. 6th National Passive Solar Conference*, Portland, OR, September 8–12, 1981. Newark, DE: American Section of the International Solar Energy Society, pp. 590–594.

RTL, Inc. 1975. *Mobile Home Solar Energy Systems Development* (unpublished report).

Sandia Laboratories. 1979. *Passive Solar Buildings*. SAND 79-0824, Albuquerque, NM: Sandia Laboratories.

Saunders, N. B. 1976. Weston residence. *Proc. Passive Solar Heating and Cooling Conference and Workshop*, Albuquerque, NM, May 18–19, 1976. LA-6637-C. Los Alamos, NM: Los Alamos Scientific Laboratory, pp. 90–100.

Schiff, M. 1978. House at Crooked Creek. *Proc. 2d National Passive Solar Conference*, Philadelphia, PA, March 16–18, 1978. Newark, DE: American Section of the International Solar Energy Society, pp. 38–42.

Schmitt, E. A. 1976. The Sun House. *Proc. Passive Solar Heating and Cooling Conference and Workshop*, Albuquerque, NM, May 18–19, 1976. LA-6637-C. Los Alamos, NM: Los Alamos Scientific Laboratory, pp. 111–118.

Solar Energy Research Institute. 1980. *The Design of Energy Responsive Buildings: Building Case Studies*. Golden, CO: SERI.

Solar Energy Research Institute. 1983. Passive Solar Manufactured Buildings: Design, Construction, and Class B Results, Golden, CO: SERI.

Solar Energy Research Institute. 1984. *Passive Solar Homes: 20 Case Studies*. Golden, CO: SERI.

Shibley, R. G., and D. F. Weaver. 1982. A monograph on passive and low energy alternatives in the United States. *Proc. First International Passive and Low Energy Alternatives I Conference*. pp. 15.71–15.104.

Shore, R. 1976. Pitkin County's airport terminal; or, where did all the natural gas go. *Proc. Passive Solar Heating and Cooling Conference and Workshop*, Albuquerque, NM, May 18–19, 1976. LA-6637-C. Los Alamos, NM: Los Alamos Scientific Laboratory, pp. 129–131.

Smeallie, P. H. 1980. Design & Energy: results of a national student design competition. *Proc. 5th National Passive Solar Conference*, Amherst, MA, October 19–26, 1980. Newark, DE: American Section of the International Solar Energy Society, pp. 804–808.

Stoops, J. L., J. J. Deringer, S. Moreno, and H. P. Misuriello. 1984. *Summary Report: The BEPS Redesign of 168 Commercial Buildings*. Pacific Northwest Laboratory.

Stoops, J. L. 1983. A baseline for energy design. *Proc. 8th National Passive Solar Conference*, Santa Fe, NM, September 7–9, 1983. Boulder, CO: American Solar Energy Society, pp. 407–412.

Stuby, F. C. 1983. A few examples of passive solar architecture in Switzerland. *Proc. 8th National Passive Solar Conference*, Santa Fe, NM, September 7–9, 1983. Boulder, CO: American Solar Energy Society, pp. 977–982.

Sullivan, P. W., and K. R. Malcolm. 1979. An analysis of the incremental construction costs of passive solar buildings. *Proc. 4th National Passive Solar Conference*, Kansas City, MO, October 3–5, 1979. Newark, DE: American Section of the International Solar Energy Society, pp. 670–673.

Systems Consultants, Inc. 1976a. *National Program for Solar Heating and Cooling of Buildings, Project Data Summaries.* Vol. 1: *Commercial and Residential Demonstrations.* Washington, DC: Energy Research and Development Administration.

Systems Consultants, Inc. 1976b. *National Program for Solar Heating and Cooling of Buildings, Project Data Summaries.* Vol. 2: *Demonstration Support.* Washington, DC: Energy Research and Development Administration.

Systems Consultants, Inc. 1976c. *National Program Plan for Solar Heating and Cooling of uildings, Project Summaries.* Vol. 3: *Research and Development.* Washington, DC: Energy Research and Development Administration.

Taylor, R. D. 1979. Energy savings and incremental costs for conservation and passive solar construction. *Proc. 4th National Passive Solar Conference*, Kansas City, MO, October 3–5, 1979. Newark, DE: American Section of the International Solar Energy Society, pp. 674–676.

Ternoey, S., L. Bickle, C. Robbins, R. Busch, and K. McCord. 1985. *The Design of Energy-Responsive Commercial Buildings.* New York: Wiley.

Terry, K. 1976. The Karen Terry house. *Proc. Passive Solar Heating and Cooling Conference and Workshop*, Albuquerque, NM, May 18–19, 1976. LA-6637-C. Los Alamos, NM: Los Alamos Scientific Laboratory, pp. 132–136.

Theis, C. C. 1982. The Waugh residence: a passive solar retrofit. *Proc. 7th National Passive Solar Conference*, Knoxville, TN, August 30–September 1, 1982. Newark, DE: American Section of the International Solar Energy Society, pp. 197–202.

Thompson Hancock Witte & Associates, Inc. 1980. *Passive Retrofit Handbook.* Washington, DC: U.S. Department of Energy.

Total Environmental Action. 1974. *Solar Energy Home Design in Four Climates.* Washington, DC: U.S. Department of Energy.

Total Environmental Action. 1979. *The Brookhaven House.* Washington, DC: U.S. Department of Energy.

Total Environmental Action. 1980. *Passive Solar Design Handbook.* Vol. 1: *Passive Solar Design Concepts.* DOE/CS-0127/1. Washington, DC: U.S. Department of Energy.

University of Detroit School of Architecture. 1975. *Solar Energy Proposal for Mobile Homes* (unpublished report).

U.S. Department of Energy. 1978. *Solar Heating and Cooling: Research and Development Project Summaries.* Washington, DC: USDOE.

U.S. Department of Energy. 1979. *Passive Solar Commercial Buildings Design Assistance and Demonstration: Program Opportunity Notice.* Washington, DC: USDOE.

U.S Department of Housing and Urban Development. 1976. *Solar Heating and Cooling Demonstration Program: A Description Summary of HUD Solar Residential Demonstrations Cycle 1.* Washington, DC: USHUD.

U.S. Department of Housing and Urban Development. 1977a. *Solar Heating and Cooling Demonstration Program: A Description Summary of HUD Cycle 2 Solar Residential Projects—Fall 1976.* Washington, DC: USHUD.

U.S. Department of Housing and Urban Development. 1977b. *Solar Heating and Cooling Demonstration Program: A Description Summary of HUD Cycle 3 Solar Residential Projects—Summer 1977.* Washington, DC: USHUD.

U.S Department of Housing and Urban Development. 1979. *Solar Heating and Cooling Demonstration Program: A Descriptive Summary of HUD Cycle 4 and 4A Solar Residential Projects*. Washington, DC: USHUD.

Watson, D. 1975. *Innovation in Solar Thermal House Design* (unpublished report).

Watson, D. 1978. Beyond passive design: toward an integrative conservation technology. *Proc. 2d National Passive Solar Conference*, Philadelphia, PA, March 16–18, 1978. Newark, DE: American Section of the International Solar Energy Society, pp. 30–34.

Watson, D. 1981. Bioclimatic analysis and design methods. *Passive Cooling: Proc. International Passive and Hybrid Cooling Conference*, Miami Beach, FL, November 11–13, 1981. Newark, DE: American Section of the International Solar Energy Society, pp. 597–611.

Watson, D. 1982. Winter garden atrium design: The New Canaan Nature Center. *Proc. of the First International Passive and Low Energy Alternatives I Conference*, pp. 4.13–4.21.

Whiddon, W. I., and H. T. Gordon. 1980. "Process" issues in the design of passive solar commercial buildings. *Proc. 5th National Passive Solar Conference*, Amherst, MA, October 19–26, 1980. Newark, DE: American Section of the International Solar Energy Society, pp. 176–180.

Windheim, L. S., and K. V. Davy. 1981. The substitution of daylighting for electric lighting in a large office building. *Proc. 6th National Passive Solar Conference*, Portland, OR, September 8–12, 1981. Newark, DE: American Section of the International Solar Energy Society, pp. 875–879.

Wray, W. O., J. D. Balcomb, and R. D. McFarland. 1979. A semi-empirical method for estimating the performance of direct-gain passive solar heated buildings. *Proc. 3d National Passive Solar Conference*, San Jose, CA, January 11–13, 1979. Newark, DE: American Section of the International Solar Energy Society, pp. 41–47.

Yanda, W. F. 1976. Considerations for retrofitting an attached solar greenhouse. *Proc. Passive Solar Heating and Cooling Conference and Workshop*, Albuquerque, NM, May 18–19, 1976. LA-6637-C. Los Alamos, NM: Los Alamos Scientific Laboratory, pp. 160–164.

Zimmerman, D. R. 1984. Tennessee Valley authority technical and design assistance data base. *Proc. 9th National Passive Solar Conference*, Columbus, OH, September 1984. Boulder, CO: American Solar Energy Society, pp. 314–319.

9 Performance Monitoring and Results

Donald J. Frey

9.1 Introduction

This chapter contains a discussion of performance monitoring in full-size passive and hybrid residential and commercial buildings and the major results from these monitoring activities. Most of the buildings discussed were occupied during the performance evaluation period, though some were full-size test buildings and were unoccupied during all or part of the monitoring period. The literature contains a number of accounts of performance monitoring that would fall within this scope; however, this information has been screened and only the most significant projects are included. The information presented in this chapter was current in 1987. Readers are cautioned that some of the methods for performance monitoring have been improved in more recent years. The term *performance* is used in this chapter to describe the thermal and energy-related aspects of buildings. In the broad sense, performance can include building function, economics, environmental impact, and other factors, but these uses of the term are not intended.

Performance monitoring is a commonplace activity in the traditional disciplines of science and engineering. The methods of measurement, data collection and analysis, applied to buildings all evolved out of these traditional fields of research. Since buildings are complicated systems that received little scientific study prior to the interest in passive solar and conservation technologies, an evolutionary period resulted during which the thermal dynamics of buildings were investigated and methods of testing buildings were developed. Work is still in progress in these areas, though many techniques have been well established and utilized. Monitoring projects are discussed chronologically in this chapter so that the history of this development can be clearly presented.

Researchers have engaged in performance monitoring for a variety of reasons. The variety makes this a broad topic and one with diversified monitoring approaches and results. For example, many researchers have been interested in quantifying energy flows for either individual components or systems. Others have been interested in energy balances on whole buildings. Performance data are useful for improving designs or operation routines and for identifying desirable properties of innovative materials and components. Results of performance monitoring are also used to calculate the economic benefit of passive design options. Professionals interested in using performance information range from researchers and designers to building owners and contractors to lenders and public officials.

The thread that ties together many performance monitoring projects is the fact that they comprise the same two main activities, observations and analysis. Observations frequently consist of data collection that ranges in complexity from manual reading of thermometers and utility meters to automatic data collection with sophisticated sensors and electronic equipment. Data analysis methods range from numerical manipulation of data to calculate averages and totals, to statistical methods used to establish relationships among variables, to engineering methods using mathematical models to predict unknown quantities not directly measured.

The goal of this chapter is to present an overview of the performance monitoring that has been performed in full-scale buildings. This includes a discussion of monitoring goals and objectives, levels of monitoring detail, descriptions of monitoring programs, and results from these programs. Most of the projects described in this chapter were sponsored by the U.S. government and performed by government laboratories or contractors.

9.2 Monitoring Methods

Before beginning the discussion of performance monitoring programs, it is worthwhile to look at some of the approaches to monitoring and the levels of monitoring detail that have been utilized. This discussion is divided into data collection methods, types of data collected, data analysis techniques, and five basic approaches used to monitor passive and hybrid solar buildings. The information presented in this section is central to the descriptions of monitoring programs in section 9.3. A system of nomenclature is established in this section for identifying data collection methods, types of data collected, data analysis techniques, and monitoring approaches used in section 9.3 to provide information about each of the projects discussed.

9.2.1 Data Collection Methods

The following methods have been used to collect data. For each method, a letter is given in parentheses next to its name to be used in section 9.3 to describe the data collection approach used by each of the projects. As a general note, the sensors used increase in sophistication as the methods for collecting data become more automated. For example, common thermometers are sufficient for manual data collection but are not adequate for automated data collection.

Manual Data Collection (M) Individual meters, thermometers, or other instruments are read manually, and the data are recorded by hand. The interval between readings depends on how frequently the devices are read. Using this method, average values for certain parameters, such as temperature, cannot be accurately calculated, and the accuracy with which any mathematical average of readings approximates the long term average of measured values depends on the interval between readings and the rate at which the value of the measured quantity changes. Maximum and minimum values of the measured quantity during the interval between readings are often indicated by manually read meters. For other parameters, such as fuel consumed, averages over time will be correct.

Utility Bills (U) Energy consumption values are derived from meter readings provided in utility bills. The interval depends on the frequency of the meter reading. The consumption values are the total for the interval. Average energy consumption values are easily derived by dividing by the length of time in the interval. Some meters record both the total energy consumed and the peak energy demand during the interval.

Strip-Chart Recorders (S) Data is continuously plotted on paper tape. Strip-chart recorders are most commonly used to measure temperatures. Because the plots are continuous, totals and averages can easily be derived from the data, though the data points must be read by hand.

Early Data Loggers (E) Early data loggers sampled channel values at regular intervals but could not store average or total values for the interval. The frequency of reading and printing was adjusted to compensate for this behavior.

Computerized Data Loggers (C) Computerized data loggers overcome all the shortcomings of the other methods. They can be programmed to produce total or average values over specified intervals, derived from frequent sampling of some sensors and continuous sampling of others. Intervals between sampling and data storage can usually be selected by the user and controlled from the keyboard. Many of these systems also allow data to be retrieved over telephone lines.

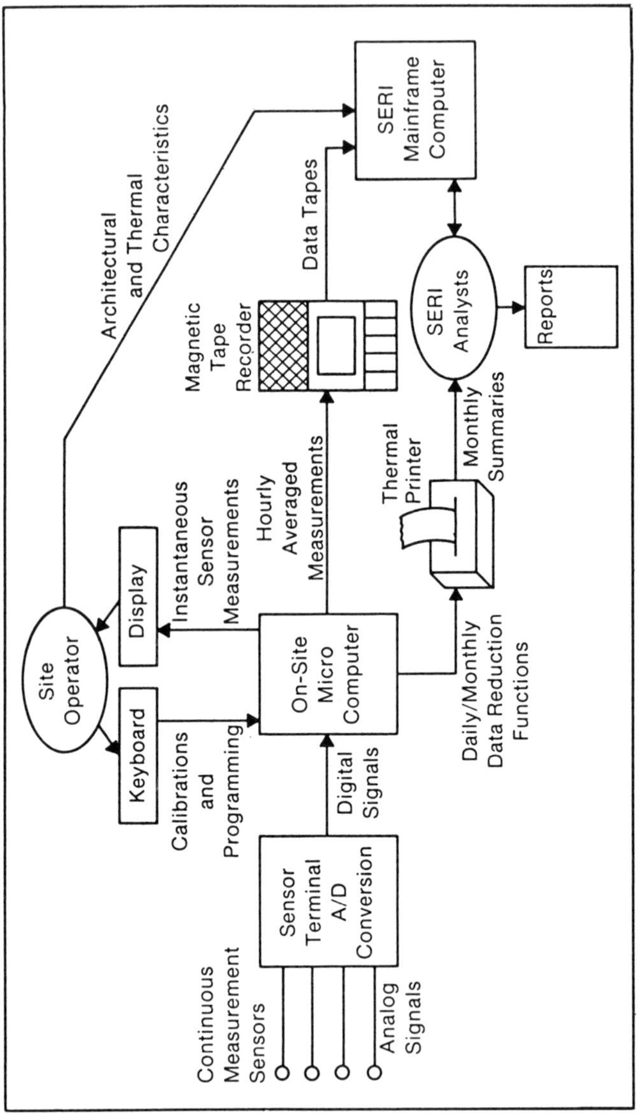

Figure 9.1
Class B data-handling process.

Figure 9.1 is a diagram of the data-handling process used by the Solar Energy Research Institute (SERI) in their Class B program for performance evaluation of residential passive buildings. At the heart of this system was the microcomputer-controlled data acquisition system.

9.2.2 Types of Data Collected

The main categories of data collected are temperatures, weather, energy use, building configuration, and occupancy. Each of these is subdivided below and arranged so that the most frequently collected information is presented first. The roman numerals are used to indicate which categories of data were collected in each of the projects described. These are followed by the arabic numerals to provide additional detail. For example, *I 1,3*; *II 1,2*; *III 1,2* means that the data collected were inside air and building surface temperatures, ambient temperature and solar radiation, and whole building electricity and gas.

I. Temperatures
1. Inside air (dry bulb)
2. Inside mean radiant temperature (MRT)
3. Building surface
4. Inside air (wet bulb) or humidity
5. Cross-section through components

II. Weather
1. Ambient temperature
2. Solar radiation
3. Wind speed
4. Wind direction
5. Relative humidity (wet bulb temperature)

III. Energy
1. Whole building electricity
2. Whole building gas (or other fossil fuel)
3. Submetered electricity
4. Submetered gas (or other fossil fuel)
5. Heat flux

IV. Building Components and Systems
1. Position of ventilation devices (windows open or closed)
2. Position of movable devices (insulation)

3. Status of lighting (on/off)
4. Air flow rate (supply duct)

V. Occupants

1. Surveys
2. Occupied vs. unoccupied

Programs that evaluated a large number of buildings using standardized equipment and techniques also employed a standard set of measurements for each building. Examples of programs with standard measurements are the SERI Class B program, the National Solar Data Network (NSDN) passive monitoring program, and the Passive Solar Commercial Buildings Program.

9.2.3 Data Analysis Techniques

Three basic techniques are used for data analysis: numerical manipulation, statistical analysis, and engineering analysis. Each of these is explained in greater detail below. The letter in parentheses is used to designate which of these methods was employed by each of the projects discussed.

NUMERICAL MANIPULATION (A) This is the most basic level of data analysis and is used to derive totals, averages, maximums, minimums, and the time of day when events occurred. Numerical manipulation of data is very useful for understanding what happened but not necessarily why it happened.

STATISTICAL ANALYSIS (S) Using statistical analysis, relationships between two or more variables are determined. This approach provides a means to identify correlations among system parameters to derive cause-and-effect relationships and performance lines or curves.

ENGINEERING ANALYSIS (E) These methods use engineering equations and advanced engineering analysis methods. Equations are used to interpret the data and to relate data values to each other. The equations may be thermal networks or other models that describe energy balances on components, systems, or whole buildings, or to describe physical processes, such as conductive, convective, and radiative heat transport. Advanced engineering analysis methods are used to identify parameters of the thermal behavior that are not measured directly and to predict behavior over periods of time longer than the data period. Advanced engineering analysis methods may be applied to both component and whole building analysis.

It is important to emphasize the distinction between component analysis and whole-building analysis. Some of the monitoring programs were aimed at component performance, without a specific interest in what was happening with the rest of the building. Other programs addressed whole-building performance but were not concerned with component performance. Each project is described as being concerned with component performance or whole-building performance.

9.2.4 Five Monitoring Approaches

The objectives of the monitoring programs are to gain new information about buildings, whether it is to quantify performance or more fully understand the physical mechanisms that control their behavior. What does seem to vary is the hypothesis on which the monitoring program is based. Below are descriptions of five approaches to performance monitoring, each based on a different hypothesis.

Each of the five approaches involves the methods of data collection, the various types of data, and the data analysis techniques that are discussed at the beginning of this section. As will be seen in the discussions of the projects, the data collection and analysis procedures used within any general approach can be quite different.

9.2.4.1 Approach 1 The first approach, and probably the simplest, is one in which temperatures and energy flows are measured to compare the energy use of the building to that of a reference building, such as other actual buildings, or to calculated building performance. The hypothesis is that the difference between the energy use of the test building and the reference building is attributable to the differences between them, for example, their passive solar features.

This approach is subject to numerous sources of error. Some of the largest uncertainties are interior temperatures, internal heat gains, occupancy, weather, and construction differences. If temperatures are not similar for the test and the reference buildings, then the seasonal loads are quite different. Internal gains can contribute substantially to reducing auxiliary heating requirements or to increasing cooling loads. Occupants can influence heating and cooling loads by leaving doors and windows open or by using sources of heat other than those being measured, such as wood stoves and fireplaces. In a single location, weather can be substantially different from one year to the next. Assumed insulating values may not be realized in the actual buildings. It is very difficult to use this approach to understand cause and effect.

9.2.4.2 Approach 2 The second approach investigates the functional performance of a building or its passive systems. Building and passive component temperatures are monitored and then analyzed to determine that all the components are responding qualitatively or quantitatively in the manner intended by the designer. The hypothesis is that correct temperature response of the building or the passive system components means that passive heat is being delivered to the building.

A fundamental question not answered by this approach is whether the building consumes less auxiliary energy than conventional buildings because of its passive features. Current design practice is for passive buildings to have greater total glazing area than conventional buildings and, as a result, increased envelope conductance. It is possible for the passive heat to simply make up for an increased load without reducing auxiliary energy.

9.2.4.3 Approach 3 The third approach is to use simulation analysis jointly with measured data to calculate building or component performance. This is a powerful approach when the inputs to the simulation are derived from short-term monitoring. The changes in building or component performance resulting from occupant effects (thermostat setpoints and internal gains), weather, and other driving forces can be studied. The hypothesis is that the building or component will perform as predicted by the simulations.

The advantage of this approach is that buildings and components do not have to be monitored for long periods of time, and the uncertainty in the data as a result of occupants or the properties of the building may be eliminated. The short-term monitoring has to be sufficient to derive the inputs to the simulations, such as envelope conductance, thermal mass, effective aperture area, seasonal average infiltration rate, and mechanical system efficiency.

9.2.4.4 Approach 4 In the fourth approach, an energy balance is derived for an entire building. The hypothesis is that passive performance can be determined if all or most of the energy flows into and out of the building can be measured or calculated from measurements. This approach produces numbers that include total load, heat supplied by auxiliary, heat from internal sources, and heat from passive solar.

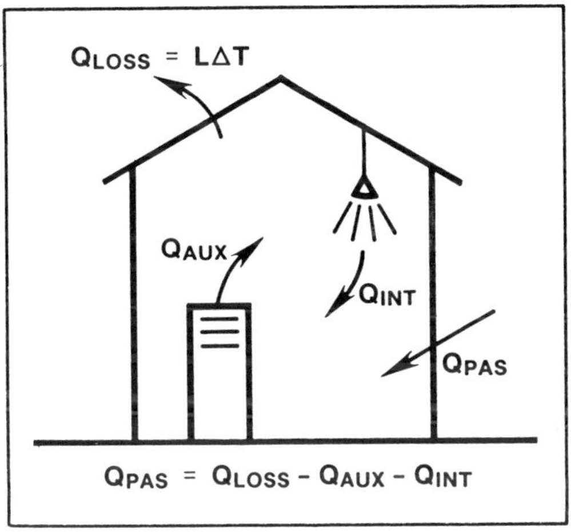

Figure 9.2
Class B energy balance method.

The major assumption built into this approach is that the primary building energy flows can be measured or calculated from measurements and thereby determine the terms in an energy balance equation. Two techniques are used to calculate the passive contribution. One is referred to as an *additive method*, in which gains and losses through glazing, energy into storage, heat from auxiliary sources, etc., are summed to directly calculate the passive contribution, and the other is a *subtractive method*, in which the passive contribution is found by subtracting auxiliary heat and heat from internal gains (lights and appliances) from the total building load based on the indoor-outdoor temperature difference. The individual values are more difficult to derive for the additive approach, but once they are known, the total gains can be compared to the total losses to verify that they are equal. In the subtractive approach, all the error is lumped into the passive contribution and verification of accuracy is not possible.

Figure 9.2 is a schematic drawing of the whole-building energy balance approach that was used in the SERI Class B program. The contribution of passive solar was calculated by the subtractive method.

9.2.4.5 Approach 5 The fifth approach is similar to the fourth one, but instead of deriving an energy balance for an entire building, it is derived for individual components or subsystems of the building. This approach can include detailed monitoring of the envelope and mechanical systems, such as infiltration and air flow rates. This approach may involve direct measurement of the thermal and physical properties of materials. The hypothesis is that component or subsystem performance can be determined if all or most of the energy that flows into and out of the component or subsystem can be measured or calculated from measurements. The purpose of such studies usually is to validate simulation models and to understand basic physical behavior.

9.3 Monitoring Programs

This section contains descriptions of the major programs in the United States for monitoring passive solar buildings. The majority of these programs were sponsored by the U.S. Department of Energy (DOE) or by other organizations within the government. Work done in the private sector, such as by utilities and individuals, comprises only a small part of the overall work.

However, some of the significant earliest projects were performed abroad—the first one described in England and the second one in France. They show a high level of sophistication in their monitoring and analysis approaches that was not seen in programs in the United States until much later. All the remaining projects took place in the United States and are examples of work from 1974 through 1985.

9.3.1 The Wallasey School

Organization: British Department of the Environment and University of Liverpool (Morris Davies)
Number of buildings: 1
Type: school
Location: Wallasey, England
Monitoring period: January 1969 to July 1970
Components or whole buildings: both
Data collection methods: E
Type of data collected: I 1, 2, 3, 4; II 1, 2; III 1, 3; IV 1, 3; V 1
Data analysis technique: A, S, and E
Monitoring approach: 2, 4, and 5
Building occupancy: occupied

Significance and results: Data were collected at the Wallasey School using an automated data acquisition system, and analysis was performed on a digital computer. Davies and his team statistically analyzed the data to discern the functional performance of the passive heating elements in one wing of the school. They created a thermal network to solve for unknown ventilation rates, and they applied statistical techniques to investigate relationships among the measured parameters. They formed an energy balance on the school wing and determined the percentages of the building's heat input attributable to sun, lights, and students. Methods similar to these are used over and over again in subsequent monitoring programs. They recognized that ventilation is an unknown that is hard to measure but which must be determined to properly describe a building energy balance.

On a seasonal basis, the direct-gain solar wall was found to be a net gainer of energy even though it showed a net loss for the coldest months. An approximate determination of heat input to the building showed 70% from the sun, 22% from lighting, and 8% from people.

Reference: Perry 1976.

9.3.2 Trombe Wall Houses In France

Organization: CNRS Solar Energy Laboratory
Number of buildings: 3
Type: residential
Location: Odeillo and Chauvency-Le-Chateau, France
Monitoring period: December 1974 through February 1975
Components or whole buildings: components
Data collection methods: E
Type of data collected: I 1, 3, 5; II 1, 2; IV 4
Data analysis technique: A and E
Monitoring approach: 5
Building occupancy: occupied
Significance and results: The researchers in France had as their objectives to evaluate the energy delivered to the room by the passive system, which is known today as the Trombe wall, and to calculate the average efficiency of the collector. These concepts are identical to concepts used to evaluate active solar collectors. Temperatures, solar radiation, and air flow rates were measured using sophisticated equipment, and analysis was performed using rigorous heat transfer equations. No attempt was made to calculate

whole-building performance or to investigate the theoretical performance of the wall using a thermal network. One of the major problems identified by the research team was the difficulty with measuring volumetric air flow in ducts, particularly over the laminar, transition, and turbulent air flow regimes.

Reference: Trombe et al. 1976.

9.3.3 The Atascadero House

Organization: Harold Hay and California Polytechnic University
Number of buildings: 1
Type: residential
Location: Atascadero, California
Monitoring period: 1974
Components or whole buildings: both
Data collection methods: E
Type of data collected: I 1, 2, 3, 5; II 1, 2; III 1; IV 2
Data analysis technique: A, S, and E
Monitoring approach: 3, 4, and 5
Building occupancy: occupied
Significance and results: The Atascadero house contains a passive system for both heating and cooling. This was the first significant project to investigate passive cooling performance. It was also one of the first projects for which the performance of the building was simulated. A subtractive method was used to calculate the heating and cooling performance of the passive system.

Reference: Hay, 1976.

9.3.4 Los Alamos National Laboratory

Some of the earliest monitoring work was performed by researchers at the Los Alamos National Laboratory (LANL). LANL instrumented twenty-one full-scale residential and commercial passive solar buildings during the period 1976 through 1980. The first report containing data on these buildings (Stromberg and Woodall 1977) covered five buildings, three in New Mexico, one in New Jersey, and one in California. Monitoring was not the purpose of the report, though some functional performance, including plots of thermal comfort and solar response, and comparisons of actual and calculated utility consumption are reported. The second report (Hoskins and Stromberg 1979) contains information about fifteen buildings, nine in New Mexico, two each in Vermont and New Hampshire, one in New Jersey, and one in Virginia. Like the first report, monitoring was not the primary purpose. Functional performance is reported for some of the buildings along with more in-depth engineering analysis of thermal process. The third report in this group (Jones 1982) concentrates on detailed monitoring of six buildings. Thermal networks are developed for a number of the buildings or their components. All six of these buildings are located in New Mexico. The remainder of this section focuses on performance monitoring of the most noteworthy buildings.

9.3.4.1 The Hunn Residence

Organization: Los Alamos National Laboratory
Number of buildings: 1
Type: residential
Location: Los Alamos, New Mexico
Monitoring period: 1977 to 1979
Components or whole buildings: both
Data collection methods: E
Type of data collected: I 1, 2, 3, 5; II 1, 2; III 1
Data analysis technique: A and E
Monitoring approach: 4 and 5
Building occupancy: occupied
Significance and results: The Hunn residence has a Trombe wall with 240 ft^2 of aperture area and 140 ft^2 of windows for direct solar gain. The Trombe wall does not vent to the occupied space in winter but is connected to a fan and a rock bed system for heat storage and active discharge. Hunn (1978) reports that the rock bed was not used after initial testing because storage temperatures were cooler than anticipated.

Hunn (1978) solved for the Trombe wall's overall solar heating contribution to the house load for the period between February 1, 1977, and January 31, 1978, using a subtractive energy balance method.

Balcomb and Hedstrom (1980) calculated the heating contribution of the Trombe wall for the period between November 1, 1978, and April 16, 1979, using a seven-node solution of the diffusion equation driven by temperatures measured hourly within the wall (see figure 9.3). By comparison, in the original analysis of a Trombe wall (cited above), Trombe et al. (1976) calculated the contribution of the wall by using surface tempera-

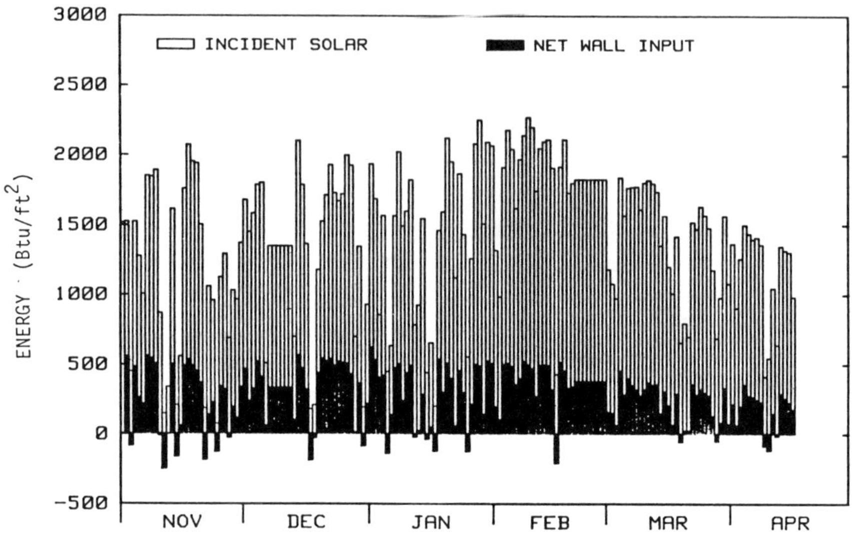

Figure 9.3
Hunn house: daily Trombe-wall performance—November 1, 1978, to April 16, 1979.

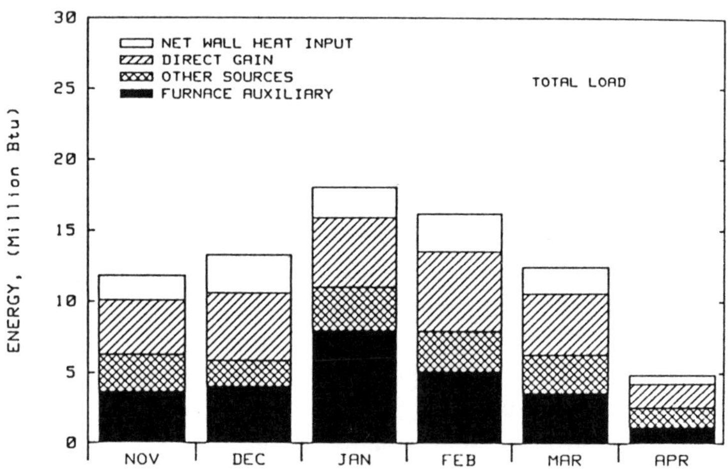

Figure 9.4
Hunn house: monthly solar, internal sources, and auxiliary energies—November 1, 1978, to April 16, 1979.

tures and solving separately for conductive, convective, and radiative heat transport. Figure 9.4 contains graphs of the energy sources contributing to the load on a monthly basis. The long-term efficiency of the Trombe wall during this data period was 20% relative to incident radiation, and the passive solar heat from the Trombe wall and direct gain provided a solar fraction of 59%.

Comfort is reported (Balcomb and Hedstrom 1980) as a histogram of living-room air temperatures and a calculated value of 50.9 of the discomfort index defined by Carroll (1980). The discomfort index used here and in all the buildings reported on by Los Alamos measures the extent to which effective temperatures calculated from measurements departed from preferred effective temperatures. The calculation of effective temperature includes air temperature, mean radiant temperature, and relative humidity.

References: Hunn 1978, 1979, 1981, Balcomb and Hedstrom 1980, Jones 1982.

9.3.4.2 The Jones House

Organization: Los Alamos National Laboratory
Number of buildings: 1
Type: residential
Location: Santa Fe, New Mexico
Monitoring period: 1979–80 heating season
Components or whole buildings: component
Data collection methods: E
Type of data collected: I 1, 5; II 1, 2
Data analysis technique: E
Monitoring approach: 2 and 5
Building occupancy: occupied
Significance and results: The solar heating system on this house consists of a natural-convection, air-heating collector connected to a rock bin. Heat is discharged from the bin by a fan. The rock bin on this house was found to be quite leaky, seriously effecting performance. In one of the first examples of passive system diagnosis using simulation analysis, a finite difference model was created to simulate the measured temperature profiles (Perry 1980). The best agreement with the measured data was found by assuming reverse thermosiphon flow at night through the rock bin-collec-

tor loop. Leaks in the dampers permitted the reverse flow. Though auxiliary energy usage of the house was low and heat was evenly distributed through the house in the ducted system, the full potential of the heating system was not realized because of the reverse thermosiphon flow.

References: Hunn and Jones 1978, Perry 1980, Jones 1982.

9.3.4.3 The Ghost Ranch Greenhouse

Organization: Los Alamos National Laboratory
Number of buildings: 1
Type: residential (dormitory)
Location: Ghost Ranch (Abiquiu), New Mexico
Monitoring period: 1979 to 1980
Components or whole buildings: both
Data collection methods: E
Type of data collected: I 1, 3, 5; II 1, 2
Data analysis technique: A and E
Monitoring approach: 2, 3, and 5
Building occupancy: occupied
Significance and results: Three separate monitoring experiments were performed at the Ghost Ranch greenhouse. First, a thermal-network model of the whole building was created based on the PASOLE framework (Jones and McFarland 1979). The thermal-network diagram of the model is shown in figure 9.5. It was used to calculate the building's performance, which was then compared to measured data. This work is significant because it provided confidence in using simulation models to predict the long-term thermal performance of buildings.

Second, temperature and air flow measurements were taken to analyze the thermocirculation between the greenhouse and the adjoining room. Air flow velocity and temperature measurements in the greenhouse were plotted using two different expressions for the temperature relationships. Neither model showed significantly less scatter than the other as shown in figure 9.6.

Third, the heat storage capacity of the dirt floor of the greenhouse was calculated from temperature measurements and the diffusion equation. The results showed the floor to be thermally equivalent to a container of water 1.3 in. thick covering the same area.

References: Jones and McFarland 1979, Jones 1982.

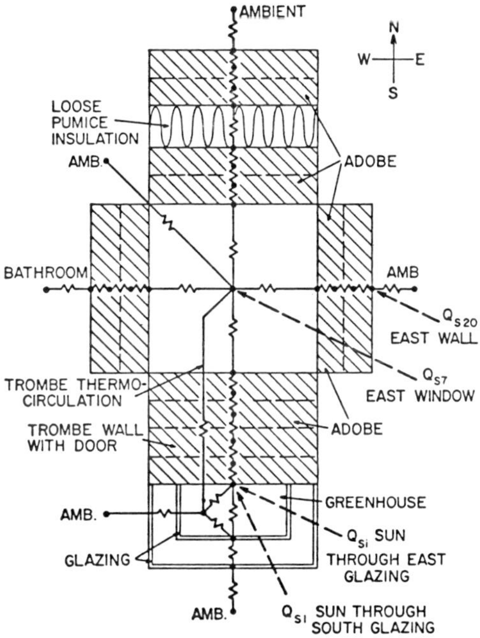

Figure 9.5
Ghost Ranch greenhouse: mathematical thermal network.

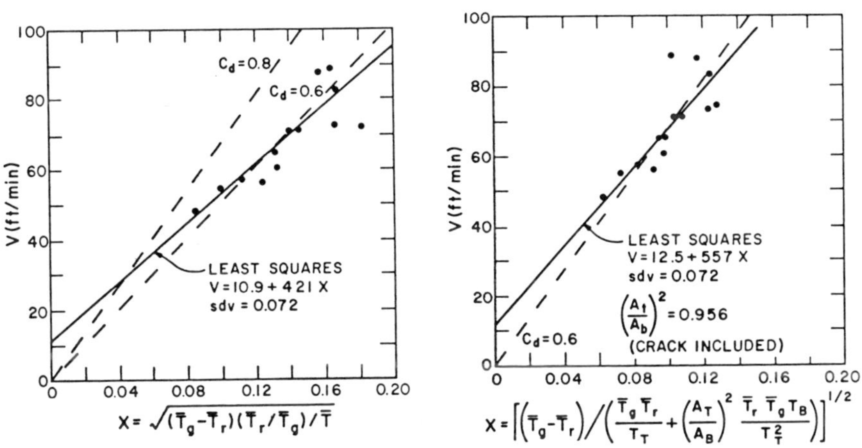

Figure 9.6
Ghost Ranch greenhouse, correlation of thermocirculation air flow velocity to average temperatures (on the left) and to inlet and average temperatures (on the right).

9.3.4.4 The Gunderson Residence

Organization: Los Alamos National Laboratory
Number of buildings: 1
Type: residential
Location: Santa Fe, New Mexico
Monitoring period: February 1979
Components or whole buildings: component
Data collection methods: E
Type of data collected: I 1, 3, 5; II 1, 2
Data analysis technique: E
Monitoring approach: 5
Building occupancy: occupied
Significance and results: This work reports on the performance of a water-loaded Trombe wall. The heat conversion efficiency of the wall is reported to be 38%. Details of the analysis technique are not reported. The water-loaded Trombe wall appears to provide a steady delivery of heat throughout a diurnal cycle, even during partly cloudy weather.

References: Jones 1982.

9.3.4.5 Balcomb Residence

Organization: Los Alamos National Laboratory
Number of buildings: 1
Type: residential
Location: Santa Fe, New Mexico
Monitoring period: 1978–1981
Components or whole buildings: Both
Data collection methods: E
Type of data collected: I 1, 2, 3, 4, 5; II 1, 2, 3, 4; III 3; IV 4
Data analysis technique: A, S, and E
Monitoring approach: 2, 3, 4, and 5
Building occupancy: occupied
Significance and results: The Balcomb house is probably the most thoroughly analyzed passive solar home in the United States. Owner and occupant J. Douglas Balcomb and his colleagues at Los Alamos have studied many of the performance aspects of this house in great detail and developed interesting and useful methods for analyzing building performance.

The building has a two-story sunspace, heat from which is stored in the sunspace floor and in the massive adobe wall that separates the sunspace from the occupied area of the house. Heat is also stored in two rock beds that are fan-charged with warm air from the top of the sunspace. Sources of heat gain, heat loss, and heat transport mechanisms have been quantified from measurements during several monitoring periods. Balcomb has used an additive methodology to form an energy balance on the building. Details of this methodology included a detailed solar-gain calculation for six glazing orientations, using measured hourly solar radiation data and determining the building loss coefficient from data collected during a twenty-day period in which no venting occurred and the fireplace was not used. The air infiltration rate of 0.6 air changes per hour was derived from blower-door tests. The overall building heat loss coefficient agrees well with values calculated using ASHRAE steady-state techniques. Figures 9.7 and 9.8 contain summaries of the energy flows during a 175-day period in the 1978–79 winter season. During normal operation, heat vented and heat from a wood-burning fireplace had to be inferred. Actual auxiliary energy consumed agrees well with estimated values using the SLR correlations.

Balcomb, Hedstrom, and Perry (1980) used hourly humidity data to calculate the transpiration of water from plants and found that a significant amount of energy was required for this process. The evaporation of water in the greenhouse also provides benefits for the building: it helps to keep down the greenhouse temperatures on sunny days and provides needed humidity to the house. The heat loss designated as *evaporation* in figure 9.7 results from exfiltration of water vapor generated in the greenhouse.

Balcomb, Hedstrom, and Perry (1981) report histograms of temperatures and values of the discomfort index (Carroll 1980) for each room of the house. The discomfort indices range from 23.1 $°F^2$ for the central bedroom (the most comfortable room) to 99.8 $°F^2$ for the sunspace. The greenhouse has temperature swings much larger than the adjacent living spaces. The sunspace temperatures are acceptable, however, to plants and for some daytime activities. A histogram of temperatures in the dining room is shown in figure 9.9.

The mass wall separating the greenhouse from the living space provides most of its stored heat back to the greenhouse at night. Calculations of heat flux into the massive adobe wall of the greenhouse (Balcomb, Hedstrom, and Moore 1979) and the greenhouse floor (Balcomb, Hedstrom, and Perry

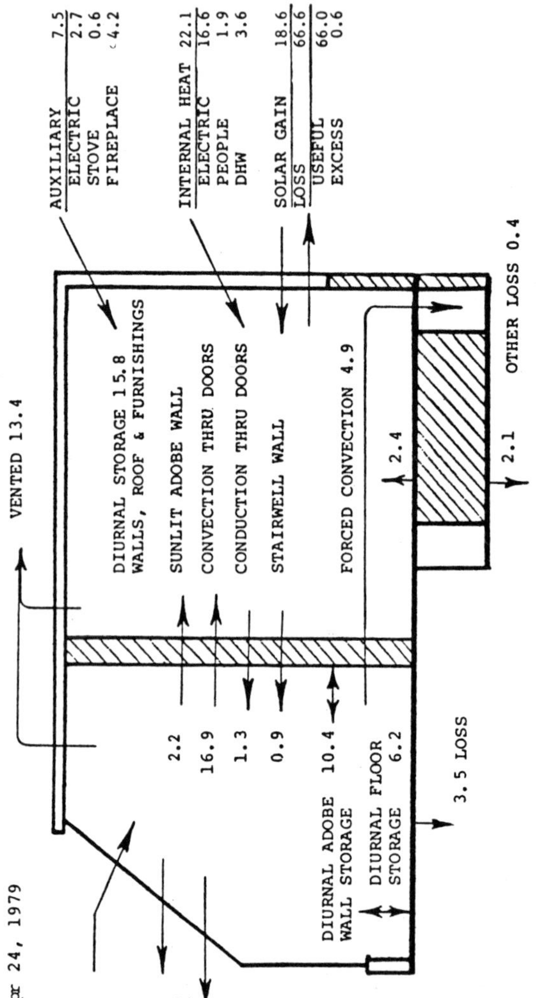

Figure 9.7
Balcomb house: energy flows, November 1, 1978, to April 24, 1979.

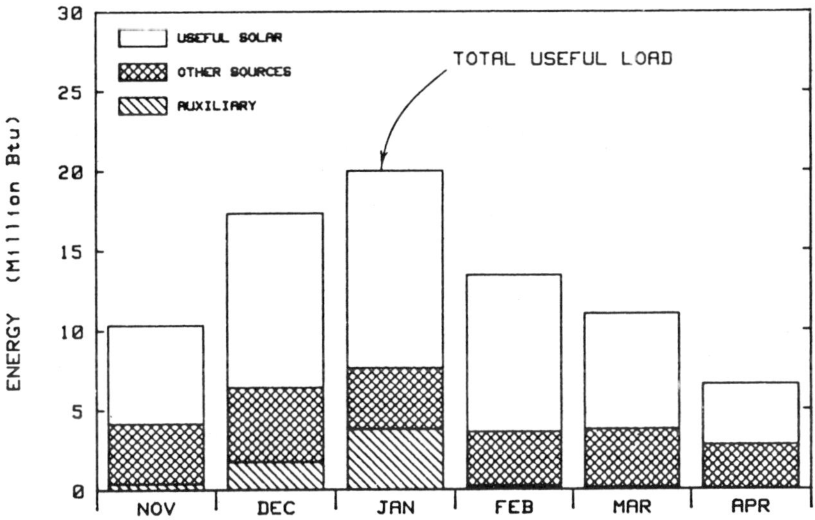

Figure 9.8
Balcomb house: monthly useful solar, internal sources, and auxiliary energies, November 1, 1978, to April 24, 1979.

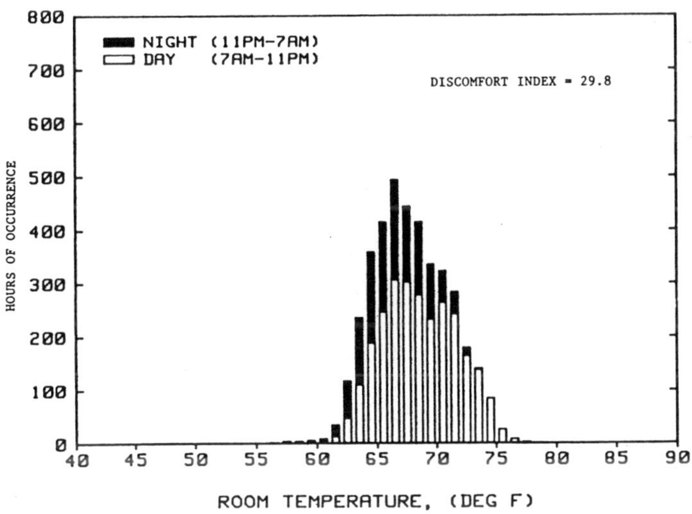

Figure 9.9
Balcomb house: dining-room temperature histogram for the period November 1, 1978, to April 24, 1979. Discomfort index unit is in °F^2. Source: Balcomb, Hedstrom, and Perry 1981.

1980) were performed using measurements of temperatures within the massive elements and the diffusion equation. These two studies quantified the heat stored in the wall and floor that is returned to the greenhouse at night. Figure 9.10 contains plots of the greenhouse floor temperatures and calculated heat flows for a four-day period.

The performance of the two rock beds and their contribution to comfort was investigated (Balcomb, Hedstrom, and Moore 1979, Perry 1980). The rock beds contribute substantially to the temperature stability and comfort of the greenhouse and living space. Ducts that deliver air from the greenhouse to the rock beds were retrofitted with backdraft dampers to improve overall effectiveness.

In summary, the building is substantially solar-heated—it has a solar fraction of 83% and a system efficiency of 27%.

References: Balcomb 1979, 1980, 1981a, 1981b, Balcomb, Hedstrom, and Moore 1979, Balcomb, Hedstrom, and Perry 1980, 1981, Perry 1980, Jones 1982.

9.3.4.6 Mobile/Modular Home Unit 2

Organization: Los Alamos National Laboratory
Number of buildings: 1
Type: residential
Location: Los Alamos, NM
Monitoring period: 1979–80 heating season
Components or whole buildings: both
Data collection methods: E
Type of data collected: I 1, 3; II 1, 2; III 2; IV 2
Data analysis technique: A and E
Monitoring approach: 3, 4, and 5
Building occupancy: unoccupied
Significance and results: This is one of the first, if not the first altogether, passive solar mobile/modular home to be monitored. The design, combining direct-gain windows and a roof pond, is adaptable to a wide range of buildings. An extremely detailed evaluation was done on this building, using a complex PASOLE thermal network model (see figure 9.11). PASOLE was used to form monthly energy balances, to reconcile the data, and to derive some of the information about the building that was not measured directly, such as infiltration rate. PASOLE was also used to

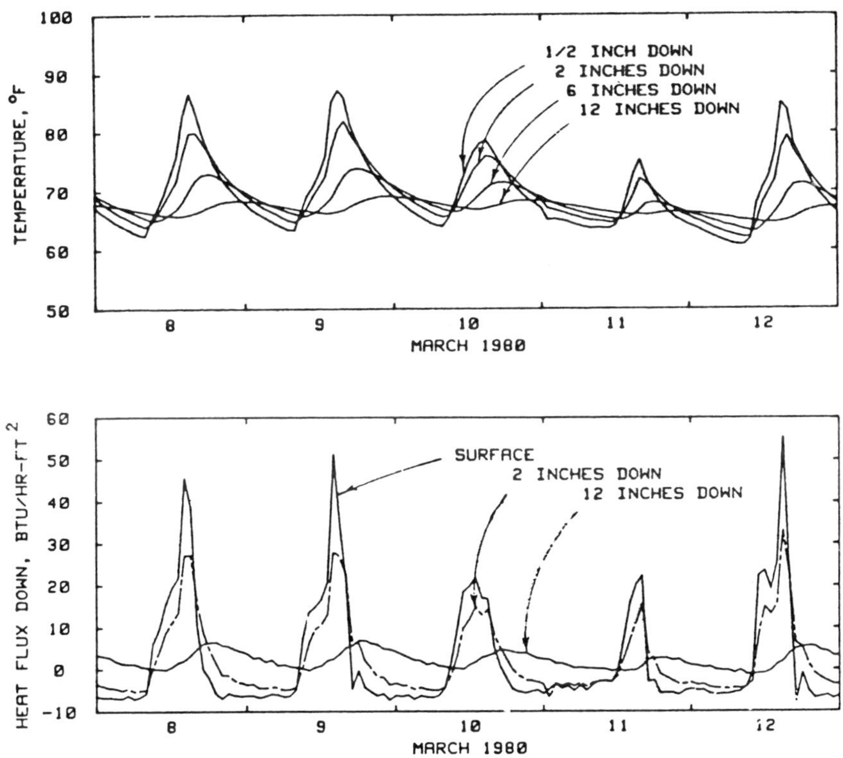

Figure 9.10
Balcomb house: greenhouse floor temperatures and heat fluxes, March 8–10, 1980.

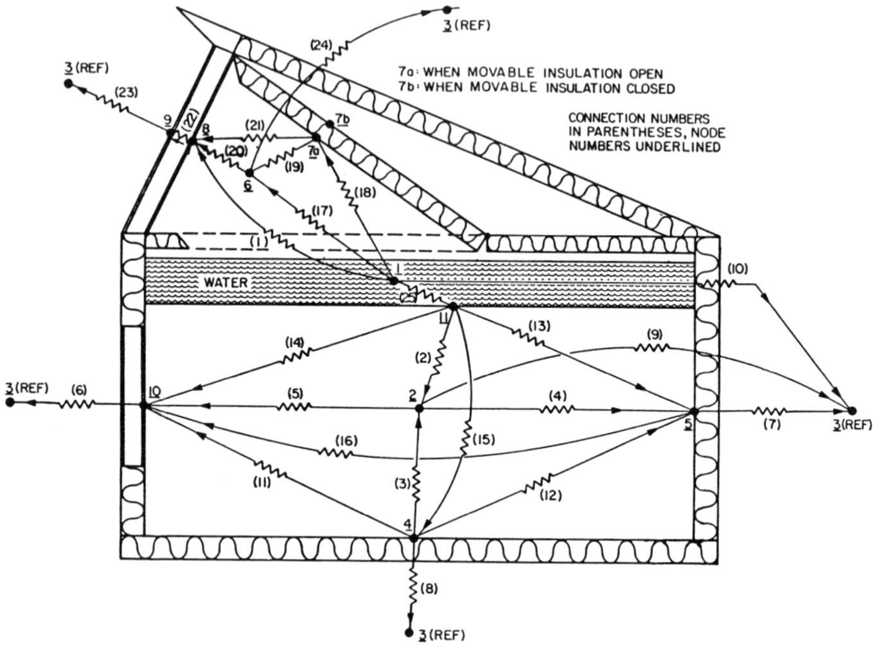

Figure 9.11
Mobile/Modular Home Unit 2: simulation model thermal network.

predict the performance of the building under a range of conditions and in a variety of locations.

The building was a good performer with a solar fraction of 80% and an overall efficiency of approximately 27%. The roof-pond system was found to have poor optical efficiency, and its usefulness was dependent on the amount of heat provided to the living space from direct gain. The direct gain could overheat the space, which reversed the heat flow from the ceiling. Movable insulation panels providing a measure of thermal control improved the thermal efficiency of the roof pond.

Reference: Jones 1982.

9.3.5 National Solar Data Network

The National Solar Data Network (NSDN) was established to fulfill one of the objectives of the National Solar Heating and Cooling Demonstration Act of 1974, which was to collect and evaluate information about solar

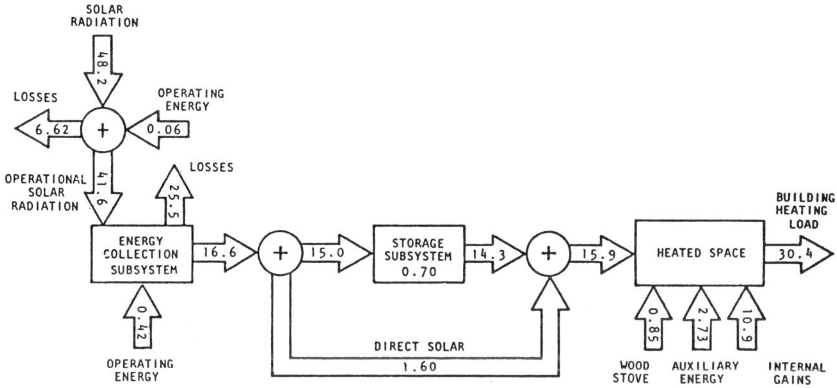

Figure 9.12
Standard NSDN passive space heating flow diagram.

energy systems and disseminate reports to potential users. The NSDN was originally established to monitor the performance of active solar space and water-heating systems, and as a result developed a methodology that concentrates on solar energy collected, energy delivered to storage, and energy delivered to the load, as shown in figure 9.12. This basic philosophy carried over into the the energy balance approach used by the NSDN to monitor passive solar buildings.

The NSDN was originally operated by IBM. IBM developed the equipment that was used, which consisted of the site data acquisition system (SDAS), telecommunications equipment to connect the SDAS to the host computer, and a central data-processing system (CDPS). These are shown in figure 9.13. The data collected at each site was stored for one or two days in the memory of the SDAS. The host computer would call each SDAS, and the data would be transferred via telephone lines to the host computer. The data were processed in the CDPS, and monthly reports for each site were produced. In 1979, the responsibility for the NSDN was transferred to the Vitro Corporation of Silver Spring, Maryland. Monthly performance of each of the buildings in the program was calculated and reports were prepared.

Organization: Vitro Corporation
Number of buildings: 20
Type: residential and commercial

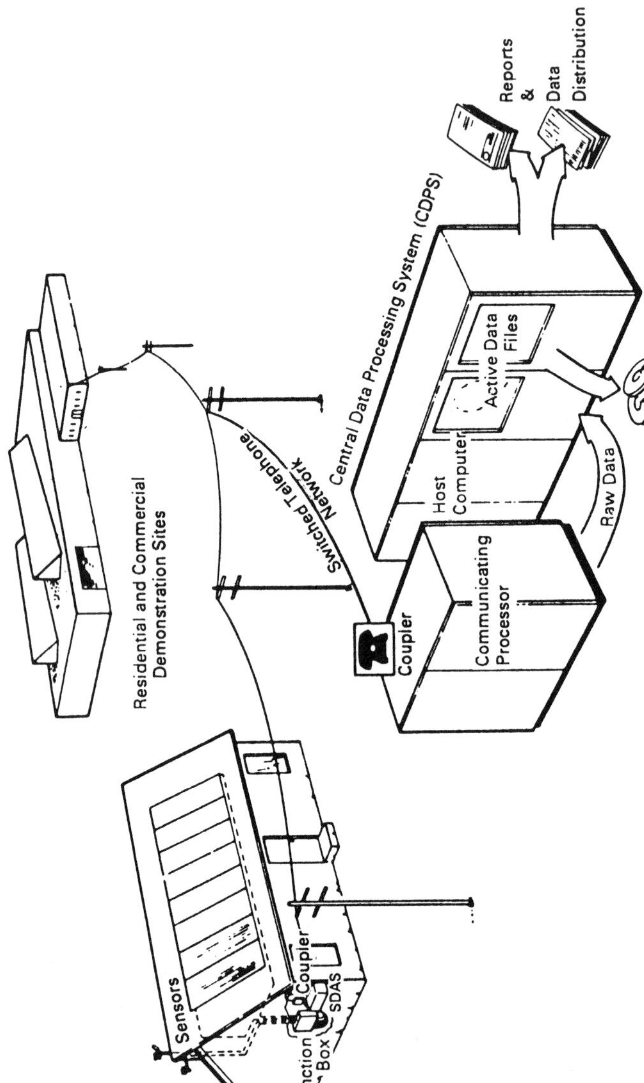

Figure 9.13
Diagram of NSDN data-collection process.

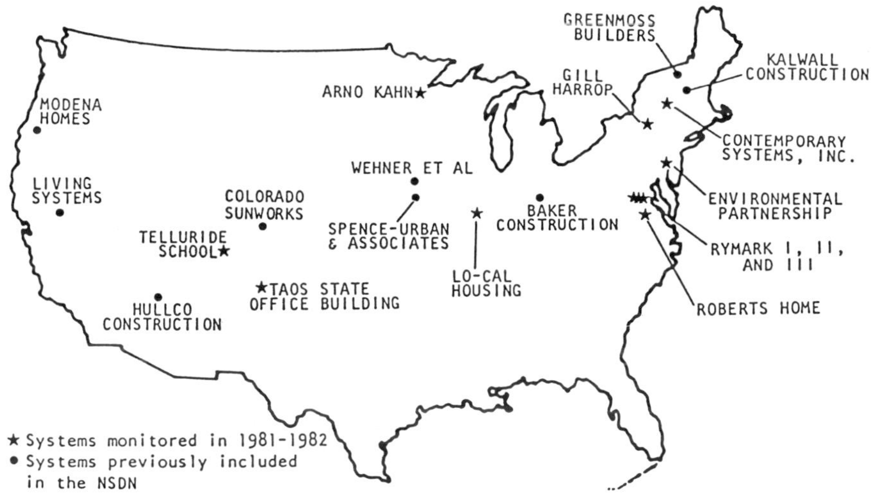

Figure 9.14
Location of passive solar buildings monitored by NSDN.

Location: throughout the United States (see figure 9.14)
Monitoring period: 1977–1982
Components or whole buildings: both
Data collection methods: C
Type of data collected: I 1, 4; II 1, 2, 3, 4, 5; III 1, 2, 3, 4; IV 3, 4
Data analysis technique: A, S, and E
Monitoring approach: 2, 3, 4, and 5
Building occupancy: occupied
Significance and results: The NSDN program was the first to collect data from a large number of homes throughout the United States using standardized equipment and procedures (see figure 9.14). It was also the first to use computer-controlled data acquisition equipment with data retrieval via the telephone line. Many of the buildings were monitored with as many as ninety sensors. The data collected was intended to produce performance assessments at the system level rather than at the subsystem or component level.

As the work by the NSDN progressed, the data analysis methodology used became similar to the Class B methodology that had been developed at the Solar Energy Research Institute (explained below). Analysis of passive system performance was based on a subtractive method, with the passive solar heating performance being the unknown. A significant difference between the NSDN approach and the Class B approach is the technique each used to calculate the total building heat load. The NSDN calculated the conductive heat loss by summing the conductive losses for each envelope component using ASHRAE conductance values and the measured temperature difference across the component. The sum of these losses for each component was added to the infiltration rate, calculated using ASHRAE procedures, to obtain the total building heat load. At four of the sites, electric coheating and tracer-gas tests were used to measure the building heat load and infiltration rate. These values were subsequently used in the analysis.

Another approach was a determination of energy flows between components of the building, the environment, the passive system, and auxiliary systems. This approach developed from NSDN's primary work, namely, the evaluation of active solar heating systems.

References: Howard and Pollack 1981, Pollack 1980, 1981a, 1981b, 1981c, Spears 1981a, 1981b, 1981c, 1981d, 1982a, 1982b, 1982c.

9.3.5.1 Roberts Home Performance results for the Roberts home in Reston, Virginia (Spears 1982a) are presented to illustrate the type of information produced by the NSDN. This three-story house has a total of 2,300 ft^2 (210 m^2) of floor area. It has a two-story Trombe wall with 663 ft^2 (62 m^2) of single glazing and a sunspace located on the third floor with 99 ft^2 (9.2 m^2) of double glazing tilted 48 degrees from the horizontal. The 12-in.(30-cm)-thick Trombe wall, the hollow-core north wall, and two centrally located fireplaces provide 264,000 lbm (120,000 kg) of solar heat storage. The mass is illustrated in the drawing in figure 9.15. During the winter, four fans draw air from the space between the Trombe wall and the glazing, and force it through the cores of the north wall. Movable insulating curtains are used to reduce nighttime losses through the Trombe wall. Walls of the house are insulated to R-24 (4.2 m^2 °C/W) and the roof to R-32 (5.6 m^2 °C/W). Summer cooling is provided by a combination of heat avoidance, earth tubes, and ventilation induced by thermal chimneys. The sunspace glazing is shaded on the outside during the summer with a rolldown shading device. Auxiliary space heating is provided by electric baseboard heaters and the fireplace.

Figure 9.16 contains a heating energy flow diagram for the months of February, March, and April 1982. During this period, 52% of the space

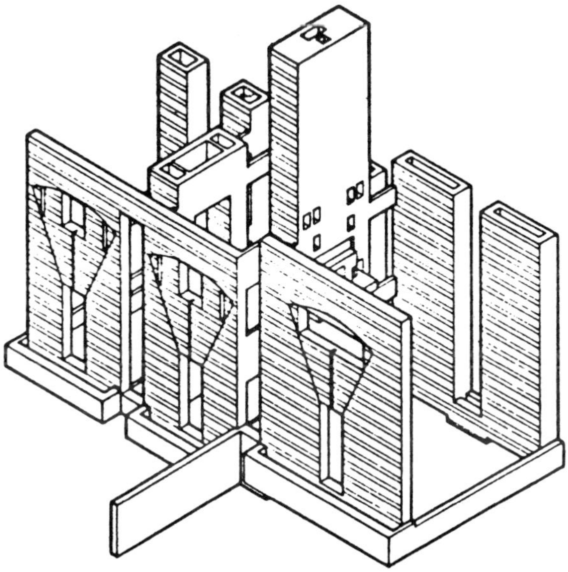

Figure 9.15
Roberts home: mass walls configuration.

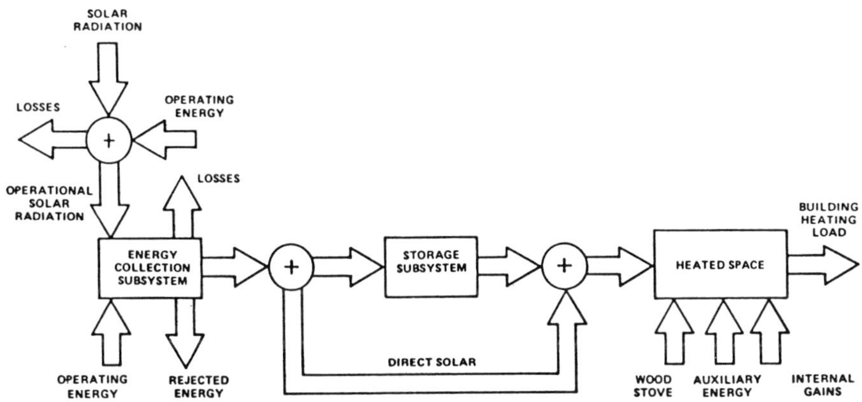

Figure 9.16
Heating energy flow diagram for Roberts home, February 1982 through April 1982 (values in million Btu).

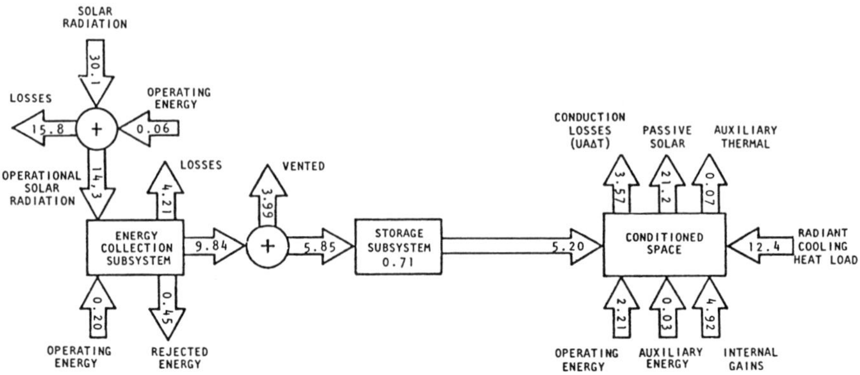

Figure 9.17
Cooling energy flow diagram for Roberts home, May 1982 through July 1982 (values in million Btu).

heating needs were provided by passive solar and 36% from internal gains (lights, appliances, and people). Of the incident solar radiation, 33% was delivered to the heated space. The sum of the purchased sources of heat (auxiliary and internal gains), normalized to the floor area and the heating degree-days, was 3.34 Btu/ft^2 °F day (18.9 Wh/m^2 °C day).

Figure 9.17 contains a cooling energy flow diagram for the Roberts home during the months of May, June, and July 1982. During this period, the total cooling load was 24.76 million Btu (7.25 MWh). Over 85% of this load was rejected from the building by passive means (ventilation and earth tubes). Another 14% was conduction through the building envelope. There was virtually no remaining sensible load. The passive systems in this building work well in both the cooling and heating seasons.

9.3.5.2 General Conclusions The NSDN predicted auxiliary energy requirements for a number of the buildings they monitored using the solar load ratio (SLR) method. The predicted auxiliary values were based on measured solar radiation in the plane of the collector along with measured ambient and interior air temperatures. Hourly corrections were made for the effects of shading. Corrections were also made for measured internal gains.

Table 9.1 contains a comparison of predicted and measured values of auxiliary energy for eight residential and three commercial buildings. For most of the buildings, the agreement between the predicted and measured

Table 9.1
Predicted vs. actual auxiliary energy usage (all values in million Btu)

Site	Predicted auxiliary energy usage	Actual auxiliary energy usage
Residential low-mass		
Lo-Cal Housing	26.3	25.8
Rymark I	28.6	27.2
Rymark II	33.0	32.1
Rymark III	33.3	30.8
Residential high-mass		
Arno Kahn	46.2	61.2
Environmental Partnership	32.3	27.3
Gill Harrop	12.1	12.6
Roberts Home	6.83	3.38
Commercial high-mass		
Contemporary Systems, Inc.	72.6	65.0*
Taos State Office Building	226	224
Telluride School	341	300

* Energy from active solar system.

values is good. Notable exceptions are the Arno Kahn house located in Duluth, Minnesota, for which the actual auxiliary value is 50% more than the predicted value, and the Roberts home for which the actual value is 50% less than the predicted value. The NSDN notes that "on a month-to-month basis, there was a larger amount of difference, and use of long-term weather data and average values of internal gains will result in a larger variance" (Raymond and Pollock 1984). The dependence of auxiliary energy use on thermostat setpoint, internal gains, and weather has been examined by a number of researchers (Bishop and Frey 1984, Frey and Holtz 1985). The NSDN conclusions are consistent with these other observations.

Most of the sites monitored by the NSDN were first-generation passive solar designs with large glazing areas. NSDN results stressed the importance of adequate conservation features and the sensitivity of building performance to occupant effects, especially the use of operable components such as movable insulation. NSDN found that the buildings with the best net gains had triple-glazed windows or were equipped with movable insulation but that movable insulation would have been more beneficial if its control systems were automated.

NSDN noted that all types of passive systems delivered energy but that direct-gain systems had slightly better performance than the others. Analyzing energy and construction costs, it concluded that direct gain/sun-tempered systems are the most cost-effective. NSDN also said that sunspaces are usually not cost-effective unless they provide needed additional floor space. It observed very little air flow in the outer loop of a double-envelope house and concluded that it actually performed like a well-insulated sunspace system. NSDN also concluded that overhangs reduce system performance and are unnecessary in space heating dominated climates. It noted that temperature stratification was a problem in buildings that were higher than one story.

NSDN noted that vented mass walls with backdraft dampers performed better than unvented walls and that systems with thermal-storage walls located close to the collector glazing should have multiple layers of glazing or movable insulation to reduce storage losses.

The NSDN was also one of the first programs to monitor the performance of commercial buildings. NSDN noted that commercial buildings that are unoccupied at night are extremely good candidates for passive solar energy for heating and daylight when it is directly available. Incremental passive solar costs can be kept low since the internal mass of commercial buildings is often adequate to provide all the thermal storage required. Existing south-facing brick walls on commercial buildings can be cost-effectively retrofitted with exterior glazing and vents to create mass wall systems.

References: Adams 1982, Greenmoss 1978, Howard and Pollock 1981, Kennedy 1980, Miller and Pollock 1980, Pollock 1980, 1981a, 1981b, 1981c, 1982, Raymond and Pollock 1984, Shippee 1979, Sparkes and Raman 1981, Spears 1981a, 1981b, 1981c, 1982a, 1982b, 1982c, Weston 1979a, 1979b.

9.3.6 Solar Energy Research Institute

In 1980, the Solar Energy Research Institute (SERI), funded by the U.S. Department of Energy, embarked upon three programs to evaluate the performance of passive solar residential buildings. Each program had a different level of detail and monitoring expense, and had its own specific audience and purpose. The first level, called Class A, was directed toward the research community, and its purpose was to collect high-quality data from a small number of buildings for component research and validation of computer codes. The second level, Class B, was initially directed toward Congress and industry for the purpose of demonstrating the extent to which passive homes save energy. The third level, Class C, was directed toward consumers and was intended to provide marketing and performance information from surveys, audits, and fuel-bill information.

9.3.6.1 The SERI Class A Program

Organization: Solar Energy Research Institute
Number of buildings: 1
Type: residential
Location: Golden, Colorado
Monitoring period: 1981–1983
Components or whole buildings: both
Data collection methods: C
Type of data collected: I 1, 2, 3, 4, 5; II 1, 2, 3, 4, 5; III 1, 3
Data analysis technique: A, S, and E
Monitoring approach: 2, 3, 4, and 5
Building occupancy: unoccupied
Significance and results: The Class A building at SERI was monitored in extreme detail. In addition to the data noted above, infiltration rates were measured zone by zone. Thermography was performed to investigate uniformity of the envelope insulation and to locate unusual heat loss sources. The configuration of the building was changed from one set of tests to another.

Information was obtained of high-enough quality to use in computer code validation. The thermophysical properties of envelope components were measured and, in many cases, found to differ from assumed handbook values.

References: Hunn, Turk, and Wray 1982, Judkoff, Wortman, and Burch 1983, Judkoff and Wortman 1984.

9.3.6.2 The SERI Class B Program

Organization: Solar Energy Research Institute
Number of buildings: 60
Type: residential
Location: throughout the United States (see figure 9.18)
Monitoring period: 1981 through 1983
Components or whole buildings: whole buildings

Data collection methods: C
Type of data collected: I 1; II 1, 2, 3; III 1, 2, 3, 4; IV 2
Data analysis technique: A and E
Monitoring approach: 4
Building occupancy: occupied
Significance and results: The Class B program went a step beyond what was being done by NSDN in the use of microcomputer based data acquisition equipment. The Class B equipment not only collected the data, it also performed on-site analysis, printed results on paper tape, and stored hourly data summaries on magnetic cassette tape (see figure 9.1). The goal of the program was to provide a consistent measure of the thermal performance of different types of passive buildings in different climates. This was the first monitoring program to use a standardized data acquisition system, standardized data acquisition methodology, and standardized data analysis methodology to derive performance factors that were comparable among the group of monitored buildings. Instrumentation was limited to that needed to calculate a monthly building energy balance, separating passive solar heating from the other building energy flows. Thermal storage and other individual components were not monitored, and no attempt was made to determine the details of the thermal processes. The Class B methodology is a good example of the fourth approach to performance evaluation.

9.3.6.2.1 GENERAL RESULTS Figure 9.18 is a map with the locations of the Class B sites. The sites in the states of Washington and Oregon were monitored by the Bonneville Power Administration (BPA) using the Class B methodology and equipment. The BPA project is discussed in a subsequent section of this chapter. Table 9.2 contains a summary of the distribution of the Class B houses and their passive systems.

One of the unique features of the Class B program was the one-time tests performed to determine heat loss coefficients, heating system efficiencies, and effective leakage areas for infiltration. These values were input to the computerized data logger for the on-site data analysis. The Class B analysis included a calculation of the infiltration rate by the use of the Sherman-Grimsrud equation (Grimsrud, Sherman, and Sonderegger 1983).

Major problems in the Class B program resulted from the data acquisition equipment. Early in the program, the data loggers were unreliable

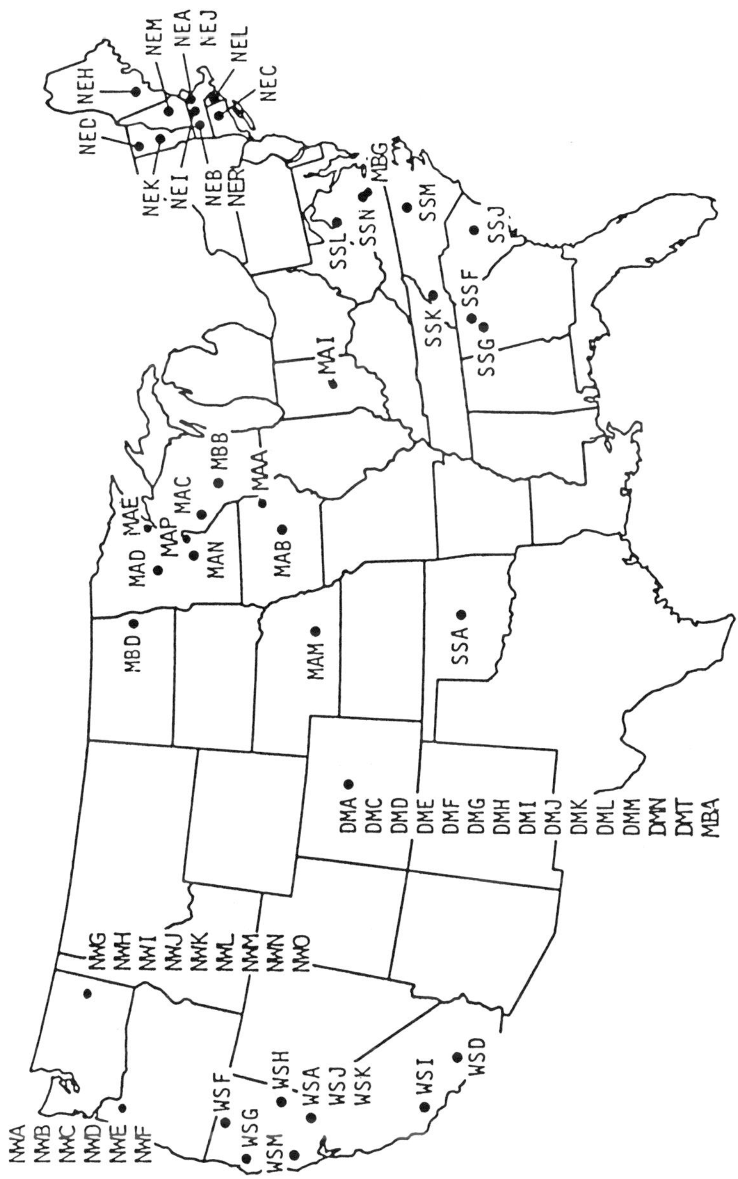

Figure 9.18
Locations of Class B homes.

Table 9.2
Distribution of Class B houses and system types*

	Number of buildings	Sunspace	Trombe wall	Direct gain	Water wall
Northeast	11	6	1	10	0
Midwest	10	3	2	10	0
South	10	2	2	9	2
California	14	2	1	12	0
Denver	15	5	3	12	0
Total	60	18	9	53	2

* Many buildings had more than one type of passive system.

because of design flaws in the electronics. The flaws were corrected and the reliability substantially increased. Another problem resulted from the cassette recorders used to store data. Recovering data from the tapes was extremely time-consuming, and often the recording quality was so poor that the data could not be recovered. The printed summaries proved to be valuable for salvaging information.

The overall performance of forty buildings in the Class B program for the 1981–1982 heating season is shown in figure 9.19. The height of each bar represents the total heating load. It is divided into passive solar, internal, and auxiliary heating components. The energy quantities are normalized to the conditioned floor area and the heating degree-days. The degree-days are based on measured indoor-outdoor temperature differences. The buildings are ordered from left to right according to increased purchased heating energy, auxiliary plus internal. At the top of each bar is an indication of the passive heating type (DG = direct gain, SS = sunspace, TW = Trombe wall, WW = water wall) and the ratio of the building's south glazing to floor area.

Each building is identified by a three-letter site code. The first two letters denote the region of the country and the third denotes the specific building in the region (DM = Denver, MA = Mid-America, SS = South, WS = California). In addition to these, site code prefix MB identifies buildings in the SERI Passive Solar Manufactured Buildings Program (MBA is in Denver and MBB is in Wisconsin). Two of the buildings on this list are not passive buildings (WSE and WSJ) but conventional buildings that were monitored as a basis with which to compare the passive performance.

Results from the Class B program for the period 1981–1983 show that the buildings have low auxiliary heating needs. Auxiliary heating is gener-

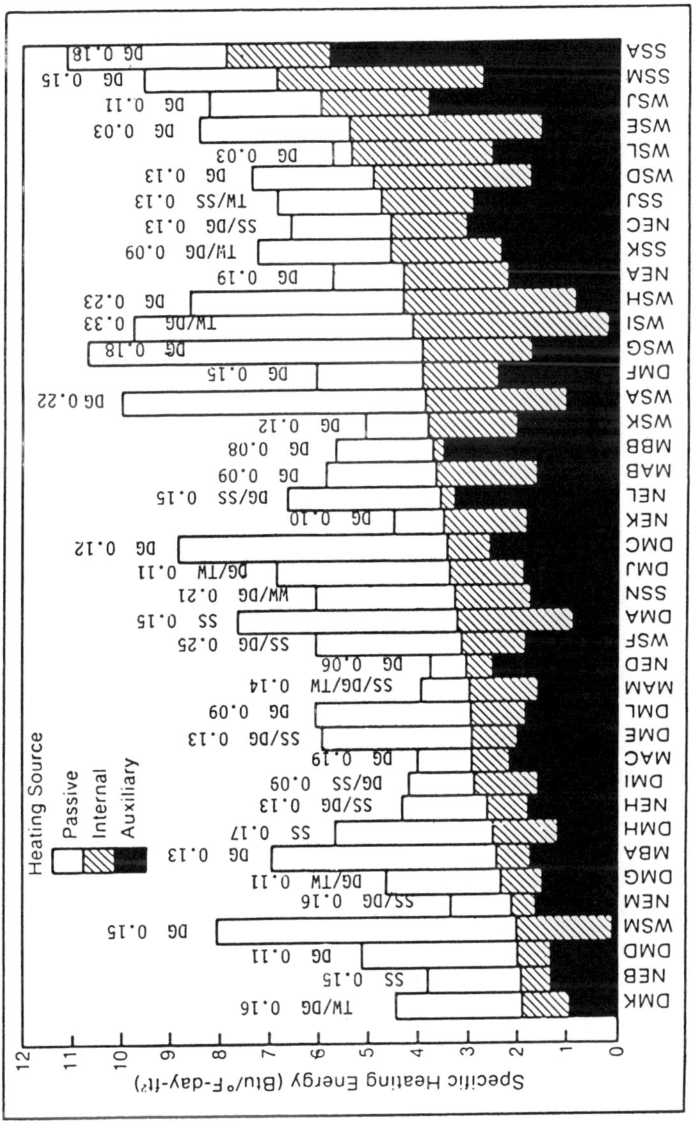

Figure 9.19
Normalized heating-season energy summaries for forty passive residences in the Class **B** program, the 1981–82 heating season.

ally less than 3.0 Btu/°F day ft^2 (17 Wh/°C day m^2) with and average of 2.4 Btu/°F day ft^2 (13.6 Wh/°C day m^2). Conventional buildings typically use 11 Btu/°F day ft^2 (62 Wh/°C day m^2) or more of total purchased energy.

The solar performance is variable. The passive solar systems contribute an average of 39% of the total heating load, or 55% of the net heating load (total load minus internal heat). Several of the buildings in figure 9.19 have large passive contributions, but this does not necessarily translate into energy savings. Site WSG, for example, has a large passive fraction and still has purchased energy above the average for the group.

Insulation and weatherization are important to the thermal performance of passive systems. Tight, well-insulated buildings tend to use less purchased energy than loose buildings. In figure 9.19, the largest purchased-energy users (right side of the graph) have relatively high total heating loads; and, with the exception of a few very heavily solar driven buildings in Denver and Northern California, the buildings with higher heat loss coefficients required more purchased heat. Some of the smallest purchased-energy users (left side of the graph) were buildings in the Northeast and Mid-America that had a moderate amount of passive heating and very low heat loss coefficients.

All three major types of passive systems, direct gain, sunspace, and thermal-storage wall, performed about the same. Several homes with sunspaces used very little auxiliary heat but did not have especially large solar contributions. This suggests that the sunspace, besides being a solar collector, acts as a thermal buffer that reduces heat loss.

SERI has examined the performance of each of the houses it monitored in detail. As an example, site DMH is a two-story townhouse in Denver, Colorado, with an attached sunspace and rock bed for thermal storage. Figure 9.20 shows a plot of building heat loss coefficient per floor area (Btu/°F day ft^2) versus specific total purchased heat (same unit) for all the sites monitored during the 1981–82 heating season (Swisher and Cowing 1983), showing site DMH having a below-average heat loss coefficient and a below-average purchased-energy requirement. Figure 9.21 shows solar aperture per unit of floor area for the same group versus specific total purchased heat. Site DMH has above-average solar aperture to floor area. The sunspace in this building is contributing to both decreased heat loss and increased solar aperture area. Table 9.3 contains descriptive information for the building, which is part of the SERI summary. Also included in

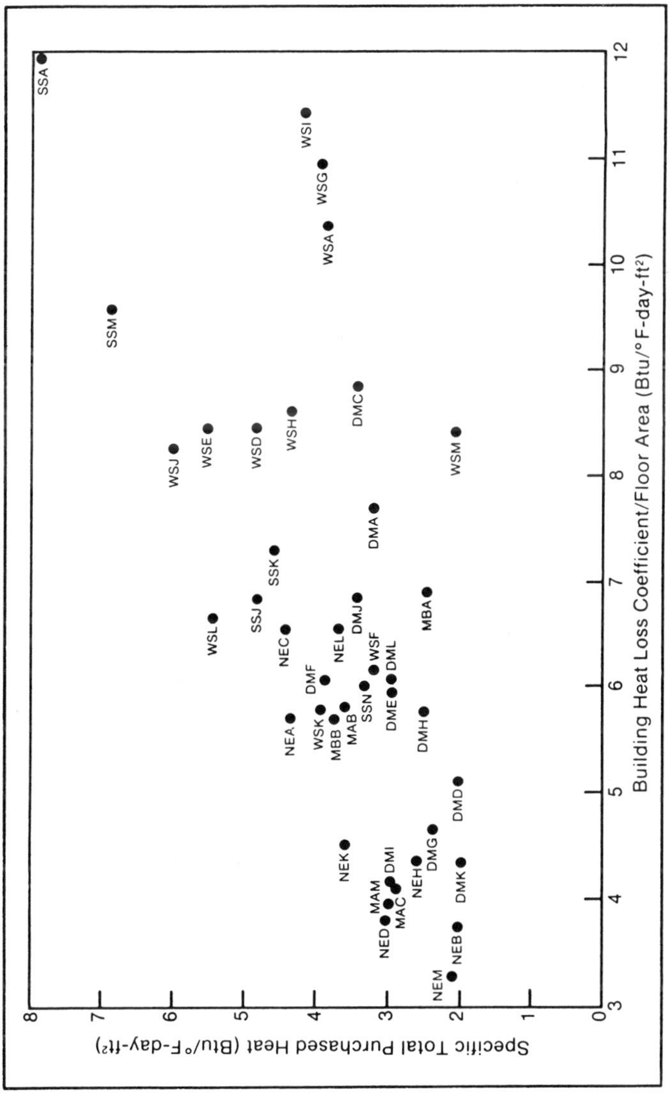

Figure 9.20
Purchased heating energy vs. building heat loss for forty Class B buildings.

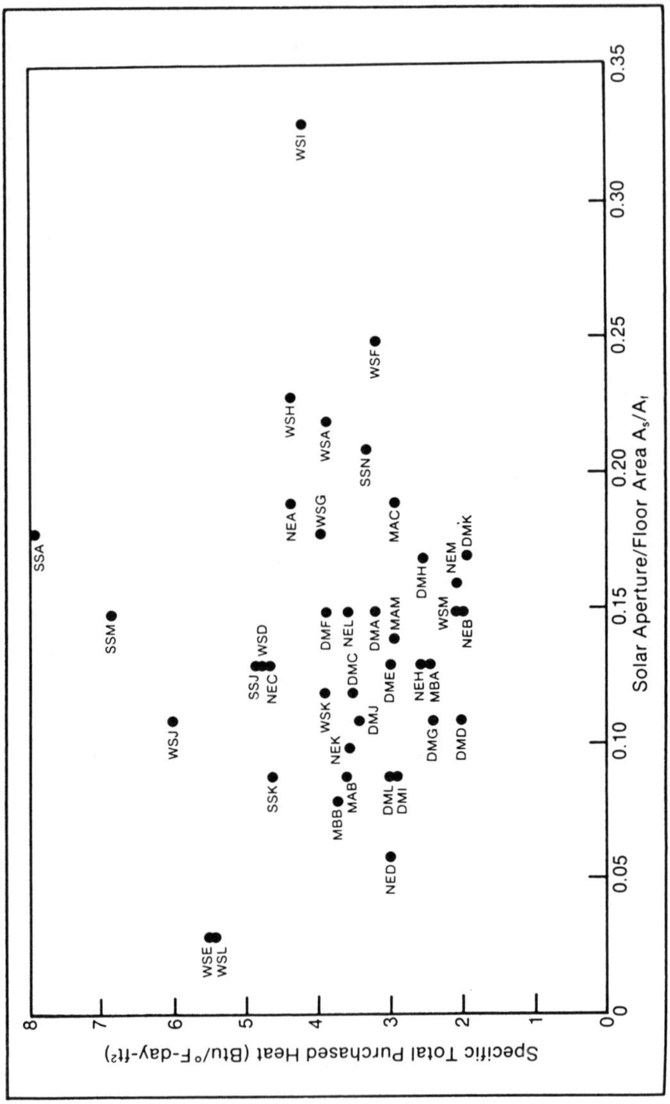

Figure 9.21
Purchased heating energy vs. solar aperture-floor area ratio for forty Class B sites.

Table 9.3
Summary information for SERI site DMH

Location:	Denver, Colo., latitude 39° 45′ N longitude 104° 52′ W, elevation 5280 (1609 m)
Nearest ETMY site:	Denver, Colo.
Building	
Type of building:	multifamily
General type of construction:	frame, two stories
Ground coupling:	slab on grade
Passive/hybrid heating system(s)	
Primary heating system type:	sunspace, rock bed
Primary heating system area:	191 ft^2 (17.8 m^2)
Primary heating system tilt angle:	90°, 71 ft^2 (6.6 m^2) 45°, 120 ft^2 (11.2 m^2)
Other system type:	none
Movable insulation:	none
Auxiliary heating system	
System and distribution type:	natural gas, forced air
Net system delivery efficiency:	0.73
Day thermostat setting:	65°F (18.3°C)
Night thermostat setting:	60°F (15.6°C)
Living zone floor area:	1088 ft^2 (101.1 m^2)
Passive solar aperture area:	191 ft^2 (17.8 m^2)
Building heat loss coefficient:	260 Btu/h °F (137 W/°C)
Air infiltration rate*:	0.77 air change/h

* Based on average winter temperature of 37°F; wind speed of 7 mph.

this summary are photographs of each building, sketches of floor plans and sections, and a list of monitored data points. Figure 9.22 is the monthly performance of the building broken down into the internal, auxiliary, and passive heat categories. Table 9.4 contains a summary of this information along with statistics about weather and building temperatures. Figure 9.23 contains a plot of winter data for the building, intended to show the operation of the building under different ambient conditions. During the first day of the four-day period, temperatures are cool and sunshine is plentiful. Ambient temperatures drop throughout the second and third days, and the fourth day is cloudy but warmer. The responses of the building, sunspace, and auxiliary systems can be clearly seen in the plot.

9.3.6.2.2 CALIFORNIA SITES Significant contributions were made to the Class B program by sites in California, performed at the California State

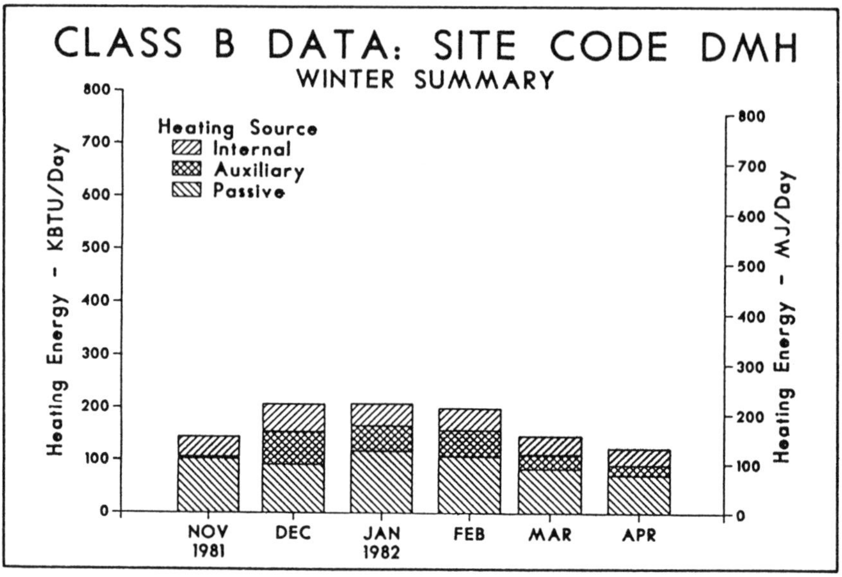

Figure 9.22
Heating season summary for the SERI site DMH.

University, Sacramento, for the California State Energy Commission with guidance and instrumentation from SERI. The California group made some major changes that eventually became part of the standard procedures (Shea et al. 1983). One of the most noteworthy changes was the adoption of the Sherman-Grimsrud infiltration model (Grimsrud, Sherman, and Sonderegger 1983), in which infiltration was calculated as a function of indoor-outdoor temperature difference and wind speed, separating infiltration from the conductive heat loss by performing both blower-door and tracer-gas measurements on eleven of their test houses. A plot of air changes per hour (ACH) predicted from blower-door measurements and application of the Sherman-Grimsrud equation versus tracer-gas measurements is shown in figure 9.24 (Mahajan et al. 1983b).

Other contributions include the direct measurement of natural gas consumed by furnaces and water heaters (instead of inferring natural gas consumption from measurements of flue gas temperatures) and an improved process of data transcription from the cassette tapes.

The researchers in California also compared measured auxiliary heating and cooling energy consumed in three of their monitored houses to simu-

Table 9.4
Performance summary for the SERI site DMH

British units

Month	Number of days of complete data	Radiation (kBtu/ft² day) Horizontal	Radiation Incident	Average temperature (°F) Outdoor	Average temperature Indoor	Degree-days (65°F) (°F day)	Total load	Heating energy balance (kBtu/day) Passive	Heating energy balance Auxiliary	Heating energy balance Internal	Specific purchased heat (Btu/°F day ft²) Total purchased	Specific purchased heat Auxiliary
Nov	14	0.89	1.65	46.2	70.7	563	152	107 (71%)	4 (3%)	40 (27%)	1.68	0.16
Dec	24	0.59	1.50	32.1	67.0	1021	218	96 (44%)	66 (30%)	55 (25%)	3.19	1.74
Jan	31	0.72	1.47	30.0	65.2	1085	219	124 (57%)	50 (23%)	44 (20%)	2.45	1.31
Feb	28	0.97	1.56	31.3	65.1	944	210	113 (54%)	53 (25%)	43 (21%)	2.61	1.44
Mar	31	1.34	1.35	41.0	66.0	744	156	88 (57%)	29 (19%)	38 (24%)	2.46	1.07
Apr	30	1.73	1.17	47.7	68.8	519	131	76 (57%)	20 (16%)	34 (26%)	2.35	0.88
	Average	1.04	1.45	38.1	67.1	—	181	101 (56%)	37 (20%)	42 (23%)	2.50	1.17

SI units

Month	Number of days of complete data	Radiation (MJ/m² day) Horizontal	Radiation Incident	Average temperature (°C) Outdoor	Average temperature Indoor	Degree-days (18.3°C) (°C day)	Total load	Heating energy balance (MJ/day) Passive	Heating energy balance Auxiliary	Heating energy balance Internal	Specific purchased heat (kJ/°C day m²) Total purchased	Specific purchased heat Auxiliary
Nov	14	10.1	18.7	7.9	21.5	313	160	113 (71%)	4 (3%)	42 (27%)	34.3	3.2
Dec	24	6.7	17.1	0.0	19.5	567	230	102 (44%)	70 (30%)	58 (25%)	65.2	35.5
Jan	31	8.1	16.7	−1.1	18.4	603	231	131 (57%)	53 (23%)	46 (20%)	50.0	26.8
Feb	28	11.0	17.7	−0.4	18.4	524	222	119 (54%)	56 (25%)	46 (21%)	53.3	29.4
Mar	31	15.2	15.3	5.0	18.9	413	164	93 (57%)	31 (19%)	40 (24%)	50.2	21.9
Apr	30	19.7	13.3	8.7	20.4	288	138	80 (576%)	22 (16%)	36 (26%)	48.0	17.9
	Average	11.8	16.4	3.4	19.5	—	191	106 (56%)	39 (20%)	45 (23%)	51.1	23.9

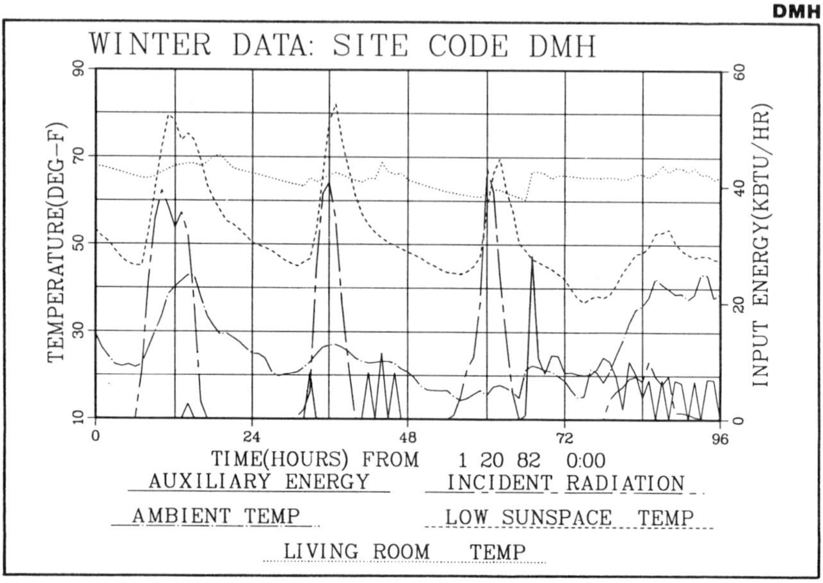

Figure 9.23
Data plots for the SERI site DMH.

lated consumption using the DOE 2.1 and CALPAS 3.01 computer codes (Mahajan et al. 1983b). Discrepancies between predicted values and measurements were most commonly attributed to the actual weather being different than the values used in the simulations and to setpoint temperatures being different than assumed in the simulations.

9.3.6.2.3 COMPARISON WITH SLR Duffy and Odegard (1984) compared auxiliary energy predictions made using SLR correlations (see Jones and Wray, chapter 4 of this volume) with Class B estimates of auxiliary and internal gains. They used information from fifty-four sites, comprising sixty-nine site-years of heating-season data. Their work presents three different methods for performing SLR predictions of auxiliary energy, each using a different set of heating degree-days. One is based on measured degree-days calculated from measured interior and exterior temperatures. The second is based on balance point temperatures, and the third is based on degree-days from historical weather records. The first method minimizes the dependence of the predicted auxiliary on the heat loss coefficient that is used along with internal gains to estimate the balance-point temper-

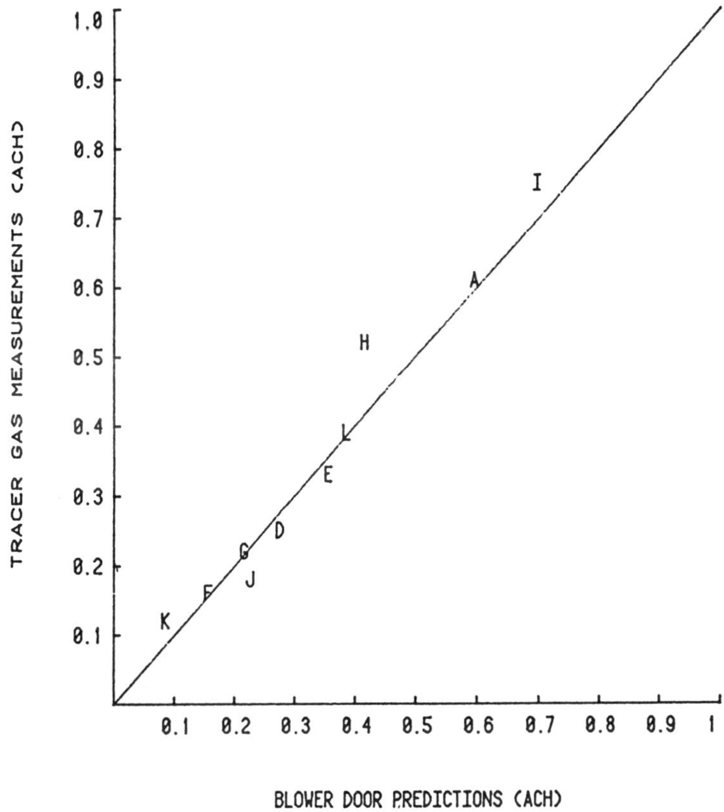

Figure 9.24
Comparison of the infiltration rate (ACH) in eleven Class B houses in California using tracer-gas measurements and the blower-door methodology.

ature. The second produces an SLR prediction that can be compared directly to measured auxiliary energy. The third method was used to estimate the differences in predictions that result from using long-term weather data instead of the measured weather from specific monitoring periods.

In addition to the three different sets of degree-days for making the SLR predictions, the solar savings fraction (SSF) sensitivity curves were applied in three different ways to account for the specific construction and geometries of the buildings that affect its thermal characteristics. The result is eight unique predictions of auxiliary (or auxiliary plus internal gains) energy.

Duffy and Odegard found that the difference between the predicted and measured auxiliary energy is small when the averages over all the houses are compared, but that the difference between predicted and measured for individual houses is relatively large on the average. They note that the averages over all the houses range from near zero to 11% of the mean measured auxiliary energy for the eight different predictions. The SLR method using measured temperatures produced the most accurate predictions of average energy consumption. The difference between average measured and predicted auxiliary energy using balance point temperatures and the long term weather data produced results of approximately the same level of error. The different methods for correcting the predicted auxiliary energy values based on the sensitivity curves tended to increase the error for the averages over all houses, though differences for some individual houses decreased. Using the historical weather data introduces little additional random error between the predicted and measured auxiliary energy values, relative to the other uncertainties in both the SLR method of prediction and the Class B monitoring and analysis process. There is a systematic tendency for underprediction for houses with relatively small auxiliary requirements and a tendency for overprediction for houses with large auxiliary requirements.

To investigate possible causes of discrepancies between measured and predicted values, Duffy and Odegard divided the population of houses into subsamples according to geographic area, passive system type, and relative amount of thermal mass. They found that for direct gain houses, auxiliary is overpredicted for buildings with low mass and underpredicted for buildings with high thermal mass. On the average, however, no clear pattern is evident in the comparison of predicted and measured auxiliary for houses with different thermal-mass levels. The sample of sunspace houses was not large enough for meaningful comparison. They also found that no clear pattern is evident between predicted and measured auxiliary for various regions of the country.

Duffy and Odegard also investigated the measured data and noted that a probable source of some of the discrepancy between measured and predicted auxiliary values may be due to discrepancies between the measured and actual heat loss coefficient values for the buildings. Predicted minus measured auxiliary energy versus the absorbed solar minus useful solar energy are plotted in figure 9.25. Useful solar energy, calculated using the Class B methodology, depends on the measured loss coefficients. Absorbed solar energy was calculated independently from pyranometer measurements. Since the discrepancy in auxiliary energy appears to be correlated with this difference, it is argued that errors in the loss coefficient accounts for much of the discrepancy.

Duffy and Odegard mention additional sources of error as being differences between the reference designs of the SLR method and the actual buildings, and the way the SLR method treats multiple passive system types in a single building. Sensitivities on these parameters are not reported.

References: Acorn Structures 1982, Duffy 1982, Dynamic Homes 1982, Fowlkes 1981, 1982, Frey, Holtz, and Swisher 1982, Frey McKinstry, and Swisher 1982, Gustashaw 1983a, 1983b, Halperin and Swisher 1982, Hamilton and Sachs 1982, Holtz and Hamilton 1980, Holtz and Frey 1982, Holtz and Swisher 1982, Holtz et al. 1985, Lofchie, Noble, and Duffy 1983, Mahajan et al. 1981, 1983a, 1983b, MASEC 1981, Palmiter Hamilton, and Holtz 1979, SERI 1984, Shea et al. 1983, Solar Age 1983, Swisher 1981, 1982a, 1982b, Swisher and Cowing 1983, Swisher and Duffy 1983, Swisher, Frey, and Holtz 1984, Thornton 1982, 1983.

9.3.6.3 The SERI Class C Program The Class C program at SERI was quite different from the Class A and B programs. Where Class A and Class B relied almost exclusively on direct measurements made in buildings, Class C was based on information gathered by a trained interviewer who obtained subjective nontechnical information by assisting building occupants with completing a questionnaire, by gathering technical information about the buildings, and by receiving energy information in the form of fuel bills and weather data from the nearest weather station. The Class C

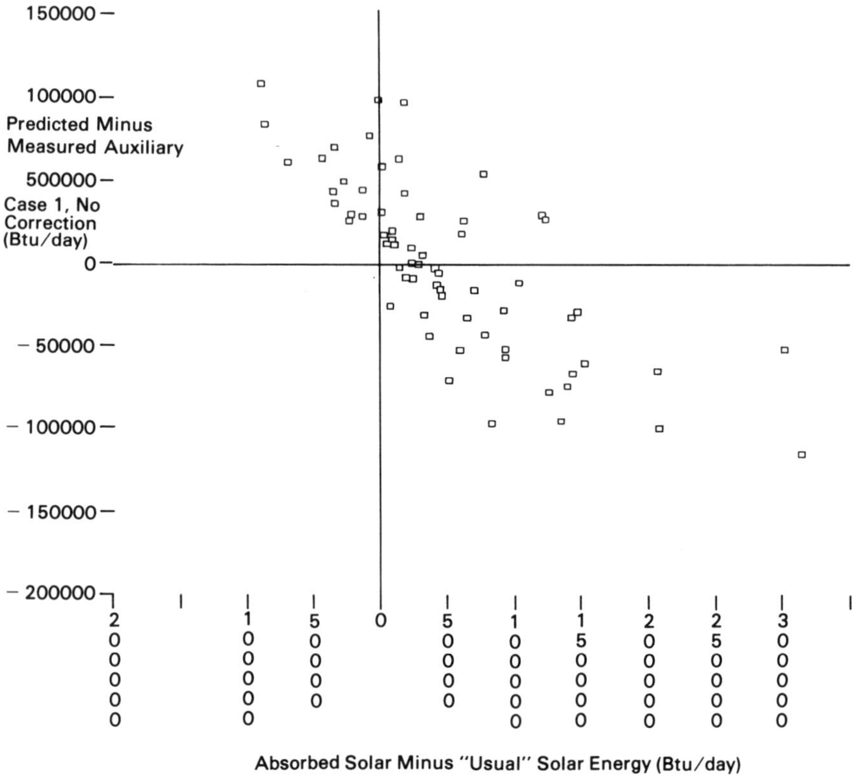

Figure 9.25
Predicted minus auxiliary energy compared to absorbed solar minus useful solar energy.

program, though directed by SERI, was conducted primarily by the Regional Solar Energy Centers (RSECs).

Organization: Solar Energy Research Institute and Regional Solar Energy Centers
Number of buildings: 335
Type: residential
Location: throughout the United States
Monitoring period: 1980 and 1981
Components or whole buildings: whole buildings
Data collection methods: M and U
Type of data collected: I 1; II 1; III 1, 2; V 1
Data analysis technique: A and S
Monitoring approach: 1
Building occupancy: occupied
Significance and results: The Class C program was the first large-scale program to gather information about passive solar buildings, occupant perceptions of these buildings, their energy consumption, and costs. This effort was valuable for identifying the most common problems in passive buildings, as shown in table 9.5, and for identifying areas for future research and development. This program produced information about energy consumption as well as building costs, interior temperatures, and occupant comfort.

References: Hamilton et al. 1981, Saras 1981.

9.3.7 Bonneville Power Administration

Organization: Bonneville Power Administration (BPA)
Number of buildings: 16
Type: residential
Location: Spokane, Washington, and Portland, Oregon
Monitoring period: 1982 and 1983
Components or whole buildings: whole buildings
Data collection methods: C
Type of data collected: I 1; II 1, 2, 3; III 1, 2, 3, 4; IV 2
Data analysis technique: A and E
Monitoring approach: 4
Building occupancy: occupied and unoccupied
Significance and results: The Bonneville Power Administration (BPA) pat-

Table 9.5
Problems in passive homes identified by Class C methods

Problem	Number of sites where this has ever been a problem	Percentage of total number of sites audited
Overheating	181	42.7
Keeping windows or other glass clean	124	37.0
Condensation on windows	114	34.0
Glare	94	28.1
Not warm enough	62	18.5
Weather-stripping or caulking	62	18.5
Fading of furniture, walls, or coverings	61	18.2
Excess humidity	59	17.6
Drafts	58	17.3
Rooms cool down too fast	55	16.4
Extreme temperature swings	52	15.5
Lack of privacy	47	14.0
Mechanical/electrical system failure	38	11.3
Stagnant odors	25	7.5
Covering sloped windows	23	6.9
Assured solar access	14	4.2
Zoning restrictions	9	2.7
Other	33	9.9

terned their residential passive solar monitoring program after the SERI Class B program. The methodology and equipment used are identical. The buildings were located in two cities, Spokane, Washington, and Portland, Oregon. They are identified in figure 9.18 with the prefix NW.

All the buildings were frame construction. Most of them had 2 × 6-in., R-19 (5 × 15-cm, 3.4 m^2 °C/W) walls and R-38 (6.7 m^2 °C/W) ceilings. Windows were double glazed, and some buildings had air-lock entries. Careful attention was paid to caulking and weather-stripping. The average overall heat loss coefficient was 322 Btu/h °F (170 W/°C). Many of the houses were electrically heated. The heating system efficiency of the buildings with gas furnaces was measured. The average overall heating system efficiency of the gas systems was 46%. This low efficiency was due in part to furnaces being located in unconditioned spaces with uninsulated ducts.

Ten of the houses had sunspaces, two had masonry thermal-storage walls, and four used direct gain only. BPA participated in design reviews on all the homes so that the buildings would be exemplary homes. Most of the sunspaces had a transfer fan to move heat into the living area during the heating season, and some had exhaust fans to vent heat in summer.

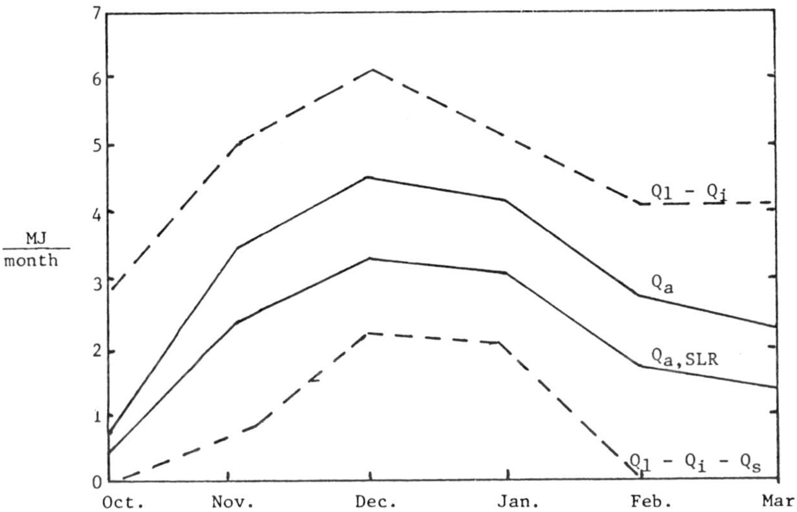

Figure 9.26
Monthly auxiliary heating: measured, predicted, and bounded for the BPA Class B homes.

BPA looked at the bounds for the auxiliary energy in their group of houses. The upper bound is the heat loss from the living zone minus the internal heat gains, $Q_l - Q_i$, representing no utilization of available solar energy. The lower bound is the heat loss from the living zone minus the internal heat gains minus the available solar energy, $Q_l - Q_i - Q_s$, representing full utilization of solar energy. The actual auxiliary heat is expected to be between these two bounds.

BPA then calculated a predicted auxiliary value using solar load ratio (SLR) correlations (Balcomb et al. 1984), measured values for degree-days and insolation, and internal gains were used. Monthly building loss coefficients were derived from measured loads, corrected for wind, and measured interior temperatures were used for Tset. The predicted auxiliary heating is designated as $Q_{a,SLR}$. Results of this analysis and the bounds described above are shown in figure 9.26, which shows these parameters averaged over all the houses. The measured auxiliary exceeds the predicted values in all months. Looking at system types, BPA found that direct-gain systems performed very close to predicted values but that the sunspace systems performed significantly worse. In fact, sunspaces used almost twice as much auxiliary energy as predicted in the BPA analysis.

BPA discusses possible sources of error in the analysis. One possible source is identified as the determination of the loss coefficient and the monthly adjustment for wind. A second is the determination of auxiliary energy in some houses based on a single heating system efficiency value. Another is the measurement of internal gains and adjustments applied for the amount of heat delivered to the space. The variation between the predicted values and the actual values was only significant for sunspace systems and was apparently due to poor utilization of available solar energy. The main reason identified is poor transport of heat to the living space. BPA notes that fans specified by designers for this purpose were not always correctly, or even completely, installed by HVAC contractors. Some were tied to the backup heating systems, but delivery temperatures were too cool for delivery at furnace flow speeds. It should be noted that there are no SLR correlations for sunspaces with fan delivery and that sunspaces with fan delivery are usually less efficient than those with natural-convection delivery (Jones, McFarland, and Lazarus 1983).

References: McKinstry, et al. 1983.

9.3.8 Brookhaven National Laboratory

Brookhaven monitored in detail two very-low-energy-use houses in the New England area of vastly different designs.

9.3.8.1 The Brookhaven House

Organization: Brookhaven National Laboratory
Number of buildings: 1
Type: residential
Location: Upton, Long Island, New York
Monitoring period: 1980 and 1981
Components or whole buildings: both
Data collection methods: C
Type of data collected: I 1, 3, 4, 5; II 1, 2, 3, 4; III 1, 3
Data analysis technique: A, S, and E
Monitoring approach: 3, 4, and 5
Building occupancy: unoccupied
Significance and results: The first building discussed was called the Brookhaven Natural Thermal-Storage House or, simply, the Brookhaven House. It is a two-story, 2000-ft^2 (186-m^2) house located on the grounds of the Brookhaven National Laboratory. Superinsulation and passive solar sys-

tems are incorporated into its traditional New England design. It has two types of passive systems. One is an unvented Trombe wall with 196 ft^2 (18.2 m^2) of aperture area and the other a sunspace with a 245-ft^2 (22.8-m^2) aperture. The sunspace has a mass wall between it and the living zone, which is identical in size and construction to the Trombe wall. The building is well insulated with total thermal resistance values of R-27 (4.8 m^2 °C/W) in the walls, R-38 (6.7 m^2 °C/W) in the attic, R-10 (1.8 m^2 °C/W) around and under the foundation and basement slab, and R-5 (0.9 m^2 °C/W) under the crawl space. Heat from the backup system is supplied only to the first floor and returned from the second floor. Heat rises to the second floor through the stairway and through ceiling grills that are connected to registers in each room on the upper level.

One method employed to analyze the performance of the house was a whole-building energy balance. Coheating was used to measure conductive losses, and both tracer gas and pressurization were used to determine infiltration rates. This house is extremely conserving of auxiliary heating energy. Purchased auxiliary energy during the 1981–82 heating season was 2.1 Btu/ft^2 °F day (12 Wh/m^2 °C day), 25% of the normal usage for conventional houses in this region. The maximum daily solar heating fraction was 94%. The minimum value was 60%, occurring on a cloudy day. Temperature measurements throughout the house indicate uniform comfort. The second floor receives adequate distribution of heat through the stairway and thermal-transfer grills.

Because the Trombe wall and sunspace thermal-storage masses are of similar construction and equal area, their performance was compared by the researchers at Brookhaven. Heat gains and losses, hourly temperature profiles, and peaks were examined. They found that the maximum solar contributions of the Trombe and sunspace walls were 15 and 7 Btu/h ft^2 (47 and 22 W/m^2), respectively, which demonstrates a twofold superiority of the Trombe wall over the sunspace. Suggested improvements to the sunspace included removing snow when it occurs and decreasing losses through the glazing with night insulation during nonsolar hours.

References: Adams 1982, Ghaffari and Jones 1982, Temple 1980.

9.3.8.2 The Mastin House

Organization: Brookhaven National Laboratory
Number of buildings: 1
Type: residential

Location: Middletown, Rhode Island
Monitoring period: 1979–1980
Components or whole buildings: both
Data collection methods: E
Type of data collected: I 1, 4; II 1, 2, 5; III 1, 3; IV 4
Data analysis technique: A, S, and E
Monitoring approach: 2 and 4
Building occupancy: occupied
Significance and results: The Maştin house is a double-envelope building. The house has three stories, including the basement, with 2,000 ft^2 (186 m^2) of living area. The house is of wood-frame construction with single walls of 2 × 6-in. (5 × 15-cm) studs and R-19 (3.4 m^2 °C/W) insulation on the east and west. The double north wall is constructed with 2 × 4-in. (5 × 9-cm) studs on the inside and 2 × 6-in. (5 × 15-cm) studs on the outside, with an airspace of 8 in. (20 cm) in between. The inner and outer walls have R-11 (1.9 m^2 °C/W) and R-19 (3.4 m^2 °C/W) insulating, respectively. The top-floor ceiling is similarly constructed. There are 520 ft^2 (48 m^2) of double-glazed, south-facing windows, two-thirds of which are on a 45° sloped roof of the sunspace. The other three elevations have minimal glazing. Below the basement floor is a 4-in. (10-cm) concrete slab poured on the excavated earth. The basement floor itself is supported by wood joists supported by 8-in. (20-cm) concrete block. The floor joist space is insulated by R-11 (1.9 m^2 °C/W) fiberglass batts.

The performance-monitoring of the Mastin house is the first rigorous scientific study of the performance of a double-envelope house. The Mastin house is a low user of auxiliary energy, 2.3 Btu/ft^2 °F day (13 Wh/m^2 °C day), about one-quarter the amount of energy of typical older houses with the same floor area.

The loop in a double-envelope house is supposed to distribute heat from the sunspace by thermocirculation, storing energy in the ground under the crawlspace. Reverse flow is supposed to deliver stored heat at night. The researchers at Brookhaven found that no significant storage and release of heat takes place in the house. They also found such extreme temperature stratification that there is a need for auxiliary heat on the lower level when the upper level is overheated.

Brookhaven introduced an interesting and useful method of plotting the data. Shown in figure 9.27 is a plot of auxiliary energy requirement normalized to the heating degree-days (Btu/°F day) versus weighted insolation

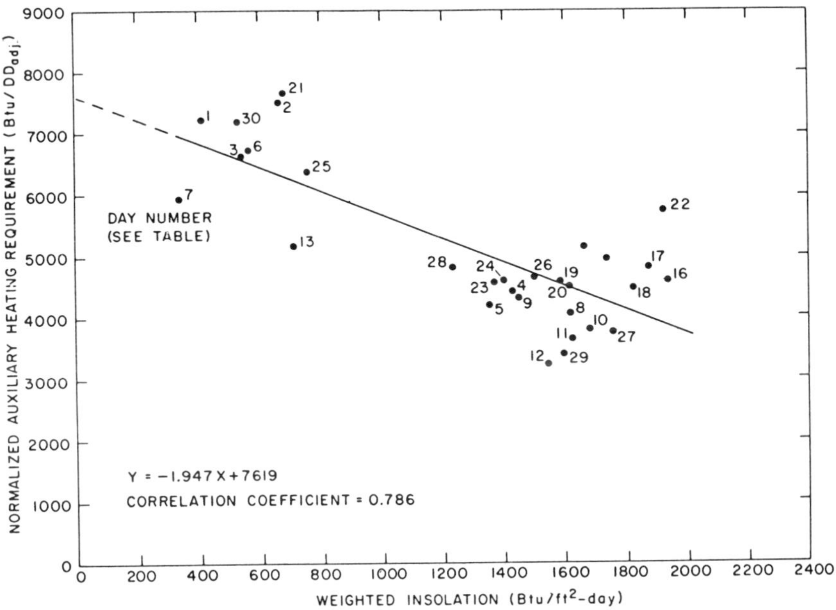

Figure 9.27
Auxiliary heating requirements for the Mastin house vs. insolation, January 17 to February 16, 1980.

(Btu/ft^2-day). The data are in two main clusters corresponding to sunny and cloudy days. The *y*-intercept value is the heat requirement for no solar input. The data in figure 9.27 were taken with the house operating as designed. During a subsequent period of time, the loop was blocked to prevent any air flow. If the convective loop actually stored heat for later release to the building, a significant change in the performance line should result. The second performance line is shown in figure 9.28 (marked "*convective loop*" blocked) along with the original line. Brookhaven conclude that the change in performance after blocking the loop is not large enough to be conclusive, considering experimental errors, but it did not harm the performance.

In another set of tests, the upper row of windows was blocked with insulation to decrease the aperture area. During this time, there was clearly an improvement in the performance of the building. Brookhaven concluded that most or all of the benefit derived from solar gain through the

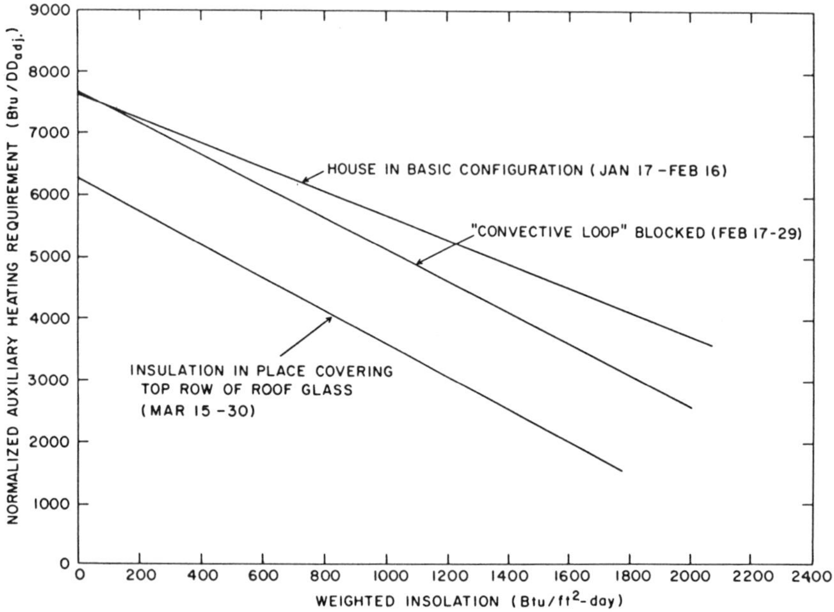

Figure 9.28
Comparison of auxiliary heating requirements for three configurations of the Mastin house.

upper row of roof glass is offset by heat loss through the glass and that the sloped glazing causes daytime overheating.

References: R. F. Jones 1981.

9.3.9 NAHB Research Foundation

The NAHB Research Foundation, Inc., has participated in a number of the programs that were discussed above. They were involved with the design of a number of the houses that were monitored by the NSDN, and they were the recipient of the Class B monitoring program when it was transferred from SERI. They assumed responsibility for the Class B program in 1983, when sixteen buildings were being monitored by SERI. The NAHB Research Foundation completed the monitoring responsibilities at these buildings. In addition, they reviewed data that had been collected by SERI and produced a collection of results that reflect the requirements for information voiced by members of an advisory panel representing the views of various segments of the housing industry. The results presented here are from the NAHB Research Foundation's work with the Class B data base.

Organization: NAHB Research Foundation, Inc.
Number of buildings: 16
Type: residential
Location: throughout the United States
Monitoring period: 1983–1985
Components or whole buildings: whole buildings
Data collection methods: C
Type of data collected: I 1; II 1, 2, 3; III 1, 3, 4; IV 2
Data analysis technique: A, S, and E
Monitoring approach: 4
Building occupancy: occupied
Significance and results: The work done by the NAHB Research Foundation, Inc., (NAHB/RF) is significant for a number of reasons. NAHB/RF is the research arm of the National Association of Home Builders. Their interest in performance evaluation of housing constitutes an major shift from government laboratories to private industry in performing these activities. NAHB/RF also worked closely with an advisory group known as the Residential Passive Solar Performance Evaluation Council (RPSPEC) to define the types of data analysis and reporting that were useful to industry groups. Two activities came from this work. One was to analyze the existing Class B data base to derive information for the industry that had not yet been presented to them, and the second was to standardize whole-building testing and evaluation procedures, and then identify equipment that was suitable at reasonable cost to implement them. The first activity is complete and the second is ongoing. Highlights of these studies are presented below.

One issue studied was the effect of movable insulation for improving the operation of passive buildings. The data from seventy buildings, of which twenty-six had some type of movable insulation, was examined. Two statistics, the building performance index (BPI) and the passive heating fraction (PHF) were compared for the two groups. The BPI is the total purchased heat (from the auxiliary system, lights, and appliances) delivered to the living zone, divided by the living zone area and the number of degree-days (Btu/ft^2 °F day). The PHF is the utilized passive heat divided by the total load. The BPIs were calculated as 3.55 ± 1.32 for the group with movable insulation and 3.73 ± 2.83 for the group without it. Similarly, the PHFs were calculated as 0.35 ± 0.13 and 0.38 ± 0.16, respectively. The 5% difference in BPI between the two groups is negligible. Possible reasons for the small differences in performance are that the actual operation of the movable insulation is so irregular that it negates much of the insulation's potential performance benefit, or that the insulation does not perform as it is predicted to, or both. Hourly data on the position of night insulation was analyzed from eight buildings to examine if the insulation was being used correctly on south windows. Using an analytical model, it was found that insulation was operated well in some houses, very poorly in some, and that on the average had no net benefit.

A second study examined the efficiencies of the gas-fired, forced-air auxiliary heating systems. The data for this analysis came from the original one-time tests consisting of building load measurements using coheating and heating system efficiency measurements. The heating systems efficiencies ranged from 33% to 73%, with an average value of 49%. Though all the furnace systems were essentially new, these numbers are lower than one might expect. Combustion efficiency of eleven furnaces was measured, and all values were approximately 80%. Further examination of the data showed that the larger the furnace size relative to design-condition heat loss, the lower the measured efficiency. This result suggests that a large reason for the low efficiencies was the frequent on/off cycling of oversized furnaces.

Poor transfer of heat from sunspaces to living spaces was noted from the Class-B data. This problem was also noted by the Bonneville Power Administration. For example, the operation of an isolatable sunspace at building DME in Arvada, Colorado, is illustrated by the plot of measured temperatures in figure 9.29. During the first day, a Friday, the occupants are at work, and the door between the sunspace and living zone is closed. The day is sunny and temperatures in the sunspace are quite high. Auxiliary energy is needed most of the night. During the next two sunny days, the doors to the sunspace are open. Sunspace temperatures are cooler, living space temperatures are warmer, and less auxiliary is needed. Obviously, natural convective heat transfer through the door opening cannot occur if the door is closed. The example may be seen as illustrating an operational error on the part of the occupants, or it may be seen as a design error if it is unrealistic to expect occupants to operate the sunspace door correctly.

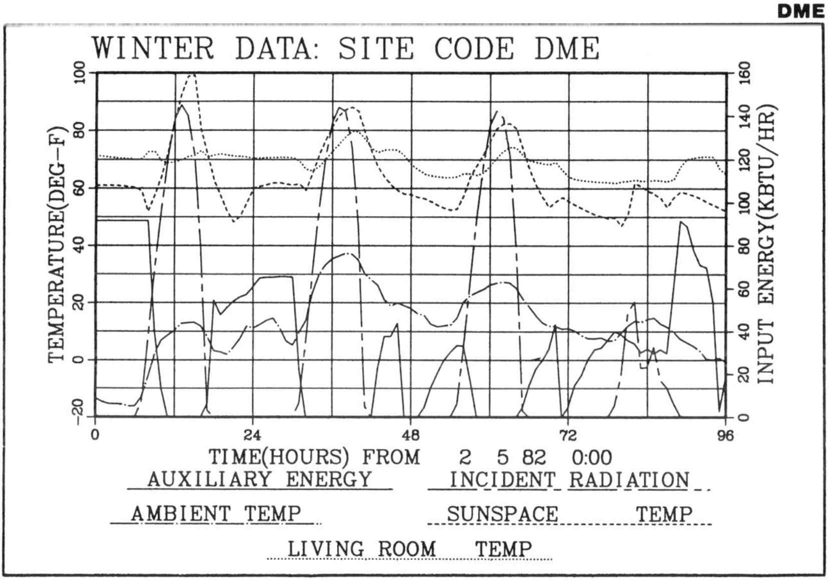

Figure 9.29
Hourly data plots for Class B site DME.

Other problems noted by NAHB/RF were homes with actively charged rock beds in which heat loss from the bed significantly decreased the performance of the bed, similar to the problems noted by Los Alamos. NAHB/RF also noted an earth-sheltered house with a continuous, though small, requirement for auxiliary heat and numerous houses with poor distribution of heat, with some zones too warm and other requiring, auxiliary heat to maintain comfort, such as was discussed by Brookhaven for the Mastin house. These issues point out the need for proper design for thermocirculation in passive buildings and the need to properly integrate auxiliary heating into them.

NAHB/RF looked at the cooling season impact of passive designs but did not form strong conclusions. Some of the houses were consistently warm, while others remained cool. There was no systematic tendency to overheat, and a number of the buildings performed very well in summer, such as the Roberts house reported by NSDN.

NAHB/RF went beyond the original Class B scope to examine some consumer, marketing, and economic issues, using a sample of twenty-two home owners and builders. Twenty of the owners were very satisfied with their houses. The two who were not had significant flaws in their homes. One was earth-sheltered and was not well insulated from the ground, which lead to large heat losses and high energy bills in the winter. The other was the owner of a Trombe-wall house who did not like to operate the overly complicated controls for vents and wall insulation, though heating bills were very low. NAHB/RF also found that the passive homes sold for about as much (per unit of floor area) as conventional homes in the same area and that the tendency was for them to be bought as fast, if not faster, as conventional houses.

References: Bishop and Frey 1984, 1985, Frey and Bishop 1984a, 1984b, Frey and Holtz 1985, Frey 1986.

9.3.10 Tennessee Valley Authority

Organization: Tennessee Valley Authority (TVA)
Number of buildings: 35
Type: residential
Location: throughout the Tennessee Valley
Monitoring period: 1980–1983
Components or whole buildings: both
Data collection methods: C
Type of data collected: I 1, 3, 4, 5; II 1, 2, 3, 4, 5; III 1, 3, 5
Data analysis technique: A, S, and E
Monitoring approach: 3 and 4
Building occupancy: occupied
Significance and results: The Tennessee Valley Authority (TVA) investigated the performance of passive solar houses to determine their performance and potential for offsetting electrical load on their system. TVA produced eleven different passive solar house designs, and thirty-five houses were built in twelve communities throughout the Tennessee Valley. Between two and five homes were built of each design. TVA designated four levels of performance evaluation, though many of the houses were monitored at a level of detail slightly greater than the Class B homes. Results of the TVA work are not widely published.

Some early results show two identical houses that perform quite differently. The performance differences are attributed to differences in the way the occupants use the buildings, such as the setting of thermostats, use of lights and appliances, and opening of windows. Lifestyle differences and ways of accounting for them in performance evaluation have also been reported by Bishop and Frey (1985). Though the energy consumption of these two houses is different, one uses one-fifth and the other about half the energy of a conventional building.

In another building, results showed a mass wall to be a net loser of energy. The building was inspected and it was found that the glazing covering the wall was open at the top and bottom, allowing ambient air to cool the surface of the mass wall.

References: Suhs and Wessling 1980, Robertson 1982.

9.3.11 Passive Solar Commercial Buildings Program

Organization: U. S. Department of Energy
Number of buildings: 19
Type: commercial
Location: throughout the United States (see figure 9.30)
Monitoring period: 1982–1984
Components or whole buildings: whole buildings
Data collection methods: M, U, and C
Type of data collected: I 1; II 1, 2, 3, 5; III 1, 2, 3, 4; IV 2, 3; V 1
Data analysis technique: A, S, and E
Monitoring approach: 1, 2, 3
Building occupancy: occupied
Significance and results: The Passive Solar Commercial Buildings Program was extremely important because it demonstrated that passive solar strategies can save significant amounts of energy in a variety of commercial building types. The buildings in this program were all designed by architects chosen by the building owners, but the designs were all reviewed by experts under contract to DOE. As a result, designs were often modified and performance was optimized so that this was a group of exemplary buildings. Performance predictions were made by the designers, and measured performance was compared to the predicted values. The locations of the buildings are shown in figure 9.30.

Performance evaluation of the buildings was designed to measure weather conditions and to submeter energy consumption by end use. Seven of the buildings had manually read submeters, and twelve had automated data acquisition systems. Procedures for monitoring the buildings and reporting the data were standardized. In addition to the energy and weather data, information was collected from the occupants on a

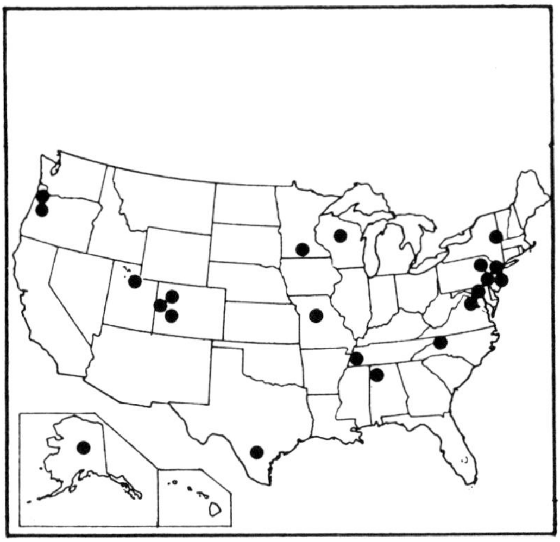

Figure 9.30
Locations of commercial buildings.

regular basis regarding their thermal and visual comfort, their interaction with the building, and deviations from their normal building-use patterns.

A variety of results came out of this program with information about building design, performance, and occupant interaction. One of the lessons about design was that passive strategies used for residential buildings are not directly applicable to nonresidential buildings. Residential passive strategies applied to these buildings often exaggerated loads rather than decreasing them. In these buildings, lighting rather than heating is often a major portion of the energy cost, and the timing for heating needs is different than in residential buildings. Table 9.6 contains information about each of the buildings as well as the primary energy costs of the base-case building (with no passive solar features). The final design solutions for a number of buildings incorporated high, light-diffusing vertical glass clerestories and roof monitors for daylighting that also functioned as solar collectors but were properly shaded to prevent summer and fall overheating. As many buildings had perimeter daylighting as had clerestories, but reports of glare by building occupants were more frequent in these buildings. There was no evidence that the contribution of light shelves was significant.

Table 9.6
Base-case building energy costs

Project #	Building type	Location	Primary energy use: Heat	DHW	Cool	Lights	Other
1	Medium office	Salt Lake City, UT	**		*	***	
2	School	Fairbanks, AK	***	*		**	
3	Senior citizens	Baltimore, MD	***		**	*	
4	School	Bessemer, AL	*		***	**	
5	Library	Mt. Airy, NC			*	***	**
6	Auto maintenance	Philadelphia, PA	***			**	*
7	College	Glenwood Springs, CO	***			*	**
8	Workshop	New Braunfels, TX	**		***	*	
9	Church addition	Columbia, MO	***		*	**	
10	Recreation	Philadelphia, PA	***	*		**	
11	Airport	Gunnison, CO	***			**	*
12	Retail store	Wausau, WI	**		*	***	
13	College	Princeton, NJ	***			**	*
14	Office	Princeton, NJ	***		*	**	
15	Visitors' center	Troy, NY	***			*	**
16	Bank	Wells, MN			**	***	*
17	Greenhouse	Memphis, TN	***		*		**
18	Airport	Grand Junction, CO	**			***	**
19	Gymnasium	Alexandria, VA	**		*	***	

Key: *** Highest energy cost
** Second highest energy cost
* Third highest energy cost
(All costs based on local utility rates.)

A significant philosophy that emerged from the design process for these buildings was that simple designs that serve multiple purposes work the best. This means that that energy-related features of the buildings had to be integrated during the design process rather than tacked on as separate features. Some preliminary designs that were highly automated or controlled were revised to be more simple. Some occupants went to extremes to override thermostats or lighting controls.

Figure 9.31 shows the aggregated results of the monitoring of these buildings. The base-case building information derives from simulations of energy conserving buildings but without the solar features. These buildings conform to the Building Energy Performance Standards (BEPS) that were being developed at about the same time that these buildings were being designed, built, and monitored. The passive solar buildings use 45% less energy than the base-case requirements and 14% less than the BEPS

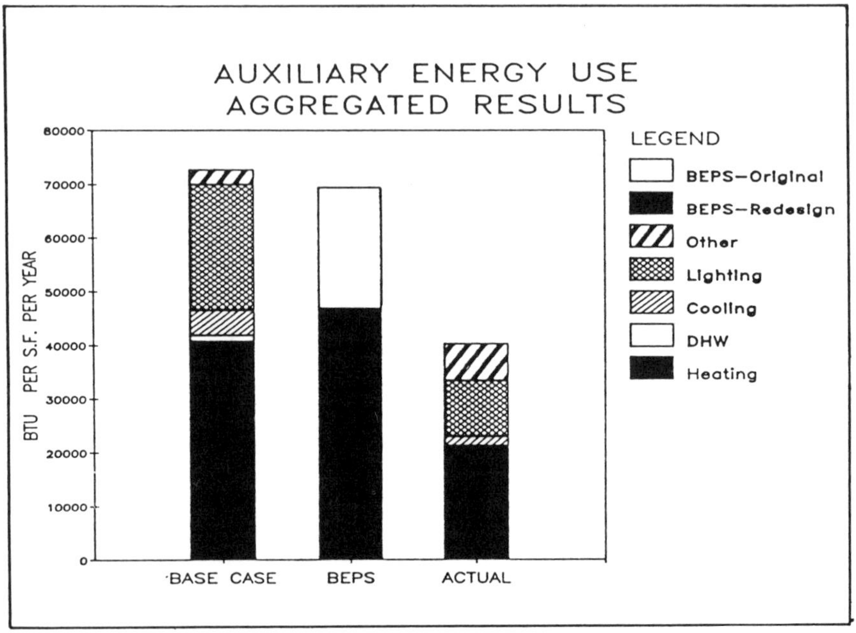

Figure 9.31
Aggregated results of auxiliary energy use.

redesign buildings. The greatest reductions were in cooling (64%), followed by lighting (56%) and heating (49%). Figure 9.32 shows auxiliary energy use on a project-by-project basis. For retrofit buildings, total energy savings averaged 60% over the base-case buildings.

These results dispel the notion that nonresidential passive solar heated buildings will incur cooling penalties that offset the savings. They also show that daylit buildings do not suffer high heating and cooling loads. Indeed, the energy for all three purposes was reduced by over one-half. Evidence from these buildings points to the hypothesis that good daylighting design reduces internal heat gains from artificial lighting to achieve a net cooling reduction. Additional reduction in load is achieved by natural ventilation strategies.

An important component of the monitoring of these buildings was the monthly questionnaires that were completed by both regular and part-time users of the buildings. Occupant-related issues that influenced energy

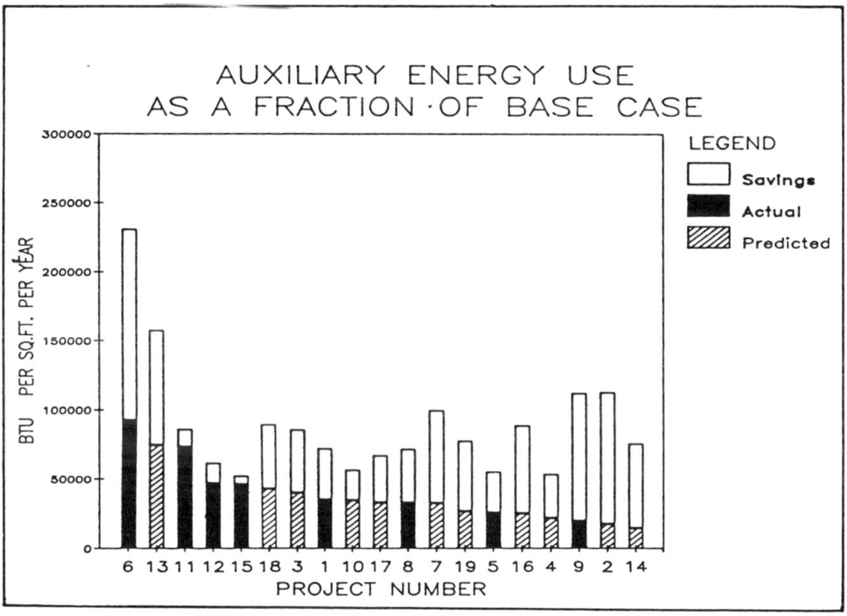

Figure 9.32
Auxiliary energy use as a fraction of base-case requirements.

consumption were identified as changed use (extended hours of operation, greater number of occupants than initially planned, and different activities than planned), complex instructions for building operation, conflicts between solar strategies (natural ventilation blocked by shading devices, insulating shades raised to admit daylight, etc.), and new building start-up problems. The majority of the users liked the buildings, and 80% reported infrequent problems. The most frequently reported problems were the building being cool in the morning and warm in the afternoon. Significant problems other than thermal ones were related to visual glare and the acoustics of hard, sound reflective surfaces. Very rarely were there complaints of poor quality or insufficient light for either manually or automatically controlled lighting systems.

An analysis of the roof aperture daylighting systems on the Mt. Airy Library and the Community United Methodist Church was performed by researchers at Lawrence Berkeley Laboratory. Their work included data acquisition at the buildings, data acquisition in models of the buildings, and computer simulations using BLAST (Andersson et al. 1985). They found that manual control of the electric lighting performs as well or better than simple automatic controls would perform on these same buildings. Reasons for this conclusion are observations of frequent occupant satisfaction with daylighting levels well below current standards for electric lights, large apertures allowing lights to remain off for extended periods of time, and the usage patterns of these buildings (occupancy does not begin until after sufficient daylight is available). Researchers noted that the potential energy reduction from the use of roof apertures appears far greater for lighting than it does for heating or cooling and that with proper shading the roof apertures do not aggravate the cooling energy requirement.

References: Andersson et al. 1985, BHKR 1982a, 1982b, 1984, Frey et al. 1983, Frey and Yager 1984, Frey, Swisher, and Yager 1984, Gordon, Hart, and Kantrowitz 1984, Gordon et al. 1984, 1985, Kantrowitz 1984, Swisher, Frey, and Holtz 1984.

9.4 Recent Monitoring Activities

A significant number of passive solar residential and nonresidential buildings were monitored in the United States during the first half of the 1980s. A number of organizations were involved in these activities and a significant number of results were produced. The monitoring of passive buildings in the United States has decreased steadily since 1984, and today only a small number of buildings are being monitored. This section contains a brief discussion of recent monitoring activities, mostly sponsored by the U.S. Department of Energy.

9.4.1 Solar Energy Research Institute—Short-Term Monitoring

The Solar Energy Research Institute (SERI) collected data from both residential and commercial buildings in support of its short-term method for predicting energy performance (Subbarao 1984). SERI collected data during the winter of 1987–88 from three buildings as follows:

- The Sargent home, a passive solar residence in Golden, Colorado. Data were collected in the house over a three-day period.
- The Carr home, a conventional residence in Fredericksburg, Virginia. Data were collected in the house over a three-day period.

• The Smith house in Madison, Wisconsin. Data were collected over short periods of time before and after the house was retrofitted with passive solar features.

In addition to these houses, SERI analyzed data collected by Architectural Energy Corporation from the Wonderland Hills home, which is being continuously monitored as part of the IEA Task VIII activities.

Also in support of the short-term monitoring project, SERI collected data in two commercial buildings. One is the Washington Association of Counties building in Olympia, Washington, and the other is the Enerplex building in Princeton, New Jersey, owned by the Prudential Insurance Company.

SERI has created a short-term monitoring procedure consisting of the following steps.

1. Inventory the areas of all components of the building (walls, windows, roof, etc.) and the materials that each is constructed from to arrive at an initial model for the building. (SERI refers to this as the model developed from a building audit.)

2. Perform short-term tests on the building and collect data continuously. These tests include a period of steady coheating, a period when the temperatures are allowed to decay with no heating, and a measurement of the efficiency of the mechanical system. The results of these tests are parameters, determined from regression, that are used to renormalize the initial model of the building.

3. The renormalized model is used to generate the results of the monitoring, such as annual heating and cooling energy requirements and peak loads.

Some of the conclusions that have been drawn by the SERI research team are as follows:

• The building load coefficient derived from the audit, including modeled infiltration, has always been higher than the renormalized value by approximately 20%. Two possible reasons for this are that the audit model ignores radiant heat exchanges and that a heat exchange effect takes place in the walls of the actual building, which tends to moderate the infiltration heat loss.

• The calculated solar gains from the audit description are 10–15% too large. This appears to be a result of ignoring screens, shading, and other window obstructions that occur in the actual building.

• The calculated internal mass values are generally lower by approximately 20% when the furnishings of the building are ignored.

• The procedure requires that the externally coupled mass be computed through the renormalization process. It is too small to approximate accurately from the audit information.

• The procedure does not reliably separate the thermal coupling of the house to the basement from the coupling to the outside air. SERI has not been able to derive repeatable values for this coupling. The analysis generally keeps the basement at a fixed temperature.

• On average, the SERI method has been able to predict short-term energy values to within 10% of measured values. No comparisons have been made between measured long-term energy and predictions using short-term data.

References: Burch et al. 1986, Subbarao 1980, 1982, 1984.

9.4.2 Lawrence Berkeley Laboratory

9.4.2.1 Daylighting Performance Evaluation Methodology Monitoring work at the Lawrence Berkeley Laboratory (LBL) has centered on daylighting. Historically, the need for daylighting evaluation arose from the congressionally mandated Solar in Federal Buildings Program (SFBP). The SFBP resulted in the installation of a large number of active solar heating systems on federally owned buildings, the construction of a somewhat smaller number of passive solar buildings, and an even smaller number of passive systems that were retrofitted to existing buildings. The passive systems provide the buildings not only with heat but lighting as well. The performance evaluation requirements for SFBP, published in the Federal Register, are very specific regarding the information that must be obtained and reported about these systems. Since the solar systems that were funded and installed under the SFBP were predominantly active space heating systems, the performance evaluation specifications are written primarily for these active systems, calling for such information as solar energy collected, solar energy stored, solar energy delivered to the load, solar system efficiency, and fossil fuel energy saved.

At the time that the SFBP performance evaluation requirements were written, there were no standardized evaluation methods that would provide the level of information required. The responsibility to develop such an evaluation was placed by DOE at Rockwell International/ETEC in Canoga Park, California. ETEC contracted with the Solar Energy Research Institute (SERI) to develop a thermal performance evaluation methodology and with LBL to develop a daylighting performance evaluation methodology. The work at SERI is described above and became additional work on the SERI short-term method as applied to nonresidential buildings. The work at LBL resulted in the development of the daylighting performance evaluation methodology (DPEM).

DPEM is a method used to extrapolate the annual effectiveness of a daylighting system from data collected over a limited time period. This process uses short-term measurements to calibrate a simulation model. The simulation model represents the energy performance of the daylighting system. The model used in the current version of DPEM is SuperLite. This method is an improvement over simulation alone, which may not accurately model the specific situation in the real building. In addition, it eliminates the need for long-term data collection to perform annual performance evaluations.

9.4.2.2 Application to the South Staffordshire Water Company Building
A major activity in the U.S./U.K. Bilateral agreement involving LBL in the United States and Data Build in the United Kingdom has been the refinement of DPEM and its application to the evaluation of a building in Walsall, England. The building is owned by the South Staffordshire Water Company (SSWaCo). This project was launched in mid-1986, monitoring began in March 1987, and the evaluation was completed in the fall of 1987.

The building was monitored for exterior irradiation (diffuse sky and global sky values), interior illumination levels, and power usage for electric lights. All measurements on the interior of the building were made in second-floor areas (two floors above the ground floor). A data logger on the second floor recorded and stored all data.

The building is five stories high and has 35,000 ft^2 (3250 m^2) of floor area. The shape of the building is a cube that is modified so that upper floors overhang the floors below to give the building the appearance of an inverted pyramid that provides shading to lower floors. Ambient lighting is provided by daylight, which is supplemented when necessary by ceiling-mounted fluorescent fixtures. Task lighting is provided to each work station and can be controlled by the occupants. The building has continuous strip windows on all elevations for daylighting. These have specularly reflective light shelves to shade the lower portions of the windows and to reflect light deep into the room. In addition, the building has higher-than-normal ceilings, partially reflective deep window sills, high-performance glazing, and manually operated window shades to provide control.

The evaluation of the daylighting performance of the SSWaCo building is complete, and reports are pending. Plans are being made to continue the collaboration between LBL and Data Build and extend the usefulness of DPEM to the evaluation of atria.

Reference: Andersson, Erwine, and Kammerud 1986.

9.4.3 NAHB National Research Center—Short-Term Monitoring

Over the last several years, the NAHB National Research Center (NAHB/NRC) has been working to develop methods for extrapolating results from short-term tests to predictions of long-term building performance. The approach developed by Duffy and Saunder (1987) is based on an analytical model to predict building performance that uses building thermal parameter inputs estimated from the results of several short-term tests. Possible analytical models that this technique can be used with include the SERI-RES simulation program, the SLR correlation method, or even the ASHRAE steady-state methods. The accuracy achieved for a given house will depend on whether the house is strongly solar driven and on how massive it is.

The short-term tests include electric coheating estimate of the heat loss coefficient in conjunction with blower-door and tracer-gas tests to quantify the infiltration portion of the heat loss coefficient. To improve the accuracy of the estimate of heat loss coefficient, mass temperatures are measured and the heat input or removed from the building as a result of charging or discharging the mass is accounted for in the analysis.

Also part of the short-term test protocol is a cool-down test in which the temperature of the building is allowed to fall without any heat input to the building. Analysis of the results of this test yields the effective time constant of the building. Combined with the measured value of the loss coefficient, the time constant yields the effective thermal capacitance of the building. The efficiency of the auxiliary heating system can be determined as part of the short-term tests. The solar aperture is measured, and the effect of overhangs and shading devices to decrease this area is estimated.

The information derived from the short-term tests is input to the analytical model to derive estimates of auxiliary heating energy requirements and to calculate performance indices, similar to those calculated in the Class B program. Duffy and Saunders (1987) describe the best procedures and analysis methods to use for different types of houses, depending if they are strongly solar driven and/or ground coupled.

The Duffy-Saunders method includes the determination of occupant effects on energy performance. A given building will have a range of auxiliary energy requirements depending on the way in which the house is operated. Occupants determine thermostat setpoints, allowable interior temperature swings, added infiltration, internal gains, and management of controls and night insulation. Duffy and Saunders suggest standard input assumptions for a family with young children, a family with older children, and an older couple. Using these standard operating assumptions, three separate predictions of auxiliary energy may be derived.

The tests and procedures proposed by Duffy and Saunders differ in a number of significant ways from the tests and procedures proposed by Subbarao at SERI. Included in these is the nature of the heat inputs to the building during the short-term tests. Duffy and Saunder's method uses steady-state and step inputs, while Subbarao commonly uses sinusoidal inputs. The Subbarao method lumps infiltration with the conductive losses, while Duffy and Saunders construct an infiltration model that is wind and temperature driven, then tunes the coefficients of the model with measured data. The Subbarao method produces system vectors in the frequency domain, while the coefficients derived using the Duffy and Saunders method have significance in the time domain.

The developmental work by NAHB/NRC led to monitored results from three houses sponsored by New England Electric (NEE). These houses have provided significant opportunities for experimenting with testing procedures and for comparing long-term performance predictions with measured performance data.

Some of the conclusions of this work are as follows:

• Estimates of total heating-season average heat loss coefficient derived from short-term tests compared to heat loss coefficients derived from ASHRAE handbook values were within 2% for one house, 10% for another, and 40% for a third. The first two house were relatively new, having been constructed in 1984, and the third house was old, constructed in 1898. Differences are attributed to discrepancies between *as designed* and *as built* characteristics, the conservative nature of the ASHRAE estimates, and differences between estimated and actual infiltration rates.

• Comparisons between actual and predicted energy consumption, using the ASHRAE handbook values and the values measured in the short-term tests and applying three prediction methods, indicate that the predictions are better using the values from short-term tests. The prediction methods were the SLR method, the unutilizability method, and the ASHRAE degree-day method. Surprisingly, the ASHRAE method resulted in smaller differences between predicted and actual than the other two methods.

• The experimentally derived thermal parameters appear to be repeatable. Numerous data sets from some of the houses were analyzed. For the "worst-case" house, the estimates of conductive heat loss had a standard deviation of 8% of the mean. The standard deviation for time constants estimated from cool-down tests were less than 8% of the mean.

9.4.4 Architectural Energy Corporation—IEA Task VIII

Architectural Energy Corporation (AEC) monitored two residential buildings as part of their work under a contract with DOE to participate in the International Energy Agency (IEA) Task VIII activities. The broad objectives of this portion of IEA Task VIII were to design, construct, and evaluate the performance of prototype low-energy-use housing in the United States. To meet these objectives, two designer/builder teams were selected and participated in the activities of the project. One team was in Virginia and the other was in Colorado. By working with the designer/builder teams, designs were created that incorporated features for low energy consumption and utilization of solar energy that were tempered with the experience of the builder team and their knowledge of what design features are necessary to meet the requirements of the home-buyer market.

One of the buildings is a 4,000-ft^2 (372-m^2) home located in Oakton, Virginia. The other house is a duplex in Boulder, Colorado. Each side of the duplex has approximately 2,000 ft^2 (186 m^2) of floor area. Monitoring activities in these houses are complete and the analysis of the data is in progross.

The data analysis yielded two types of information. First, the performance of the houses was calculated using the Class B methodology. The short-term tests were conducted in these houses using methods quite similar to the ones suggested by Duffy and Saunders. The results of the long-term monitoring were compared to the Class B data base. Second, computer models of each house were created, and the models were run using the weather conditions measured at each site. The results of these computer models were compared with the measured energy consumption, and the models were adjusted until the predicted consumption matched the actual values. Using models that agree with the measured values, features of the building were modified to create a reference building. The reference building had construction features, insulation levels, and glazing distribution selected to be typical for the particular area of the country. The simulations of the solar building and the reference building were used to determine the savings attributable to the features of the passive building.

Results show that the value of the building performance index for total purchased energy in the house in Virginia is 3.36 Btu/ft^2 °F day (19 Wh/m^2 °C). The computer simulations that were done during the design studies indicated that the home would require 3.2 Btu/ft^2 °F day, so these results agree well with the predictions. The total annual energy consumption of this house has also been compared to the consumption of three neighboring houses of similar size. The results are that the IEA house uses 29% less energy than the average of the other three houses and that its total utility costs are approximately 23% less.

9.5 Summary

This chapter contains a summary of monitoring and results, primarily in the United States from 1969 to 1987, for full-size passive solar residential and commercial buildings. These activities range from detailed heat transfer studies that provide an understanding of the thermodynamics of passive solar buildings to monitoring programs that investigate the energy consumption characteristics and energy savings of passive buildings. The objective of this chapter is to present an overview of significant and innovative methods used to evaluate the performance of passive solar buildings and to summarize significant results.

References

[Some entries are not cited in the chapter but describe significant work or add details to the information presented.]

Abrams, D. W., C. C. Benton, and J. M. Akridge. 1980. Simulated and measured performance of earth cooling tubes. *Proc. 5th National Passive Solar Conference*, Amherst, MA, October 19–26, 1980. Newark, DE: American Section of the International Solar Energy Society, pp. 737–741.

Acorn Structures. 1982. *DOE/SERI Solar-Passive-Hybrid Manufactured Buildings Program, Final Report*. Concord, MA: Acorn Structures.

Adams, J. A. 1982. Solar performance: results from the field. *Solar Age*, July 1982, p. 44–50.

AIA Research Corporation. 1978. *The Survey of Monitored Passive/Hybrid Buildings*. Washington, DC: AIA Research Corporation.

Andersson, B., M. Adegrah, T. Webster, W. Place, R. Kammerud, and P. Albrand, 1985. *Effects of Daylighting Options on the Energy Performance of Two Existing Passive Commercial Buildings*. Berkeley, CA: Lawrence Berkeley Laboratory.

Andersson, B., B. Erwine, and R. Kammerud. 1986. *Daylighting Performance Evaluation Methodology*. Berkeley, CA: Lawrence Berkeley Laboratory.

Architectural Energy Corporation. 1985. *Installation Manual for Automated Data Collection Systems in Buildings; Passive Solar Commercial Buildings Program*. Washington, DC: U.S. Department of Energy.

Arumi-Noe, F., and D. O. Northrup. 1979. A field validation of the thermal performance of a passively heated building as simulated by the DEROB system. *Engineering Education 2(1)*, January 1979.

Balcomb, J. D., R. D. McFarland, and S. W. Moore. 1978. Passive testing at Los Alamos. *Proc. 2d National Passive Solar Conference*, Philadelphia, PA, March 16–18, 1978. Newark, DE: American Section of the International Solar Energy Society, pp. 602–609.

Balcomb, J. D., and W. O. Wray. 1978. Evaluation of passive solar heating. *Proc. Annual Meeting of the American Section of the International Solar Energy Society*, Denver, CO, August 28–31, 1978. Newark, DE: American Section of the International Solar Energy Society, pp. 64–68.

Balcomb, J. D. 1979. Summary review: solar passive space heating systems. *Proc. 2d. Annual Solar Heating and Cooling Systems Operational Results Conference*, Colorado Springs, CO. November 27–30, 1979. Los Alamos, NM; Los Alamos Scientific Laboratory, pp. 275–277.

Balcomb, J. D., J. C. Hedstrom, and S. W. Moore. 1979. Performance data evaluation of the Balcomb solar home (SI units). *Proc. 2d. Annual Solar Heating and Cooling Systems Operational Results Conference*, November 27–30, 1979. Colorado Springs, CO. Los Alamos, NM: Los Alamos Scientific Laboratory.

Balcomb, J. D. 1980. Passive solar space heating. *Proc. Annual Meeting of the American Section of the International Solar Energy Society*, Phoenix, AZ, June 2–6, 1980. Newark, DE: American Section of the International Solar Energy Society, pp. 696–705

Balcomb, J. D., D. Barley, R. McFarland, J. Perry, Jr., W. Wray, and S. Noll. 1980. *Passive Solar Design Handbook*, Vol. 2: *Passive Solar Design Analysis*. DOE/CS-0127/2. Washington, DC: U.S. Department of Energy.

Balcomb, J. D., and J. C. Hedstrom. 1980. Determining heat fluxes from temperature measurements made in massive walls. *Proc. 5th National Passive Solar Conference*, Amherst, MA,

October 19–26, 1980. Newark, DE: American Section of the International Solar Energy Society, pp. 136–140.

Balcomb, J. D., J. C. Hedstrom, and J. E. Perry. 1980. Performance evaluation of the Balcomb solar house. *Proc. International Colloquium on Passive Solar House Testing*, Nice, France, December 11–12, 1980. Los Alamos, NM: Los Alamos Scientific Laboratory.

Balcomb, J. D., J. C. Hedstrom, and J. E. Perry. 1981. Performance summary of the Balcomb solar house. *Proc. Annual Meeting, American Section of the International Solar Energy Society*, Philadelphia, PA, May 27–30, 1981. Newark, DE: American Section of the International Solar Energy Society.

Balcomb, J. D. 1981a. Heating remote rooms in passive solar buildings. *Proc. International Solar Energy Society, Solar World Forum*, Brighton, England, August 23–28, 1981. Oxford, England, Pergamon Press. pp. 1835–1840.

Balcomb, J. D. 1981b. Heat storage duration. *Proc. 6th National Passive Solar Conference*, Portland, OR, September 8–12, 1981, Newark. DE: American Section of the International Solar Energy Society, pp. 44–48.

Balcomb, J. D. 1983. Evaluating the performance of passive-solar-heated buildings. *Proc. 6th Annual ASME Technical Conference on Commercial Building Heating and Cooling Application*, Orlando, FL, April 19–21, 1983. Los Alamos, NM: Los Alamos National Laboratory.

Balcomb, J. D., R. W. Jones, R. D. McFarland, and W. O. Wray. 1984. *Passive Solar Heating Analysis: A Design Manual.* Atlanta, GA: ASHRAE.

Bier J. 1979. Performance of a low-cost owner-built home using vertical solar louvers. *Proc. 3rd National Passive Solar Conference*, San Jose, CA, January 11–13, 1979. Newark, DE: American Section of the International Solar Energy Society, pp. 643–647.

Bishop, R., and D. Frey. 1984. Quick methods for accurately predicting building thermal performance from short-term measurements. *Proc. 9th National Passive Solar Conference*, Columbus, OH, September 1984. Boulder, CO: American Solar Energy Society, pp. 244–249.

Bishop, R., and D. Frey. 1985. Occupant effects on energy performance of monitored houses. *Proc. 10th National Passive Solar Conference*, Raleigh, NC, October 15–20, 1985. Boulder, CO: American Solar Energy Society, pp. 395–400.

Burch, J., K. Subbarao, C. Christensen, E. Hancock, M. Warren, and M. Krarti. 1986. *Macrodynamic Methods for Building Thermal Monitoring.* SERI/TR-3044. Golden, CO: Solar Energy Research Institute.

Burt, Hill, Kosar, Rittelmann Associates (BHKR). 1982a. Performance evaluation manual for automated data collection. *Passive Solar Commercial Buildings Program.* Washington, DC: U.S. Department of Energy.

Burt, Hill, Kosar, Rittelmann Associates (BHKR). 1982b. Performance evaluation manual for submetered data collection. *Passive Solar Commercial Buildings Program.* Washington, DC: U.S. Department of Energy.

Burt, Hill, Kosar, Rittelmann Associates (BHKR). 1984. *Passive Solar Commercial Buildings Program: Anthology of Articles and Papers.* Chicago, IL: U.S. Department of Energy.

Carroll, J. A. 1980. An index to quantify thermal comfort in homes. *Proc. 5th National Passive Solar Conference*, Amherst, MA, October 19–26, 1980. Newark, DE: American Section of the International Solar Energy Society, pp. 136–140.

Chen, B., and J. W. Mitchell. 1981. Performance results of the Dennis Demmel "double shell" home, Hartington, Nebraska. *Proc. 6th National Passive Solar Conference*, Portland, OR, September 8–12, 1981. Newark, DE: American Section of the International Solar Energy Society, pp. 832–836.

Converse, A. D., and M. Hall-Martindale. 1981. *Monitoring and Design Study of the White Mountain School Hybrid Solar Building*. Final Report, June 1, 1978—January 31, 1981. Hanover, NH; Thayer School of Engineering, 1981.

Dubin, S., and T. Tucker. 1979. Performance of HUD solar demonstration program instrumented passive houses. *Proc. 3d National Passive Solar Conference*, San Jose, CA, January 11–13, 1979. Newark, DE: American Section of the International Solar Energy Society, pp. 541–546.

Duffy, J. J. 1982. Class B passive solar monitoring results for the northeast. *Proc. 7th National Passive Solar Conference*, Knoxville, TN, August 30—September 1, 1982. Newark, DE: American Section of the International Solar Energy Society, pp. 789–794.

Duffy, J. J. 1983. *Estimation of the Uncertainty in Class B Monitoring Results*. SERI/STR-254-2274. Golden, CO: Solar Energy Research Institute.

Duffy, J. J., and D. Odegard. 1984. *Solar Load Ratio Design Tool Predictions Compared to Level B Monitoring Data*. SERI/STR-254-2251. Golden, CO: Solar Energy Research Institute.

Duffy, J. J., and D. Saunders. 1987. *Short Term Test Method for Measuring the Thermal Performance of Buildings*. Upper Marlboro, MD: NAHB Research Foundation, Inc.

Duffy, J., D. Saunders, and J. Spears. 1987. Low-cost method for evaluation of space heating efficiency of existing homes. *Proc. 12th National Passive Solar Conference*, Portland, OR, July 11–16, 1987. Boulder, CO: American Solar Energy Society, pp. 296–300.

Dynamic Homes, Inc. 1982. A modular passive solar home. *Proc. Passive and Hybrid Solar Energy Update*, Washington, DC, September 15–17, 1982. Washington, DC: U.S. Department of Energy, pp. 286–289.

Fate, R. E. 1980. *Analysis of the New Mexico State University Passive Solar House*. Master's thesis, New Mexico State University, Las Cruces.

Fleischacker, P., G. Clark, and P. Giolma. 1983. Geographic limits for comfort in unassisted roof pond cooled residences. *Proc. 8th National Passive Solar Conference*, Santa Fe, NM, September 7–9, 1983. Boulder, CO: American Solar Energy Society, pp. 835–838.

Fleming, W. S., J. Brown, and T. M. Griffin. 1983. Energy performance monitoring of low cost passive solar homes: results and conclusions. *Proc. ASME Winter Annual Meeting*, Boston, MA, November 13, 1983. New York, NY: American Society of Mechanical Engineers.

Fowlkes, C. W. 1979. Measured performance of a passive solar residence in Bozeman, Montana. *Proc. Solar 79 Northwest Conference*, Seattle, WA, August 10, 1979. Springfield, VA: National Technical Information Service, pp. 113–116.

Fowlkes, C. W. 1979. Measured thermal performance of a passive solar residence in Bozeman, Montana. *Proc. 4th National Passive Solar Conference*, Kansas City, MO, October 3–5, 1979. Newark, DE: American Section of the International Solar Energy Society, pp. 704–707.

Fowlkes, C. W. 1981. Class B passive cooling monitoring program. *Preconference proc. Passive and Hybrid Solar Energy Program Update*. Washington, DC, August 1981. Washington, DC: U.S. Department of Energy, pp. 2B-40–2B-42.

Fowlkes, C. W. 1982. Passive cooling monitoring at the Class B level. *Proc. ASME Solar Conference*, Albuquerque, NM, April 1982. New York: American Society of Mechanical Engineers, pp. 578–584.

Freeborne, W. E., G. Mara, and T. Lent. 1979. Performance of solar energy systems in the residential solar demonstration program. *Proc. 2d. Annual Solar Heating and Cooling Systems Operational Results Conference*, Colorado Springs, CO, November 27–30, 1979. Golden, CO: Solar Energy Research Institute, pp. 137–145.

Frey, D., M. Holtz, and J. Swisher. 1981. *Class B Performance Monitoring of Passive/Hybrid Solar Buildings.* SERI/TP-254-1492. Golden, CO: Solar Energy Research Institute.

Frey, D., J. Swisher, and M. Holtz. 1981. Class B performance monitoring of passive/hybrid solar buildings. *Proc. ASME Solar Energy Division Fourth Annual Conference*, Albuquerque, NM, April 26–30, 1982. Golden, CO: Solar Energy Research Institute.

Frey, D., M. McKinstry, and J. Swisher. 1982. *Installation Manual for the SERI Class B Passive Solar Data Acquisition System.* SERI/TR-254-1671. Golden, CO: Solar Energy Research Institute.

Frey, D., M. Holz, J. Morgan, and J. Vest. 1983. Monitored heating season performance of the Mount Airy library building. *Proc. 8th National Passive Solar Conference*, Santa Fe, NM, September 7–9, 1983. Boulder, CO: American Solar Energy Society, pp. 391–396.

Frey, D., and R. Bishop. 1984a. Industry based residential building thermal performance evaluation program. *Proc. Passive and Hybrid Solar Energy Update*, Washington, DC, September 5–7, 1984. Washington, DC: U.S. Department of Energy, pp. 222–226.

Frey, D., and R. Bishop. 1984b. Lessons learned in passive solar homes. *Proc. 9th National Passive Solar Conference*, Columbus, OH, September 1984. Boulder, CO: American Solar Energy Society, pp. 517–522.

Frey, D., J. Swisher, and A. Yager. 1984. Results and lessons learned from monitoring four passive solar commercial buildings. *Proc. 9th National Passive Solar Conference*, Columbus, OH, September 1984. Boulder, CO: American Solar Energy Society, pp. 162–167.

Frey, D., and A. Yager. 1984. Field instrumentation for the passive solar commercial buildings monitoring program. *Proc. Passive and Hybrid Solar Energy Update*, Washington, DC, September 5–7, 1984. Washington, DC: U.S. Department of Energy, pp. 247–251.

Frey, D., and M. Holtz. 1985. Use of microcomputers for performance evaluation of residential buildings. *Proc. 1st National Conference on Microcomputer Applications for Conservation and Renewable Energy*, Tucson, AZ, February 26–28, 1985. Tucson, AZ: University of Arizona.

Frey, D. 1986. Occupant effects and normalization of field monitoring results. *Proc. National Workshop on Field Data Acquisition for Building and Equipment Energy-Use Monitoring*, Dallas, TX, October 16–18, 1985. Oak Ridge, TN: Oak Ridge National Laboratory.

Ghaffari, H., and R. Jones. 1982. Heating season thermal performance of the Brookhaven house. *Proc. 1982 Annual Meeting, American Solar Energy Society*, Houston, TX. Boulder, CO: American Solar Energy Society.

Gordon, H., K. Hart, and M. Kantrowitz. 1984. Non-residential buildings program design and performance overview. *Proc. Passive and Hybrid Solar Energy Update*, Washington, DC, September 5–7, 1984. Washington, DC: U.S. Department of Energy, 1984, pp. 233–239.

Gordon, H., J. Estoque, K. Hart, M. Kantrowitz. 1984. Non-residential buildings program design and performance overview. *Proc. 9th National Passive Solar Conference*, Columbus, OH, September 1984. Boulder, CO: American Solar Energy Society, pp. 155–160.

Gordon, H., et al. 1985. *Passive Overview: Passive Solar Energy For Non-Residential Buildings.* Berkeley, CA: Lawrence Berkeley Laboratory.

Greenmoss Builders, Inc. Monthly Performance Peport, December 1977, January 1978, March 1978, April 1978, May 1978, June 1978, July 1978. Waitsfield, VT: Greenmoss Builders, Inc.

Griffin, T. M. 1983. Energy performance monitoring and analysis for a passive solar residence. *ASHRAE Transactions*, No. 1B.

Grimsrud, D. T., M. H. Sherman, and R. C. Sonderegger. 1983. *Calculating Infiltration: Implications for a Construction Quality Standard.* LBL-9416. Berkeley, CA: Lawrence Berkeley Laboratory.

Grondzik, W., L. Boyer, and T. Johnston. 1981. Variations in earth covered roof temperature profiles. *Passive Cooling: Proc. International Passive and Hybrid Cooling Conference*, Miami Beach, FL, November 11–13, 1981. Newark, DE: American Section of the International Solar Energy Society, pp. 146–150.

Gustashaw, D. H. 1983a. *A Comparison of the Analytically Developed and Field Measured Overall Conductive Heat Transfer Coefficients in Passive Solar Residential Construction.* Draft report. Atlanta, GA: ESG, Inc.

Gustashaw, D. H. 1983b. *Class B Monitoring Program Handbook for Southeastern Sites with Data Based on INIT2 Software (Applicable to Heating Season).* Atlanta, GA: ESG, Inc.

Gustashaw, D. H., and R. A. Jacobs. 1983. *Class B Monitoring Program Handbook for Southeastern Sites with Data Based on SUM83 Software (Applicable to Cooling Season).* Atlanta, GA: ESG, Inc.

Hagan, D. A., and R. F. Jones. 1983. Linear regression techniques in analysis of the Blouin superinsulated house. *Proc. 8th National Passive Solar Conference*, Santa Fe, NM, September 7–9, 1983. Boulder, CO: American Solar Energy Society, pp. 259–264.

Halperin, D. and J. Swisher. 1982. Microcomputer based system for monitoring passive solar homes. *Proc.* 13th Annual Conference on Modeling and Simulation, Pittsburgh, PA, April 22–24, 1982. Research Triangle Park, NC: Instrument Society of America, pp. 803–807.

Hamilton, B., B. Sachs, R. Alward, and A. Lepage. 1981. Non-instrumented performance evaluation of 335 passive solar homes. *Proc. 6th National Passive Solar Conference*, Portland, OR, September 8–12, 1981. Newark, DE: American Section of the International Solar Energy Society, pp. 3–7.

Hamilton, B., M. J. Holtz, and L. S. Palmiter. 1979. *Low Cost Performance Evaluation of Passive Solar Buildings.* SERI/RR-63-223. Golden, CO: Solar Energy Research Institute.

Hamilton, B., and B. Sachs. 1982. Side by side comparison of a conventional house and a passive solar superinsulated redesign. *Proc. 7th National Passive Solar Conference*, Knoxville, TN, August 30–September 1, 1982. Newark, DE: American Section of the International Solar Energy Society, pp. 21–26.

Hamilton, B., B. Sachs, J. Duffy, and A. Persilly. 1982. Measurement-based calculation of infiltration in passive solar performance evaluation. *Proc. 8th National Passive Solar Conference*, Santa Fe, NM, September 7–9, 1983. Boulder, CO: American Solar Energy Society, pp. 295–300.

Hay, H. 1976. Atascadero residence. *Proc. Passive Solar Heating and Cooling Conference and Workshop*, May 18–19, 1976. LA-6637-C. Los Alamos, NM: Los Alamos Scientific Laboratory, pp. 101–107.

Hayes, J. W. 1978. Performance of a passive solar house designed for New England. *Proc. Annual Meeting of the American Section of the International Solar Energy Society*, Denver, CO, August 28–31, 1978. Newark, DE: American Section of the International Solar Energy Society, pp. 207–210.

Hodges, L. 1980. The Hodges residence: performance of a direct gain passive solar home in Iowa. *Proc. 5th National Passive Solar Conference*, Amherst, MA, October 19–26, 1980. Newark, DE: American Section of the International Solar Energy Society, pp. 304–308.

Holtz, M., and B. Hamilton. 1980. *Program Plan: Performance Evaluation of Passive/Hybrid Solar Heating and Cooling Systems.* SERI/PR-721-788. Golden, CO: Solar Energy Research Institute.

Holtz, M., and D. Frey. 1982. Preliminary Class B results. *Proc. Passive and Hybrid Solar Energy Update*, Washington, DC, August 9–12, 1981. Washington, DC: U.S. Department of Energy, February 1982, pp. 2B-62–2B-64.

Holtz, M., and J. Swisher. 1982. Performance of passive solar buildings in the U.S. *Proc. International Solar Architecture Conference*, Cannes, France, December 1982. New York: American Solar Energy Society, pp. 31–43.

Holtz, M., D. Frey, R. Bishop, and J. Swisher. 1985. The future of passive solar design. *Solar Age*, October 1985, pp. 49–56.

Hoskins D., and R. Stromberg. 1979. *Passive Solar Buildings*. SAND 79-0824. Albuquerque, NM: Sandia Laboratories.

Howard, B. D., and E. O. Pollock. 1981. *Performance of Passive Solar Space Heating Systems, Comparative Report, 1980–1981 Heating Season*. SOLAR/0022-82/39. Chicago, IL: U.S. Department of Energy.

Hunn, B. 1978. A hybrid passive/active solar house: first year performance of the Hunn residence. *Proc. 2nd National Passive Solar Conference*, Philadelphia, PA, March 16–18, 1978. Newark, DE: American Section of the International Solar Energy Society, pp. 247–251.

Hunn, B. D. 1979. Performance and cost of a hybrid passive/active solar house. *Proc. ASHRAE Symposium on Air Infiltration*, Philadelphia, PA. January 1979. New York, NY: American Society of Heating, Refrigerating and Air Conditioning Engineers.

Hunn, B. 1981. Long-term performance of the Hunn passive solar residence. *Proc. 6th National Passive Solar Conference*, Portland, OR, September 8–12, 1981. Newark, DE: American Section of the International Solar Energy Society, pp. 64–68.

Hunn, B. D., and M. M. Jones. 1978. An air thermosiphon solar heating system: the Jones house, Santa Fe, New Mexico. *Proc. Annual Meeting of the American Section of the International Solar Energy Society*, Denver, CO, August 28–31, 1978. Newark, DE: American Section of the International Solar Energy Society, pp. 230–234.

Hunn, B., W. Turk, and W. Wray. 1982. The DOE solar Class A performance evaluation program: preliminary results. *Proc. Passive and Hybrid Solar Energy Update*, Washington, DC, September 15–17, 1982. Washington, DC: U.S. Department of Energy.

Hyatt, T., and D. W. Abrams. 1982. Homeowner's monitoring program: a low cost study of passive system performance initial results. *Proc. ASME Solar Energy Conference*, Albuquerque, NM, April 1982. New York: American Society of Mechanical Engineers, pp. 585–596.

Jones, R. F. 1981. *Case Study of the Mastin Double-Envelope House*. BNL 51460. Upton, NY: Brookhaven National Laboratory.

Jones, R. W. 1982. *Monitored Passive Solar Buildings*. LA-9098-MS. Los Alamos, NM: Los Alamos National Laboratory.

Jones, R. W., and R. D. McFarland. 1979. Simulation of the Ghost Ranch greenhouse-residence. *Proc. 3d National Passive Solar Conference*, San Jose, CA, January 11–13, 1979. Newark, DE: American Section of the International Solar Energy Society, pp. 35–40.

Jones, R. W., R. D. McFarland, and G. S. Lazarus. 1983. Mass and fans in attached sunspaces. *Proc. 3d Energy Conserving Greenhouse Conference*, Hyannis, Massachusetts, November 19–21, 1982. New York: American Solar Energy Society, pp. 81–90.

Judkoff, R., D. Wortman, and J. Burch. 1983. *Measured Versus Predicted Performance of the SERI Test House: A Validation Study*. SERI/TP-254-1953. Golden, CO: Solar Energy Research Institute.

Judkoff, R., and D. Wortman. 1984. *Validation of Building Energy Analysis Simulations Using 1983 Data from the SERI Class A Test House*. Golden, CO: Solar Energy Research Institute.

Kantrowitz, M. 1984. Report on occupancy evaluation from the Passive Solar Commercial Buildings Program. *Proc. Passive and Hybrid Solar Energy Update*, Crystal City, VA, September 5–7, 1984. Washington, DC: U.S. Department of Energy, pp. 252–256.

Kedzierski, P., and A. Lau. 1982. Short term testing to determine the thermal characteristics of as-built passive solar houses. *Proc. Passive and Hybrid Solar Energy Update*, Washington, DC, August 9–12, 1981. Washington, DC: U.S. Department of Energy, February 1982, pp. 2B-55–2B-57.

Kelbaugh, D. 1978. Kelbaugh house: recent performance. *Proc. 2d National Passive Solar Conference*, Philadelphia, PA, March 16–18, 1978. Newark, DE: American Section of the International Solar Energy Society, pp. 69–75.

Kennedy, M. 1980. *Rural Housing Research Unit (RHRU), Clemson, South Carolina: Solar Energy System Performance Evaluation, November 1979–April 1980.* Richland, WA: Automation Industries, Inc.

Kunz, Jr., W. S. 1981. The thermal performance of the peterson residence: a double shell envelope house. *Proc. 6th National Passive Solar Conference*, Portland, OR, September 8–12, 1981. Newark, DE: American Section of the International Solar Energy Society, pp. 49–53.

Tennessee Valley Authority. 1984. *Lakeland Wesley Village Monitoring Results for June 1982–May 1983.* Chattanooga, TN: Tennessee Valley Authority.

LaVigne, A. B., and M. A. Schuldt. 1981. Thermal performance of an earth-sheltered passive solar residence. *Proc. 6th National Passive Solar Conference*, Portland, OR, September 8–12, 1981. Newark, DE: American Section of the International Solar Energy Society, pp. 54–58.

Lofchie, H., E. Noble, and J. J. Duffy. 1983. The Massachusetts multi-family passive solar program: a comparison of monitoring data to design tool predictions. *Proc. 8th National Passive Solar Conference*, Santa Fe, NM, September 7–9, 1983. Boulder, CO: American Solar Energy Society, pp. 229–234.

McCulley, M. T., R. O'Mears, and C. O. Pederson. 1982. Energy monitoring results of a superinsulated passive solar building. *Proc. ASHRAE/DOE Conference on Thermal Performance of the Exterior Envelopes of Buildings II.* Las Vegas, NV, December 6–9, 1982. Champaign, IL: U.S. Army Construction Engineering Research Laboratory.

McCulley, M. T., J. J. Siminovitch, and D. E. Bergeson. 1981. Correlation studies of passive solar buildings based on instrumented field and testing computer simulation studies. *Proc. 4th International Conference on Alternative Energy Sources*, Miami, FL, December 14–16 1981. Coral Gables, FL: Clean Energy Research Institute, University of Miami, pp. 120–122.

McKinstry, M., P. Busse, L. Lambert, and P. Gillen. 1983. Heating season results from the Bonneville Power Administration Class B solar monitoring program. *Proc. 8th National Passive Solar Conference*, Santa Fe, NM, September 7–9. 1983. Boulder, CO: American Solar Energy Society, pp. 211–216.

Maeda, B. T., and P. W. Grant. 1980. Comparison of simulation and measured performance of the Suncatcher house design using SOLSIM and SOLEST. *Procc. Systems Simulation and Economics Analysis Conference*, San Diego, CA, January 23, 1980. Golden, CO: Solar Energy Research Institute, pp. 323–329.

Maeda, B. T., P. W. Grant, and R. D. Anson. 1981. *Suncatcher Monitoring Project.* Final Report, March 1, 1978–June 30, 1981. Davis, CA: Davis Alternative Technology Associates.

Mahajan, B. M., and S. T. Liu. 1983. Results of the NBS passive test building: a status report. *Proc. Passive and Hybrid Solar Energy Update*, Washington, DC, September 26–28, 1983. Washington, DC: U.S. Department of Energy, pp. 70–78.

Mahajan, S., C. Newcomb, M. Shea, D. Woodford, E. Hodapp, W. Armaes. 1981. Cooling performance data and analysis for three passive/hybrid homes in Davis, California. *Proc.*

Solar Engineering 1981, Reno, NV, April 27–May 1, 1981. New York, NY: American Society of Mechanical Engineers, pp. 616–625.

Mahajan, S., C. Newcomb, M. Shea, D. Mort, and P. Morandi. 1983a. One time measurement of the infiltration rate and conductive loss coefficient for houses in California—Class B sites. *Proc. 8th National Passive Solar Conference*, Santa Fe, NM, September 7–9, 1983. Boulder, CO: American Solar Energy Society, pp. 241–246.

Mahajan, S., C. Newcomb, M. Shea, and D. Mort. 1983b. *Performance of Passive Solar and Energy Conserving Houses in California*. Sacramento, CA: California State University at Sacramento; SERI/STR-254/2017. Golden, CO: Solar Energy Research Institute, 1983.

Mancini, T. R., and P. R. Smith. 1979. *Economical Solar Heated and Cooled Residence for Southern New Mexico, Final Report*. Las Cruces, NM: New Mexico Energy Institute, New Mexico State University.

Mancini, T. 1982. The New Mexico State University passive solar house. *Proc. Passive and Hybrid Solar Energy Update*, Washington, DC, September 15–17, 1982. Washington, DC: U.S. Department of Energy, November 1982, pp. 89–92.

Miller, P. C., and E. Pollock. 1980. *Hullco Construction, Prescott, AZ: Solar Energy System Performance Evaluation, October 1979–May 1980*. Chicago, IL: U.S. Department of Energy.

Morris, W. S. 1983. Performance evaluation of a thermosiphon/blockbed system. *Proc. 8th National Passive Solar Conference*, Santa Fe, NM, September 7–9, 1983. Boulder, CO: American Solar Energy Society, pp. 265–269.

Nisson, N., and E. Williams. 1980. Preliminary monitoring of ten south wall solar retrofits in the Northeast. *Proc. 5th National Passive Solar Conference*, Amherst, MA, October 19–26, 1980. Newark, DE: American Section of the International Solar Energy Society, pp. 313–317.

Palmiter, L., B. Hamilton, and M. Holtz. 1979. *Low Cost Performance Evaluation of Passive Solar Buildings*. SERI/RF-63-223. Golden, CO: Solar Energy Research Institute.

Passive Solar Class B Monitoring Reports, September 1981. Minneapolis, MN: Mid-American Solar Energy Complex (MASEC).

Peck, J. F., and H. J. Kessler. 1980. Integrating optimized evaporative cooling systems with passive homes, concepts and results. *Proc. Annual Meeting of the American Section of the International Solar Energy Society*, Phoenix, AZ, June 2–6, 1980. Newark, DE: American Section of the International Solar Energy Society, pp. 952–956.

Peck, J. F., T. L. Thompson, and H. J. Kessler. 1981. *Off-Peak Power Use in Passive Solar Homes: Performance, Monitoring and Analysis of Periodic Heating and Cooling in High Mass Homes*. EPRI-EM-1966. Palo Alto, CA: Electric Power Research Institute.

Perry, J. E. 1980. Rock bed behavior and reverse thermosiphon effects. *Proc. 5th National Passive Solar Conference*, Amherst, MA, October 19–26, 1980. Newark, DE: American Section of the International Solar Energy Society, pp. 106–110.

Perry, Jr., J. E. 1976. The Wallasey School. *Proc. Passive Solar Heating and Cooling Conference and Workshop*, May 18–19, 1976. LA-6637-6. Los Alamos, NM: Los Alamos Scientific Laboratory, pp. 223–237.

Pfister, P. J., and D. T. Dixon. 1980. Results of an instrumented passive retrofit in a cold climate: Pfister residence. *Proc. Annual Meeting of the American Section of the International Solar Energy Society*, Phoenix, AZ, June 2–6, 1980. Newark, DE: American Section of the International Solar Energy Society, pp. 762–766.

Pollock, E. O., S. J. Sersen, and P. C. Miller. 1980. *Performance of Passive-Solar Space Heating Systems, Comparative Report, 1979–1980 Heating Season*. SOLAR/0022-81/39. Chicago, IL: U.S. Department of Energy.

Pollock, E. O. 1981a. NSDN passive program. *Proc. Passive and Hybrid Solar Energy Update*, Washington, DC, August 9–12, 1981. Washington, DC: U.S. Department of Energy, pp. 2B-74–2B-76.

Pollock, E. O. 1981b. Comparison of monitored passive buildings. *Proc. 6th National Passive Solar Conference*, Portland, OR, September 8–12, 1981. Newark, DE: American Section of the International Solar Energy Society, pp. 69–73.

Pollock, E. O. 1981c. Comparison of Monitored Passive Buildings. Silver Spring, MD: Automation Industries, Inc., pp. 3–7.

Pollock, E. 1982. NSDN passive program. *Proc. Passive and Hybrid Solar Energy Update*, Washington, DC, September 15–17, 1982. Washington, DC: U.S. Department of Energy, pp. 207–212.

Raymond, M. G., and E. O. Pollock. 1984. *Comparative Report: Performance of Passive Solar Space Heating Systems, 1981–1982 Heating Season.* SOLAR/0022-84/39, National Data Program. Washington, DC: U.S. Department of Energy, 1984.

Reid, R. L., et al. 1981. *Evaluation of Passive-Solar Modular Dwelling, June 30, 1980–August 30, 1981, Final Report.* Knoxville, TN: Tennesse University.

Robertson, R. 1982. An overview of TVA's residential passive solar monitoring. *Proc. Passive and Hybrid Solar Energy Update*, Washington, DC, August 9–12, 1981. Washington, DC: U.S. Department of Energy, pp. 2B-67–2B-70.

Rogers, B. T. 1976. Some performance estimates for the Wright house—Santa Fe, New Mexico. *Proc. Passive Solar Heating and Cooling Conference and Workshop*, Albuquerque, NM, May 18–19, 1976. *Proc.* LA-6637-C. Los Alamos, NM: Los Alamos Scientific Laboratory, pp. 189–199.

Rogers, B. T. 1978. Preliminary report on Sundwelling passive units at the Ghost Ranch. *Proc. Annual Meeting of the American Section of the International Solar Energy Society*, Denver, CO, August 28–31, 1978. Newark, DE: American Section of the International Solar Energy Society, pp. 46–52.

Saras, A. 1981. The passive solar residential Class C monitoring program in the northeast region. *Proc. 6th National Passive Solar Conference*, Portland, OR, September 8–12, 1981. Newark, DE: American Section of the International Solar Energy Society, pp. 15–19.

Schramm, D. 1978. Passive solar remodeling in northern climates. *Proc. Annual Meeting of the American Section of the International Solar Energy Society*, Denver, CO, August 28–31, 1978. Newark, DE: American Section of the International Solar Energy Society, pp. 9–15.

Schramm, D. 1979. Low cost performance testing procedures for a passive solar retrofit. *Proc. 3d National Passive Solar Conference*, San Jose, CA, January 11–13, 1979. Newark, DE: American Section of the International Solar Energy Society, pp. 553–560.

Shea, M., D. Mort, S. Mahajan, and C. Newcomb. 1983. *Documentation of Data Processing Procedures and Extension of Class B Data Analysis.* SERI/STR-254-2085. Golden, CO: Solar Energy Research Institute.

Shippee, P. 1979. Performance of the Sunearth house. *Proc. 2d Annual Solar Heating and Cooling Systems Operational Results Conference*, Colorado Springs, CO, November 27–30, 1979. Golden, CO: Solar Energy Research Institute, pp. 163–166.

Solar Age. 1983. The best passive heating data yet. *Solar Age*, July 1983.

Solar Energy Research Institute. 1984. *Summary of 1982–1983 Class B Results.* SERI/SP-281-2362. Golden, CO: Solar Energy Research Institute.

Solar Energy Research Institute Passive Technology Program and The Memphremagog Group. 1980. *Program Area Plan—Performance Evaluation of Passive/Hybrid Solar Heating and Cooling Systems.* SERI/PR-721-788. Golden, CO: Solar Energy Research Institute.

Sonderegger, R., P. Condon, and M. Modera. 1980. *In-situ Measurements of Residential Energy Performance Using Electric Co-heating.* LBL-10117. Berkeley, CA: Lawrence Berkeley Laboratory.

Sparkes, H. R., and K. Raman. 1981. *Performance of instrumented passive system in the HUD Solar Residential Demonstration Program.* ASHRAE Transactions New York, NY: American Society of Heating, Refrigerating, and Air Conditioning Engineers, Vol. 87, Part 2. 1981, pp. 667–685.

Spears, J. 1981a. Performance analysis methodology for passive heating systems in the NSDN. *Proc. 6th National Passive Solar Conference*, Portland, OR, September 8–12, 1981. Newark, DE: American Section of the International Solar Energy Society, pp. 8–12.

Spears, J. W. 1981b. *Baker Construction, Cincinnati, OH, Solar Energy System Performance Evaluation, October 1980–May 1981.* Silver Spring, MD: Automation Industries, Inc.

Spears, J. W. 1981c. *Solar-Energy System Performance Evaluation, Rymark I and Rymark II, Frederick, MD, January 1981–April 1981.* Silver Spring, MD: Automation Industries, Inc.

Spears, J. W. 1981d. *Performance Analysis Methodology for Passive Heating Systems in the NSDN.* Silver Spring, MD: Automation Industries, Inc.

Spears, J. W. 1982a. *Solar Energy System Performance Evaluation: Roberts Home, Reston, Virginia, February 1982–July 1982.* Silver Spring, MD: Vitro Laboratories.

Spears, J. W. 1982b. Side-by-side testing of three production type passive solar homes. *Proc. 1982 Annual Meeting, American Solar Energy Society*, Houston, TX, Boulder, CO: American Solar Energy Society.

Spears, J. W. 1982c. Side-by-side testing of three production type passive solar homes. *Proc. Mid-Atlantic Energy Conference*, Baltimore, MD: December 7, 1982. Silver Spring, MD: Vitro Laboratories.

Starr, G., and B. Melzer. 1978. The Soldyne house. *Proc. 2d National Passive Solar Conference*, Philadelphia, PA, March 16–18, 1978. Newark, DE: American Section of the International Solar Energy Society, pp. 317–320.

Stromberg, R., and S. Woodall. 1977. *Passive Solar Buildings: A Compilation of Data and Results.* SAND77-1204. Albuquerque, NM: Sandia Laboratories.

Subbarao, K. 1980. *Scaling Relations for the Performance of Passive Systems.* Golden, CO: Solar Energy Research Institute.

Subbarao, K. 1982. *The Dynamic Response of Thermal Masses and Their Interactions Using Their Vector Diagrams.* SERI/TP-254-1632. Golden, CO: Solar Energy Research Institute.

Subbarao, K. 1984. *BEVA (Building Element Vector Analysis)—A New Hour-by-Hour Building Energy Simulation with System Parameters as Inputs.* SERI/TR-254-2195. Golden, CO: Solar Energy Research Institute.

Subbarao, K., and J. Anderson. 1982a. *A Frequency-Domain Approach to Passive Building Energy Analysis.* SERI/TR-254-1544. Golden, CO: Solar Energy Research Institute.

Subbarao, K., and J. Anderson. 1982b. *A Fourier Transform Approach to Building Simulations with Thermostatic Controls.* Golden, CO: Solar Energy Research Institute.

Subbarao, K., D. Mort, and J. Burch. 1984. *Short Term Tests for Determining Long Term Building Performance.* Golden, CO: Solar Energy Research Institute.

Suhs, N. E., and F. C. Wessling, Jr. 1980. Initial monitoring results of TVA's Solar House no. 2. *Proc. 5th National Passive Solar Conference*, Amherst, MA, October 19–26, 1980. Newark, DE: American Section of the International Solar Energy Society, pp. 356–360.

Swisher, J. 1981. Summertime results from the Class B solar performance monitoring program. *Proc. 6th National Passive Solar Conference*, Portland, OR, September 8–12, 1981. Newark, DE: American Section of the International Solar Energy Society, pp. 13–14.

Swisher, J. 1982a. *Measured Passive Solar Performance from New Residences in Denver*, Colorado. SERI/TP-254-1682. Golden, CO: Solar Energy Research Institute.

Swisher, J. 1982b. Performance results from passive solar residences in Denver, Colorado. *Proc. 7th National Passive Solar Conference*, Knoxville, TN, August 30–September 1, 1982. Newark, DE: American Section of the International Solar Energy Society, pp. 783–788.

Swisher, J., and T. Cowing. 1983. *Passive Solar Performance: Summary of 1981–1982 Class B Results*. SERI/SP-281-1847. Golden, CO: Solar Energy Research Institute.

Swisher, J., and J. J. Duffy. 1983. Measured performance of 50 passive solar residences in the United States. *Proc. 8th National Passive Solar Conference*, Santa Fe, NM, September 7–9, 1983. Boulder, CO: American Solar Energy Society, pp. 223–227.

Swisher, J., D. Frey, and M. Holtz. 1984. Results and lessons learned from monitoring three passive solar commercial buildings. *Proc. 11th Annual Wattec Conference and Exposition* February 21–24, 1984, Knoxville, TN. Knoxville, TN: Ltm Consultants, pp. 31–32.

Teller, R., and M. Holtz. 1985. IEA Task VIII residential design-build projects: preliminary energy analysis. *Proc. 10th National Passive Solar Conference*, Raleigh, NC, October 15–20, 1985. Boulder, CO: American Solar Energy Society, pp. 721–726.

Temple, P. 1980. Class A monitoring of the Brookhaven House. *Proc. 5th National Passive Solar Conference*, Amherst, MA, October 19–26, 1980. Newark, DE: American Section of the International Solar Energy Society, pp. 351–355.

Thornton, R. K. 1982. Performance of a major passive solar retrofit on a turn-of-the-century multi-family brick row house. *Proc. 7th National Passive Solar Conference*, Knoxville, TN, August 30–September 1, 1982. Newark, DE: American Section of the International Solar Energy Society, pp. 773–776.

Thornton, R. 1983. Performance and evaluation of a passive solar superinsulated residence in the Northeast. *Proc. 8th National Passive Solar Conference*, Santa Fe, NM, September 7–9, 1983. Boulder, CO: American Solar Energy Society, pp. 271–276.

Trombe, F., J. F. Robert, M. Cabanat, and B. Sesolis. 1976. *Some performance characteristics of the CNRS solar house collectors. Proc. Passive Solar Heating and Cooling Conference and Workshop*, May 18–19, 1976. LA-6637-6. Los Alamos, NM: Los Alamos Scientific Laboratory, pp. 201–222.

Tsongas, G., J. Peterson, M. McKinstry, and P. Gillen. 1983. Measured versus SLR-predicted performance of a monitored direct gain home. *Proc. 8th National Passive Solar Conference*, Santa Fe, NM, September 7–9, 1983. Boulder, CO: American Solar Energy Society, pp. 321–326.

Vieira, R., G. Clark, and J. Faultersack. 1983. Energy savings potential of dehumidified roof pond residences. *Proc. 8th National Passive Solar Conference*, Santa Fe, NM, September 7–9, 1983. Boulder, CO: American Solar Energy Society, pp. 829–834.

Walsh, E., J. Duffy, and S. Frye. 1987. Cost-benefit analysis of monitored passive solar houses in New England. *Proc. 12th National Passive Solar Conference*, Portland, OR, July 11–16, 1987. Boulder, CO: American Solar Energy Society, pp. 79–83.

Welch, J. H., K. R. Olson, and D. J. Mackinnon. 1979. Thermal performance of a totally passive solar greenhouse in Flagstaff, AZ. *Proc. 3d National Passive Solar Conference*, San

Jose, CA, January 11–13, 1979. Newark, DE: American Section of the International Solar Energy Society, pp. 674–677.

Weston, M. W. 1979a. Project monitor: executive summary; introduction; and summary and recommendations. Final report. *Proc. DOE Regional Solar Update Conference*, Dearborn, MI, July 11 1979. Washington, DC: U.S. Department of Energy.

Weston, M. W. 1979b. Results of performance analysis of passive systems in the National Solar Data Network. *Proc. 2d. Annual Solar Heating and Cooling Systems Operational Results Conference*, Colorado Springs, CO, November 27–30, 1979. Golden, CO: Solar Energy Research Institute, pp. 85–97.

Wexler, A., and B. Wilcox. 1979. Preliminary results from performance measurements of a passive house in Stockton, CA. *Proc. 3d National Passive Solar Conference*, San Jose, CA, January 11–13, 1979. Newark, DE: American Section of the International Solar Energy Society, pp. 603–605.

10 Design Tools

John S. Reynolds

Design tools are devices that focus attention on one aspect of design; they enable the designer to improve that aspect of the design's performance. The user of a design tool must meet a very challenging task: how to resolve the potential conflict between the process of design and the act of analysis. Design involves both intuition, which grows out of experience, and information provided by specific acts of analysis. The potential conflict arises when the information from analysis is too detailed, arrives too soon, and tends to overwhelm the designer's intuition. (This is especially likely when a designer lacks experience and is therefore more confident with finite-element analysis than with vague intuition.) In the "too much too soon" scenario, the design process is influenced more by the tool available than by the design program. The result can be a design product less suitable for its intended purpose, rather like designing a bench instead of a chair, because the available tools favor benches. Therefore, one of the earliest issues in design tool selection is to choose the right tool at the right time.

A simple distinction can be made (Balcomb 1986) between tools used to guide a designer toward a result and tools used by the designer to evaluate the result once achieved. One problem here is that evaluation at one stage becomes guidance at another. In addition, for some design processes, a rule of thumb is sufficient for evaluation ("it's close enough"). For many other designers, rules of thumb are only a guide as to where to start. It is too easy to fall into the assumption that architects use guidance tools, while engineers use evaluation tools.

A more thorough distinction (Brown, Reynolds, and Ubbelohde 1982) presents tools that correspond to four stages of design: conceptual, schematic, developmental, and final.

Conceptual tools are used at the earliest stage of design. They are ideas that a designer brings to his or her task, perhaps out of intuition and past experience, and are not necessarily specific to a single type of building or a particular site. A sample of such a conceptual tool is the idea of elongating a building on its east-west axis in order to expose a relatively large wall area to the south.

Schematic tools include rules of thumb and simple charts that provide the first sizing, shaping, and placement information for elements of a building at a specific site. Many architects and builders use such tools as their most advanced experience with evaluation; if a more detailed analysis

Table 10.1
Sample of rules of thumb for solar-glazing area

	Area of solar glazing as ratio of floor area		Approximate SSFs			
			No night insulation		With R9 night insulation	
Location	Low	High	Low	High	Low	High
Birmingham, Alabama	0.09	0.18	22	37	34	58
Mobile, Alabama	0.06	0.12	26	44	34	60
Montgomery, Alabama	0.07	0.15	24	41	34	59
Phoenix, Arizona	0.06	0.12	37	60	48	75
Prescott, Arizona	0.10	0.20	29	48	44	72
Tucson, Arizona	0.06	0.12	35	57	45	73
Winslow, Arizona	0.12	0.24	30	47	48	74
Yuma, Arizona	0.04	0.09	43	66	51	78
Fort Smith, Arkansas	0.10	0.20	24	39	38	64
Little Rock, Arkansas	0.10	0.19	23	38	37	62
Bakersfield, California	0.08	0.15	31	50	42	67
Daggett, California	0.07	0.15	35	56	46	73
Fresno, California	0.09	0.17	29	46	41	65
Long Beach, California	0.05	0.10	35	58	44	72
Los Angeles, California	0.05	0.09	36	58	44	72

Source: Balcomb et al. 1980.

is necessary, they hire a consultant to use more advanced tools. A sample of such a tool: a residence in a given climate should have an area of south-facing glass that is a certain percentage of its total floor area, providing that night insulation is utilized (see table 10.1).

Developmental tools allow comparisons between whole systems over a typical heating or cooling season and are likely to begin to distinguish between what is best for the particular building program and what is best for thermal performance. For example, in a particular climate, a bowling alley provided with a percentage of its floor area in south-facing direct-gain glass area might yield a somewhat higher annual solar savings fraction than a Trombe wall of the same area (see table 10.2). However, the Trombe wall is much more compatible with the hardwood floors of the bowling alley lanes; it may be impossible to provide the necessary area of thermally massive surface for a successful direct-gain system in such a building.

Final tools, usually detailed, hour-by-hour simulations, provide the ultimate relationship between size, system type, building program, and de-

Table 10.2
Sample of solar savings fraction (SSF) vs. load collector ratio (LCR)

Montgometry, Alabama									2269 DD
SSF =	.10	.20	.30	.40	.50	.60	.70	.80	.90
WW-A3	340	162	98	66	47	34	25	18	12
WW-A6	287	162	106	74	54	40	30	22	15
WW-B2	303	161	103	71	51	38	28	20	14
WW-B4	303	183	124	89	66	50	38	28	19
WW-C3	320	206	144	106	80	61	46	35	24
TW-A3	334	144	84	55	39	28	20	15	10
TW-E3	356	211	142	101	75	56	42	31	21
TW-F3	245	117	71	48	34	25	18	13	9
TW-F4	201	104	66	45	32	24	18	13	8
TW-G3	180	91	56	38	27	20	15	11	7
TW-G4	144	75	47	32	23	17	12	9	6
TW-I2	248	128	80	55	40	29	22	16	10
TW-I3	262	140	90	62	45	33	25	18	12
TW-I4	259	149	99	70	51	38	29	21	14
TW-J2	282	159	104	73	53	40	30	22	15
DG-A1	247	109	62	39	24	14	—	—	—
DG-A2	261	119	72	49	34	24	16	10	5
DG-A3	312	145	90	63	46	34	25	18	10
DG-B2	268	124	76	52	38	29	21	14	8
DG-C3	361	173	106	75	56	44	34	26	18
SS-A1	687	214	113	70	47	33	24	16	11
SS-A2	606	260	152	99	69	50	36	26	17
SS-A5	1155	215	103	62	40	28	20	14	9
SS-B1	473	163	88	55	38	27	19	13	9
SS-B3	442	152	81	51	35	24	17	12	8
SS-B5	618	147	73	44	29	20	14	10	6
SS-C4	313	149	90	61	43	31	23	17	11
SS-D1	633	255	144	93	64	46	33	23	15
SS-D2	535	282	178	122	87	64	47	34	23
SS-E1	479	202	116	75	52	37	27	19	12

Source: Balcomb et al. 1983.

tailed performance. Architects and builders often avoid using such tools, preferring to hire a consultant to do so. These tools can be time consuming to use, especially if such usage is infrequent and the process less familiar. An example of applying a final tool is that, for the above-mentioned bowling alley, an unvented 12-in. (30-cm) thick Trombe wall is found to deliver more heat during a typical winter night between 2 and 10 A.M. than will a vented 10-in. (25-cm) thick Trombe wall of the same material. Since the internal heat gains of the bowling alley (from electric lights and bowlers) meet the heating needs in this climate until 2 A.M., the unvented thicker wall is selected, even if the thinner, vented wall cost less to construct and promised a slightly higher annual SSF at the earlier, developmental design stage. When the typical nightly pattern was analyzed, the real "winner" emerged.

The process is described elsewhere (Jones and Wray, chapter 4 of this volume) for the development and verification of some of the most common early-stage design tools. What follows here is a description of many design tools and an indication of their popularity with designers. The relationship between popularity and veracity is not entirely clear, especially in the intuition-dominated early-tool examples. Sometimes, what is easy to use or understand is preferred to what is most accurate.

10.1 Conceptual Tools

10.1.1 Site Assessment

One of the first design tasks is site selection; specifically, the placement of a building on a given site. The designer faced with a wide range of site choices must immediately begin a balancing act between three potentially disparate influences: what is best for the aesthetics of the building-site relationship, what is best for the building program, and what is best for overall energy consumption. The question of energy consumption quickly separates into three categories: daylighting (supplemented by electric lighting), solar heating (supplemented by the range of available auxiliary heating fuels at the site), and passive cooling (supplemented by hybrid or mechanical cooling).

Until recently, daylighting was not seen as very important; the purpose of early design tools was to distinguish between the relative importance of heating and of cooling. However, in many climates, a high percentage of daylighting—with proper summer shading—will tip the balance from a cooling-dominated building (due to heat from its many electric lights) to a heating-dominated one (due to heat loss through increased window area and greatly reduced heat from electric lights). A popular, simple early tool is provided in Dubin and Long (1978), where for various building functions and for many locations across the United States, the likely ranking of dominance is listed for lighting, heating, cooling, service hot water, and miscellaneous electric power usages (see table 10.3).

The popular tools that emphasize heating versus cooling begin with Olgyay's (1963) Bioclimatic Charts, upon which typical monthly average days are plotted. These charts for a given location show the relative amount of time during which certain months are overheated (and how much wind might relieve the overheating), just right (within the "comfort zone"), or underheated (and how much solar radiation might relieve the underheating). These charts are well known and widely taught to architecture students. The major problem with them is that they emphasize comfort in exterior environments, which is useful for terraces and courtyards but not for interior design tending to predict a much greater reliance upon heating than is, in fact, the case in our insulated buildings of the 1980s.

Later, less famous tools include Mazria's (1979) patterns for site selection, beautifully illustrated and very clearly presented for maximum appeal to architectural students as well as to builders. Brown, Reynolds, and Ubbelohde (1982) present an extensive set of tools that are less appealingly presented but more thoroughly integrated in a process of stating goals, setting criteria, and evaluating design results. In addition, these tools are separated into site, building-cluster, building, and component scales, which helps the designer to focus on site-scale tools for site-scale design tasks. A more general set of tools (Niles and Haggard 1980) invites the designer to consider a site's major heat sources and major heat sinks, and then to site the building accordingly. Because they are less tied to a specific design act, these tools may encourage creativity among the more confident, experienced designers.

10.1.2 Building Shape

Closely following the task of site selection is the design task of selecting building shape. At its most basic level, this choice is "tall or short, thick or thin"; a matrix relating these basic choices to daylight, heating, and cooling appears in Stein and Reynolds (1992). Again, the influences of aesthetics,

Table 10.3
Ranking the energy use in buildings

Building types	Heating and ventilating	Cooling and ventilating	Lighting	Power and process[a]	Domestic hot water
	Categories of energy usage				
Schools					
A	4	3	1	5	2
B	1	4	2	5	3
C	1	4	2	5	3
Colleges					
A	5	2	1	4	3
B	1	3	2	5	4
C	1	5	2	4	3
Office buildings					
A	3	1	2	4	5
B	1	3	2	4	5
C	1	3	2	4	5
Stores					
A	3	1	2	4	5
B	2	3	1	4	5
C	1	3	2	3	5
Religious buildings					
A	3	2	1	4	5
B	1	3	2	4	5
C	1	3	2	4	5
Hospitals					
A	4	1	2	5	3
B	1	3	4	5	2
C	1	5	3	4	2

Rank 1 = most energy used in typical building of this type, relative to other uses.
Source: F. Dubin and G. C. Long, *Energy Conservation Standards for Building Design, Construction and Operation*, © 1978, McGraw-Hill.
a. Extensive use of computer terminals in workplaces may increase the ranking of this category.
A = climate with fewer than 2500 heating degree days.
B = climate with 2500 to 5500 degree days.
C = climate with 5500 to 9500 degree days.

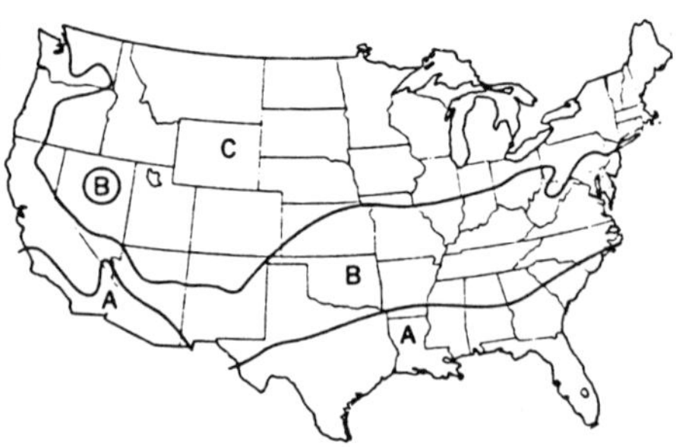

program, and energy performance must immediately be sorted through. And again, the potentially conflicting influences of daylighting (thin buildings favored), solar heating (south glass favored) and cooling (north glazing favored) must be resolved.

Olgyay (1963) presented recommendations for building plan (east-west dimension as a ratio to north-south dimension) that are widely taught. These studies assumed a rather low contribution of internal heat sources (typical of residences) and assumed the 1960 levels of insulation. Thus, they are somewhat dated, unduly favoring heating as the dominant design task in many climates. Mazria (1979) continues his appealing presentation of patterns: here, the patterns concern building shape and orientation, the treatment of the north side, and a guide to the location within the plan of various indoor activities (see figure 10.1). A wider set of patterns, linked to regional climate zones and analysis techniques, is found in AIA (1978). Following a similar format and accompanied by a supporting calculation process is Watson and Labs (1983).

In the more complex tool organization of Brown, Reynolds, and Ubbelohde (1982), the designer will be lead to design tools at the building-cluster and building scales. Perhaps the most evocative of these early tools are the whole-building concepts in Niles and Haggard (1980): "tin boxes, thermos bottles, chicken coops and pyramids," which achieve distinctions between combinations of light-weight construction, lots of glass, lots of mass, and lots of insulation.

A more recent illustrated set of siting and building-shape guidelines appeared in Brown and Cartwright (1985); drawn from the earlier work (Brown, Reynolds, and Ubbelohde 1982), the organization of the tools is less complex.

10.2 Sun Charts

As part of the consideration of location and shape of a building, the question arises as to the direction of the sun. Solar gain available to a building is important to daylighting, solar heating, and passive cooling, where avoidance of such gain is a typical objective. (See also Burns, chapter 2 of this volume.)

A procedure for plotting the skyline from a given location was presented by Mazria and Winitsky (1976), and widely published in Mazria (1979) and

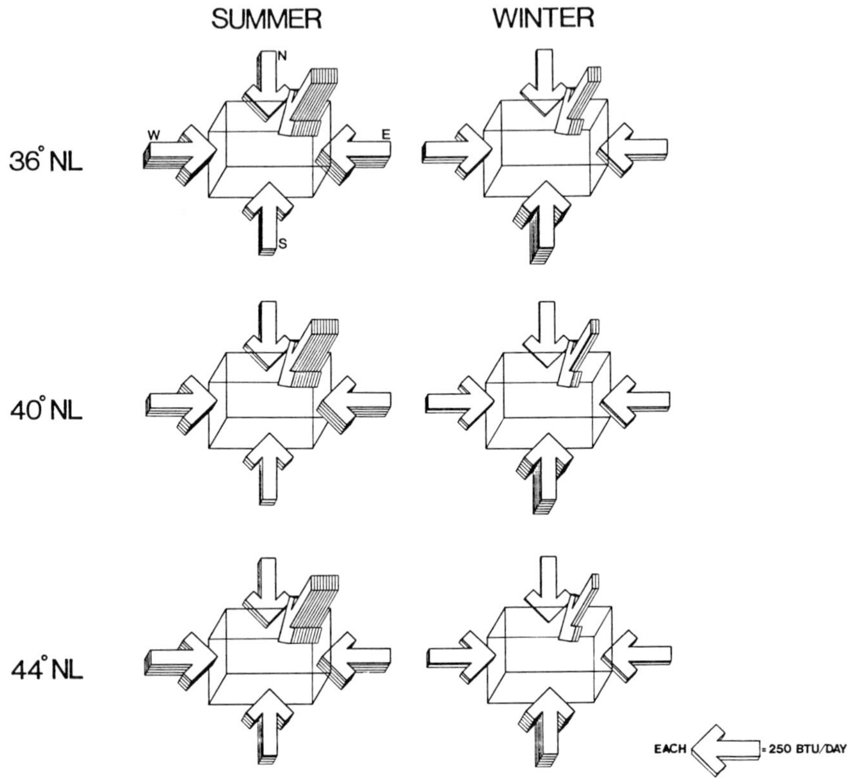

Fig. 10.1
Design pattern for building shape and orientation. Source: Mazria 1979. Reprinted from *Passive Solar Energy Book* © 1979 by Edward Mazria—permission granted by Rodale Press, Inc., Emmaus, PA 18098.

McGuinness, Stein, and Reynolds (1980)—see figure 2.16. This process is somewhat tedious because a separate skyline must be constructed for each location and each building surface, where sun and shadows must be assessed. But because it requires on-site usage of the tool, it encourages observation by the designer of other on-site influences. For sites where obstructions are remote (distant hills or tall buildings), this tool is ideal.

Related tools have been developed as devices that are carried to the site: they include a range from the inexpensive Solar Card (Solarvision, Inc., Harrisville, NH), requiring a steady hand, to the much more costly tripod-mounted Solar Pathfinder (Solar Pathways, Inc., Glenwood Springs, CO). Like the skyline plotting procedure in Mazria, the Solar Card uses a vertical projection: the sun's seasonal paths through the sky are presented as a wrap-around elevation relative to the location being analyzed. In contrast, the Solar Pathfinder uses a horizontal projection favored both by Olgyay (1963) and the popular Solar Calculator of Libbey-Owens-Ford (1974). In the horizontal projection, the sun's seasonal paths through the sky are presented on a plan whose center corresponds to the location being analyzed. In this author's experience, the vertical projection is somewhat easier for less experienced designers to grasp. Each projecting technique, however, has its advantages, and architecture students are generally taught to understand both the vertical and horizontal projection systems.

A third projection technique, again a horizontal one in which the location being analyzed is at the center of a plan, is called the sun-peg chart (Brown, Reynolds, and Ubbelohde 1982, Stein and Reynolds 1992). Here, in a procedure reminiscent of the earliest solar observations, shadows are cast by a stick (peg) of given height, inserted in the ground (see figure 10.2). The ground is the base of an architectural model; the peg is at the center of a horizontal graph. Upon this graph are a north arrow and various lines whose intersections represent various combinations of dates and times of day. When the model (with peg and graph) is carried out into the sun and tilted so that the end of the peg's shadow coincides with a given date/time intersection, the shadows cast on the model exactly correspond to that date and time. This tools has many uses for the designer, the earliest of which is to assess a given location for sun and shadows. The tool's great value is the combination of its three-dimensional model and the fourth dimension of time. A designer can consider a site, building, and interior space through a typical June day, a typical December day, or a series of,

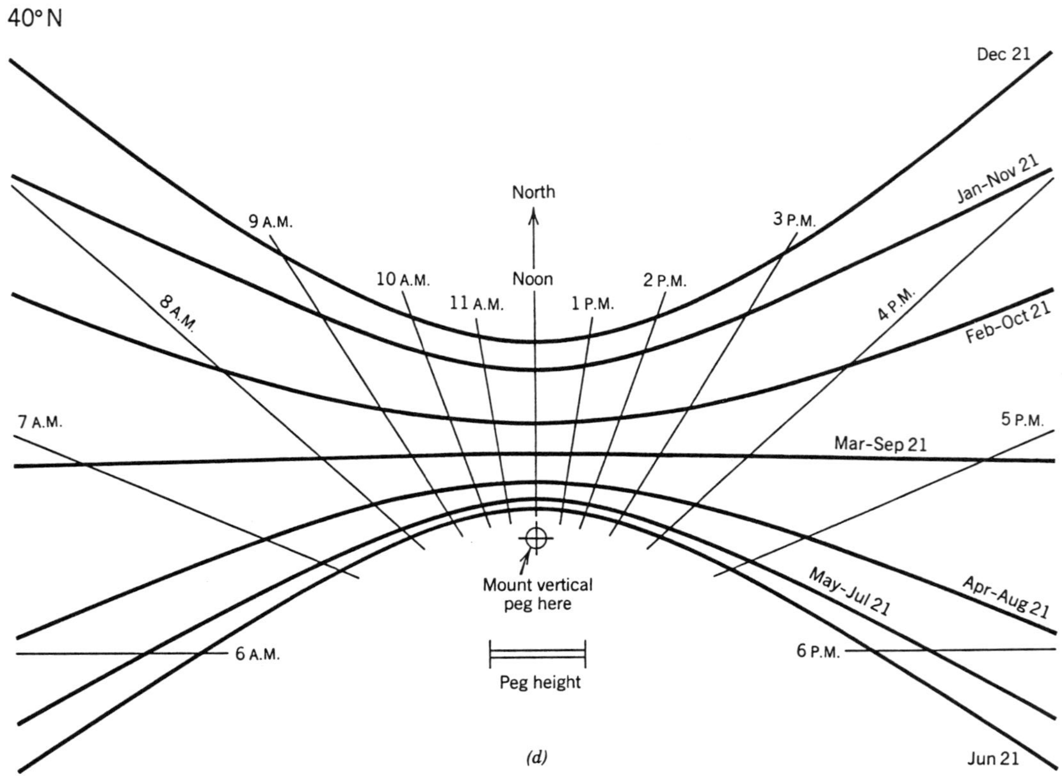

Fig. 10.2
Sun chart for 40° NL, based on shadow of vertical peg. Source: Stein, Reynolds and McGuinness 1986.

say, 11 A.M. sun patterns throughout the year (as for a church service starting at such an hour). Teachers of design have thus found this tool to be among the low-cost, more powerful energy-related tools. Its primary disadvantage is that any object that potentially casts shadows must be included in the model, which is unreasonable for distant hills or tall buildings.

For designers who prefer to work at night or indoors—or who live in cloudy climates—an electric light source can be substituted for the sun. Such electric lamps should be of special construction so that they emit rays that are essentially parallel (more a spotlight than a floodlight). If the sun-peg chart is too crude, specially constructed, though expensive, tilting tables are available that are cranked into positions corresponding to latitudes, dates, and times. At the other extreme is the heliodon, a room in which tracks are installed, enabling a traveling light source to emulate the sun's paths relative to a model.

Shading masks are a tool that helps design shading devices to transmit sun in the heating season and shade it in the cooling season. The Olgyay brothers, Victor and Adalar, developed an extensive series of such sun masks (Olgyay 1963, Olgyay and Olgyay 1957). As a result of their work, the Solar Calculator (which is perhaps the single most widely used energy design tool to date) was developed and distributed by Libbey-Owens-Ford (1974)—see figure 10.3. It enables the designer to determine the solar altitude and azimuth at any given date and time, the profile angle (the altitude of the sun relative to a given orientation), the angle of incidence (helpful to assessing how much of the incident solar radiation will penetrate the glass for a given orientation), and the portion of the year when at least 500 footcandles of daylight will be available. This elegant, multipurpose device encourages a three-dimensional approach to solar control where each facade has a set of shading devices most appropriate to its orientation.

10.3 Rules of Thumb

Whenever a complex set of relationships may be characterized by a simple ratio, a "rule of thumb" emerges. As a starting point in sizing (such as of windows and mass), rules of thumb are very useful to designers. As a final goal for such sizing, such rules are commonly used by builders for whom

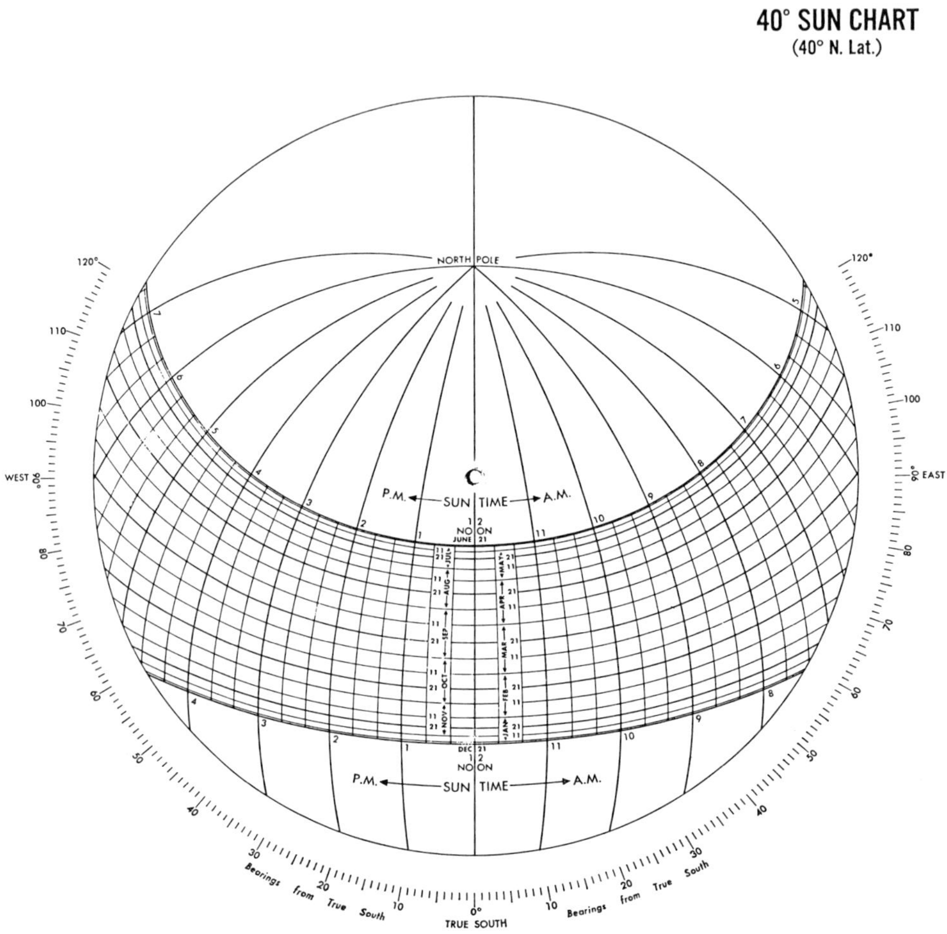

Fig. 10.3
The Libbey-Owens-Ford sun chart for 40° NL with transparent overlays for altitude, profile angle, and daylight. Source: Libbey-Owens-Ford 1974.

Table 10.4
Overall heat loss criteria

	Maximum heat loss (BTU/DD ft^2)	
Annual heating degree days (base 65°F)	Conventional buildings	Passively solar heated buildings, exclusive of solar wall
Less than 1000	9	7.6
1000–3000	8	6.6
3000–5000	7	5.6
5000–7000	6	4.6
Greater than 7000	5	3.6

Source: Balcomb et al. 1980.
Reprinted with permission from Stein, Reynolds, and McGuinness 1986.

solar heating (or daylighting) is a welcome amenity rather than a central feature. One of the reasons for the enduring popularity of rules of thumb is that for some they are guidance tools, for others they are evaluation tools.

10.3.1 Overall-Building Envelopes and Design Strategies

There are a few overall-building rules of thumb that are useful as points of departure, or points of early evaluation. For heating, there are suggested overall heat loss coefficients (Balcomb et al. 1980) that depend upon the severity of the climate (see table 10.4). The next level of complexity in heating guidelines is the U-overall criteria of the ASHRAE-90A Standard (ASHRAE 1980), which set U-overall criteria by component (roof, wall, or floor), climate, and building type (residential or nonresidential).

For cooling, there is a graphic technique for establishing the applicable passive cooling strategies developed by Milne and Givoni (1979) and subsequently published in Stein and Reynolds (1992)—see figure 10.4. The next cooling issue is the rate of heat gain per unit of floor area, which is heavily dependent on both the extent of daylighting and the rate of internal heat gain that varies by function. These heat gains are combined in estimates presented in Stein and Reynolds (1992)—see table 10.5. With an estimated heat gain rate, a designer may then enter other rules-of-thumb described in section 10.3.4. A more particular rule of thumb is the overall thermal transfer value (OTTV) from the ASHRAE Standard 90A, which relates window area to wall area, mass of construction, shading devices, and latitude. A separate OTTV is published for roofs. The OTTV is an

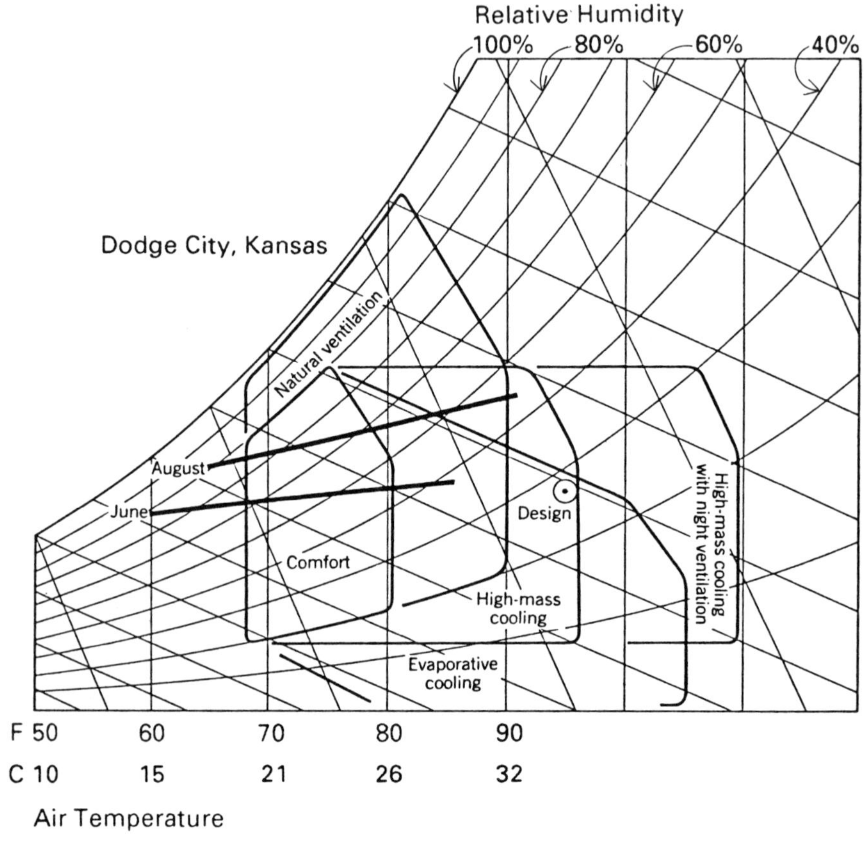

Fig. 10.4
Passive cooling strategies. The typical days for two months in Dodge City, Kansas, are plotted as lines; the ASHRAE summer design condition is shown as a point. Source: Milne and Givoni 1979. Reprinted with permission from Stein, Reynolds, and McGuinness 1986.

Table 10.5
Rules of thumb for heat gains

Part A. Internal heat sources—people and equipment

Function	Area per person (ft^2)	Sensible heat gain (Btu/h ft^2 of floor area)		
		People[a]	Equipment[b]	Total
Office	100	2.5	3.4	5.9
School: elementary	100	2.5	3.4	5.9
School: secondary, college	150	1.7	3.4	5.1
Hospital	100	2.5	Varies	2.5 plus
Clinic	50	5.0	Varies	5.0 plus
Assembly: theater[c]	15	15.3	—	15.3
Assembly: arena[c]	15	16.7	—	16.7
Restaurant	25	11.0	Varies	11.0 plus
Mercantile	50	5.0	Varies	5.0 plus
Warehouse	1000	0.4	—	0.4
Hotels, nursing homes	300	0.8	3.4	4.2
Apartments[d]	300	0.8	(see note d)	(see note d)

Part B. Internal heat sources—lighting, daylight, and electric

Function	Sensible heat gain (Btu/h ft^2 of floor area)[e]		
	$DF < 1$	$1 < DF < 4$	$DF > 4$
Office	5.1	2.0	0.5
School: elementary	6.3–6.8	2.5–2.7	0.6–0.7
School: secondary, college	6.3–6.8	2.5–2.7	0.6–0.7
Hospital	6.8	2.7	0.7
Clinic	6.8	2.7	0.7
Assembly: theater[c]	3.8	1.5	0.4
Assembly: arena[c]	3.8	1.5	0.4
Restaurant	6.3	2.5	0.6
Mercantile	5.1–6.8	2.0–2.7	0.5–0.7
Warehouse	2.4	1.0	0.2
Hotels, nursing homes	6.8	2.7	0.7
Apartments[d]	Up to 6.8	Up to 2.7	Up to 0.7

Table 10.5 (cont.)

Part C. Heat gains through envelope (Btu/h ft^2 of floor area)		Outdoor design temperature	
		90 F	100 F
I. Gains through externally shaded windows:			
Find $\frac{\text{total window area}}{\text{total floor area}}$,	then multiply by	16	21
II. Gains through opaque walls:			
Find $\frac{\text{total opaque wall area}}{\text{total floor area}} \times (U_{wall})$,	then multiply by	15	25
III. Gains through roofs:			
Find $\frac{\text{total opaque roof area}}{\text{total floor area}} \times (U_{roof})$,	then multiply by	35	45

Part D. Summary gains (Btu/h ft^2 of floor area)

I. Passive cooling systems for "open" buildings:
- Cross ventilation
- Stack ventilation
- Nighttime or "open" hours of thermal mass/night ventilation
- Total: add Parts A, B, and C gains to obtain total cooling load

II. Passive cooling system for "closed" buildings:
- Roof ponds
- Evaporative cooling
- Daytime or "closed" hours of thermal mass/night ventilation
- Total: add Parts A, B, C and Part E (below) gains, to obtain total cooling load

Part E. Gains from infiltration/ventilation of "closed" buildings (Btu/h ft^2 of floor area)	Outdoor design temperature	
	90 F	100 F
Find $\frac{\text{total window + opaque wall area}}{\text{total floor area}}$, then multiply by	1.0	1.9
OR		
Find $\frac{\text{known total cfm of outdoor air}}{\text{total floor area}}$, then multiply by	16.0	27.0

a. Adapted from Buehrer (1978).
b. The usual load of 1 W/ft^2 is assumed here. However, heavy use of computers can produce loads of up to 6 W/ft^2.
c. Gains listed for these functions are only for the seating areas, not for lobbies, stage areas, kitchens, and so on.
d. Residential internal gains often assumed at 225 Btu/h per occupant plus 1200 Btu/h total from appliances.
e. Adapted from Northwest Power Planning Council, *Maximum Lighting Standards*, 1983.
Source: Stein, Reynolds, and McGuinness 1986.

overall, averaged U-value. It allows for the designer to meet a criterion by manipulating the sizes of openings and the construction of walls and roofs; for example, the larger the window, the more insulated the wall must be. The effects of a wall's thermal mass and window shading devices are included in these procedures.

10.3.2 Daylighting

One of the most form-influencing rules of thumb in architectural design is that promulgated by Hopkinson and Kay (1969). It has been repeated many times since: that portion of a building's floor area that is eligible for consideration as daylighted by windows must be within 2 to 2.5 times the height of the windows from the exterior wall. This is a rule of thumb of singular importance because rather than dealing directly with the quantity of daylight, it sets an early limit on the ultimate amount of floor area that can even be considered for daylighting. As such, it establishes a fundamental relationship between ceiling height and building thickness: "thick or thin, tall or short."

Next is a familiar set of daylighting rules of thumb that relate overcast-sky daylight factor (the percentage of the available daylight outdoors that will be provided indoors) to the ratio of the building's window or skylight areas to the floor area served by such daylight apertures. Pioneered by Hopkinson and Kay (1969), this set of simple formulas has also been repeated many times (Millet and Bedrick 1980, Brown and Cartwright 1985, Stein and Reynolds 1992). Three problems arise with these rules: they do not deal with clear-sky conditions, they do not specify how much floor area should be served by the skylight, and they provide little guidance on an appropriate target for the winter daylight factor. The references indicated above also include suggested daylight factor levels, but a choice remains: does the designer aim for lots of winter daylight indoors and provide seasonal shading to control summer heat gain, or does he or she aim for adequate summer daylighting and accept more extensive levels of electric lighting in winter? Since a typical availability of summer daylight outdoors is about six times that of winter, this question is of rather large importance. More recently, the relationship of skylight area to floor area for atrium-type buildings has been expressed as rules of thumb (Cartwright 1985).

10.3.3 Solar Heating

The next window-sizing rules of thumb concern solar heating. The well-known early rules recommend ratios of south-facing window area to total solar-heated floor area. Mazria (1979) published such ratios for seven climate zones based on avoidance of overheating on clear winter days. Separate ratios are given for direct gain, Trombe wall, and sunspace. Next were the location-specific ratios in Balcomb et al. (1980)—see table 10.1. These widely used rules state a range of ratios of south-facing glass area to floor area for each of the listed 219 U.S. and Canadian locations based on avoidance of overheating on clear winter days. The passive system is assumed to be half direct gain, half water wall. Also stated are the corresponding ranges of approximate solar savings fractions: one range for windows without and another range for windows with night insulation. Los Alamos authors have since developed rules of thumb based on considerations of life-cycle cost (Balcomb et al. 1984). In this set of guidelines, two recommendations are made for each location, one based upon low fuel cost, one upon high fuel cost. For each case, there are recommended values for the load collector ratio (LCR) and the conservation factor (CF). CF is a multiplier used to determine optimum R-values for roof, walls, etc. LCR is used to determine the optimum passive solar glazing area.

Although rules of thumb based upon life-cycle cost would seem most useful, the different format of these two approaches makes the former more useful in the earliest stages of design. This is because a very early design decision involves a ratio between floor area and south glass, a readily understood quantity. In contrast, most designers have no idea of the design impact of the abstract LCR when beginning design. Furthermore, most building codes already state required minimum R-values based upon both conservation and construction influences.

Window sizing is closely related to the sizing of thermal mass for solar-heating as well as some passive cooling techniques. The earliest solar-heating-mass sizing guidelines appeared in Mazria (1979). More specific rules of thumb tied to ranges of solar savings fractions appeared in Balcomb et al. (1980); these are simple recommendations on the pound mass of masonry or water per unit of south-facing window area (lbm/ft^2). In Balcomb et al. (1984), there is a more detailed set of mass guidelines that are specific to each passive system type. For direct gain, there is more emphasis upon mass area than upon mass weight; the recommendation is a minimum mass to aperture area ratio of 6:1. Mass sizing for passive cooling will be discussed in a later paragraph.

For the architect or engineer, the next step beyond these simple overall-building, window, and mass-sizing rules of thumb is to consider the sizing variations for specific systems. For the builder of homes, the calculation of energy savings is somewhat of a luxury and is likely to exceed his or her experience in dealing with heat transfer factors. A set of builder guidelines that go beyond basic rules of thumb was developed at Los Alamos National Laboratory and can be found in Balcomb (1987).

10.3.4 Cooling

Where window openings are used for natural ventilation, rules of thumb are available for the ratio of openable windows to floor area (e.g., Brown, Reynolds, and Ubbelohde 1982). In cooling, both the solar exposure of windows and the rate of internal heat generation are typically more influential than inside-outside temperature difference. Floor area must therefore be related to both the extent of daylighting and the function (rate of internal gain), as is demonstrated in Stein and Reynolds (1992). After these introductory steps, window sizing for cross or stack ventilation can be done with more confidence.

Relatively simple sizing rules of thumb for cross and stack ventilation are found in Brown and Cartwright (1985), and rules of thumb that relate rate of heat gain to wind speed, window area, and indoor-outdoor temperature difference are found in Stein and Reynolds (1992)—see figure 10.5, also for stack effect rules of thumb. A more detailed cross-ventilation design procedure allowing for architectural projections, neighboring buildings, terrain, and insect screens is found in Chandra (1983).

Stein and Reynolds (1992) also list rules for night ventilation of thermal mass; one enters the graphs with rate of heat gain, summer design temperature, and the average diurnal outdoor temperature swing; separate performance curves are presented for average-mass and high-mass buildings.

Roof pond design guidelines are found in Mazria (1979), who published a simple ratio relating roof pond area to floor area. This was superceded by Fleischhacker, Clark, and Giolma (1983), who published the expected rates of heat loss for roof pond on summer nights. This shifted the emphasis to the roof pond as a cooling device and presupposed that the designer knew the rate of heat gain per unit floor area. Fleischhaker's work formed

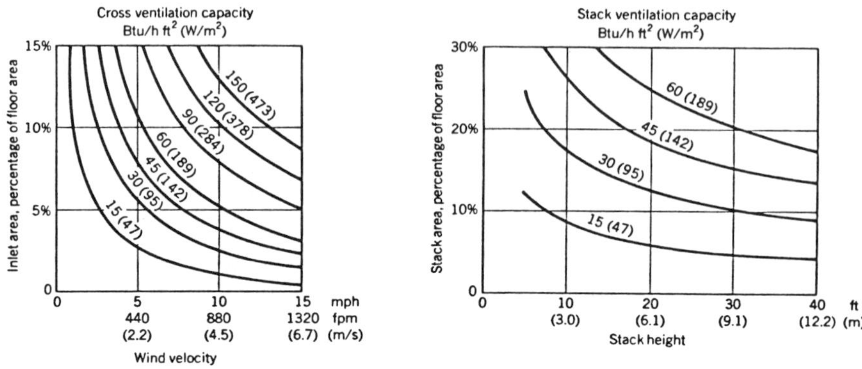

Fig. 10.5
Rules of thumb for cross ventilation and stack ventilation. The graphs assume an interior-exterior temperature difference of 3°F (1.7°C). Source: Stein, Reynolds, and McGuinness 1986.

the basis for the simple rules for roof ponds in Stein and Reynolds (1992) that relate the rate of heat gain, pond depth, diurnal outdoor temperature swing, and pond area.

Evaporative cooling rules of thumb are also presented in Stein and Reynolds (1992), based upon climate and heat gain. The same reference presents rules of thumb for earth tubes in which air velocity and rate of heat exchange per linear foot per degree temperature difference are fixed. This allows only the depth of the tube (hence the earth-air temperature difference) and the tube diameter (affecting air volume) to be adjusted by the designer. Somewhat more extensive rules of thumb for earth tubes are found in Abrams (1986).

The designer is now reaching the limit of design tools used primarily for guidance; performance evaluation tools emerge as the range of generic system types under consideration narrows, and the detailed variations within each system type increase. It is also at this point in design that questions of quantity (such as energy consumption) begin to be considered in a different way than questions of quality (such as thermal and visual comfort). It is perhaps here that the conflict is most apparent between the need for quick, simple, and easy-to-use design tools—which got the designer this far—and the need for sophisticated tools that yield annual energy balances and that may show that the designer took the wrong path early on. Several attempts are underway to harness the sophistication of computer analysis with the simplicity of early design choices in the style of the rule of thumb; an example is the software for Macintosh systems called Energy Scheming, under development by G. Z. Brown, Tomoko Sekiguchi, and Barbara-Jo Novitski (1989) at the University of Oregon. It includes a graphic-design aid for sizing windows and a single-zone building energy use analysis for four typical days representing four seasons.

10.4 Advanced Daylighting Design Tools

When the designer moves beyond simple sizing ratios for windows and skylights, several quite different design tools emerge that are likely to lead the subsequent design development in somewhat different directions. We begin with the tools more likely to be used by architects and end with tools for the consultant-engineer.

In the approach described by Moore and Anderson (1985), the emphasis is placed upon surfaces that are either the sources or the primary reflectors of daylighting. This approach deals far more with a geometric than a numeric concept and thus appeals more to the visually oriented architect than to the engineer who is left wondering "how much," while the architect has just been satisfied with "how." For the architect, this approach is highly evocative and encourages creative approaches to those critical surfaces in the vicinity of windows and skylights. Quality rather than quantity is the issue.

In an approach that gives the architect a visual pattern and the engineer a numerical distribution of daylight, Millet and Bedrick (1980) published a series of "footprints" that translate window shape and placement to daylighting contours on a workplane (parallel to floor level). Since multiple windows produce overlapping contours and corrections should be made for internal reflectance, this method quickly becomes tedious when numerical accuracy is desired.

For a single opening's impact upon a space—such as the oculus above the center of the Pantheon in Rome—this method gives a quick illustration of quality and quantity. For multiple openings or those of unusual shape, the architect is more likely to turn to the daylighting scale model.

The daylighting model is potentially one of the most powerful of all the design tools, combining studies in space and time, quantity and quality. Its primary drawbacks are the tediousness and expense involved in its construction and in the level of design decisions required before the model can be built. Both these are drawbacks because they discourage the designer from making subsequent changes, which would necessitate rebuilding the model. It can be a beautifully integrative guideline tool because it suggests three- and four-dimensional solutions to whatever problems the model may illuminate—if the designer has the patience, time, and money to take such changes seriously. The daylight model, when combined with the sunpeg chart, clearly relates sun patterns to locations of thermal mass. With considerably less accuracy, general patterns of air distribution can be studied in a wind tunnel for passive cooling. Yet another of the model's attractive features is that it so successfully communicates the nature of the building-to-be to a client.

Primarily an evaluation tool, the model deals successfully with both visual comfort and quantity of daylight. Perhaps no other design tool so readily illustrates the difference between direct light through clear glass and diffused light achieved by a filter that can be so simply added to the same opening. As the size of the model increases, it is easier for the designer to place herself within the space and thus to become aware of differential brightness that can lead, at best, to sparkle, at worst to glare. For measuring daylight quantity, the model's interior can be compared directly with the daylight on the exterior, yielding the daylight factor. This measurement process can be somewhat tediously done throughout a building, or representative readings can be taken for use in computer calculations (discussed later on). Guidelines to successful model construction and measurements are found in several sources, notably Evans (1981).

As the design task moves more toward quantity than quality of daylight, numerical techniques become attractive. Some of these retain a degree of graphic presentation that encourages the designer to see larger patterns of daylight distribution, such as in CIE (1970). With the advent of the personal computer, the "number crunching" associated with daylighting and its numerous reflectances has become much faster and far less tedious. Programs in widespread use include MICROLITE 1-2-3 (Bryan and Fergle 1986), DAYLITE (Ander, Milne, and Schiler 1986), and SUPERLITE (Kim et al. 1986).

For the mainframe computer, the daylighting calculation subroutine associated with DOE 2 seems particularly promising because it will either calculate the daylight distribution from simple windows in simple spaces, or representative values from a daylight model (with more complex geometries) can be used as the basis for the program's calculations (see the summary at the end of this chapter).

10.5 Tools for Estimating Thermal Performance

When a building design has developed to the point where its thermal envelope, aperture, and mass size can be tightly defined, a numeric analysis is usually made that predicts the building's need for auxiliary heating, lighting, or cooling under average climate conditions. This can be done for any time period, although a heating season (or calendar year) is most common.

There are two common approaches to solar heating performance prediction: the unutilizability method (Duffie and Beckman 1974) and the solar load ratio (SLR) method (Balcomb et al. 1980, 1982, 1984). (See Jones and Wray, chapter 4 of this volume.) Of the two, the SLR method has gained wide acceptance and will be discussed later on.

Early in the advent of passive solar heating, a partly graphic performance prediction technique was used in the U.S. Department of Housing and Urban Development (HUD) Passive Residential Design Competition in 1978. Although it was crude in its lack of detailed distinction between systems (rather than distinguish between direct gain, Trombe wall, and sunspace, it gave the choice of masonry or water, night insulated or not), it yielded both monthly and annual solar heating fractions. Because the solar heating fraction was presented graphically, the designer could quickly see the approximate effect of a change in aperture size or increased insulation in construction. This relatively quick and easy tool entered architectural textbooks with McGuinness, Stein, and Reynolds (1980) and introduced a large number of today's designers to the techniques for predicting solar performance.

One of the most extensive graphic performance examples, though limited to sites in California, was the *Passive Solar Handbook* (Niles and Haggard 1980), which showed how several reference homes would perform with a variety of passive systems in a range of south aperture areas and overall U-values, in fourteen locations representative of California's climate zones. Included were supplementary curves showing sensitivity to such variations as mass wall thickness and material composition.

A primarily manual graphic technique appears in Johnson (1981), who recommended that the month of March be used as a representative of the entire heating season. This allows a much quicker manual approximation of yearly energy savings; the estimated error is within 15% of the annual performance, providing the building was not unusually well insulated or the winter cloud cover not unusually heavy.

Up to this date, most designers have been using manual performance prediction tools that yielded the solar fraction, that is, the percentage of the building's total annual heating energy that would be provided by the sun. Two major changes occurred about this time: the programmable calculator became a common item in offices, and the concept of the solar savings fraction was advanced. The benefit of the solar savings fraction was that it compared the solar building to a very similar one that was nonsolar. Before this, designers could readily tell a client that 70% of their heating energy would come from solar energy but were unable to easily compare the actual backup energy used to that of a more conventional, nonsolar design. The solar savings fraction (SSF) first appeared in Balcomb et al. (1980) along with the load collector ratio (LCR). (Unfortunately, the LCR is larger for less-solar, smaller for more-solar buildings, which makes even more bewildering the array of acronyms and number systems that designers must master.) In this widely used work, six system types (water wall, Trombe wall, and direct gain, each either with or without night insulation) were listed for their combination of LCR and resulting SSF for over two hundred locations in the United States and Canada. A number of commercial products appeared that allowed the designer to use a programmable calculator to get monthly solar savings fractions. Two popular products were PASCALC (developed by Total Environmental Action) and PEGFIX (developed by the Princeton Energy Group).

The limitation of the SLR method to only six system types soon led to a much wider variety presented first in Balcomb et al. (1982) and refined in Balcomb et al. (1984). Sunspaces and numerous variations of direct-gain and thermal-storage wall are added to the system types for a total of ninety-four system variations. Sensitivity curves are included to allow the designer to examine many departures from the ninety-four systems. Adapted by many computer programs, quickly done manually (for the annual SSF), and abridged for inclusion in standard textbooks, such as Stein and Reynolds (1992), this has become probably the most widely used energy prediction technique of all.

While Balcomb's group was refining their techniques for utilizing the solar load ratio method, the personal computer was rapidly replacing the programmable calculator. In the late 1970s, only designers with some ties to a university or a large research facility could readily utilize the power of the mainframe computer to achieve, with great speed and accuracy, comparison of an array of design alternatives. Most building designers, with varying degrees of patience, sat as programmable calculators eked out their results. By 1986, the personal computer (microcomputer) has become commonplace in designers' offices, and Rittelmann and Ahmed (1985) list 230 energy-related design tools, of which 100 were for microcomputers (mainframe computers had 59, programmable calculators 30, and manual methods 41). A summary of the 1985 Rittelmann report reveals certain patterns of hardware, user types, and uses of the tools. Architects had slightly more manual tools available than engineers, engineers slightly more microcomputer tools than architects. Almost all the 230 tools surveyed dealt with heating, almost half with cooling, but other energy usages were much less often included. In almost every energy-use category, there were many more tools for microcomputer than for mainframes, programmable calculators, or manual tools. Tools for residential and small commercial applications outnumbered those for large commercial ones. For passive solar heating, 121 tools were listed; of these, 23 relied upon the SLR method, 31 upon thermal network analysis (see Niles, chapter 3 of this volume), 9 upon response factors, and 58 used other algorithms. Again, most of these tools were for microcomputers.

10.6 Tools for Estimating Thermal Comfort

As the fraction of a building's heat that is provided by solar energy increases, so does the potential for that building to overheat on a clear day. At least two undesirable results are uncomfortable people and opened windows that waste the solar energy being collected and lower the predicted performance. As was the case for estimating the need for auxiliary heating, tools for estimating comfort began with more simple graphic techniques and developed into faster, more accurate ones suitable for microcomputers.

A caution here is that comfort is much harder to quantify than auxiliary heat; it is a factor that is part physiological, part psychological. A crackling fire may be wasting far more energy up the chimney than it is contributing to the house, but it looks, sounds, smells, and feels more successful than the Trombe wall nearby. Comfort varies with air temperature, surface temperature, air motion, and relative humidity; air quality also enters in (as in "stuffiness"). Whether a given measured thermal environment is "comfortable" will depend on the age, activity rate, and clothing of the occupant, and may vary with both the season and the cultural setting. Yet, almost all the tools for designers measure but one aspect: air temperature. An interior temperature of 83°F on a January day in Albuquerque may readily cause annoyance to occupants; in normally cloudy Seattle it may be cause for rejoicing.

The early and graphic methods for thermal comfort estimation include the delta-T-solar approach in Balcomb et al. (1980) and the sunny March day analysis advocated by Johnson (1981). The delta-T-solar method begins with data for the average outdoor January temperature. Curves for various latitudes and passive system types then establish delta-T-solar, the indoor temperature rise attributable to solar energy (see figure 10.6). After adding temperature rise due to internal sources, an average indoor January temperature is obtained. Diurnal temperature swings for different passive systems may also be estimated.

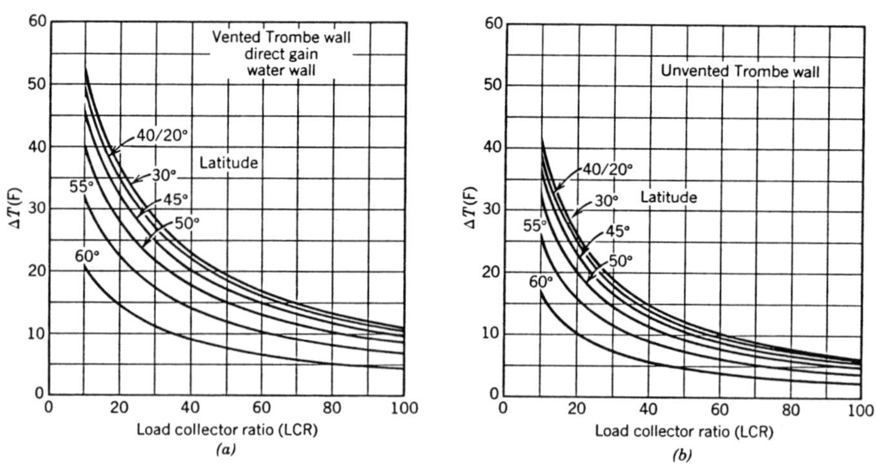

Fig. 10.6
Estimating delta-T-solar for a clear January day. Source: Balcomb et al 1980. Reprinted with permission from Stein, Reynolds, and McGuinness 1986.

Johnson's method, based on the work of Niles (1979), combines graphs and calculations that yield an internal equilibrium temperature, and then diurnal temperature swings that depend upon whether the thermal mass is coupled convectively or radiatively to the solar energy.

Programmable calculators took some of the tediousness out of the thermal network approach; a popular program was TEANET, developed by Total Environmental Action, Harrisville, NH. TEANET calculates hourly temperatures for a variety of locations within a thermal network set up by the designer. It allows for the movement of night insulation and for changes in internal gains due to building schedule.

10.7 Comprehensive Tools

One tendency is toward comprehensive analysis tools that integrate the effects of daylighting, passive solar heating, passive cooling, and auxiliary and internal heat sources. Because of the complexity of the factors, these tools are primarily designed for use on mainframe computers. Because of the volume of inputs required to describe the building's characteristics, these tools are rarely used in the early stages of design and therefore are far more likely to change later details than to guide earlier concepts.

There are a number of detailed programs that include a wide range of factors and that yield annual energy predictions broken down by end usages, such as heating, cooling, lighting, service hot water, etc. Popular available programs include the CALPAS series (primarily residential and small commercial applications, weaker on daylighting), AXCESS (comparative energy uses of alternative methods for meeting energy requirements), BLAST (a particularly strong program for simulating the effect of thermal mass but weaker on daylighting), MICROPAS (heating, cooling, and ventilating; includes thermal mass but weak on daylighting), and TRACE (primarily for HVAC application, weak on daylighting).

Perhaps the most widely used and comprehensive of these programs is the DOE series. DOE 2.1C includes heating, cooling, daylighting, and other energy usages within buildings. Although the daylighting calculation variations are quite limited (light shelves not included, at this time), there is a provision for the use of values from a daylight model. One of DOE's weaker features is an inability to deal in detail with thermal mass, which is of particular importance for both passive solar heating and several approaches to passive cooling.

10.8 Future Directions

Recent years have seen a tendency toward sophisticated, numeric-based design tools that are strong on the evaluation of a highly evolved building design but weak on the guidance of concept formation in the early design stages. If the development of graphic-based design tools can approach that of numeric-based ones, there is a better chance of influencing the building designer early enough that the fundamentally right choices can be made. Exciting as it is to see how today's computer analyses can be used to patch up a design to overcome its energy inadequacies, wouldn't it be far more exciting to produce a design that is correctly conceived in the first place?

References

Abrams, D. W. 1986. *Low Energy Cooling.* New York: Van Nostrand Reinhold.

AIA Research Corporation. 1978. *Regional Guidelines for Building Energy Conserving Homes.* Washington, DC: U.S. Department of Housing and Urban Development.

American Society of Heating, Refrigerating and Air-Conditioning Engineers. 1980. *ASHRAE Standard 90A-1980, Energy Conservation in New Building Design.* Atlanta, GA: ASHRAE.

Ander, G. D., M. Milne, and M. Schiler. 1986. Fenestration design tool: a microcomputer program for designers. *Proc. 1986 International Daylighting Conference I.* Long Beach, CA: International Daylighting Conference, pp. 187–193.

Balcomb, J. D., D. Barley, R. McFarland, J. Perry, W. Wray, and S. Noll. 1980. *Passive Solar Design Handbook.* Vol. 2: *Passive Solar Design Analysis.* DOE/CS-0127/2. Washington DC: U.S. Department of Energy.

Balcomb, J. D., C. E. Kosiewicz, G. S. Lazarus, R. D. McFarland, and W. O. Wray. 1983. *Passive Solar Design Handbook.* Vol. 3: *Passive Solar Design Analysis.* Edited by R. W. Jones. DOE/CS-0127/3. Washington, DC: U.S. Department of Energy; Boulder, CO: American Solar Energy Society, 1983.

Balcomb, J. D., R. W. Jones, R. D. McFarland, and W. O. Wray. 1984. *Passive Solar Heating Analysis: A Design Manual.* Atlanta, GA: ASHRAE.

Balcomb, J. D. 1986. Design tools for passive solar applications. *Proc. Annual Meeting, American Solar Energy Society,* Boulder, CO, June 11–14, 1986, Boulder, CO: American Solar Energy Society, pp. 2–7.

Balcomb, J. D. 1987. Builder guidelines. *Proc. 12th Passive Solar Conference,* Portland, OR, July 11–16, 1987. Boulder, CO: American Solar Energy Society, pp. 403–405.

Brown, G. Z., J. S. Reynolds, and M. S. Ubbelohde. 1982. *InsideOut: Design Procedures for Passive Environmental Technologies.* New York: Wiley.

Brown, G. Z., and V. Cartwright. 1985. *Sun, Wind, and Light.* New York: Wiley.

Brown, G. Z., T. Sekiguchi, and B. J. Novitski. 1989. *Energy Scheming, Version 1.0.* Eugene, OR: Department of Architecture, University of Oregon.

Bryan, H. J., and R. J. Fergle. 1986. MICROLITE 1-2-3: a coherent microcomputer environment for daylighting design. *Proc. 1986 International Daylighting Conference I.* Long Beach, CA: International Daylighting Conference, pp. 173–181.

Cartwright, V. 1985. The use of lightwells as a daylight strategy. *Proc. 10th National Passive Solar Conference*, Raleigh, NC, October 15–20, 1985. Boulder, CO: American Solar Energy Society, pp. 115-118.

Chandra, S. 1983. A design procedure to size windows for naturally ventilated rooms. *Proc. 8th National Passive Solar Conference*, Santa Fe, NM, September 7–9, 1983. Boulder, CO: American Solar Energy Society, pp. 105–110.

Commission Internationale de l'Eclairage (CIE). 1970. *Daylight: International Recommendations for the Calculation of Natural Daylight.* Paris: Commission Internationale de l'Eclairage.

Dubin, F. S., and C. G. Long. 1978. *Energy Conservation Standards for Building Design, Construction, and Operation.* New York: McGraw-Hill.

Duffie, J. S., and W. A. Beckman. 1974. *Solar Energy Thermal Processes.* New York: Wiley.

Evans, B. E. 1981. *Daylight in Architecture.* New York: McGraw-Hill.

Fleischhacker, P., G. Clark, and P. Giolma. 1983. Geographic limits for comfort in unassisted roof pond cooled residences. *Proc. 8th National Passive Solar Conference*, Santa Fe, NM, September 7–9, 1983. Boulder, CO: American Solar Energy Society, pp. 835–838.

Hopkinson, R. G., and J. D. Kay. 1969. *The Lighting of Buildings.* New York: Praeger.

Johnson, T. E. 1981. *Solar Architecture: The Direct Gain Approach.* New York: McGraw-Hill.

Kim, J. J., K. M. Papamichael, S. Spitzglas, and S. Selkowitz. 1986. Determining daylight illuminance in rooms having computer fenestration systems. *Proc. 1986 International Daylighting Conference I.* Long Beach, CA: International Daylighting Conference, pp. 204–208.

Libbey-Owens-Ford. 1974. *Sun Angle Calculator.* Toledo, OH: Libbey-Owens-Ford Glass Co.

Mazria, E., and D. Winitsky. 1976. *Solar Guide and Calculator.* Eugene, OR: Center for Environmental Research, University of Oregon.

Mazria, E. 1979. *The Passive Solar Energy Book.* Emmaus, PA: Rodale Press.

McGuinness, W. J., B. Stein, and J. S. Reynolds. 1980. *Mechanical and Electrical Equipment for Buildings.* 6th ed. New York: Wiley.

Millet, M. S., and J. R. Bedrick. 1980. *Manual Graphic Daylighting Design Method.* Berkeley, CA: Lawrence Berkeley Laboratory.

Milne, M., and B. Givoni. 1979. Architectural design based upon climate. In *Energy Conservation Through Building Design.* New York: McGraw-Hill.

Moore, F., and G. Anderson. 1985. *Concepts and Practice of Architectural Daylighting.* New York: Van Nostrand Reinhold Co.

Niles, P. W. B. 1979. Graphs for direct gain house performance prediction. In *Passive Systems '78.* Newark, DE: American Section of the International Solar Energy Society, pp. 76–81.

Niles, P. W. B., and K. L. Haggard. 1980. *Passive Solar Handbook.* Sacramento, CA: California Energy Commission.

Olgyay, V. 1963. *Design with Climate.* Princeton, NJ: Princeton University Press.

Olgyay, A., and V. Olgyay. 1957. *Solar Control Shading Devices.* Princeton, NJ: Princeton University Press.

Rittelmann, P. R., and S. F. Ahmed. 1985. *Design Tool Survey.* Washington, DC: International Energy Agency and U. S. Department of Energy.

Stein, B., J. S. Reynolds, and W. J. McGuinness. 1986. *Mechanical and Electrical Equipment for Buildings.* 7th ed. New York: Wiley.

Stein, B., and J. S. Reynolds. 1992. *Mechanical and Electrical Equipment for Buildings.* 8th ed. New York: Wiley.

Watson, D., and K. Labs. 1983. *Climatic Design: Energy Efficient Building Principles and Practices.* New York: McGraw-Hill.

Contributors

J. Douglas Balcomb

Principal engineer at the Solar Energy Research Institute, Dr. J. Douglas Balcomb has been working on the quantification and evaluation of passive solar heating techniques since 1975. He organized the solar research team at the Los Alamos National Laboratory and has been very active in the leadership of both the International Solar Energy Society and the American Solar Energy Society. He has monitored numerous buildings, conducted original research on natural convective air-flow, published 120 technical papers and several books, lectured extensively including 51 seminars in 26 countries, consulted on the design of about 60 buildings, participated on International Energy Agency and United Nations projects, and received five awards for his contributions to the field. He did his undergraduate work at the University of New Mexico and received his PhD in nuclear engineering from the Massachusetts Institute of Technology.

Patrick J. Burns

Patrick J. Burns is a Professor in the Department of Mechanical Engineering at Colorado State University, where he has pursued the study of solar and thermal systems since 1978. His research interests include passive solar systems simulation and design, experimental investigations of passive solar systems, parameter estimation in building heat transfer, and the calculation of radiative loads in general thermal systems. He holds MS and PhD degrees in Mechanical Engineering from the University of California at Berkeley, and a BS in Mechanical Engineering from Tulane University. He holds membership in the American Societies of Mechanical, and Heating, Refrigerating and Air-conditioning Engineers.

Donald J. Frey

Donald J. Frey has been involved with passive solar energy research since 1975. Mr. Frey is professionally affiliated with Architectural Energy Corporation in Boulder, Colorado where he is the Executive Vice President. Prior to cofounding Architectural Energy Corporation in 1982, Mr. Frey worked for the Solar Energy Research Institute. Mr. Frey's major accomplishments in solar energy research include managing the Class B passive solar residential monitoring program, and participating in the performance evaluation aspects of the Passive Solar Commercial Buildings Program, Solar In Federal Buildings Program, and IEA Task VIII. Mr. Frey also participated in IEA Task XI and the U.S./U.K. Bilateral on passive solar research. In addition to numerous project reports and articles, Mr. Frey was a contributor to and editor of the book *Sun/Earth: How to Use Solar and Climatic Energies*. He was also a major contributor to the REM/DESIGN residential energy analysis software developed by Architectural Energy Corporation.

Michael J. Holtz

Michael J. Holtz is President of Architectural Energy Corporation of Boulder, Colorado, a contract research, development, and design consulting firm specializing in energy and the environment. He has been involved with solar buildings research since 1973. Mr. Holtz has been associated with the AIA Research Corporation as Director of Solar Energy Programs and the National Renewable Energy Laboratory (formerly the Solar Energy Research Institute) as Chief and Acting Manager of Building Systems Research. Mr. Holtz's major accomplishments in the field of solar research include development of performance analysis and testing procedures, implementation of residential and commercial solar demonstration programs, and development of simplified solar design tools. In 1975 he authored the book *Solar Dwelling Design Concepts.* He holds both a Bachelor and Masters of Architecture.

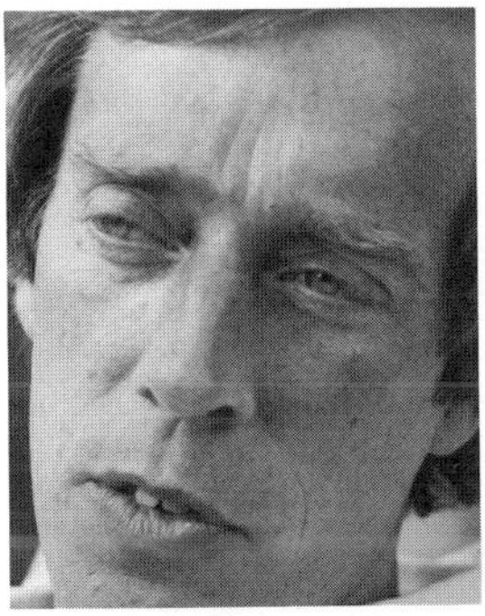

Timothy E. Johnson

Timothy E. Johnson is a Principal Research Associate and teacher in the Department of Architecture at MIT. He is also principal of Johnson & Johnson Design/Build, an architectural and software services firm that has designed and built several solar residences in the United States and Europe. He has been involved with passive solar energy research and applications since 1974, when he began work on high-performance windows. Mr. Johnson is director of the MIT Solar Building No. 5 project, which demonstrated the first application of low-emissivity windows for passive space heating in 1978. He holds a patent on phase change floor and ceiling tiles for storing solar heat. His books include *Solar Architecture: The Direct Gain Approach* and the *Low-e Glazing Design Guide*. He holds degrees in mechanical engineering from MIT and the University of Michigan.

Robert Jones

Robert Jones has been involved with passive solar energy research since 1978. Dr. Jones was a staff member at Los Alamos National Laboratory of most of that time, and was leader of the solar energy program after 1983. His major accomplishments in solar energy research include modeling and simulation of passive solar systems. He was editor and co-author or the DOE manual, *Passive Solar Design Handbook, Volume 3*, and the ASHRAE manual, *Passive Solar Heating Analysis*. Other publications include *The Sunspace Primer: A Guide for Passive Solar Heating*. He is former editor of the *Passive Solar Journal* and *SunWorld*. He holds a PhD in physics from the University of Colorado.

Fuller Moore

Fuller Moore has been involved with passive solar research since 1975. He is Professor of Architecture at Miami University, where he founded the Center for Building Science Research. He has researched the design and experiment use of passive solar test modules as a guest scientist with the Solar Group Q-11 at Los Alamos National Laboratory. Major accomplishments in solar energy research include three USHUD Passive Solar Competition Design Awards, co-authorship (with J. McFarland) of *Passive Solar Test Modules*, authorship of *Concepts and Practice of Architectural Daylighting*, and the award of a Senior Fulbright-Hayes Fellowship to the Soviet Union to lecture on solar heating and daylighting. He has been a Director of the Passive Division of American Solar Energy Society, and is Chair-Elect of both the Society of Building Science Educators and the Daylighting Network of North America.

Philip W. B. Niles

Philip W. B. Niles has been involved with passive solar energy research since 1973. Professor Niles teaches in the mechanical engineering department of the California Polytechnic State University at San Luis Obispo.

His major accomplishments in solar energy research include helping design and test the first roof-pond passive house, and writing the State of California's residential energy simulation program, CALPAS. He has also been active in solar crop dehydration research and simplified calculation methods. He has served as an energy consultant in Africa for USAID. His publications include *Passive Solar Handbook* (with Kenneth Haggard).

John S. Reynolds

Professor of Architecture at the University of Oregon since 1967, John Reynolds teaches design, environmental control systems, passive solar heating, and passive cooling. He is also Director of the Solar Energy Center. His architectural firm, Equinox Design, has won National Awards for Energy Innovation in 1984 and 1988. He is coauthor of the widely used texbook *Mechanical and Electrical Equipment for Buildings* in its 6th (1980), 7th (1986), and 8th (1992) editions. He has written many articles and research papers on passive solar design, and has taught at North and South American universities. His memberships include the AIA and the American Solar Energy Society. His professional degrees are from the University of Illinois (Urbana) and the Massachusetts Institute of Technology.

William O. Wray

William O. Wray has been involved with passive solar energy research since 1978. Dr. Wray is professionally affiliated with Los Alamos National Laboratory. His major accomplishments in solar energy research include the development of quantitative design tools for direct gain buildings, which he later wrote up for inclusion in *Passive Solar Heating Analysis*, a design manual published jointly by Los Alamos and ASHRAE. He also developed and wrote a design manual for the U.S. Navy entitled *Passive Solar Design Procedures for Naval Installations*, which specifically addressed the concrete block and metal buildings typical of Navy construction. Additionally, for four years he represented the United States as an expert on computer modeling of passive solar buildings in the NATO-sponsored Committee on the Challenges to Modern Society.

Index

Absorption
in direct gain analysis, 255–256
distributed, 82–83
by glass, 73, 79–82, 201, 211
local, 83–84
by opaque external objects, 90–95
in sunspace analysis, 275
in thermal storage wall analysis, 264–265
by water, 133
Additive monitoring methods, 407, 417
Admittance in thermal storage modeling, 123–126, 225
Adobe, sensitivity curves for, 255, 263
Aerogel in windows, 208–210
Aesthetics
in building integration, 335, 344, 347–348
in early passive buildings, 342
and shape of building, 489–491
in site assessment, 488
Air
convection modeling for, 139–144
emissivity of, 157
sensors for movement of, 303
in windows, 76–77, 207–208
Air gaps in direct gain analysis, 257
Air-glass interfaces, 76–77
Alberta, University of, 321, 323
All-solar subdivisions, 12
Ames Laboratory, 316–317
Analytical results, 235
of auxiliary heat (*See* Auxiliary heat systems and measurements)
of backup heating demand, 244, 280–281
building variables for, 236–237
of mixed systems, 279–280
of radiant panels, 279
of roof ponds, 278
of thermal comfort criteria, 244–247, 281–288
of thermosiphoning air panels, 279
of transwalls, 278–279
Antireflection glazing treatments, 201–203
Architectural Energy Corporation (AEC), 470–471
Architecture
in building integration, 335–336, 345–348, 359–360
regionalism in, 14
Argon-filled windows, 200, 207
ASHRAE Handbook of Fundamentals, 47
Atascadero House, 410
Atria in commercial buildings, 14
Auxiliary heat systems and measurements, 237–244, 304, 307
in building tests, 302
direct gain, 248–257
efficiency of, 456
for sunspaces, 267–278
for thermal storage walls, 257–267
for Trombe walls, 247–248
AXCESS program, 511

Backup heating demand, 244, 280–281
Backward-difference algorithm, 119–120
Backward-looking methods, 181
Baer, Steve, 341
Balanced systems, 13, 334–335, 359–360, 406–407
Balcomb residence, 416–420
Basements, heat transfer through, 126–131
Bateson State Office Building, 384
Beam radiation, 50–51
and overhangs, 64
and tree transmittance, 57, 60
Bermed walls, heat transfer through, 128, 131
Between-the-glazing window insulation, 219–220
BEVA (Building Energy Vector Analysis), 33–34
Bioclimatic Charts, 489
Biot number, 121
Black chrome for thermal storage, 223–224
BLAST program, 28, 511
for radiation, 155
for roof apertures, 464
Blinds, 218–219
Bonneville Power Administration (BPA), 432, 447–450
Brookhaven house, 450–451
Brookhaven National Laboratory, 450–454
BuilderGuide, 36
Building Energy Performance Standards (BEPS), 461–462
Building integration, 331
aesthetics in, 335, 344, 347–348
architecture in, 335–336, 345–348, 359–360
in commercial buildings, 364–387
economics in, 335, 360
future of, 387–389
goals of, 334–336
and passive subsystems, 332–336
in residential buildings, 336–364
Building performance index (BPI), 455–456
Buildings
components and systems in, 403–404
test, 301–302, 317–324
variables in, 236–237

Bulk transparent insulation, 208–210
Butler Manufacturing Company, 378

Calibration of test modules, 306–307
California
  office-building construction program in, 380–381
  SERI monitoring programs in, 439–442
CALPAS programs, 119–120, 511
Capacitance in thermal network analysis, 112
Carey Arboretum, 384
Carpeting, 256
Case study of design guidelines, 18–21
Caudill, William W., 366–367
Ceilings in convection modeling, 144
Central data-processing systems (CDPS), 423
Central difference method, 118–119, 121
Cholesteric crystals, 218
Circumsolar radiation, 42
Class A program (SERI), 431
Class B program (SERI), 431–445
Class C program (SERI), 445–447
Clear sky radiance, 160–161
Coatings for glazings
  antireflection, 202–203
  electrochromic, 216–218
  insulation, 203–210
Code validation, 293, 300–301
  calibration for, 306
  instrumentation for, 305
  test modules for, 311–313, 324–325
Comfort. *See* Thermal comfort
Commercial buildings, 364
  atria in, 14
  daylighting in, 6, 13–14, 376, 386, 460
  design process in, 14
  energy and cost-savings potential for, 26
  glazings in, 210–218
  lessons learned from, 386–387
  milestones in, 368–386
  monitoring of, 430, 459–464
  in 1972, 365–368
  thermal balancing in, 14
Comparative testing, 293–300, 307
Components. *See* Materials and components
Comprehensive design tools, 511–512
Computations with solar gain modeling, 95–98
Computer-aided design and drafting (CADD), 387
Computerized data loggers, 401
Computers, 16, 21–22, 28. *See also* Code validation
Conceptual tools, 485, 488–491
Concrete, 225, 254–255, 263
Condensation with window insulation, 220
Conductance in thermal network analysis, 112
Conservation
  balanced systems for, 13, 334–335, 359–360
  categories of, 381–382
  guidelines for, 193–194, 243
  in optimum mix methods, 192–194
Construction
  in building integration, 333
  importance of, 364
  quality of, 13
Containers for phase-change materials, 229–230
Convection modeling
  for air and surfaces, 139–144
  for outside of buildings, 144–149
  between rooms, 149–152
  test boxes for, 309
  for windows, 207–208
Convective heat transfer coefficient, 93–95
Cooling
  in building integration, 334
  by phase-change materials, 228–229
  rules of thumb for, 503–505
  and windows, 211
Coordinate systems
  for shading calculations, 66
  for solar radiation geography, 46–48
Correlation methods, 182
  degree-day, 183–185
  solar load ratio, 185–187
  unutilizability, 187–189
Costs, 6, 10–11
  in commercial buildings, 375–376, 386
  potential savings in, 26
  in residential buildings, 363
Counting sensors, 304
Creativity, importance of, 375
Crystals for electrochromic coatings, 216–218
Curtains, 221

Data
  analysis of, 404–405
  types of, 403–404
Data acquisition, 304–305
  methods for, 400–403

sensors for, 302–304
in SERI program, 432–434
Data loggers, 304, 401, 432, 434
Daylighting
benefits of, 10
in commercial buildings, 6, 13–14, 376, 386, 460
design tools for, 505–507
from roof apertures, 463–464
rules of thumb for, 501
and shape of building, 491
in site assessment, 488–489
Daylighting performance evaluation methodology (DPEM), 467
DAYLIGHT program, 506
Deciduous trees
shading by, 54–57
transmittance by, 57–61
Degree-day method, 183–185, 237
Delta-T-solar method, 510
Demonstration programs, 343, 368–370, 389
Denver Metropolitan Solar Homebuilders Program, 351–352
DEROB program
for absorptance, 256
for radiation, 155
for thermal storage mass, 253
Design
for building integration, 332–334
for commercial buildings, 14
competitions for, 364, 389
costs of, 11
flexibility in, 362
guidelines for, 16–21, 35–36, 363, 387–388
handbook for, 358
performance evaluation of, 7
rules of thumb for, 497–501
Design & Energy competition, 378–379
Design of Energy-Responsive Commercial Buildings, 376–377
Design tools, 4, 485–487
categories of, 21–23
comprehensive, 511–512
conceptual, 488–491
for daylighting, 505–507
evolution of, 16–18
rules of thumb, 495–505
vs. simplified methods, 181
sun charts, 491–495
for thermal comfort estimation, 509–511
for thermal performance estimation, 507–509
use of, 23–25
Developmental tools, 486
Diagnostic procedures, 388
Diffuse solar radiation component, 42, 44, 51–54, 59
Direct-comparison experiments, 293–294
Direct gain and direct-gain systems, 235
in auxiliary heat systems, 248–257
in mixed systems, 279–280, 287
thermal comfort in, 281–285
Directional distribution of solar radiation, 41–46
Direct normal solar radiation, 44–45
Direct-reading thermometers, 303
Discomfort index (DI), 246–247
Distributed absorption, 82–83
Distributed masses
response functions for, 167
spatial discretization of, 121–123, 125–126
Diurnal heat capacity (DHC) method, 194–195
Diurnal temperature swings, 283–284
DOE programs, 506, 511
Doors, 149–152
Doped semiconductors, 205–206
Double-envelope houses, 451–454
Duffy-Saunders method, 468–469

Economics, 6, 10–11
in building integration, 335, 360
of commercial buildings, 375–376, 386
in early passive buildings, 342
life-cycle, 21, 335
potential savings, 26
of residential buildings, 363
of retrofits, 358
Education, 15–16, 294
Effective capacity, 254
Effective U-value, 95
Ehrenkrantz Group study, 371–372
Electric power, monitoring, 304
Electrochromic materials, 216–218
End walls, 151
Energy Conscious Design Awards Program, 381–382
Energy crisis, 4, 342
Energy savings, 10, 26
Energy usage, collecting data on, 304, 401, 403, 405
Enerplex office buildings, 385
Engineering analysis, 404–405
Environmental quality, 26
Environmental Research Laboratory, 320
Equivalent constant thermostat setpoint, 190–192

Eric Meng Associates, 359
Euler integration method, 116–118, 120
Evacuated windows, 207–208
Evaluation tools, 21–24. *See also* Analytical results; Performance evaluation
Evaporation
heat transfer algorithms for, 162–164
rules of thumb for, 504
Event sensors, 304
Explicit finite-difference algorithms, 116–118
External film coefficient, 94
External objects, absorption on, 90–95
External surfaces, reflection from, 62
Extinction coefficient for glass, 76

Fans
in BPA monitoring, 450
in Roberts home, 426
in sunspace systems, 271–272, 286–287
in thermal storage wall analysis, 262, 285
Farralones Institute, 323–324
Fast solar load ratio (FSLR) method, 186–187
Feedback loops, 22
Film coatings for glazings
antireflection, 202–203
insulation, 203–210
Final tools, 486, 488
Finite-difference algorithms, 116–120
First Village, 343–344
Floors
heat transfer through, 126–131
modeling of, 87
Florida Solar Energy Center, 320–322
Fluoropolymer binders, 224
Footprints for daylighting, 505
Forward-difference algorithms, 116–117, 120, 122–123
Forward-looking methods, 181
Fourier transforms, 167
Free-running experiments, 307
FREHEAT program
Biot number with, 121
for finite-difference methods, 118–119
for radiation modeling, 159
Frequency-domain methods
for admittance, 125
for heat storage modeling, 164–171
for temperature swings, 283
for thermal storage analysis, 225
for Trombe walls, 168, 285, 288
Functional building performance, 406
Funding, 5–6
Furniture, 82–83

Gas fills for glazing, 206–207
Ghost Ranch greenhouse, 414–415
Glazings, 199–200
absorption by, 73, 79–82, 201, 211
in building integration, 333
in direct gain analysis, 252–253, 256–257
high transmission approaches to, 201–203
insulation approaches to, 203–210
properties of, 73–79
reflection by, 74, 76–77, 201–203
in sunspace analysis, 275–278
in thermal storage wall analysis, 265–267
types of, 73
Globe temperature, 303
Graded-index glazing treatments, 201–202
Ground
heat transfer through, 126–131
reflection from, 44, 53–54, 59–60
temperature of, in radiation modeling, 159–160
Guidance tools, 21–24
Guidelines
for conservation, 193–194, 243
for design, 16–21, 35–36, 363, 387–388
in direct gain analysis, 250
for home builders, 35–36
vs. simplified methods, 181
in sunspace analysis, 269
in thermal storage wall analysis, 259
Gunderson residence, 416

Habitat Center, 315–316
Harmonic analysis, 7, 225. *See also* Frequency–domain methods
Hay, Harold, 294, 308, 343
Heating systems
backup, 244, 280–281
in building integration, 334
Heat loss
degree-day method for, 183–185
rules of thumb for, 497
Heat storage modeling
convection heat transfer algorithms for, 139–152
evaporation heat transfer algorithms for, 162–164
frequency domain methods for, 164–171
liquid phase storage, 132–139
radiation heat transfer algorithms for, 152–162
solid phase storage, 121–131
Heat transfer rates, 39, 293

Heliodons, 495
Hermetically sealed buildings, 366
High-iron glass, 211
High-quality construction, 13
High transmission approaches to glazings, 201–203
Historic buildings, 361
History, 3–6, 27–29. *See also* Building integration
Home builders, guidelines for, 35–36
Hooker Chemical and Plastics office building, 385
Horizontal brightening, 42
Hours of discomfort, 284–285
Hunn residence, 411–413
Hutchinson, F. W., 337, 340

Imaging, 88
Implicit finite-difference algorithms, 118–120
Index of refraction for glass, 75–76
Indirect comparative testing, 298–300
Indoor movable window insulation, 220–222
Industrial buildings. *See* Commercial buildings
Infiltration
 in building integration, 333
 in building tests, 301
 measuring, 303
 model for, 440
Input-boundary conditions, 122
Instability errors, 117
Instantaneous energy balance, 91–93
Instrumentation for test modules, 302–307
Insulation. *See also* Movable insulation systems
 in building integration, 333
 in early passive buildings, 341
 and glazings, 203–210, 219–220
 importance of, 436
Integration. *See* Building integration
Intensity of solar radiation, 41
Internal building mass
 in sunspace analysis, 275
 in thermal storage wall analysis, 265
International Energy Agency (IEA), 361, 470
International Expert Group Meeting, 355
International House, 379
International Passive and Hybrid Cooling Conference, 355–356

Jones House, 413–414
Justice Department building, Sacramento, 385

Kahn house, 429
Keck, Fred, 337, 339–342
Keck, George, 337, 339–342
Kelbaugh, D., 355
Krypton gas, 207

Lakehead University, 310
Landmark Building System, 378
Larkin Building, 366
Latent thermal storage, 226–230
Lawrence Berkeley Laboratory, 466–468
Layers, glazing
 in direct gain analysis, 256–257
 in sunspace analysis, 276
 in thermal storage wall analysis, 266–267
Leddy, William, 359
Life-cycle economics, 21, 335
Light green glass, 211
Lighting systems, 334
Lightweight surfaces, 275
Liquid crystals, 216–218
Liquid phase storage modeling
 phase-change storage, 134–136
 rockbed storage, 136–139
 roof ponds, 133–134
 water walls, 132–133
Liquid temperatures, 303
Liquid-to-solid phase-change materials, 134
Load collector ratio (LCR) method and tables, 189–190, 236, 239–243, 508
 development of, 349
 in direct gain analysis, 249
 and movable insulation, 251, 260, 270
 in sunspace analysis, 268
 in thermal storage wall analysis, 258–259
Local absorption, 83–84
Local solar radiation, 46–54
Local solar time, 48–49
Location of mass
 in direct gain analysis, 256
 in sunspace analysis, 272
Lockheed Company office building, 386
Longwave emittance
 and absorption, 93
 and mass surface characteristics, 264
Longwave radiation
 inside buildings, 152–159
 outside buildings, 159–162

Los Alamos National Laboratory
monitoring program at, 410–422
test boxes at, 309–310
test rooms at, 311–312, 349
Low absorption glazing, 201
Low-emissivity windows, 200, 203–213
Low-iron glass, 73, 201, 265–266
LOWTRAN program, 161
Lumped parameters, 112, 121–122, 125–126

McFarland, Robert, 19
Manual data collection, 401
Masonry for thermal storage, 225
Massachusetts Institute of Technology
early solar house by, 340
test rooms at, 310–311
Mass location
in direct gain analysis, 256
in sunspace analysis, 272
Mass material
in direct gain analysis, 255
in sunspace analysis, 275
in thermal storage wall analysis, 263–264
Mass surfaces. *See* Surfaces
Mass thickness
in direct gain analysis, 254–255
in sunspace analysis, 273
and thermal comfort, 284
in thermal storage wall analysis, 262–263
Mastin house, 451–454
Materials and components
assessment of, 388
commercial glazings, 210–218
in direct gain analysis, 255
movable window insulation, 218–222
residential glazings, 199–210
in sunspace analysis, 275
testing, 301
for thermal storage, 223–230, 263–264
Mazria, E., 489, 491
Mean radiant temperature (MRT), 80, 89–90, 140, 156–158
Mechanical antireflection glazing treatments, 201–202
Michel, Jacques, 340
Microgrids, 206
MICROLITE 1-2-3 program, 506
MICROPAS program, 511
Mid-America Solar Energy Complex (MASEC), 352
Mixed systems
analytical results for, 279–280
thermal comfort in, 287–288
Mobile/modular homes, 420, 422
Moment method, 45–46
Monitoring programs, 8–11, 399
approaches to, 405–408
Architectural Energy Corporation, 470–471
for Atascadero House, 410
Bonneville Power Administration, 447–450
Brookhaven National Laboratory, 450–454
data collection and analysis for, 400–405
Lawrence Berkeley Laboratory, 466–468
Los Alamos National Laboratory, 410–422
for mobile/modular homes, 420, 422
NAHB Research Foundation, 454–464, 468–470
National Solar Data Network, 422–430
Passive Solar Commercial Buildings Program, 459–464
short-term, 33–35, 464–466, 468–470
Solar Energy Research Institute, 430–447, 464–466
Tennessee Valley Authority, 458–459
for Trombe wall houses, 409–410
for Wallasey School, 408–409
Monte Carlo techniques, 88–89
Morris, Scott, 310
Movable insulation systems, 10, 218–222
in direct gain analysis, 251–252
in Roberts home, 426
in sunspace analysis, 270
in thermal network analysis, 115
in thermal storage wall analysis, 260–261
MRT network method, 156–158
Multi-Family Passive Solar Program, 354
Multilayer walls, 168
Multiple glazings, 78–79

NAHB Research Foundation, 454–464, 468–470
National Bureau of Standards, 317
National Center for Appropriate Technology (NCAT), 313
National Passive Solar Conferences, 27, 343–344, 347, 350, 355–356, 359–361
National Passive Solar Design Awards Program, 379–380
National Passive Solar Design Competition, 359
National security, 26
National Solar Data Network (NSDN), 404, 422–430

Natural light. *See* Daylighting
NBSLD program, 28, 154
Nebraska, University of, 313–314
Nematic crystals, 218
Net load coefficient (NLC), 236
Net reference load, 237
Networks. *See* Thermal network modeling
New Mexico State Office Building, 369–370
Nighttime. *See also* Movable insulation systems
 in solar gain modeling, 95–96
 thermostat setback during, 250
 ventilation during, 504
Normalization, testing for, 294
Normalized collector area, 236
Northeast Solar Energy Center (NESEC), 353
Numerical manipulation, 404

Occupants, data from, 404, 462–463
Oil dependence, 342
Olgyay, V., 489, 491
One-dimensional RC equivalent circuits, 129–130
Opaque external objects, absorption on, 90–95
OPEC oil embargo, 342
Operating modes for test modules, 294–302
Operational testing, 293
Optically switching windows, 213–218
Optimum mix simplified methods, 192–194
Outdoor movable window insulation, 218–219
Overall thermal transfer value (OTTV), 497, 501
Overcast-sky daylight factor, 501
Overhangs, 260
 in early passive buildings, 337, 341
 shading from, 62–65
 and sun charts, 67–69
 and tree transmittance, 60
Overheating problems, 15, 463

Pacific Gas and Electric Company, 324
Packed beds, 136
Paint for thermal storage, 224
Pala Passive Solar Project, 318–320
Participating medium method, 86
PASCALC program, 508
PASOLE program, 19
 for Ghost Ranch greenhouse, 414
 for mobile/modular homes, 420, 422
 phase-change materials in, 135
 validation of, 311–313, 349

Passive and Hybrid Solar Manufactured Buildings Program, 351
Passive heating fraction (PHF), 455–456
Passive Retrofit Handbook, 358–359
Passive Solar Commercial Building Program (PSCP), 372–376, 459–464
Passive Solar Design Analysis, 349
Passive Solar Design Concepts, 348–349
Passive Solar Design Handbook, 72, 348–349. *See also* Analytical results
Passive Solar Design Strategies: Guidelines for Home Builders, 35
Passive Solar Energy Book, 348
Passive Solar Handbook, 507. *See also* Analytical results
Passive Solar Heating Analysis, 72, 189, 239, 350. *See also* Analytical results
Passive Solar Manufactured Buildings Program, 377
Passive Solar Program Research and Development Announcement, 346
Passive Solar Retrofit Guidebook, 358
Peak demand, 11, 244
PEGFIX program, 508
Pennsylvania State University, 316
Perception problems, 15–16
Performance evaluation, 6–7, 238–239, 399. *See also* Analytical results
 in direct gain analysis, 249
 methods for, 400–408
 monitored buildings for, 8–11
 programs for (*See* Monitoring programs)
 in residential buildings, 362
 in sunspace analysis, 268
 test modules for, 7–8
 in thermal storage wall analysis, 258–259
 tools for estimating, 507–509
Phase-change storage, 134–136, 226–230
Photochromic materials, 214
Pitkin County airport terminal, 382, 384
Plants, 82–83
Plastic glazing films, 201, 203–210
Polarization of solar radiation, 46
Pole-and-zero procedure, 125
POOLS program, 164
Preliminary design phase, 24
Problems, 463
 identification of, 375
 in passive solar development, 15–16
Programming design stage, 23–24
Program Opportunity Notices, 368–369, 372
Projected area, 236
Proof-of-concept demonstrations, 293–294, 307

Prudential Insurance Company buildings, 385
PSTAR (Primary and Secondary Terms Analysis and Renormalization), 34–35
Purdue University, 317–318
Pyranometers, 302
Pyrheliometers, 302
Pyrolytic coating for glazing, 206

Questionnaires, 462–463

Radiant panels, 279–280
Radiation
 beam, 50–51
 geography of, 46–49
 local, 50–54
 longwave, 152–162
 modeling of, 41–46
 networks for, 84–89
 reflected (*See* Reflected radiation)
 sensors for, 302
 shielding of, 303
Radiation energy density, 90
Radiosity, 84–86
RC equivalent circuits, 129–131
Recent developments
 guidelines for home builders, 35–36
 short-term energy monitoring, 33–35, 464–466
Reflected radiation, 44, 53–54
 in evacuated windows, 208
 from external surfaces, 62
 by glazing, 74, 76–77, 201–203
 in radiation modeling, 154
 specular, 87–89
 and tree transmittance, 59–60
Regionalism in architecture, 14
Relationships. *See* Correlation methods
Research
 facilities for, 388
 future of, 25
Residential buildings, 12–13, 336
 energy and cost-savings potential for, 26
 glazings for, 199–210
 lessons learned from, 362–364
 milestones in, 342–361
 movable window insulation for, 218–222
 in 1972, 337–342
Residential Passive Solar Performance Evaluation Council (RPSPEC), 455
Residential Solar Demonstration Program, 343
Residential Solar Heating and Cooling Program, 342
Resistance in glazing, 203–210
Response surfaces, 243
Roberts home, 426–428
Rockbed storage, 136–139
 in Balcomb residence, 417, 420
 in early passive buildings, 341
 in NAHB/RF monitoring, 457
Roof aperture systems, 378, 463–464
Roof ponds
 analytical results for, 278
 evaporation in, 162–164
 modeling, 133–134
 proof-of-concept experiments for, 294
 radiant losses from, 160
 rules of thumb for, 504
Rooms, convection modeling between, 149–152
Room temperature and thermal comfort, 281–283
Roughness and convection, 93–94, 148
Rules of thumb, 22, 485, 495
 for cooling, 503–505
 for daylighting, 501
 for design strategies, 497–501
 for solar heating, 502–503
Runga-Kutta methods, 118, 120

St. Louis, MO, LCR table for, 240–242
Salt hydrates, 226
Sandia Laboratories, 350
Saunders, Norman, 341
Scaling laws, 360
Scattered solar radiation, 41–42
Schematic design phase, 24
Schematic tools, 485–486
Second-order Runga-Kutta method, 118
Selective absorbers, 223
Selective transmitters, 211–212
Sensible heat storage, 225–226
Sensitivity data, 243, 254–255, 263
Sensors, data acquisition, 302–304
SERI-RES program
 phase-change materials in, 135
 for rockbeds, 138
Shades, 220, 222
Shading
 in building integration, 333
 coordinate mapping for, 66
 in direct gain analysis, 250–251
 from overhangs, 62–65
 in sunspace analysis, 270
 in thermal storage wall analysis, 260, 285
 from trees, 54–62
Shading calculator, 69
Shading coefficient of glass, 79
Shading masks, 68, 495

Shape of buildings, 489–491
in building integration, 333
and radiation, 86–87
Shell Oil Company office complex, 385
Sherman-Grimsrud infiltration model, 440
Short-term energy monitoring (STEM), 33–35, 464–466, 468–470
Showcase of Solar Homes, 361
Shutters, 219
Simplified methods, 181
correlation, 182–189
diurnal heat capacity, 194–195
optimum mix, 192–194
supplemental, 189–192
Simulation analysis, 7
as evaluation tools, 24–25
evolution of, 16–17
heat storage modeling, 121–139
monitoring, 406
thermal network modeling, 111–120
validation of, 362–363
Single glazings, 76–78
Site
assessment of, 488–489
in building integration, 332–333
planning and developing, 12
Site data acquisition systems (SDAS), 423
Sky diffuse solar radiation, 44, 51–53, 59
Sky temperature, 159–161
Skytherm processes, 343
Skytherm Processes and Engineering, 311
Slabs
heat transfer through, 126–131
sensitivity curves for, 254–255
Smectic crystals, 218
Soda-lime glass, 73
Soil, heat transfer through, 126–131
Solar Calculator, 493, 495
Solar Card, 493
Solar Dwelling Design Concepts, 343
Solar Energy Research Institute (SERI), 351–352, 430
Class A monitoring program, 431
Class B monitoring program, 431–445
Class C monitoring program, 445–447
recent programs, 464–466
test boxes at, 310
test buildings at, 324
test rooms at, 313
Solar gain modeling, 39–40
and comfort, 89–90
computations with, 95–98
of distributed absorption, 82–83
of external surface absorption, 90–95
geography for, 46–49
of glazings, 73–82
of local absorption, 83–84
of local solar radiation, 50–54
radiation networks for, 84–89
for shading, 54–66
for solar radiation, 41–46
sun charts for, 66–73
Solar heat gain factor (SHGF), 79
Solar heating, rules of thumb for, 502–503
Solar heating fraction (SHF), 238, 243
Solar in Federal Buildings Program (SFBP), 370–371, 466
Solar Load Ratio (SLR) methods, 7, 10, 19, 185–187, 507
with BPA monitoring, 449–450
in direct gain analysis, 249–250
introduction of, 347
and movable insulation, 251, 260, 270
with NDSN monitoring, 428–430
with SERI monitoring, 442–445
in sunspace analysis, 268
in thermal storage wall analysis, 258–259
Solar Pathfinder, 493
Solar radiation. *See* Radiation
Solar Room Company, 317
Solar savings, 237
Solar savings fraction (SSF), 238
with SERI monitoring, 444
for thermal performance, 508
Solar Utilization Network (SUN), 353–354
Solid-liquid/solid-solid phase-change materials, 226–229
Solid phase heat storage modeling
for heat transfer through ground, 126–131
network reduction in, 123–126
spatial discretization with, 121–123
Solid-state electrochromic coatings, 216
Solid temperature, 303
South Staffordshire Water Company building, 467–468
Spatial arrangement in building integration, 333
Spatial discretization, 87, 121–123
Spectral distribution of solar radiation, 41, 43–44
Specular reflections, 87–89
Sputter-coated indium-tin-oxide, 205–206
Stacked coatings for glazing, 204–205
Stack effect, 149–150, 303
State-variable equations, 115–116
Statistical analysis, 404
Status sensors, 304
Strip-chart recorders, 401

Subbarao method, 469
Subdivisions, all-solar, 12
Subsystems
monitoring, 407–408
passive, 332–336
Subtractive monitoring methods, 407
Sullivan, Louis, 365
Sulphur hexafloride gas, 207
Sun charts, 66–73, 491–495
Sundwellings Demonstration Center, 323
Sun-peg charts, 493–495, 506
Sunrise and sunset calculations, 95–96
Sunspaces
in auxiliary heat performance, 267–278
in Balcomb residence, 417
in BPA monitoring, 450
in Brookhaven house, 451
fans in, 271–272, 286–287
in First Village, 343
gains in, 235
performance of, 436
in residential buildings, 12
in retrofits, 358
in Roberts home, 426
in SERI monitoring, 445
thermal comfort in, 285–287
SUNSPOT program, 87, 349
Sun-tempered designs, 353
SUPERLITE program, 506
Supplemental simplified methods, 189–192
Surfaces
absorption by, 90–95, 256
conductance by, 145
convection modeling for, 139–144
in direct gain analysis, 252–256
reflection from, 62
response, 243
roughness of, 93–94, 148
in sunspace analysis, 272, 275
temperature of, 303
and thermal comfort, 282, 284
for thermal storage, 223–224, 264–265
Survey of Passive Solar Buildings, 346

TARP program, 156–158
TEANET program
for finite-difference methods, 119
for thermal networks, 511
Technology-transfer program, 350
Temperatures
collecting data on, 403, 405
ground, 159–160
mean radiant, 80, 89–90, 140, 156–158
room, 281–283
sensors for, 303
sky, 159–161
Temperature swings
and thermal comfort, 283–287
and thermal mass, 15
Tennessee Valley Authority
monitoring by, 458–459
office complex at, 385
test rooms by, 317
Test modules, 7–8
calibration of, 306–307
classification of, 293
for code validation, 324–325
instrumentation for, 302–307
objectives with, 293
operating modes for, 294–302
test boxes, 307–310
test buildings, 317–324
test rooms, 310–317
Thermal balancing, 14
Thermal comfort, 13
in building integration, 335
criteria for, 244–247
in direct gain analysis, 281–285
in mixed systems, 287–288
and solar gain modeling, 89–90
in sunspaces, 285–287
in thermal storage wall systems, 285
tools for estimating, 509–511
Thermal network modeling, 6, 111–112
and absorption, 91–92
in solid phase modeling, 123–126
solution of equations for, 116–120
system equations for, 113–116
Thermal storage, 235
in auxiliary heat performance, 257–267
in building integration, 333
bulk materials for, 225–230
in direct gain analysis, 252–254
in early passive buildings, 337
in mixed systems, 279–280, 287
and overheating, 15, 463
surface treatment for, 223–224
thermal comfort in, 285
Thermochromic materials, 214–216
Thermosiphon air panels (TAP), 279, 378
Thermostat settings
in direct gain analysis, 250
equivalent constant thermostat setpoint, 190–192
in sunspace analysis, 269–270
in thermal storage wall analysis, 259
Thick films for glazing, 205–206

Thickness of mass
in direct gain analysis, 254–255
in sunspace analysis, 273
and thermal comfort, 284
in thermal storage wall analysis, 262–263
Thin films for glazing, 204–205
Time
corrections for, 48–49
in solar gain modeling, 96–98
Tin-oxide for glazing, 205–206
Titanium oxynitride for glazing, 202
Total incident energy, 42–43
Total load coefficient (TLC), 183–184, 236
TRACE program, 511
Translucent glazing, 252–253
Transmittance
of glass, 73–74, 76
of trees, 57–61
Transparent insulation, 208–210
Transwalls, 278–279
Trees
shading by, 54–57
transmittance by, 57–61
Trinity University, 320
TRNSYS program, 138
Trombe wall systems
auxiliary heat performance with, 247–248
in Brookhaven house, 451
fans in, 262, 285
frequency-domain analysis for, 168, 285, 288
vs. glazing, 253
glazing layers for, 266
in Gunderson residence, 416
heat transfer through, 150
in Hunn residence, 411–413
invention of, 340–341
lumped parameters for, 126
mass surface for, 264
mass thickness in, 263
in mixed systems, 279–280, 287
modeling of, 19
monitoring programs for, 409–410
in Roberts home, 426
shading and ventilation in, 260
testing, 298
in test rooms, 312–313
vents in, 150, 247–248, 261–262, 285
Truncation errors, 117, 121–123

Unutilizability method, 10, 187–189, 507
U-overall criteria, 497
Utility bills, 401
U-value method, 95
UWENSOL program
for finite-difference methods, 119
for radiation modeling, 155

van Dresser, Peter, 341
Venetian blinds, 219
Ventilation
in building integration, 334
in direct gain analysis, 250–251
rules of thumb for, 504
in sunspace analysis, 270
in thermal storage wall analysis, 260, 285
Vents
convection modeling with, 149–152
for sunspaces, 270–271
in Trombe wall systems, 150, 247–248, 261–262, 285
View factors, 155
Village Homes, 348
Vitro Corporation, 423–428

Wainwright Building, 365–366
Wallasey School, 367–368, 408–409
Wall vents for sunspaces, 270–271
Warehouses, 369
Water
absorption coefficient of, 133
in sunspace analysis, 273–274
for thermal storage, 225–226
Water-wall systems
in Gunderson residence, 416
modeling, 132–133
shading and ventilation in, 260
simulation of, 18–19
in sunspace analysis, 273–274
thermostat settings in, 259
transwalls, 278–279
vs. vented Trombe walls, 248
Watt-hour meters, 304
Wavelength-sensitive plastic film, 203
Weather
collecting data on, 403
and design, 35
Weatherization, importance of, 436
Weighing functions, 6–7
Wessling Consulting, 314–315
Whole-building concepts, 376–377
in Brookhaven house, 451
and shape of building, 491
Wind
in Bioclimatic Charts, 489
in BPA monitoring, 450
and convection, 144–148

Wind (cont.)
and infiltration, 303
and surface conductance, 94
Windows. *See also* Glazings
air space in, 207–208
in building integration, 333
with bulk transparent insulation, 208–210
for daylighting, 13–14
evacuated, 207–208
movable insulation for (*See* Movable insulation systems)
optically switching, 213–218
size of, 503
in Wallasey School, 367–368
Wing wall reflectors, 87–88
Worksheets, design, 36
Wright, Frank Lloyd, 366

Zero-purchased-energy buildings, 388
Zome wall-panel system, 341
Zomeworks, 308–309